Exploring Pillars of Sustainability for Modern Age Improvements

Farhan Sabir Ujager
De Montfort University, Dubai, UAE

Hassan Ali
University of Birmingham, Dubai, UAE

Sayed Sayeed Ahmad
De Montfort University, Dubai, UAE

IGI Global
Scientific Publishing
Publishing Tomorrow's Research Today

Vice President of Editorial Melissa Wagner
Managing Editor of Acquisitions Mikaela Felty
Managing Editor of Book Development Jocelynn Hessler
Production Manager Mike Brehm
Cover Design Phillip Shickler

Published in the United States of America by
IGI Global Scientific Publishing
701 East Chocolate Avenue
Hershey, PA, 17033, USA
Tel: 717-533-8845
Fax: 717-533-8661
E-mail: cust@igi-global.com
Website: https://www.igi-global.com

Library of Congress Cataloging-in-Publication Data

Names: Ujager, Farhan, 1985- editor.
Title: Exploring pillars of sustainability for modern age improvements / Edited by Farhan Ujager, Hassan Ali, Sayed Sayeed Ahmad.
Description: Hershey, PA : IGI Global Scientific Publishing, [2025] | Includes bibliographical references and index. | Summary: "This book aims to explore the vital interface of contemporary pillars of sustainability, including technology, finance, policies, politics, and related domains"-- Provided by publisher.
Identifiers: LCCN 2024049396 (print) | LCCN 2024049397 (ebook) | ISBN 9798369357484 (hardcover) | ISBN 9798369357491 (paperback) | ISBN 9798369357507 (ebook)
Subjects: LCSH: Sustainable development. | Technology--Environmental aspects. | Sustainable design. | Green movement.
Classification: LCC HC79.E5 E984 2025 (print) | LCC HC79.E5 (ebook) | DDC 338.9/27--dc23/eng/20241213
LC record available at https://lccn.loc.gov/2024049396
LC ebook record available at https://lccn.loc.gov/2024049397

British Cataloguing in Publication Data
A Cataloguing in Publication record for this book is available from the British Library.

Table of Contents

Detailed Table of Contents

This chapter explores the concept of sustainable sustainability, underscoring the critical necessity of integrating environmental, social, and economic factors at the core of human society to ensure its long-term continuation. The interdependence of critical societal components, such as climate, economics, trade, technology, governance and many others, is highlighted, all of which are dependent upon finite natural resources. In addition, this chapter discusses a critical assessment of contemporary sustainable technologies, with a particular focus on electric vehicles (EVs) and artificial intelligence (AI). Despite the potential to reduce environmental impacts, these technologies are met with challenges such as the ecological footprint of battery manufacturing, energy-intensive AI processes, and ethical and societal concerns associated with AI implementation.

This work aims at analyzing the sustainable practices of 5-star hospitality in Algarve (Portugal). It explores how these hotels communicate sustainability, which practices they adopt and main drivers/constraints. First observations were made in their own websites and in Booking.com. Then an online survey was used to collect more information. It was perceived a higher level of sustainable behavior in internal processes. Main areas are energy management, water/waste management and social responsibility. Most popular practices are the least expensive, and main drivers are environmental responsibility, strategic vision, and efficiency gains. Although the main obstacle referred is the investment involved, lack of qualified personnel and technical issues are other issues. The growth of new businesses, personnel training, and specialized communication in the area can overcome them. The originality of this study lies on: the 5-star category in Algarve filling a gap in the literature; the need of multiple research methods; and application of the sustainability triad approach.

The space industry has expanded rapidly over several decades, with a significant increase in the number of launches and satellites in orbit. However, this expansion has generated sustainability concerns in space, including increased greenhouse gas emissions from satellite launches and orbital debris polluting the atmosphere. This raises concern about the sustainable use of the Earth's orbits as space debris accumulates. The amount of space debris, made up of different leftovers from objects like defunct satellites and rocket fuels, has been increasing for many years. In addition to contaminating the environment, this floating orbital garbage increases the chance of crashing into current satellites. Unless interventions are contemplated, the likelihood of more orbital garbage is expected to rise in the coming years. However, eliminating space waste is fraught with legal, political, economic, and regulatory issues as well as practical technological challenges. This chapter examines this confluence of factors with a view to contemplating remedies to solve the orbital debris problem.

Effective leadership focused on sustainability is key to achieving operational excellence in a supply chain. Leaders with the right qualities implement sustainable practices throughout the chain, considering environmental, social, and economic factors. Success is influenced by company culture, stakeholder engagement, regulations, and data use. This approach leads to operational excellence with better efficiency, quality, and resilience, along with a competitive advantage and happy stakeholders. The framework is a roadmap for businesses to leverage leadership for a sustainable and successful supply chain.

Digital arts, guided by New Media Design principles, are transforming architectural heritage conservation. This chapter explores how digital tools, including 3D mapping, VR/AR, and simulations, enhance sustainability in preservation practices. By creating immersive experiences and fostering community engagement, digital art not only documents but reimagines historical sites, ensuring their accessibility and relevance for future generations. This multidisciplinary approach tackles challenges and embraces opportunities, showcasing the potential of digital technology to reshape heritage conservation towards a more sustainable and interactive future.

Passive design is a sustainable approach to architecture that prioritizes natural elements and energy efficiency. It encompasses strategies such as heat recovery ventilation, spatial openness for natural light and airflow, energy-efficient HVAC systems, and heat control mechanisms. Effective insulation, high-performance windows, proper solar orientation, and an airtight envelope are crucial components. By minimizing environmental impact and maximizing occupant comfort, passive design offers a holistic solution for high-performance architecture. This chapter will provide detailed insights into these strategies and their real-world applications.

Green Financial Markets play a significant role in directing capital towards environmentally friendly projects, renewable energy, and other green initiatives. The demand for green financial products such as green bonds and sukuks, green derivatives, as well as green investment funds, is expected to increase phenomenally as consumer awareness of environmental issues grows. However, as the sector expands, it encounters challenges such as greenwashing, identifying and managing green risks, and ensuring data availability and quality. Effective regulations and supportive policies harmonizing global standards and reporting will ensure transparency, reliability, and accountability in financial flows directed towards green investments. The ongoing evolution of the green markets will support moderating the impact of climate change and undoubtedly offer new opportunities for investors and businesses to thrive in a green economy.

The demand for geospatial information is growing rapidly across diverse sectors, including urban planning, agriculture, disaster management, and public health. Geospatial information has become a crucial requirement for sustainable development across all countries and sectors. This chapter aims to investigate and establish the rationale of geospatial data for sustainable development with the help of various use cases. The study also provides recommendations for enabling access to geospatial data.

Numerous challenges are being faced by the healthcare system of Pakistan. It lacks universal access to quality healthcare, an inadequate supply of healthcare human resources, and healthcare sector inconsistencies. Incorporating the Spatial Data Infrastructure (SDI) in Pakistan can be a possible solution to tackle most of these problems. SDI is a platform that organizes and centralizes geographical information, such as the population, the placement of healthcare facilities, and disease incidence. By employing SDI, these challenges may be met efficiently by the planners. When implemented, spatial data infrastructure can support optimizing human resource development, investment in resources, and evidence-based decision-making. Furthermore, SDI can ensure the delivery of actual and timely information to stakeholders thus enhancing the aspect of accountability. This chapter discussed how SDI can help in coping with relative issues concerning the Pakistani health sector and thereby presented a way of creating a sustainable healthcare sector.

Preface

This book focuses on the concept of sustainability in a broad and inclusive way, exploring its significance across various aspects of modern life beyond just environmental concerns. It covers a wide range of fields such as artificial intelligence (AI), architecture, interior design, education, space exploration, accounting, smart cities, energy consumption through smart technologies, climate challenges, leadership, and health technology. By offering insights from such diverse areas, the book aims to give readers a different perspective on sustainability, allowing them to understand it from multiple angles rather than through a singular lens. One of the key messages of this book is that sustainable projects or initiatives, while often well-intentioned, can sometimes lead to unsustainable outcomes if a holistic approach is not adopted. It is crucial to understand the broader, often hidden, impacts of these so-called sustainable projects, as they may, in fact, create new problems when viewed narrowly or executed without considering their long-term and interconnected effects.

In providing a well-rounded understanding of sustainability, this book emphasizes the importance of adopting sustainability in every domain of life. The motivation behind this work is to show that sustainability should not be treated as a short-term solution or trend, but rather as a comprehensive approach that addresses the complex global challenges we face today. Through the presentation of modern concepts, the latest research findings, case studies, and in-depth analyses, readers will be able to see how sustainability plays a vital role in shaping the future. Each chapter provides valuable insights into how these practices can be implemented effectively across different fields, demonstrating the necessity of this holistic approach.

The target audience for this book includes researchers, students, and professionals who are interested in learning about and implementing sustainable practices in their respective fields. Readers will gain an understanding of how sustainability must be viewed and practiced through an interdisciplinary lens, as no single field can achieve sustainability on its own. The interdisciplinary approach highlighted throughout the book is essential for fostering collaboration and integration across sectors, ensuring that sustainable initiatives are truly impactful.

I would like to extend my deepest gratitude to the contributors from around the world who have provided their expertise and insights to this book. Without their hard work and dedication, this comprehensive exploration of sustainability would not have been possible. I also want to thank the readers for their interest in this subject, as it is your engagement and commitment that will ultimately drive the changes necessary to create a more sustainable future.

My hope is that this book inspires readers to critically engage with the concept of sustainability, challenging them to think beyond surface-level solutions and to consider the long-term impacts of their actions. The challenges our planet faces are significant and growing, and it is more important than ever to adopt sustainable practices across all sectors of life. By applying the insights gained from this book, readers can contribute to the global effort to protect and preserve our world for future generations.

Farhan Sabir Ujager
De Montfort University, Dubai, UAE

Chapter 1
Sustainable Sustainability: A Holistic Approach

Farhan Sabir Ujager
https://orcid.org/0000-0002-7080-5141
De Montfort University, Dubai, UAE

ABSTRACT

This chapter explores the concept of sustainable sustainability, underscoring the critical necessity of integrating environmental, social, and economic factors at the core of human society to ensure its long-term continuation. The interdependence of critical societal components, such as climate, economics, trade, technology, governance and many others, is highlighted, all of which are dependent upon finite natural resources. In addition, this chapter discusses a critical assessment of contemporary sustainable technologies, with a particular focus on electric vehicles (EVs) and artificial intelligence (AI). Despite the potential to reduce environmental impacts, these technologies are met with challenges such as the ecological footprint of battery manufacturing, energy-intensive AI processes, and ethical and societal concerns associated with AI implementation.

INTRODUCTION

The fabric of human society is woven from many threads, each representing crucial aspects of life such as climate, finance, business, technology, and governance. These components are interconnected and dependent on the limited resources provided by nature. Our survival and the functionality of society hinge on the prudent utilization of these finite resources.

Sustainability, at its essence, denotes the capacity for lasting or withstanding. Thomas Malthus's theory from the late 18th century illuminates this dilemma by forecasting that population growth would surpass resource production, resulting in scarcity and famine(Malthus, 1959). This underscores the pivotal significance of sustainable practices in ensuring our ongoing existence.

Sustainability is imperative for human survival and can be characterized through diverse perspectives:

- Guiding Principles: Incorporating social and environmental standards into human endeavors, emphasizing the design and operation of systems(Liu et al., 2015).

DOI: 10.4018/979-8-3693-5748-4.ch001

- Human Objective: Striving for systems that attain environmental, social, and economic objectives (Kuhlman & Farrington, 2010).
- A Concept: Encompassing notions such as equilibrium, continuity, and adaptability(Fiksel, 2006; Schilling et al., 2018).

When examining sustainability from the viewpoint of offering tailored solutions to meet the specific needs or demands of a particular domain, researchers, innovators, or policymakers often concentrate solely on one aspect typically related to climate impact. This narrow focus may disregard broader and more comprehensive definitions of sustainability that encompass multiple interconnected dimensions (Pezzey, 2017; Rosati & Faria, 2019; Sauvé et al., 2016)

- Environmental: Safeguarding natural resources and biodiversity to ensure the vitality and functionality of ecosystems.
- Social: Securing well-being, fairness, and social equity by addressing the requirements of all community members and promoting equitable access to resources.
- Economic: Preserving economic frameworks that promote sustainable livelihoods and resilient economies, while also ensuring constant growth and future viability.
- Interdisciplinary: The recognition that genuine sustainability necessitates a comprehensive approach that integrates all of these aspects, in order to achieve a balance between ecological well-being, social justice, and economic prosperity.
- Sustainable Development: The Brundtland Commission's definition of the capacity of future generations to meet their own requirements without compromising the gratification of present needs. This theory highlights the need of strategic planning for the future and guaranteeing fairness between various generations (Borowy, 2013).

Figure 1. Interdisciplinary sustainable development

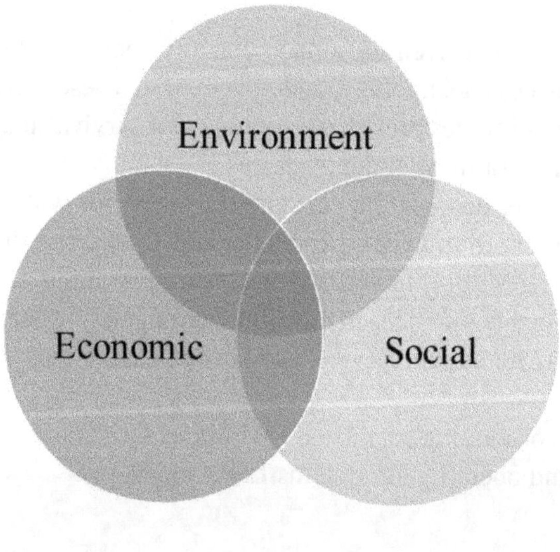

These diverse dimensions must be comprehensively examined in order to achieve a more inclusive approach to sustainability. It is imperative to integrate environmental conservation with economic stability and social equity in order to prevent sustainability initiatives from inadvertently injuring one domain while benefiting another. In order to accomplish genuine sustainability, it is essential to adopt an interdisciplinary approach that integrates economic feasibility, social equity, and environmental health, even when accounting for other critical factors. The comprehensive integration of these dimensions is crucial to ensure that these actions have a positive impact on long-term sustainability. (Pezzey, 2017). A comprehensive strategy that encompasses the environmental, social, and economic facets of existence is necessary for sustainability. It is imperative to take into account the broader implications, for instance for countries with significant populations and vulnerable economies. Inadequate planning in sustainable development initiatives may provide a number of obstacles. These are a few potential outcomes.

- Economic Disruption: The transition to sustainable practices has the potential to significantly disrupt established industries, resulting in economic instability and unemployment. In countries with weak economies, this might make inequality and poverty worse.
- Resource allocation: The transition to sustainable practices often necessitates substantial investments in new technology and infrastructure. Less developed governments may find it more difficult to finance these advancements, widening the gap between rich and developing countries.
- Social Impact: Sustainable practices have the capacity to trigger social transformations that have a diverse impact on communities. For example, the disruption of local cultures and lifestyles may occur as a consequence of deviating from conventional industries, necessitating meticulous management to preserve social unity.

Figure 2. Silo-free approach for sustainable sustainability

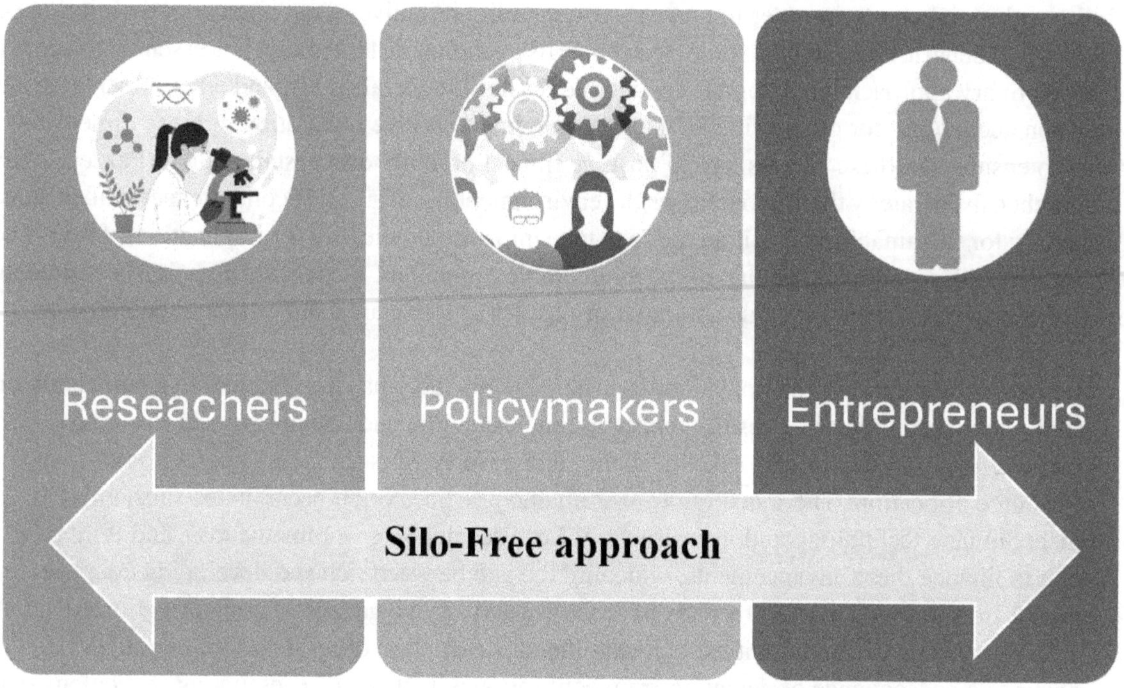

An additional obstacle to comprehending sustainability is the silo-based approach that plagues the development, implementation, and monitoring of sustainable solutions as shown in Figure.2. While endeavoring to advance sustainability in one domain, researchers, policymakers, and entrepreneurs may inadvertently cause harm in another. This outcome is the consequence of their silo-based methodologies. In some cases, researchers prioritize the development of knowledge within their respective disciplines without completely understanding the multidisciplinary implications of their sustainable solutions. The development of regulations and incentives that are intended to achieve specific sustainability objectives is frequently the primary focus of policymakers, who frequently overlook the broader economic and social implications. While entrepreneurs are instrumental in the advancement of sustainable solutions, they frequently fail to completely consider the longer-term environmental and socioeconomic consequences of their concepts. The absence of integration and holistic thinking can lead to unintended adverse outcomes, such as environmental degradation, economic upheaval, or social discord, which can undermine the overarching objective of sustainability.

In this chapter, there are two primary sections. The initial section examines specific domains that, although not explicitly linked to sustainability, can significantly influence sustainability when incorporated into societal, economic, and educational frameworks. Education, agriculture, urban design and planning, consumerism, and tourism comprise the domains. Despite the fact that other domains could potentially be relevant, the ones chosen are meant to highlight the critical role that apparently unconnected disciplines play in the goal of sustainability. The second section investigates two contemporary sustainable technologies: artificial intelligence (AI) and electric vehicles (EVs). Although these technologies may seem sustainable, they may not produce the expected benefits if not properly managed.

EXPLORING SUSTAINABLE PRACTICES

This section explores a range of sustainable approaches across several fields that contribute to reaching sustainability goals beyond net zero, decarbonisation, and climate change mitigation. In addition to addressing critical challenges such as emission reduction and decarbonisation promotion, we also endeavour to restore the disrupted equilibrium between the environment and nature that has been caused by human activities through the discussion and implementation of these approaches. The purpose of this discussion is to clarify the importance of incorporating sustainable practices into various areas of life. The goal is not simply to restrict to these practices, but to develop a mindset that is in harmony with future requirements, while utilising the resources available to us at present.

Education

The objective of sustainable educational practices is to integrate sustainability principles into the teaching methods and learning processes, thereby fostering economic viability, social responsibility, and environmental stewardship. Sustainable educational methods involve a comprehensive and multifaceted approach that includes key factors(Eagan et al., 2002; McFarlane & Ogazon, 2011).

- Interdisciplinarity: This approach integrates knowledge from a variety of fields, including environmental science, social sciences, economics, and ethics, to improve students' understanding of the interdependence of environmental, social, and economic issues.
- Focus on Experiential Learning: Exposure to practical skills and critical thinking abilities is promoted through experiential learning, which emphasises real-world projects, fieldwork experiences, and hands-on tasks, thereby enhancing students' comprehension of sustainability concepts and methods.
- Collaborative Learning: fosters communication skills, teamwork abilities, and a shared feeling of responsibility for crafting a sustainable future by encouraging teamwork, conversations, and project partnerships.
- Transformative Purpose: Sustainable education aims to inspire individuals to assess their values, convictions, and behaviours, thereby enabling them to advocate for sustainability in both personal and professional contexts. The transforming nature of this component is crucial for instigating positive changes on personal, community, and societal levels.
- Inquiry-Based Discourse: Sustainable educational methodologies encourage students to engage in meaningful conversations about sustainability issues by promoting inquiry, dialogues, and introspection. This enhances cognitive skills in analysing and evaluating complex sustainability challenges, fostering a more profound comprehension of their intricacies.
- Value and Objectives of Sustainability: The sustainable educational models align educational aims with the principles of environmental preservation, social equality promotion, and economic prosperity advancement. This serves to motivate students to develop into global citizens who are committed to fostering a more sustainable world.

Urban Designing and Planning

Sustainable urban design is an urban planning strategy that is dedicated to the creation of city environments that are economically viable, socially inclusive, and ecologically conscious. Its objective is to address the obstacles presented by accelerated urbanisation by incorporating community-centric design principles and eco-friendly construction practices(Burton & Mitchell, 2006; Díaz et al., 2024; Omole et al., 2024).

- Environmental Responsibility: Implementing practices that mitigate the ecological consequences of urban expansion by utilising renewable energy sources, energy-efficient technologies, eco-friendly building materials, waste reduction programmes, resource conservation initiatives, and decarbonisation initiatives.
- Social Inclusivity: The creation of urban spaces that are readily accessible and usable by all community members, regardless of their socio-economic status or capabilities. Sustainable urban design involves implementing inclusive design principles, engaging the community, and creating public places that promote social cohesion and connectivity.
- Economic Viability: The development of mixed-use areas that incorporate residential, commercial, and recreational spaces is essential for the promotion of economic sustainability, which reduces commuting demands, supports local businesses, and provides affordable housing solutions.
- Resilience: Through catastrophe resistance measures, climate-responsive design, and the repurposing of existing structures, resilience is a critical factor in the protection of urban environments from natural disasters and climate change.
- Community Well-being: is prioritised through the design of spaces that improve air and water quality, support physical and mental health, and contribute to a good standard of living for residents. The significance of public health is emphasised by advocating for designs that prioritise walkable neighbourhoods and verdant spaces in order to have a positive impact on individual well-being.
- Equitable Development: The objective of equitable development is to address socio-economic disparities by advocating for inclusive housing policies and strategies that prevent gentrification and guarantee universal access to essential services. Investing in effective public transport networks is crucial for reducing reliance on private vehicles, decreasing carbon emissions, and enhancing overall mobility for residents.
- Innovation and Research: Emphasis on innovation and research entails the allocation of resources to the investigation of sustainable urban design innovations, including energy-efficient technologies, eco-friendly construction materials, and innovative methods of improving community well-being.

By incorporating these elements into planning procedures, sustainable urban design aims to create cities that are both economically prosperous and ecologically conscious, while also being socially responsible, thereby ensuring a sustainable future for urban communities.

Tourism

Sustainable tourism practices entail the meticulous planning and administration of tourism activities to mitigate negative environmental effects. These practices have the potential to generate economic benefits and social advantages for all parties involved, as well as to protect cultural heritage and support local communities(Díaz et al., 2024; Mowforth & Munt, 2015).

- Environmental Conservation: Implementing safeguards to protect natural resources, reduce carbon footprint, preserve biodiversity, and reduce pollution.
- Socio-cultural Respect: Guaranteeing that tourism activities do not disrupt or degrade local ways of life; fostering cultural exchange; and upholding and endorsing local cultures, traditions, and communities.
- Economic Benefits: Ensuring that tourism enhances local economies by creating employment opportunities, fostering the growth of small businesses, and fairly dispersing economic profits among all parties involved.
- Engagement with the Community: Fostering positive interactions between travellers and residents, empowering local communities to participate in tourism development, and involving them in decision-making processes.
- Awareness and Education: Promoting responsible behaviour and increasing awareness of the significance of sustainable tourism practices among local communities, industry actors, and tourists.

Agriculture

Sustainable agriculture is the practice of farming that meets the current food demands of society while also ensuring that future generations can meet their own needs. The production of food that is economically viable for farmers, does not damage the environment, treats animals humanely, and promotes the health of consumers and producers is all part of this process(Agovino et al., 2019; Arora, 2018; Janker & Mann, 2020; Wawrzyniak, 2023).

- Environmental Conservation: The implementation of practices that protect and preserve the environment, such as soil, water, and air quality. This entails reducing the utilisation of artificial fertilisers and pesticides, preserving water, and lowering the emission of greenhouse gases.
- Economic Sustainability: The goal is to guarantee the economic prosperity of rural communities by decreasing input costs, increasing market accessibility, and diversifying income sources, all while providing producers with an equitable income.
- Social Accountability: The acknowledgment of the agricultural sector's significance and the implementation of socially fair and equitable farming practices through just labor practices, community engagement, and respect for the rights of stakeholders.
- Animal Welfare: The provision of suitable living conditions, nutrition, and healthcare to ensure the humane treatment of animals.
- Preserving Biodiversity: Maintaining and enhancing the variety of plant and animal species on farms in order to promote better crops and animals as well as more resilient ecosystems.
- Resource Efficiency: The effective utilization of land, water, and energy resources to reduce its environmental impact and minimize waste.

- Innovation and Adaptation: Utilizing new technologies such as precision agriculture, renewable energy, and conservation cultivation to improve sustainability.
- Community Engagement and Food Security: Encourage the development of local food systems to ensure that all community members have access to affordable, nutritious food.
- Long-Term Perspective: Making decisions that are based on long-term sustainability rather than short-term gains, while also taking into account the influence on future generations.

Sustainable agriculture is a flexible approach based on principles that can be tailored according to specific farm contexts.

Consumption

Sustainable consumption is the practice of making conscious decisions when purchasing commodities in order to minimise the environmental impact, improve social well-being, and promote a sustainable economy. This is achieved by acknowledging the impact of our actions on the planet's health and the welfare of society when using products (Lavuri et al., 2023; Rodriguez-Gonzalez et al., 2020; Tripathi & Singh, 2016).

- Resultant Behaviours (Actions): Sustainable consumption is a reflection of an individual's values and attitudes towards environmental protection, which includes decisions such as supporting ethical brands, reducing pollution, and choosing eco-friendly products.
- Environmental Protection Emotion (EPE): The degree of one's emotional attachment to environmental conservation can have a substantial influence on their sustainability decisions. These emotions strengthen the correlation between values, attitudes, and behaviours.
- Positive Feedback Loop (Influence of Values): An individual's pro-environment self-image, attitude, and sustainable consumption behaviours are positively influenced by altruistic and egoistic values, as well as social influences.
- Mediating Factors (Identity and Attitude): The manner in which an individual's values are translated into actions is influenced by their positive stance on sustainability and their environmental self-perception. This suggests that persons who perceive themselves as environmentally conscientious are more inclined to behave in alignment with their principles.

CRITICAL EVALUATION OF CONTEMPORARY SUSTAINABLE TECHNOLOGIES

In this section, we thoroughly examine two technological innovations that have gathered attention for their capacity to reduce environmental impact. Artificial intelligence (AI) is recognised for its capacity to enhance operational efficiency and resource utilisation in various industries, while electric vehicles (EVs) are commended for their capacity to reduce carbon emissions in the transportation sector. Nevertheless, the quest of their intended sustainability objectives necessitates more than just adoption; it necessitates a deliberate examination of design, manufacturing procedures, energy sources, infrastructure development, and end-of-life practices.

Despite the acknowledged advantages of EVs and AI technologies, they are confronted with obstacles and hazards that could potentially undermine their environmental benefits if not properly addressed. The environmental impact of battery manufacturing for EVs and the energy-intensive nature of AI algorithms only serve to emphasise the intricacies of the situation. The objective of this section is to conduct a comprehensive and strategic analysis of these technologies in order to evaluate their sustainability potential and mitigate any unintended negative consequences.

Electric Vehicles (EVs)

In numerous countries, electrified transport, including metros, electric trains, trams, is implemented extensively for public transport and is acknowledged as a well-established technology. In the past, the adoption of private electric vehicles (EVs) has been restricted by the challenges associated with battery technology, such as weight issues, high costs, brief lifespans, and lengthy charging times. EVs have become competitive with conventional gasoline-powered vehicles because to recent breakthroughs in battery technology, which have led to a significant fall in EV prices.

Charging infrastructure for electric vehicles (EVs) comprises workplace chargers, street chargers, residential home chargers, and large charging stations, all of which are supplied by the distribution network. One difficulty is successfully managing the increased energy demand to avoid negative effects on the power distribution network due to unregulated charging of electric vehicles with significant market penetration(Khalid et al., 2019). As a result, the unregulated effects of electric vehicle charging on power systems may result in an increase in peak demand, voltage fluctuations, phase imbalances, power losses, and network component overloads.

Electric vehicles (EVs) face an extra challenge due to the presence of lithium-ion batteries. The importance of lithium in the transition to renewable energy is significant, as it is used extensively in rechargeable batteries and plays a critical role in the operation of electric vehicles. Nevertheless, the process of extracting lithium has significant environmental ramifications (Kaunda, 2020; Vera et al., 2023).

- Energy Consumption: The direct lithium extraction (DLE) process typically requires a significant quantity of energy.
- Water Use: Traditional evaporative techniques used in lithium mining are well-known for their excessive water use.
- Chemical Utilisation: Certain DLE techniques utilise chemicals to improve lithium recovery, which can result in environmental hazards if not adequately controlled.
- Geothermal Energy Harvesting: The stability of geothermal fields and the adjacent ecosystem may be affected when lithium extraction is combined with geothermal energy harvesting in certain cases.
- Ecosystem Disruption: Lithium extraction activities have the potential to disrupt ecosystems and result in habitat loss, particularly in fragile environments near lithium reserves. Changes in soil composition, water levels, and ecosystem diversity can have a profound impact on local climatic patterns and the functioning of ecosystems.

The extraction of lithium not only impacts the environment but also the disposal and recycling of spent lithium-ion batteries negatively affect the environment (Costa et al., 2021; Ibrahim et al., 2020; Meshram et al., 2020).

Soil Contamination

- Disposal of lithium-ion batteries in landfills may lead to leaching of chemicals and heavy metals into the soil over time.
- Leaching of toxic metal nanooxides from batteries can contaminate soil, posing risks to plant life and ecosystems.
- Heavy metals from batteries might be carried by dumpsite leachate into surrounding soils.

Water Contamination

- Battery fires, especially when combined with landfill biogas emissions, can release toxins into the air or contaminate soil, groundwater, and surface water.
- Leaching, disintegration, and decomposition of batteries could result in the release of pollutants that contaminate groundwater sources.
- If batteries leach onto open ground, heavy metals can be washed off by rainwater into nearby rivers or lakes, further polluting these bodies of water.

Air Contamination

- Fine particles, including dust, may be emitted during battery disassembly processes or recycling activities, impacting air quality.
- Improper disposal methods, such as the incineration of batteries, can emit hazardous gases like hydrofluoric acid (HF) and hydrogen cyanide (HCN) into the air.
- Fires and explosions due to mishandling of batteries can release harmful substances, endangering both human health and the environment.

Release of Toxic Chemicals

- Spent lithium-ion batteries contain reactive salts, volatile organic electrolytes, and additives with uncertain toxicity levels.
- Battery fires, especially during thermal runaway events, can release toxins into the atmosphere or leach hazardous components into soil, groundwater, and surface water bodies.
- Toxic chemicals from batteries pose risks to human health and ecosystem integrity.

Sustainable AI

Artificial intelligence (AI) is a revolutionary technology that is indispensable for the promotion of sustainability in numerous industries. In this section, we will investigate its potential. While AI has the potential to promote sustainable projects, it is crucial to assess it from multiple perspectives using the framework of Sustainable AI. The importance of incorporating environmental implications, ethical dilemmas, and societal repercussions into the development and implementation of AI solutions is underscored in this section, which emphasises the critical convergence of AI and sustainability. Sustainable AI aims to promote responsible and sustainable practices in the development of AI in order to mitigate adverse effects on the environment and society (Van Wynsberghe, 2021).

It is crucial to distinguish between Sustainability AI and Traditional AI (Mishina et al., 2020; Wu et al., 2022).

- Holistic Perspective: Sustainable AI employs a comprehensive approach that involves the examination of the entire lifecycle of AI products, from inception to governance, to guarantee that environmental and social impacts are addressed at every developmental stage.
- Sustainability is emphasised: Sustainable AI prioritises social justice and ecological integrity, whereas conventional AI development typically prioritises efficacy and performance. Its goal is to create long-term successful AI systems that are also sustainable.
- Environmental Focus: Sustainable AI is a field that concentrates on the environmental issues that are associated with the development of AI. This encompasses the utilisation of sustainable data sources and electricity supplies, as well as the mitigation of carbon footprints associated with algorithm training. This emphasis on environmental sustainability distinguishes it from conventional methodologies that may disregard these issues.
- Ethical and Social Considerations: Sustainable AI takes into account the ethical and social ramifications of AI technologies, making sure that they uphold social norms and encourage the fair allocation of resources. Artificial intelligence development is distinguished from conventional methodologies by its ethical component.

There are numerous ways in which traditional artificial intelligence methods can significantly impact sustainability and the environment.(Chang & Ke, 2024; Chen et al., 2023; Schoormann et al., 2023; Zhao & Gómez Fariñas, 2023).

- Energy Consumption: The substantial computational capacity that traditional AI models need to operate is a direct result of their high energy consumption. Not using renewable energy sources may result in higher carbon emissions and environmental harm.
- Data Centre Impact: The extensive energy consumption necessary for cooling purposes can have a significant environmental impact on the infrastructure that supports AI systems, such as data centres. Over-reliance on resources may contribute to greenhouse gas emissions and put strain on regional ecosystems.
- E-waste and Technological Advancement: The rapid advancement of AI technology has the potential to lead to a decrease in the average lifespan of products and an increase in electronic waste (e-waste). Contaminated environments and health risks may result from the improper disposal of AI equipment.
- Concerns Regarding Inequity and Bias: The perpetuation of biases and inequities within society by conventional AI algorithms has the potential to exacerbate social and environmental injustices. Algorithms that are biassed and used in the allocation of resources or decision-making processes have the potential to exacerbate existing inequalities and have adverse effects on marginalised communities.

- Resource Depletion Issues: In the event of environmentally detrimental extraction practices, the manufacturing and maintenance of AI hardware components, such as semiconductors and rare earth minerals, may contribute to resource depletion and environmental harm.
- Transparency: In terms of transparency, traditional AI approaches frequently present challenges due to their lack of openness and accountability regarding their environmental implications.

There is a lack of defined sustainability guidelines and reporting mechanisms, which makes it difficult to assess and adequately address the environmental consequences of AI development.

- The development of energy-efficient artificial intelligence (AI) algorithms can help reduce the processing capacity necessary for AI activities, thereby reducing energy consumption and carbon emissions. This is one of the environmentally sustainable solutions that have been proposed.
- The environmental impact of AI operations can be significantly mitigated by transitioning data centres and AI infrastructure to renewable energy sources, such as solar or wind power.
- The process of conducting lifecycle assessments facilitates the identification and mitigation of environmental impacts during each stage of the lifecycle of AI products, from design to disposal. This procedure lessens the production of trash and the depletion of resources.
- Effective strategies for reducing e-waste production include the reduction of environmental impact, the promotion of responsible disposal practices, recycling, and the use of eco-friendly materials in the creation of AI hardware.
- The integration of ethical values into the design process facilitates the development of socially responsible AI systems that positively impact sustainability efforts, thereby addressing bias issues and promoting impartiality.
- The implementation of accountable mechanisms and the reporting of environmental metrics during the development of AI can increase awareness and encourage responsible practices.
- The establishment of a legislative framework for AI development that includes requirements such as undertaking environmental impact assessments and disclosing carbon footprints encourages enterprises to implement more sustainable practices.
- Collaboration and knowledge sharing among diverse stakeholders, such as researchers, policymakers, industry professionals, and community members, can foster innovation and positive transformations by providing a platform for the exchange of best practices and expertise in sustainable AI development.

CONCLUSION

Sustainable sustainability emphasises the importance of incorporating environmental, social, and economic factors into human society to ensure its long-term sustainability. This discussion acknowledges the interdependence of critical elements, including climate, finance, commerce, technology, and governance, which are dependent on finite natural resources. Sustainable practices are critical to human life because they promote social justice, economic stability, and environmental conservation. The guarantee of equitable access to these resources, the establishment of resilient economic systems, and the preservation of natural resources are among the central principles. This comprehensive viewpoint requires interdisciplinary collaboration to coordinate economic viability, equitable societal outcomes, and environmental sustainability, all while ensuring that future generations are able to meet their needs. A necessary condition for true sustainability, which goes beyond the prevention of climate change and decarbonisation, is the assessment of the wider effects of sustainable solutions on many economic and social dimensions. In order to ensure successful implementation, it is imperative to address obstacles such as economic disruption, resource allocation, and social transformations in vulnerable regions. It is essential to consider the long-term effects of sustainable practices in a number of domains, including education, consumption, urban planning, and agriculture. It is imperative to assess contemporary sustainable technologies, including electric vehicles (EVs) and artificial intelligence (AI). Despite the potential to reduce environmental impacts, the sustainability of these technologies may be compromised by obstacles such as the environmental footprint of battery production, energy-intensive AI algorithms, and ethical/social concerns associated with AI deployment. Sustainable AI prioritises the comprehensive lifecycle of AI technologies, with a particular emphasis on social equity and ecological sustainability. On the other hand, the switch to electric vehicles (EVs) calls for controlling energy use as well as the environmental effects of lithium mining and battery disposal. The design of environmentally friendly hardware, the utilisation of renewable energy sources, and the development of energy-efficient algorithms comprise a comprehensive sustainability strategy in technology. The establishment of regulatory frameworks and the promotion of collaboration among stakeholders can enhance transparency and accountability in sustainable practices. A commitment to interdisciplinary approaches that harmonise environmental, social, and economic objectives is necessary for genuine sustainability, while also ensuring that technological advancements have a positive impact on long-term sustainability objectives.

REFERENCES

Agovino, M., Casaccia, M., Ciommi, M., Ferrara, M., & Marchesano, K. (2019). Agriculture, climate change and sustainability: The case of EU-28. *Ecological Indicators*, 105, 525–543. DOI: 10.1016/j.ecolind.2018.04.064

Arora, N. K. (2018). Agricultural sustainability and food security. *Environmental Sustainability*, 1(3), 217–219. DOI: 10.1007/s42398-018-00032-2

Borowy, I. (2013). *Defining sustainable development for our common future: A history of the World Commission on Environment and Development (Brundtland Commission)*. Routledge. DOI: 10.4324/9780203383797

Burton, E., & Mitchell, L. (2006). *Inclusive urban design: Streets for life*. routledge.

Chang, Y., & Ke, J. (2024). Socially responsible artificial intelligence empowered people analytics: A novel framework towards sustainability. *Human Resource Development Review*, 23(1), 88–120. DOI: 10.1177/15344843231200930

Chen, Z., Wu, M., Chan, A., Li, X., & Ong, Y. (2023). Survey on AI sustainability: Emerging trends on learning algorithms and research challenges. *IEEE Computational Intelligence Magazine*, 18(2), 60–77. DOI: 10.1109/MCI.2023.3245733

Costa, C. M., Barbosa, J. C., Gonçalves, R., Castro, H., Del Campo, F. J., & Lanceros-Méndez, S. (2021). Recycling and environmental issues of lithium-ion batteries: Advances, challenges and opportunities. *Energy Storage Materials*, 37, 433–465. DOI: 10.1016/j.ensm.2021.02.032

Díaz, C. G., Zambrana-Vasquez, D., & Bartolomé, C. (2024). Building resilient cities: A comprehensive review of climate change adaptation indicators for urban design. *Energies*, 17(8), 1959. DOI: 10.3390/en17081959

Eagan, P., Cook, T., & Joeres, E. (2002). Teaching the importance of culture and interdisciplinary education for sustainable development. *International Journal of Sustainability in Higher Education*, 3(1), 48–66. DOI: 10.1108/14676370210414173

Fiksel, J. (2006). Sustainability and resilience: toward a systems approach. *Sustainability: Science. Practice and Policy*, 2(2), 14–21.

Ibrahim, M., Labaki, M., Giraudon, J., & Lamonier, J. (2020). Hydroxyapatite, a multifunctional material for air, water and soil pollution control: A review. *Journal of Hazardous Materials*, 383, 121139. DOI: 10.1016/j.jhazmat.2019.121139 PMID: 31520935

Janker, J., & Mann, S. (2020). Understanding the social dimension of sustainability in agriculture: A critical review of sustainability assessment tools. *Environment, Development and Sustainability*, 22(3), 1671–1691. DOI: 10.1007/s10668-018-0282-0

Kaunda, R. B. (2020). Potential environmental impacts of lithium mining. *Journal of Energy & Natural Resources Law*, 38(3), 237–244. DOI: 10.1080/02646811.2020.1754596

Khalid, M. R., Alam, M. S., Sarwar, A., & Asghar, M. J. (2019). A Comprehensive review on electric vehicles charging infrastructures and their impacts on power-quality of the utility grid. *ETransportation*, 1, 100006. DOI: 10.1016/j.etran.2019.100006

Kuhlman, T., & Farrington, J. (2010). What is sustainability? *Sustainability (Basel)*, 2(11), 3436–3448. DOI: 10.3390/su2113436

Lavuri, R., Roubaud, D., & Grebinevych, O. (2023). Sustainable consumption behaviour: Mediating role of pro-environment self-identity, attitude, and moderation role of environmental protection emotion. *Journal of Environmental Management*, 347, 119106. DOI: 10.1016/j.jenvman.2023.119106 PMID: 37806273

Liu, J., Mooney, H., Hull, V., Davis, S. J., Gaskell, J., Hertel, T., Lubchenco, J., Seto, K. C., Gleick, P., Kremen, C., & Li, S. (2015). Systems integration for global sustainability. *Science*, 347(6225), 1258832. DOI: 10.1126/science.1258832 PMID: 25722418

Malthus, T. R. (1959). *Population: The first essay*. University of Michigan Press. DOI: 10.3998/mpub.6638

McFarlane, D. A., & Ogazon, A. G. (2011). The challenges of sustainability education. *Journal of Multidisciplinary Research (1947-2900), 3*(3)

Meshram, P., Mishra, A., & Sahu, R. (2020). Environmental impact of spent lithium ion batteries and green recycling perspectives by organic acids–A review. *Chemosphere*, 242, 125291. DOI: 10.1016/j.chemosphere.2019.125291 PMID: 31896181

Mishina, N. V., Shikula, I. R., & Afanasyeva, S. A. (2020). Features of legal regulation of artificial intelligence as a guarantee of sustainable development of society. Paper presented at the *E3S Web of Conferences, 208* 06010. DOI: 10.1051/e3sconf/202020806010

Mowforth, M., & Munt, I. (2015). *Tourism and sustainability: Development, globalisation and new tourism in the third world*. routledge.

Omole, F. O., Olajiga, O. K., & Olatunde, T. M. (2024). Sustainable urban design: A review of eco-friendly building practices and community impact. *Engineering Science & Technology Journal*, 5(3), 1020–1030. DOI: 10.51594/estj.v5i3.955

Pezzey, J. (2017). *Sustainability: an interdisciplinary guide. The economics of sustainability*. Routledge.

Rodriguez-Gonzalez, P. T., Rico-Martinez, R., & Rico-Ramirez, V. (2020). Effect of feedback loops on the sustainability and resilience of human-ecosystems. *Ecological Modelling*, 426, 109018. DOI: 10.1016/j.ecolmodel.2020.109018

Rosati, F., & Faria, L. G. (2019). Addressing the SDGs in sustainability reports: The relationship with institutional factors. *Journal of Cleaner Production*, 215, 1312–1326. DOI: 10.1016/j.jclepro.2018.12.107

Sauvé, S., Bernard, S., & Sloan, P. (2016). Environmental sciences, sustainable development and circular economy: Alternative concepts for trans-disciplinary research. *Environmental Development*, 17, 48–56. DOI: 10.1016/j.envdev.2015.09.002

Schilling, T., Wyss, R., & Binder, C. R. (2018). The resilience of sustainability transitions. *Sustainability (Basel)*, 10(12), 4593. DOI: 10.3390/su10124593

Schoormann, T., Strobel, G., Möller, F., Petrik, D., & Zschech, P. (2023). Artificial intelligence for sustainability—A systematic review of information systems literature. *Communications of the Association for Information Systems*, 52(1), 8. DOI: 10.17705/1CAIS.05209

Tripathi, A., & Singh, M. P. (2016). Determinants of sustainable/green consumption: A review. *International Journal of Environmental Technology and Management*, 19(3-4), 316–358. DOI: 10.1504/IJETM.2016.082258

Van Wynsberghe, A. (2021). Sustainable AI: AI for sustainability and the sustainability of AI. *AI and Ethics*, 1(3), 213–218. DOI: 10.1007/s43681-021-00043-6

Vera, M. L., Torres, W. R., Galli, C. I., Chagnes, A., & Flexer, V. (2023). Environmental impact of direct lithium extraction from brines. *Nature Reviews. Earth & Environment*, 4(3), 149–165. DOI: 10.1038/s43017-022-00387-5

Wawrzyniak, D. (2023). Animal husbandry and sustainable agriculture: Is animal welfare (only) an issue of sustainability of agricultural production or a separate issue on its own? *Animal*, 17, 100880. DOI: 10.1016/j.animal.2023.100880 PMID: 37793705

Wu, C., Raghavendra, R., Gupta, U., Acun, B., Ardalani, N., Maeng, K., Chang, G., Aga, F., Huang, J., & Bai, C. (2022). Sustainable ai: Environmental implications, challenges and opportunities. *Proceedings of Machine Learning and Systems*, 4, 795–813.

Zhao, J., & Gómez Fariñas, B. (2023). Artificial intelligence and sustainable decisions. *European Business Organization Law Review*, 24(1), 1–39. DOI: 10.1007/s40804-022-00262-2

KEY TERMS AND DEFINITIONS

AI: Artificial Intelligence

Artificial Intelligence Sustainability: Practices that do not compromise future generations by maintaining a balance among environmental, social, and economic factors.

EV: Electrical Vehicles

Sustainable AI: Artificial Intelligence that follows sustainable practices

ML: Machine Learning

DL: Deep Learning

Smart: Intelligent

Chapter 2
Sustainable Hospitality: Lessons from a Case in Algarve, Portugal

Debora Castro
University of Algarve, Portugal

Silvia Fernandes
https://orcid.org/0000-0002-1699-5415
CinTurs & Faculty of Economics. University of Algarve, Portugal

ABSTRACT

This work aims at analyzing the sustainable practices of 5-star hospitality in Algarve (Portugal). It explores how these hotels communicate sustainability, which practices they adopt and main drivers/constraints. First observations were made in their own websites and in Booking.com. Then an online survey was used to collect more information. It was perceived a higher level of sustainable behavior in internal processes. Main areas are energy management, water/waste management and social responsibility. Most popular practices are the least expensive, and main drivers are environmental responsibility, strategic vision, and efficiency gains. Although the main obstacle referred is the investment involved, lack of qualified personnel and technical issues are other issues. The growth of new businesses, personnel training, and specialized communication in the area can overcome them. The originality of this study lies on: the 5-star category in Algarve filling a gap in the literature; the need of multiple research methods; and application of the sustainability triad approach.

INTRODUCTION

Tourism has been one of the global engines of development. In recent decades, this sector has experienced continuous and diversified growth, marked by significant impacts on economic sectors, natural resources, cultural and social relations (Batle, Sintes and Moon, 2018). With good planning and management, this sector can be a positive force to bring benefits to destinations. And the hotel industry can play an important role in its sustainability. However, there are still different reactions to this trend. For some executives, this goal is a moral requirement; for other, it is a legal requirement, and even for some other, sustainability is perceived as an inherent cost of business to maintain the company's legitimacy.

DOI: 10.4018/979-8-3693-5748-4.ch002

Some companies have begun to treat sustainability as a business opportunity, opening ways to reduce costs and risks, and increase their income and market share through innovative practices.

The present work intends to analyze sustainability in the 5-star hotels of the Algarve region (in Portugal), discovering how they communicate their sustainable behavior on their websites; which are their most adopted sustainable practices; and which are the main drivers/ obstacles. It is structured as follows: Section 1.1 presents its context and objectives; Section 2 presents the literature review on most important concepts and their relations; Section 3 presents the research framework for analyzing the sustainable practices of 5-star hotels in Algarve; Section 4 presents and discusses the results obtained in the light of the data collected and analyzed. Finally, section 5 presents the conclusions, limitations of the study, and its implications for planning this important topic of the hospitality agenda.

Context and Objectives

The concept of sustainable development was firstly discussed in the Brundtland Report (World Commission on Environment and Development - WCED, 1987). According to this report, it is considered as development that meets present needs without compromising the ability of future generations to meet their needs. In 2015, the United Nations General Assembly defined its 2030-agenda for sustainable development relating it with the United Nations SDG (Sustainable Development Goals). The expected result, by a joint work from governments, companies and citizens, is a new global model to end poverty, promote prosperity and well-being for all, protect the environment and fight climate change (UNESCO Portugal, 2020). Tourism can contribute to all these objectives (Tourism of Portugal, 2020).

Sustainable tourism development is defined as an activity that takes full account of its current condition, and its economic, social and environmental future, to address the needs of visitors, industry, environment, and host communities (Abdou, Hassan and Moustafa, 2020). In building the Strategy for Tourism 2027, Portugal decided to become one of the most competitive and sustainable destinations in the world, with a strong commitment to achieve the SDG. In the period 2010-2019, the country has registered an average annual growth rate of 7.2% in overnight stays, translated into an increase from 37 million in 2010 to 70 million in 2019 (the highest value attained). In 2020, due to Covid-19 pandemic, the weight of tourism in the national Gross Domestic Product (GDP) dropped by half (Tourism of Portugal, 2021).

Regarding the Algarve, its economic base is highly dependent on tourism. That is why it suffered a strong impact due to the pandemic: a decrease of 16.7% in its GDP (CCDR Algarve, 2020). The region has an Observatory of Sustainability for Tourism – OBSERVE - which investigates and evaluates its sustainable development (situation, trends) based on indicators that meet the strategic needs of main stakeholders involved. It aims at providing environmental, economic, sociocultural, and institutional solutions to support decision-making for a sustainable growth (Farinha, Oliveira, Silva, Lance, Pinheiro and Miguel, 2019). Some of its studies show a convergent perspective in this matter, namely in natural resources and biodiversity, security, and supply chain. However, there is a lack of uniformity in terms of sustainability initiatives (Farinha, Huertas, Pinheiro, Silva, Lance, Oliveira and Batista, 2021).

This work can contribute to know more about sustainability in a very important subsector of tourism, the hospitality. It investigates the commitment and use of sustainable practices, especially in 5-star hotel units in Algarve. More specifically, it intends to:

1) Assess their level of commitment to sustainability
2) Know which are their most popular sustainable practices

3) Identify the main drivers for implementing those practices
4) Identify the main obstacles to their implementation.

EVOLUTION OF SUSTAINABLE HOSPITALITY

The hotel industry has a strong impact on the environment due to its high consumption of water, energy, and non-durable goods along with the discharge of large amounts of waste. If these resources are not properly managed, they will have a negative impact on the environment adding operational costs to the organization (Kim, Barber and Kim, 2019). Implementing corporate sustainability in this industry provides a framework to think beyond shareholders and work with stakeholders regarding their impact on environmental protection, economic development, and social equity (Linneberg, Madsen and Nielsen, 2019). Hotel operations can promote environmental preservation and contribute to quality of life in the community (Santos, Mexas, Meirino, Sampaio and Costa, 2020).

In recent years, many sustainability practices have emerged. Such strategies can help companies become a green driver and differentiate from competitors influencing return and corporate image (Chung, 2020). Using green practices as a competitive advantage can help hotel operators make strategic decisions that will improve their technical and environmental management. Many of them recognize that the adoption of ecological practices is not only an ethical issue, but also beneficial for reducing costs, market differentiation, and social responsibility (Abdou et al., 2020).

Green marketing is a powerful promotional tool for hotels. After all, the effort made must be known by their consumers, and then a stronger bond can be established with them (Mele, Gomez and Garay, 2019). Multinational hotel corporations, which actively participate in international business, have become aware of the importance of corporate social responsibility and sustainability issues. Financial regulators around the world seem to embrace sustainability disclosures by open traded companies (Ng and Tavitiyaman, 2020). The implementation of these practices has brought positive results that are reflected in the well-being of local communities, environmental protection, and good working conditions for employees (Hassan, Shehata, El-Dief and Salem, 2020). The disclosure of issues like human rights, carbon emissions, climate risks and the SDG is increasingly relevant in the business agenda (Miranda, Ortiz and Cardona, 2019).

The Green Challenge

In the last decades, the need for aligning the business strategy with the environment made possible the expansion of competitive intelligence. This helps organizations to detect new opportunities, create value and improve their performance (Salguero, Gámez, Fernández and Palomo, 2019). The use of innovative solutions, developed for the sustainable market (green innovation), has positively affected their environmental performance. They then meet government regulations more easily and reduce waste and pollution. They also improve their competitive profile (Asadi, Pourhashemi, Nilashi, Abdullah, Samad, Yadegaridehkordi, Aljojo and Razali, 2020).

New technologies related with the interconnectivity between the hotel and tourists are examples. Virtual check-in can be mentioned, very useful in the current pandemic context (contacless technology), in addition to their role in reducing costs related to work, paper, printer, among other (Foris, Crihalmean and Panoiu, 2020). Another relevant aspect is the need to adopt green certification. There is already a number of ecological certification standards for sustainable tourism such as: Green Globe, Green Key, Travelife, Earth Check and NEPCon. Reliable certifications will contribute to consider environmental practices as a priority when choosing where to stay (Foris, Crihalmean and Foris, 2020).

On the other hand, it is necessary to involve the employees in sustainability through training, awareness, and available information (Cop, Alola and Alola, 2020). It is more beneficial to have knowledge about sustainability than just absorbing practices without understanding their inherent principles (Oriade, Osinaike, Aduhene and Wang, 2021). Hotels should not impose the use of sustainable measures, but rather create a friendly environment for that. This will enable the workforce to understand the hotel's environmental protection policies and ecological behavior (Ahmed, Guo, Qureshi, Raza, Khan and Salam, 2021).

Some studies suggest that sustainability should be a mandatory subject in the curriculum of all hospitality institutions (undergraduate and graduate), including a strong motivation among students to be personally involved with related issues and practices (Goh, Muskat and Tan, 2017). Hotel units already practice sustainable measures, but few really account for the results in their reports. There is a great need to include this information which will serve for certifications (Alvarez-Risco, Estrada-Merino and Perez-Luyo, 2020). The lack of sustainability results in the sector has been a gap that needs to be filled, so that it can be possible to identify the level of implementation. A key factor in this transition is government support in terms of incentives in recognition of the efforts toward sustainable commitment (Jaykumar, 2020). Because this issue has been more guided by efficiency gains within the existing business models, more focused on continuous growth (Jones, Hillier and Comfort, 2013).

A Triple Challenge

Sustainability is underpinned by three dimensions (figure 1) referred as the Triple Bottom Line (TBL) by John Elkington in the mid-90s. It began as an accounting framework that strove to include the environmental and social dimensions in the traditional finance-centric measurement of business performance. However, it lacks a standardized reporting method to holistically measure the three dimensions. It is a challenge to accurately assess and quantify them (Goh, Chong, Jack and Faris, 2020).

Figure 1. The Sustainability triad (TBL)

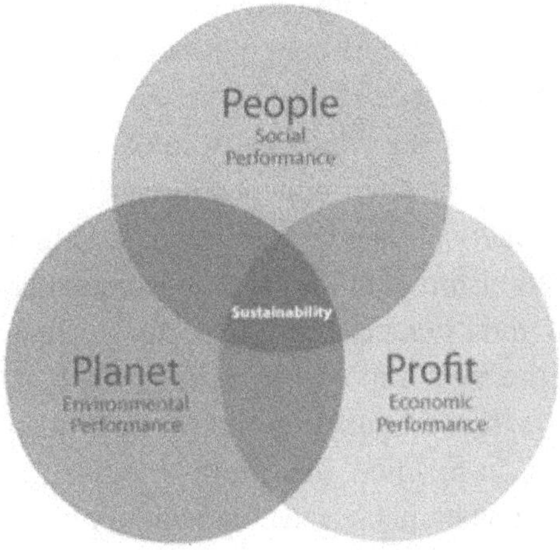

Source: own elaboration based on Gevme (2023)

Some authors refer that the organizational perception of sustainability is only linked to the environment, what shows that the concept of TBL has not yet been absorbed in the organizational context considered. This means that the full potential of these organizations is compromised (Alberton, Kieling, Lyra, Hoffmann, Lopez and Stefano, 2022). These three dimensions interact, overlap and sometimes conflict: for example, environmental sustainability can affect economic sustainability due to extra investments needed for cleaner production processes. Organizations need to act holistically for better results. Each dimension represents a necessary, but not sufficient, condition for achieving sustainability. Even if an organization manages synergies between environmental and economic dimensions, it has not yet attained sustainability (Braccini and Margherita, 2019).

There are numerous benefits for companies that adopt sustainability as a strategy, such as: identifying business opportunities, innovative solutions, increasing their corporate value and tolerance to risk, better use of resources, and strengthening relations with stakeholders (Coalition for Green Growth, 2015). Currently there is a portfolio of options for sustainable management in the hotel chain, either through disruptive or incremental innovations, or through the adoption of good practices (Ozturkoglu, Sari and Saygili, 2021). And adopting practices for sustainability is a step to strengthen the competitiveness of companies and contribute to turn Portugal into a sustainable destination, ensuring that following generations enjoy the current distinctive assets (Tourism of Portugal, 2020).

Good Practices Monitoring

Best practices in strategy are defined as a stream of actions that lead to the best change in performance and thus are the best to adopt (Laugen et al., 2005). Accordingly, the best practices in hotel green supply-chain management are streams of actions related to its dimensions which lead to the best green performance in terms of improved recycling, reduced gas emissions, water and energy consumption, and waste generated (Migdadi, 2022). The concern about the green procurement of water and energy rather than other processes is related to the fact that hotel industry is one of the most water/energy-intensive sectors (Mak and Chang, 2019). In the tertiary sector, hotels rank among the top 5 in water/energy consumption and release of high levels of CO_2 per m^2 of room floor area (Kasim et al., 2014).

Sustainability is then an increasingly important factor of decision-making in this industry. However, it is not fully assimilated into business thinking. Although some changes require additional financial effort, there are small changes or investments that can bring numerous benefits to companies. In the economic dimension, there is an increase of efficiency which helps to continue operating, generate employment and remunerate the capital invested. In the environmental dimension, hotels can reduce their ecological footprint, pollutant gas emissions, energy/water consumption, among other issues. In the social dimension, they should appreciate a relationship with the local community, suppliers, and strategic partners.

There are numerous guides of good practices for the sector (even available on the Internet) as well as for other areas of tourism. But the most important practice, which at the same time can act as a communication and monitoring tool, is the sustainability report. Its objective is to describe the economic, environmental, and social impacts. In practice, this means measuring, disclosing, and accounting for organizational performance toward sustainable development. This information is useful to different publics such as the market, workers, non-governmental organizations, and investors. This document should have a description of the firm's sustainability performance. Table 1 presents the requirements for an efficient sustainability report.

Table 1. Efficient sustainability reporting

Content	**Materiality** **Inclusion of stakeholders** **Context of sustainability** **Coverage**
Quality	Balance Comparability Accuracy Frequency Clarity Reliability
Company profile	Organization's name Operational structure Headquarters' location Type and legal nature Size of organization
Management approach	Brief overview of the organization's management regarding each indicator, to establish performance information
Economic performance	Economic performance Market presence Indirect economic impacts
Society	Community Public policies Unfair competition Compliance

continued on following page

Table 1. Continued

Content	Materiality Inclusion of stakeholders Context of sustainability Coverage
Environmental performance	Materials Energy Water Biodiversity Emissions, effluents, waste Products and services Transports
Labor practices	Employment Relation work/governance Health and safety at work Training and education Equal opportunities
Human rights	Non-discrimination Freedom of association Bargaining agreement Prevention of forced labor Complaints practices Security practices
Responsability	Customer health and safety Labeling products and services Marketing communications Customer privacy Compliance

According to leading authors in the field, sustainability reports in the hotel sector are still in their "infancy" (Medrado and Jackson, 2015; Pommier and Engel, 2021). Sustainability experts say that the employee factor (involvement, training) is rather important than cost savings. As the justification for a cost-benefit analysis in this context is unclear, further research is needed using the stakeholders' lens. For instance, customers' awareness ought to be improved through providing them more information about environmentally friendly practices they can adopt. This will lead to their concern and willingness to pay for sustainable initiatives. A low level of public environmental awareness is often due to managers' reluctance to compromise the comfort of hotel guests (Migdadi, 2022). Environmental technologies have become more mature and offer more help in dealing with these challenges. Portugal has an interesting potential in this matter because the sectors that mostly develop innovation with environmental benefits are hospitality and restaurants (figure 2). This is a contributing factor in helping our hotels become more sustainable.

Figure 2. Firms' innovation with environmental benefits

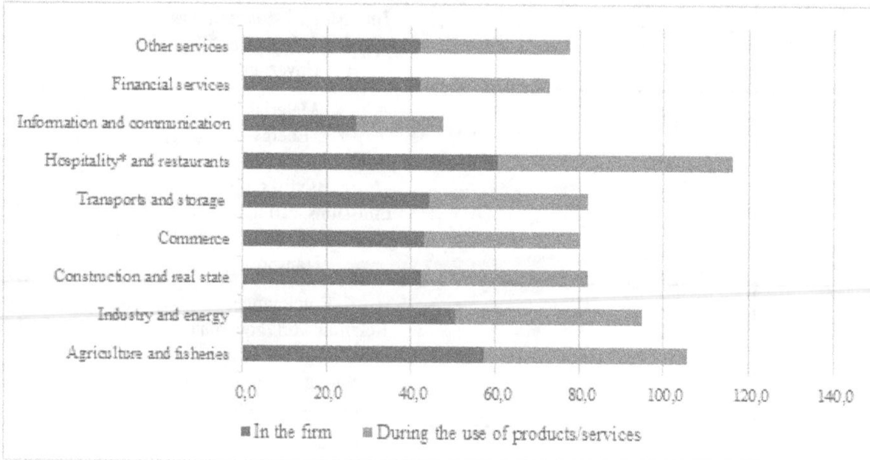

Source: CIS – Community Innovation Survey (2022), Portugal

RESEARCH DESIGN

The Methodology

The methodology applied in the first specific objective – *Assess the level of commitment to sustainability* – will be qualitative. Thus, each hotel's website will be analyzed to see if it communicates any sustainable behavior. If yes, the type of information provided will be checked according to the sustainability triad (TBL). This type of method has been used in the literature such as in the work of Moreo, Cobanoglu and DeMicco (2007). However, the present study fills some gaps that are described later in this section.

In qualitative research, instead of using a standardized measuring device, the researcher asks questions, makes interpretations, and records the observations. The main goal is to obtain an internal view of the sample under study driven by the issues raised (Kasim and Antwi, 2015; Choi, 2014). It systematically examines the content of a given body of materials with the aim of identifying patterns, themes or biases (Mohajan, 2018).

For the other three objectives (*Know which are the most popular sustainable practices*; *Identify the main drivers for implementing those practices*; *and identify the main obstacles to their implementation*), the quantitative method was applied through an online survey. Quantitative research focuses on aspects that can be quantified rather than interpreting the meanings that people give to their own actions (Rahman, 2017). The survey is based on the following relation between objectives and questions (table 2 and table 9 in Appendix 1):

Table 2. Main objectives and questions

Objectives	Questions
Which hotel departments have a higher level of good sustainable practices	1-7
What are the main drivers for implementing sustainable measures	8
What are the main challenges in implementing sustainable measures	9
Understanding the conditions of Algarve for sustainable practices	10
The answering manager's profile	11-13

Each method has its advantages and limitations, regardless of how long or widely it has been used on the subject. Thus, if more than one method is used in a single study, they can complement each other and overcome the limitations of a single method (Chu, 2015).

Comparative Analysis

The 5-star hospitality category was chosen due to the nature of its services. Since it is at the top of hotel classification, it highly considers the attributes of image/brand, modernization, and customer satisfaction. The Algarve's strategy for 2030 is following a path of change toward competitiveness and sustainability. This implies the commitment to a more informed and cohesive management, with higher levels of training and participation (even co-creation with customers/employees). Another reason for this category's choice is the kind of customer, normally with a higher purchasing power and level of desired quality. These clients tend to reveal a greater environmental awareness (Salem et al., 2022), and thus can be influencers of accommodation toward sustainability. Table 3 shows other related studies, whose comparison reflects the gaps this work helps to fill.

Table 3. Comparison with other studies for gap identification

Reference	Sample	Main method/tool	Focus
Almeida (2016)	4-star hotels in Lisbon	Sullivan matrix (diffusion of ecofriendly practices)	Obstacles to green practices
Catan (2018)	The Sheraton Lisboa Hotel & Spa	Case study (sustainability indicators)	Results (cost savings and environmental protection)
Silva (2016)	3 cases of 5-star hotels and customers	Interviews, online questionnaire (services quality measuring)	Comparing customer preferences with hotel practices in terms of sustainability
Tavares (2016)	The Neya Lisboa Hotel	Case study	Hoteliers' perception of the impact of sustainable practices on customers
Ferreira (2021)	86 hotels with the Green Key certificate in Portugal	Theory of planned behavior and application	Perceived benefits and customer's intention to adopt green hotels
Ferreira et al. (2021)	13 hotels, from 3 to 5-stars	Interviews, qualitative analysis	Perspective of managers of ecological hotels
Ornelas (2017)	28 hotels in Madeira Island	Online questionnaire and official documents	Sustainable practices and their impacts on tourism
Calheiros (2021)	4-star hotels in Porto	Interviews, online questionnaire	Sustainable practices implementation and impact

continued on following page

Table 3. Continued

Reference	Sample	Main method/tool	Focus
Alemão (2019)	1- to 5-star hotels (in Oporto, Lisbon and Algarve)	Online and emailed questionnaires	Sustainable practices in the value chain and customer perception
Preziosi et al. (2019)	EU-Ecolabel certified hotels	Survey, interviews,	Role of environmental practices/communication on guest loyalty
Pereira, Silva & Dias (2021)	Luxury hotel in the Arrábida natural park	Qualitative approach (case study)	Sustainable practices and social dimension

In resume, this comparison shows that studies on the subject focused more on 4-star hotels or on cases of unique hotels (or groups). This reflects a gap of studies around a sample of the five-star category.

The Sample

According to Tourism of Portugal (2021), the universe of five-star has 55 units located in the region. As mentioned before, this research began with a structured observation of their websites to verify whether they reveal any sustainable commitment/attitude. Also, whether they have any certification policy/strategy in terms of sustainability. To determine if each website is eligible, the hotel's structure was analyzed because in some cases it belongs to a group that uses the same website. This important aspect turned the sample of 55 units to 47 units (sample1). This sample was object of the first method referred - qualitative (website analysis).

Then the survey used for quantitative analysis was made available from 23 March to 7 June 2022, whose link was sent within a personalized email to the sample1. We obtained 13 answers (response rate=28%) which make the sample2. This can be considered a convenience sample for obtaining more detailed data like the TBL dimensions. A convenience sample is an accessible or available sample of primary data, which has limitations. Although the results cannot be generalized to the universe, it can give us some insights. Regarding the low response rate, specialists argue it is positive given both the hotels' category and the short response period (related with the dissertation deadline on which this work is based). For instance, Wu et al. (2022) conducted a comprehensive search that screened 8672 studies and 1071 online survey response rates reported in education-related research. The authors acknowledged that sending an online survey to more participants did not generate a higher response rate. Instead, contacting potential participants personally or by phone calls, to remind them about the survey, could yield a higher rate. Also, they found that the timing of these alternatives matters: when different types of survey are delivered sequentially, such as an online survey followed by a mail option, the online mode response rates improved.

Results and Discussion

The results obtained and their discussion are as follows, according to the objectives proposed.

1) Assessment of the hotels' commitment to sustainability

From the qualitative approach used, it could be observed that only 20.8% of sample1 clearly communicates a sustainable positioning on their own websites. However, it is interesting that more than half of the sample (58.3%) reveals some kind of sustainability attitude. This refers to any action in the TBL approach, such as: promote small local businesses; use local products; appreciate local cuisine; valuing local culture; encourage cycling/hiking; cultivate own vegetable garden; provide charging stations for electric cars. Another aspect observed is the lack of standardization in this communication. There is a wide variety of information making difficult to compare the practices disclosed on their websites. Around 45.8% of the sample disclose their sustainable commitment in Booking.com. This can be due to this platform's level of competition. The hotel units that participate in its sustainability filter are marked as accommodation of 'sustainable travel program'. When a client searches for a hotel with this mark, it is possible to know which actions the hotel takes to ensure a sustainable and environmentally friendly trip. When the data found in the websites was consolidated with Booking.com, 55.3% of the sample emerged as having a sustainable commitment in one or both platforms (see table 10 in Appendix 2).

2) Know their most popular sustainable practices

From this objective onward, the survey used was accessible online through an email sent to the 47 hotel units. The 13 answers received represent a response rate of ~28% (sample2). The data collected were then analyzed in SPSS, a known software for statistical analysis. The departments that stood out regarding sustainable practices are energy consumption and greenhouse-gas emission, water consumption, waste management, and social responsibility.

Energy consumption and greenhouse-gas emission

It is important to refer that in 2021 hotel units accounted for 53% of total energy consumption in Portuguese tourism sector and emitted 51% of gases (Tourism of Portugal, 2021). Table 4 shows that LED lamps and accommodation switch detectors are the most used options. A reason may be the lower investment required. More expensive options, like renewable electricity throughout the hotel, electric car charging stations, and carbon offsetting are the less applied.

Table 4. Practices related to energy/greenhouse-gas emission

LED lamps	**92%**
Electricity triggered by accommodation switch detector	85%
Double glazing	85%
Most of food is local	77%
Parking for bicycles	77%
Electric car charging stations	30%
Renewable electricity throughout the hotel	30%
Carbon offsetting	15%
Average	*61.4%*

Carbon offsetting is an important step toward environmental awareness. The trading of carbon credits (offsets) can help tourism companies meet the ambitious goal of reducing greenhouse-gas emissions (Jornal de Negócios, 2022). Yet many businesses cannot fully eliminate their emissions or lessen them quickly. It will be necessary to use carbon credits to offset emissions they can't get rid of by other means. The voluntary carbon market can help them finance climate-action projects that would not otherwise be done.

Water consumption and waste management

This issue is also a matter of concern in Algarve as this region faces a serious situation of water stress in the coming years (Sul Informação, 2021). Regarding waste in 2020, food waste represented 25% of the purchases for the hotel industry (Publituris.hotelaria, 2020). There are several programs such as *Refood*, organized by non-profit companies, where volunteers collect food from hotels, restaurants, supermarkets, and pastry shops to do its distribution to people in need. Table 5 shows the practices most applied, where elimination of plastic and efforts to reduce food waste have high adhesion. Also, the re-usage of towels and efficient water consumption in the rooms. However, a greater efficiency can be attained through investing in water re-usage, water purification systems, rainwater storage and control of its consumption.

Table 5. Practices related to water/waste issues

Re-usage of towels	**92%**
Toilets/showers with efficient water consumption	92%
Eliminated disposable plastic stirrers/straws	92%
Efforts to reduce food waste	92%
Eliminated plastic cutlery and plates	85%
Eliminated plastic bottles/cups	85%
Selective garbage/recycling	69%
Eliminated the use of water bottles	54%
Eliminated single-use packaging	46%
Average	*78.5%*

Social responsibility

This aspect reveals the company's commitment to society, a fundamental dimension of the sustainability triad. Table 6 shows the practices most applied in this matter (in the sample obtained).

Table 6. Socially responsible practices

Tours/activities made by local guides/businesses	**85%**
Investing a % of their revenue in local projects	69%
Platform for local artists to show their talents	31%
Average	*62%*

These activities are important for developing small businesses in the region, which generate jobs and provide more tourist attractions. Such efforts empower local communities and strengthen the local economy and culture.

3) Identify the main drivers for implementing those practices

The majority of responding hotels referred environmental responsibility as the main motivator for implementing sustainable practices. This expresses a commitment to carry out activities without neglecting their impacts on the environment (table 7).

Table 7. Main drivers of sustainable practices

Environmental responsibility	**92%**
Strategic vision	85%
Efficiency gains	77%
Financial savings	69%
Improvement of hotel's image	69%
Attitude of customers	62%
Public opinion	31%
Regulatory enforcement	31%
Investors' demand	23%
Average	*59.8%*

With a smaller ecological footprint and better cost-efficiency relation, a hotel can reach a sustainable competitive advantage. Be sustainable is not only investing in ecofriendly ways, but also have a good financial health and social responsibility (triad). With the current climate challenges, consumers are more aware of the need to protect the environment. On the other hand, less important drivers are public opinion and regulatory enforcement. Possible reasons can be that public opinion is difficult to quantify and there is still room for pro-sustainability legislation in the sector. Investor's demand is the least important factor, perhaps because most hotels still don't develop a sustainability report.

4) Identify the main obstacles to their implementation

Besides high costs, lack of qualified human resources and technical issues are the main constraints referred (table 8). Investing in staff training can solve part of the problem.

Table 8. Main obstacles to sustainable practices

High costs	**69%**
Lack of qualified personnel	62%
Technical/technological issues	62%
Lack of suitable partners	54%
Scarcity of financial resources	54%
Lack of available information	31%
Lack of advice on the most beneficial actions	23%
Average	*43%*

Lack of suitable partners and scarcity of financial resources are other constraints. The growth of new businesses in the area can probably overcome these issues. Lack of available information is another obstacle pointed by 31% of the respondents. For instance, Turismo de Portugal has several guides to support managers on sustainability. However, they must be interest in planning and implementing it. Lack of advice on the most beneficial actions was the next constraint referred. In fact, it is expensive to hire specialized consultants, but an adequate consultancy can suggest innovative ways to develop the sustainability triad.

Figure 3 shows a holistic view of the sample used in terms of its TBL. We can see that the percentages are similar in the three dimensions of sustainability, where the highest ones are related to the environmental pillar (energy/greenhouse-gas emissions; water/waste management). The dimensions that require more attention for future planning are related to drivers and obstacles, mainly qualified personnel and technical issues. Nevertheless, the polluting gas emissions is an issue to be continuously addressed, in terms of progressive renewable electricity and decarbonization. Financial savings from the good practices and its monitoring can offset the investment costs.

Figure 3. The sustainability triad applied

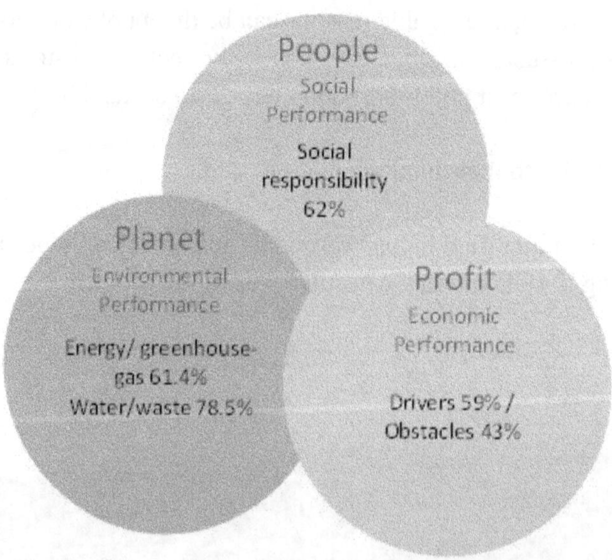

Source: own elaboration

In terms of sample characteristics, in gender ~77% are men and 23% women; in age 69% are in the interval of 30-49; followed by 15% with more than 50 years old and 15% with less than 30; in qualifications ~62% have bachelor's or undergraduate degree, followed by Master degree (15%); and in location ~54% of the respondent hotel units are situated at the east of Algarve (Loulé, Olhão) and 46% in the west (Albufeira, Portimão, Lagos, Vila do Bispo, Monchique). We can acknowledge a still weak profile of specialized graduation and a predominance of men in top/medium management positions. So, in spite of the country's awareness toward environmental technology development in hospitality (figure 2), it won't be sufficient without specializations that link this area with the global SDG and tourism trends.

This study has at least three lines of originality, compared to other related works: 1) the focus on 5-star hotels in Algarve, that fills a gap in the literature as previously referred; 2) the need it raises about using innovative research methods, like the approach used here of multiple methods (qualitative – website content analysis, and quantitative – online survey and frequency analysis). This is due to the scarcity of responses even in low season what requires other survey modes such as mail/personal contact/phone call; and 3) the sustainability triad application (figure 3) that gives a view of the most successful dimensions and those needing attention.

CONCLUSION

The present study gives more insights into the best practices in hotel green supply chain management. Most best activities here reported turn out to be related to internal green process design. This happens since hotel's internal processes generate the most emissions and engage most energy and water consumption processes. The implementation of green practices is important since they improve hotel's image through a higher level of trust and satisfaction of guests leading to long-term relationships (Moise et al., 2021). For example, various dimensions of environmental practices are positively related to tourists' stay at an environmentally friendly hotel (Rawashdeh and Al-Ababneh, 2021).

Regarding the 5-star hospitality in Algarve (Portugal), an increasing influencer category toward the sustainability triad, only 20.8% clearly communicate a sustainable positioning in their websites. Around 45.8% have disclosed their sustainable commitment in Booking.com. But when the data from their websites are consolidated with this engine, more than half are concerned with sustainable commitment in one or both platforms. Then, for a more detailed analysis, an online survey was made available. Main practices that stood out are energy consumption and greenhouse-gas emissions, water consumption, waste management, and social responsibility. Main motivators to implement sustainable practices in the sample used are: social and environmental responsibility; strategic vision; efficiency gains; financial savings and hotel's image. Main constraints referred are related to high costs; lack of qualified personnel and technical issues; lack of suitable partners and financial resources. The most limitative dimension observed is the economic, due to a remaining perception of high costs involved. However, the growth of new businesses and jobs in the subject will probably overcome this problem.

Limitations, Implications and Future Work

The present work lacks analyses to the data related with the open question, about understanding the conditions of Algarve for sustainable practices. Only four hotels participated, referring the (re)use of water as the area needing most support and solar and wind energy as strengths of the region. Nevertheless, the convenience sample used has given some insights, in spite of its limited scope. It suggests that more advanced sustainable practices are needed, which can be developed through more training in the area so that the sustainability triad becomes part of the business strategy. Also, the role of social platforms, connecting hotels to local talent and projects, and sustainability certification are increasingly important.

Sustainability issues and behavioral implications can be addressed through communication and awareness campaigns targeted at existing and potential customers (Salem et al., 2022). Publicity materials should emphasize the important role that clients can play in supporting green housing, without losing the quality of the service. The perception that green hotels have less competitive offers must be changed, fully exploiting the cost advantages of ecological practices. Online communication can be a key tool in the tourism industry (through social networks, website, newsletters, hashtags) where hotel units can disclose their sustainable practices and certificates/awards in a clear way.

According to the Booking Sustainability Report (2021), 73% of travelers would more easily choose a hotel if it implemented sustainable practices. In addition, 61% said the pandemic has made them want to travel more sustainably in the future. These data highlight how essential it is for the hotel industry to build a strategy around sustainability. For instance, Ferreira et al. (2021) refer that the pandemic has caused setbacks in environmental performance, generating more non-recyclable waste. Tourism and hospitality sectors should use this impact to reflect on new ecological measures and employee awareness. As we move toward a more conscious consumer profile, hotels must adapt offering the opportunity to travel and stay in a more sustainable way. These agents of change should treat sustainability as a holistic approach, from their operations to their choices of products, guests, local communities, and suppliers.

Future work will focus on expanding the analysis to more 5-star hotels in the region and then the 4-star category. Still in the 5-star whole dataset, analyze the data related with the open question (about understanding the conditions of Algarve for implementing sustainable practices) and with the sociodemographic profiles of the respondents.

REFERENCES

Abdou, A. H., Hassan, T. H., & Moustafa, M. E. D. (2020). A description of green hotel practices and their role in achieving sustainable development. *Sustainability (Basel)*, 12(22), 9624. DOI: 10.3390/su12229624

Ahmed, M., Guo, O., Qureshi, M. A., Raza, S. A., Khan, K. A., & Salam, J. (2021). Do green HR practices enhance green motivation and proactive environmental management maturity in hotel industry? *International Journal of Hospitality Management*, 94, 102852. DOI: 10.1016/j.ijhm.2020.102852

Alberton, A., Kieling, A. P., Lyra, F. R., Hoffmann, E. M., Lopez, M. P. V., & Stefano, S. R. (2022). Competencies for sustainability in hotels: Insights from Brazil. *Employee Relations*, 44(3), 555–575. DOI: 10.1108/ER-01-2019-0093

Alemão, C. (2019). Environmental sustainability practices in the supply chain of hotels and the consumers' perception, Master Dissertation in Management. UCP, Lisbon.

Algarve, C. C. D. R. (2020). Observatório das dinâmicas regionais. https://www.ccdr-alg.pt/site/info/observatorio-das-dinamicas-regionais

Almeida, J. (2016). Sustentabilidade na hotelaria. Uma análise de infusão/difusão em hotéis de Lisboa. Master Dissertation, Universidade Europeia. Lisbon, Portugal.

Alvarez-Risco, A., Estrada-Merino, A., & Perez-Luyo, R. (2020). Sustainable development goals in hospitality management. In Ruël, H., & Lombarts, A. (Eds.), *Sustainable Hospitality Management. Advanced Series in Management, 24* (pp. 159–178). Emerald Publishing., DOI: 10.1108/S1877-636120200000024012

Asadi, S., Pourhashemi, S. O., Nilashi, M., Abdullah, R., Samad, S., Yadegaridehkordi, E., Aljojo, N., & Razali, N. S. (2020). Investigating influence of green innovation on sustainability performance: A case on Malaysian hotel industry. *Journal of Cleaner Production*, 258, 120860. DOI: 10.1016/j.jclepro.2020.120860

Batle, J., Sintes, F. O., & Moon, C. J. (2018). Environmental management best practices: Towards social innovation. *International Journal of Hospitality Management*, 69, 14–20. DOI: 10.1016/j.ijhm.2017.10.013

Braccini, A. M., & Margherita, E. G. (2019). Exploring organizational sustainability of industry 4.0 under the Triple Bottom Line: The case of a manufacturing company. *Sustainability (Basel)*, 11(1), 36. DOI: 10.3390/su11010036

Calheiros, L. (2021). A importância das práticas sustentáveis e a sua utilização como estratégia de marketing nos hotéis de 4 estrelas da região Norte de Portugal. Master Dissertation, Politécnico do Porto, Portugal.

Catan, M. (2018). Environmental practices in hospitality industry for a more sustainable tourism. Master Dissertation, Escola Superior de Hotelaria e Turismo do Estoril. Lisbon, Portugal.

Choi, L. (2014). The strengths and weaknesses of research methodology: Comparison and complimentary between qualitative and quantitative approaches. *Journal of the Humanities and Social Sciences*, 19(4), 99–104. DOI: 10.9790/0837-194399104

Chu, H. (2015). Research methods in library and information science: A content analysis. *Library & Information Science Research*, 37(1), 36–41. DOI: 10.1016/j.lisr.2014.09.003

Chung, K. C. (2020). Green marketing orientation: Achieving sustainable development in green hotel management. *Journal of Hospitality Marketing & Management*, 29(6), 722–738. DOI: 10.1080/19368623.2020.1693471

CIS – Community Innovation Survey (2022). Principais Resultados (2018-2020). https://www.dgeec.mec.pt/np4/207/

Coalition for Green Growth. (2015). Compromisso para o Crescimento Verde. https://www.crescimentoverde.gov.pt/compromisso/

Cop, S., Alola, U. V., & Alola, A. A. (2020). Perceived behavioral control as a mediator of hotels' green training, environmental commitment and organizational citizenship behavior: A sustainable environmental practice. *Business Strategy and the Environment*, 29(3), 3495–3508. Advance online publication. DOI: 10.1002/bse.2592

Farinha, F., Huertas, D. H., Pinheiro, M. D., Silva, E. M. J., Lança, R., Oliveira, M. J., & Batista, R. (2021). Sustainable competitiveness of tourism in the Algarve region. Critical stakeholders' perception of the supply sector. *Sustainability (Basel)*, 13(11), 6072. DOI: 10.3390/su13116072

Farinha, F., Oliveira, M. J., Silva, E. M. J., Lança, R., Pinheiro, M. D., & Miguel, C. (2019). Selection process of sustainable indicators for the Algarve region - OBSERVE project. *Sustainability (Basel)*, 11(2), 444. DOI: 10.3390/su11020444

Ferreira, S. (2021). Sustentabilidade ambiental e os antecedentes da intenção comportamental dos consumidores: Uma abordagem aos hotéis ecológicos em Portugal. Master Dissertation, Universidade do Minho. Braga, Portugal.

Ferreira, S., Pereira, O., & Simões, C. (2021). Sustentabilidade ambiental na indústria hoteleira: Uma perspetiva dos gestores de hotéis ecológicos em Portugal. *International Journal of Marketing. Communication and New Media*, 9, 150–170. DOI: 10.54663/2182-9306

Foris, D., Crihalmean, N., & Foris, T. (2020). Exploring the environmental practices in hospitality through booking websites and online tourist reviews. *Sustainability (Basel)*, 12(24), 10282. DOI: 10.3390/su122410282

Foris, D., Crihălmean, N., & Panoiu, T. M. (2020). The new technologies and sustainable practices in hospitality. *Bulletin of the Transilvania University of Brasov. Series V, Economic Sciences*, 13(62), 2. DOI: 10.31926/but.es.2020.13.62.2.7

Gevme (2023). The 3 pillars of sustainability. https://www.gevme.com/en/blog/the-three-pillars-of-sustainability/

Goh, C. S., Chong, H. Y., Jack, L., & Faris, A. F. M. (2020). Revisiting triple bottom line within the context of sustainable construction: A systematic review. *Journal of Cleaner Production*, 252, 119884. DOI: 10.1016/j.jclepro.2019.119884

Goh, E., Muskat, B., & Tan, A. H. (2017). The nexus between sustainable practices in hotels and future Gen Y hospitality students' career path decisions. *Journal of Teaching in Travel & Tourism*, 17(4), 237–253. DOI: 10.1080/15313220.2017.1362971

Hassan, T. H., Shehata, H. S., El-Dief, M., & Salem, A. (2020). The social responsibility of tourism and hotel establishments and their role in sustainable tourism development in Al-Ahsa, Saudi Arabia. *Geo Journal of Tourism and Geosites*, 33(4), 1564–1570. DOI: 10.30892/gtg.334spl18-609

Informação, S. (2021). Águas do Algarve lança nova campanha de sensibilização. https://www.sulinformacao.pt/2021/08/aguas-do-algarve-lanca-nova-campanha-de-sensibilizacao

Jaykumar, P. (2020). The triumphs & challenges in the path of green hotel housekeeping. *Journal of Services Research*, 20, 21–37.

Jones, P., Hillier, D., & Comfort, D. (2013). Sustainability in the global hotel industry. Emerald Insight. *International Journal of Contemporary Hospitality Management*, 26(1), 5–17. DOI: 10.1108/IJCHM-10-2012-0180

Jornal de Negócios. (2022). O negócio do mercado de carbono. https://www.jornaldenegocios.pt/sustentabilidade/ambiente---descarbonizacao/detalhe/o-negocio-do-mercado-de-carbono

Kasim, A., Gursoy, D., Okumus, F., & Wong, A. (2014). The importance of water management in hotels: A framework for sustainability through innovation. *Journal of Sustainable Tourism*, 22(7), 1090–1107. DOI: 10.1080/09669582.2013.873444

Kasim, H., & Antwi, S. K. (2015). Qualitative and quantitative research paradigms in business research: A philosophical reflection. *European Journal of Business and Management*, 7(3), 217–226.

Kasim, H., & Antwi, S. K. (2015). Qualitative and Quantitative Research Paradigms in Business Research: A Philosophical Reflection. *European Journal of Business and Management*, 220, 217–226.

Kim, Y. H., Barber, N., & Kim, D. K. (2019). Sustainability research in the hotel industry: Past, present, and future. *Journal of Hospitality Marketing & Management*, 28(5), 576–620. DOI: 10.1080/19368623.2019.1533907

Laugen, B. T., Acur, N., Boer, H., & Frick, J. (2005). Best manufacturing practices: What do the best-performing companies do? *International Journal of Operations & Production Management*, 25(2), 131–150. DOI: 10.1108/01443570510577001

Linneberg, M. S., Madsen, M. T., & Nielsen, J. A. (2019). Micro-level translation of corporate sustainability: When strategy meets practice in the Danish hospitality sector. *Journal of Cleaner Production*, 240, 118159. DOI: 10.1016/j.jclepro.2019.118159

Mak, A. H., & Chang, R. C. (2019). The driving and restraining forces for environmental strategy adoption in the hotel industry: A force field analysis approach. *Tourism Management*, 73(1), 48–60. DOI: 10.1016/j.tourman.2019.01.012

Medrado, L., & Jackson, L. (2015). Corporate non-financial disclosures: An illuminating look at the corporate social responsibility and sustainability reporting practices of hospitality and tourism firms. *Tourism and Hospitality Research*, 16(2), 116–132. DOI: 10.1177/1467358415600210

Mele, P. R., Gomez, J. M., & Garay, L. (2019). To green or not to green: The influence of green marketing on consumer behaviour in the hotel industry. *Sustainability (Basel)*, 11(17), 4623. DOI: 10.3390/su11174623

Migdadi, Y. (2022). Identifying the best practices in hotel green supply chain management strategy: A global study. *Journal of Quality Assurance in Hospitality & Tourism*. Advance online publication. DOI: 10.1080/1528008X.2022.2065657

Miranda, D. D. P., Ortiz, J. A., & Cardona, J. R. (2019). Article determinants of CSR application in the hotel industry of the Colombian Caribbean. *Sustainability (Basel)*, 11(18), 5045. DOI: 10.3390/su11185045

Mohajan, H. (2018). Qualitative research methodology in social sciences and related subjects. *Journal of Economic Development. Environment and People*, 7(1), 23–48. DOI: 10.26458/jedep.v7i1.571

Moise, M., Gil-Saura, I., & Molina, M. (2021). The importance of green practices for hotel guests: Does gender matter? *Ekonomska Istrazivanja*, 34(1), 3508–3529. DOI: 10.1080/1331677X.2021.1875863

Moreo, A., Cobanoglu, C., & DeMicco, F. (2007). A comparative analysis of restaurant websites and hospitality school restaurant websites. *Journal of Hospitality & Tourism Education*, 19(3), 40–47. DOI: 10.1080/10963758.2007.10696896

Ng, A. W., & Tavitiyaman, P. (2020). Corporate social responsibility and sustainability initiatives of multinational hotel corporations. In: Leal Filho, W., Borges de Brito, P. & Frankenberger, F. (Eds.) International Business, Trade and Institutional Sustainability. World Sustainability Series. Springer. Doi: DOI: 10.1007/978-3-030-26759-9_1

Oriade, A., Osinaike, A., Aduhene, K., & Wang, Y. (2021). Sustainability awareness, management practices and organisational culture in hotels: Evidence from developing countries. *International Journal of Hospitality Management*, 92, 102699. DOI: 10.1016/j.ijhm.2020.102699

Ornelas, C. (2017). Sustentabilidade, razões, práticas e impactos na indústria hoteleira. Master Dissertation, Universidade da Madeira, Portugal.

Ozturkoglu, Y., Sari, F. O., & Saygili, E. (2021). A new holistic conceptual framework for sustainability-oriented hospitality innovation with Triple Bottom Line perspective. *Journal of Hospitality and Tourism Technology*, 12(1), 39–57. DOI: 10.1108/JHTT-02-2019-0022

Pereira, V., Silva, G., & Dias, A. (2021). Sustainability practices in hospitality: Case study of a luxury hotel in Arrábida Natural Park. *Sustainability (Basel)*, 13(6), 3164. DOI: 10.3390/su13063164

Pommier, B., & Engel, A. (2021). Sustainability reporting in the hospitality industry. *Research in Hospitality Management*, 11(3), 173–175. DOI: 10.1080/22243534.2021.2006937

Portugal, U. N. E. S. C. O. (2020). Objetivos do desenvolvimento sustentável. https://unescoportugal.mne.gov.pt/pt/temas/objetivos-de-desenvolvimento-sustentavel/os-17-ods

Preziosi, M., Tourais, P., Acampora, A., Videira, N., & Merli, R. (2019). The role of environmental practices and communication on guest loyalty: Examining EU-Ecolabel in Portuguese hotels. *Journal of Cleaner Production*, 237, 117659. Advance online publication. DOI: 10.1016/j.jclepro.2019.117659

Publituris.hotelaria (2020). A prioridade do grupo Accor é lutar contra o desperdício alimentar. https://www.publiturishotelaria.pt/2020/11/25/opiniao-prioridade-do-grupo-accor-lutar-desperdicio-alimentar/

Rahman, M. (2017). The advantages and disadvantages of using qualitative and quantitative approaches and methods in language "Testing and Assessment" research: A literature review. *Journal of Education and Learning*, 6(1), 102–112. DOI: 10.5539/jel.v6n1p102

Rawashdeh, A., & Al-Ababneh, M. (2021). Hotel guests' perceptions of environmentally friendly practices in Jordan. *Journal of Environmental Management and Tourism*, 12(1), 107–120. DOI: 10.14505/jemt.v12.1(49).09

Report, B. S. (2021). Looking back at the progress we made against our sustainability strategy in 2021. https://www.sustainability.booking.com/industryinsights

Salem, I., Elbaz, A., Al-Alawi, A., Alkathiri, N., & Rashwan, K. (2022). Investigating the role of green hotel sustainable strategies to improve customer cognitive and affective image. *Sustainability (Basel)*, 14(6), 3545. DOI: 10.3390/su14063545

Salguero, G. C., Gámez, M. A. F., Fernández, I. A., & Palomo, D. R. (2019). Competitive Intelligence and Sustainable Competitive Advantage in the Hotel Industry. *Sustainability (Basel)*, 11(6), 1597. DOI: 10.3390/su11061597

Santos, R. A., Mexas, M. P., Meirino, M. J., Sampaio, M. C., & Costa, H. G. (2020). Criteria for assessing a sustainable hotel business. *Journal of Cleaner Production*, 262, 121347. DOI: 10.1016/j.jclepro.2020.121347

Silva, F. (2016). Práticas ambientalmente sustentáveis nos hotéis: Perceções dos consumidores e gestores. Master Dissertation, Instituto Universitário de Lisboa (ISCTE). Lisbon, Portugal.

Tavares, B. (2016). Práticas de sustentabilidade no setor hoteleiro: Estudo de caso do Neya Lisboa Hotel. Master Dissertation, Lisbon School of Economics and Management (ISEG). Lisbon, Portugal.

Tourism of Portugal. (2020). Os objetivos do desenvolvimento sustentável. https://business.turismodeportugal.pt/pt/crescer/sustentabilidade/Paginas/objetivos-desenvolvimento-sustentavel.aspx

Tourism of Portugal. (2021). Visão geral. https://www.turismodeportugal.pt/pt/Turismo_Portugal/visao_geral/Paginas/default.aspx

WCED - World Commission on Environmental and Development. (1987). *Our Common Future*. Oxford University Press.

Wu, M., Zhao, K., & Fils-Aime, F. (2022). Response rates of online surveys in published research: A meta-analysis. *Computers in Human Behavior Reports*, 7, 100206. Advance online publication. DOI: 10.1016/j.chbr.2022.100206

KEY TERMS AND DEFINITIONS

Sustainable Hospitality: Involves integrating environmentally, socially and economically sustainable practices into every aspect of the sector's operations. It can include from reducing waste, conserving water and energy to sourcing local/organic products and adopting responsible practices.

Green Challenge: Relates to a greener and more sustainable future, i.e., green entrepreneurial practices/attitudes are needed. This encompasses eco-friendly responsible initiatives and culture. Thus, more dimensions are relevant here such as education, other industries, community, among other.

Greenhouse-Gas Emission: Carbon dioxide (CO2) makes up the majority of greenhouse gas emissions. Human emissions of greenhouse gases are the primary driver of climate change today. Tourism is responsible for around 8% of the world's carbon emissions (from flights and boat rides to lodging).

Social Responsibility: Ethical concept in which a person cooperates with other people/ organizations for the benefit of the community. Social responsibility requires that individuals and companies act in the best interests of the environment (ex.: recycling, omission practices such as "going green" by reducing greenhouse gases; etc.).

Sustainable Practices: Refer to approaches designed to reduce industry's ecological footprint and foster ecological practices that prioritize resource use efficiency, waste reduction, emission mitigation, and use of renewable energy sources. This is imperative for the industry's continuous viability.

APPENDIX 1

Table 9. The questionnaire

1. Please indicate the importance of the following aspects for your hotel unit. (Choose only one answer for each item) 1 2 3 4 5 Management of energy consumption O O O O O Management of water consumption O O O O O Management of waste O O O O O Management of gas emissions O O O O O Management of external environment O O O O O Environment-oriented purchase policy O O O O O Environment communication O O O O O 1 – Less relevant to 5 – Most relevant
2. Energy consumption. Please choose the initiatives implemented in your hotel unit. (Choose those that apply) o Differentiated monitoring of all energy consumption o Low energy consumption equipment o Master switch in guest room o Electricity activated by the accommodation key/motion detector o Energy-saving lamps o Lighting sensors o Efficient ventilation system o Solar system for water heating o Heat recovery system o Scheduled maintenance of electrical equipment o Double glazing in windows o Exploitation of renewable energies o Use of electric vehicles o Good thermal insulation o Electric car charging stations o Ecological laundry drying o Other
3. Water consumption. Please choose the initiatives implemented in your hotel unit. (Choose those that apply) o Differentiated monitoring of all water consumption o Low consumption cisterns o Water flow reducers in taps o Low intensity showers o Faucets with timer/sensor o Bath and bed linen reuse program o Optimization of laundry loads o Optimization of dishwasher loads o Automatic irrigation system with humidity control o Gray water reuse o Use of rainwater o Other

continued on following page

Table 9. Continued

4. Waste management. Please choose the initiatives implemented in your hotel unit. (Choose those that apply) o Shampoo and shower gel dispensers o Elimination of individual packaging o Selective garbage collection in rooms o Selective garbage collection in offices o Selective garbage collection in the kitchen o Collection/elimination of specific waste (used oil, batteries, etc.) o Composting of organic waste o Elimination of disposable plastic stirrers o Elimination of disposable plastic straws o Elimination of disposable water plastic bottles o Elimination of disposable plastic cups o Virtual hotel check-in o Paper reduction program o Waste water treatment o Refood o Other
5. Purchases. Please choose the initiatives implemented in your hotel unit. (Choose those that apply) o Use of environmentally friendly cleaning products o Use of organic shampoo and shower gel o Use of local products in the meals o Preference for suppliers with quality/environmental certification o Preference for using recycled products o Other
6. Social responsibility. Please choose the initiatives implemented in your hotel unit. (Choose those that apply) o Participates in social projects o Promotes the donation of materials to people in need o Supports local education (promoting study visits, insertion in labor market) o Contributes to social events (providing facilities or sponsorship) o Cooperates with local entities for a more sustainable destination o Promotes small local businesses (restaurants, delivery services, tour agencies) o Respects local socio-cultural traditions (supporting preservation of cultural heritage) o Gives training in environmental awareness (workshops, courses, lectures, campaigns) o Treats employees equally (no prejudice due to race, age, gender or religion) o Practices social inclusion (by hiring employees with some limitation/disability) o Encourages skill development among employees o Other
7. Sustainability report. Please choose the initiatives implemented in your hotel unit. (Choose those that apply) o Processes for evaluating financial results and/or performance measurement o Reports environmental, social and economic performance o Reports environmental outcomes (energy savings, carbon footprint) o Any environmental quality certificate/label o Relation with auditing companies o Other

continued on following page

Table 9. Continued

8. Indicate please motivations for adopting sustainable practices in your hotel unit. (Choose those that apply) o Efficiency gains o Financial savings from reduced use of materials, water, energy and other o Improvement of hotel's image o Social and environmental responsability o Investors' demand o Strategic vision o Customers' attitude o Public opinion o Regulations o Other
9. Indicate please constraints to the implementation of sustainable practices in your hotel unit. (Choose those that apply) o High costs o Lack of suitable partners o Lack of information o Lack of qualified staff o Technical/technological constraints (old infrastructure, etc.) o Lack of financial support o Lack of customers' interest o Lack of advice on best solutions (environmental, economic, social) o Other
10. In your opinion, does Algarve have the conditions for implementing sustainable practices? Please point out the positive, and negative aspects to be improved.
11. Indicate please your academic qualifications. (Choose only one of the options) o Secondary education o Bachelor's degree o Graduated o Master's degree o PhD degree
12. Indicate please your age: o < 30 o 30-39 o 40-49 o >= 50
13. Gender o H o M
14. Would you like to receive a copy of this work's results? o Yes o No

APPENDIX 2

Table 10. Consolidated data between own website and Booking.com

5-star hotel	own website	Booking.com	consolidated
3HB Faro		x	x
Algarve Casino Hotel			
Algarve Race Resort - Hotel			
Anantara Vilamoura Algarve Resort		x	x
As Cascatas Golf Resort & Spa			
Bela Vista Hotel & Spa - Relais & Chateaux			
Boavista Golf & Spa - Bela Colina Village			
Cascade Wellness Resort			
Conrad Algarve		x	x
Crowne Plaza Vilamoura		x	x
Dona Filipa Hotel			
Domes Lake Algarve			
Dunas Douradas Beach Club	x	x	x
EPIC SANA Algarve Hotel		x	x
Grand House - Relais & Chateaux			
Rede Grand Muthu*			
Grande Real Santa Eulalia Resort & Hotel Spa	x	x	x
Hilton Vilamoura		x	x
Hotel Quinta do Lago		x	x
Iberostar Selection Lagos Algarve	x	x	x
Longevity Health & Wellness Hotel - Adults Only			
Martinhal Sagres Beach Family Resort Hotel	x		x
Monchique Resort & Spa			
Monte Rei Golf & Country Club		x	x
Monte Santo Resort			
Palmares Beach House Hotel - Adults Only			
Penina Hotel & Golf Resort			
Pestana Alvor Praia Premium Beach & Golf Resort			
Pestana Vila Sol Golf & Resort Hotel		x	x
Rede Pine Cliffs	x	x	x
Praia do Canal Nature Resort			
Real Marina Hotel & Spa		x	x
Ria Park Hotel & Spa		x	x
Rede Nau	x		x
The Residences at Victoria by Tivoli		x	x
Tivoli Carvoeiro	x		x
Tivoli Marina Vilamoura	x	x	x
Vale da Lapa Village Resort	x	x	x
Vale d'Oliveiras Quinta Resort & Spa			
VidaMar Resort Hotel Algarve		x	x
Vila Alba Resort			
Vila Balaia - Luxury Villas Resort			
Vila Monte Farm House		x	x
Vila Valverde Design Country Hotel		x	x
Vila Vita Parc Resort & Spa	x		x
Vilalara Thalassa Resort			
Wyndham Grand Algarve		x	x
47	10	22	26

Chapter 3
Combatting Space Debris:
Reviewing Challenges and Prospects

Lucky Madikiza
University of South Africa, South Africa

ABSTRACT

The space industry has expanded rapidly over several decades, with a significant increase in the number of launches and satellites in orbit. However, this expansion has generated sustainability concerns in space, including increased greenhouse gas emissions from satellite launches and orbital debris polluting the atmosphere. This raises concern about the sustainable use of the Earth's orbits as space debris accumulates. The amount of space debris, made up of different leftovers from objects like defunct satellites and rocket fuels, has been increasing for many years. In addition to contaminating the environment, this floating orbital garbage increases the chance of crashing into current satellites. Unless interventions are contemplated, the likelihood of more orbital garbage is expected to rise in the coming years. However, eliminating space waste is fraught with legal, political, economic, and regulatory issues as well as practical technological challenges. This chapter examines this confluence of factors with a view to contemplating remedies to solve the orbital debris problem.

INTRODUCTION

Concerns about sustainability have been widely discussed in relation to the impact of human-induced pollution on the planet's ecosystems. However, there is a growing need for a focus on sustainability that extends beyond the Earth's surface and into space. The substantial growth of the space sector over several decades has resulted in a proliferation of satellites being sent into orbit. However, the thousands of satellite technologies rotating in the earth's atmosphere have raised issues about sustainability, such as air pollution caused by space debris from space-based electronics. Orbital debris is rising at a rapid rate. Thousands of space objects are in orbit, in addition to operating spacecraft with which they may collide. Due to the current situation, there is an increasing risk that a significant disaster might occur in orbit, endangering the survival of every spacecraft in orbit. This situation necessitates quick actions to remove the orbital debris that is building up in the atmosphere. The answers, though, are not so simple.

DOI: 10.4018/979-8-3693-5748-4.ch003

They necessitate several legal, economic, political, and technological factors, which makes the problem difficult to solve.

Even as the year 2030 deadline for the achievement of the Sustainable Development Goals draws closer, it should also compel much deeper contemplation within the space sector on how this major problem is to be resolved. This chapter describes the several aspects that must be considered to find a global solution to the orbital debris issue. Space debris may have implications for the achievement of other Sustainable Development Goals such as Goal 7 (Affordable and Clean Energy) given the likely impact of orbital debris on space-based renewable energy infrastructure as well as on Goal 12 (Responsible Consumption and Production) given unsustainable production patterns which need to be curtailed to re-orient the sector towards sustainability. This chapter, through an in-depth literature review as well as an overview of policy frameworks, assesses the pressing global issue of orbital debris.

The chapter will then go into the different remedies that are being thought of in relation to the political, economic, technological, and legal factors that come up when solutions to the orbital debris are considered.

In the end, the options considered require international organisations, the private sector, and governments to work together globally. The chapter will field various suggestions that can be take up to alleviate this globally pressing issue of space debris. The discussion further reflects on the implications of the orbital debris crisis in relation to the Sustainable Development Goals.

TRACING THE ORIGINS OF THE SPACE DEBRIS PROBLEM

Space debris, also known as orbital debris, refers to bits of human-made machinery left in space, such as tools used by astronauts on extravehicular missions, spent rocket stages, malfunctioning satellites, and debris from satellite fragmentation incidents which may emanate from collisions in orbit.

Nevertheless, it is crucial to be mindful that there is no clear consensus among countries throughout the world on the legal definition of space debris (Zhang, 2022, p. 19)

According to Medvedeva (2015, p. 15), despite the existence of numerous international space treaties and declarations, there is still no clear definition or adequate description of space debris, which some researchers attribute to potential confusion over the literal meaning of "debris" and uncertainty about the range of objects that can be included in the term. Indeed what falls under 'space debris' has been comprehensive including literally any junk left in space including non-functional satellites, spent rocket boosters, nuts, oxygen tanks, and bolts (Atney-Yurdin, 1991). For Medvedeva (2015, p. 16), even if there were to be a widely agreed definition of space debris, it would not address all of the difficulties surrounding it as the term would remain dependent on the phrasing in national legislations.

The first Russian satellite Sputnik 1, was launched into space in 1957, stayed in orbit for only two months, and subsequently burnt up on atmospheric re-entry a month later (Hall, 2014, p. 2). Shortly after the launch of Sputnik 1, the United States launched its own Vanguard 1 in 1958, which ceased to function approximately six years later but is still in medium-earth orbit and is anticipated to remain there for roughly the next 200 years (Hall, 2014, p. 2). The problem with human-made satellite technology is that even when they lose their functionality, they continue rotating around the Earth. Ever since the launch of the early satellite in the late 1950s, thousands of satellites have been launched into space. Kessler (1993) considers, in the context of the United States, four major periods regarding orbital debris activities. This included the era from 1966-1972 which was focused on studying collision possibilities rather than the actual number of objectives in orbit. The second period from 1974-1979, which entailed making calculations on collision probabilities and ground fragmentation tests. In the period from 1979-1988, as Kessler documents, NASA became increasingly aware of orbital debris ensuring the period starting in 1988 culminating with a realization of the need to mount monitoring and inquiry activities into orbital debris.

Although abandoned objects make up a significant portion of the debris in orbit, breakup incidents have substantially added to the litter, with the first major incident occurring in 1961, when the Thorn-Ablestar detonated in orbit, leaving more than 200 fragments (Blake, 2022, p. 2). The launch by the US and Russia catapulted the beginning of thousands of other satellites that were to be subsequently also launched into orbit from various countries. Some of the collisions were intentional rather than accidental. For example, from 1968 to 1985, the United States and the Soviet Union developed anti-satellite weapons designed to explode near targets or through direct impacts, with twelve such tests contributing 7% of classified orbital debris by 1990 (Hall, 2014, pp. 4–5). In 1966, research was undertaken to calculate estimates to characterize the orbital debris danger to evaluate the hazard for manned spacecraft by calculating the collision probability caused by other man-made objects known to be in orbit (Kessler, 1993).

Over the years, the scale of the problem of orbital debris has been increasingly mounting every year with more than 100 space launches a year worldwide (Atney-Yurdin, 1991). According to estimates made in 2020, there were more than 8,000 metric tons of debris—more than 700 school buses' worth of weight (Hosseini et al., 2021). Such scale of flying objects can be considered in relation to the severity of collisions that have been happening in space. For instance, in 2009, there were over 140 000 pieces of space debris produced by a collision of two communication satellites – Cosmos 2252 and Iridium 33- an event which has made a major contribution to the overall amount of junk fragments that have been accounted for in contemporary years (Clormann & Klimburg-Witjes, 2022, p. 961). The European Space Agency, which estimates debris to include about 34 000 objects that are less than 10cm in size, about 128 million objects that are between 1 mm and 1 cm, and 900 000 objects between 1 and 10cm in size, warns that even without launching any new satellites, these numbers will continue to escalate (Haroun et al., 2021, p. 65). The scale of the space debris problem calls for urgent intervention considering that debris is posing a threat to present and future space missions owing to severe possible collisions in space. The environmental risks for the planet Earth also always loom. For instance, in 1978, a Soviet satellite was noted to have malfunctioned and fell on Earth, with radioactive debris landing in the northern part of Canada (Hutagalung et al., 2020, p. 4).

THEORETICAL OVERVIEW: GAME THEORY

The problem of orbital debris has been approached through invoking Game Theory, among other theoretical frameworks, in several notable studies. Game Theory is associated with decision-making dynamics among role-players interacting and seeking to make strategic decisions but with competing interests. Game theory applies structured, logical approaches in the analysis of interactive decision-making and is used to identify the most effective strategies possible in contests between rational, intelligent opponents (Morgan et al., 2018). Game Theory as a concept deals with the formulation of the correct strategy that will enable an individual or entity, when confronted by a complex challenge, to succeed in addressing that challenge and premises that there will be a strategy that will allow one player to win in a circumstance or "game" (Raoof & Al-Raweshidy, 2010, p. 1). Game theory may be defined as a decision-making process in formal studies where various players make choices that directly or indirectly affect the other players' interests. According to Norozpour and Safaei (2020, p. 1), "Game Theory may be defined as a decision-making process in formal studies where various players make choices that directly or indirectly affect the other players' interests." Stated differently in another definition, "Game Theory can be also defined as the study of how the outcome of a competitive situation is dictated by interactions among the people involved in the game (also referred to as 'players' or 'agents'), based on the goals and preferences of these players, and on the strategy that each player employs"(Raoof & Al-Raweshidy, 2010, p. 2).

In terms of there being 'games,' these will involve players (the decision-makers), sets of strategies that each player may use, and a set of outcomes (the strategies selected by players to influence the outcome of the game) as well as payoffs for the players as derived from outcomes(Bhuiyan, 2018, p. 114). The notion of a 'game' is unlike what would be implied in normal parlance where people may participate in leisurely activities such as playing chess. It is hardly about any fun activity and denotes a formal strategic situation in which participants should make rational critical decisions. In effect, the notion of a 'game' points to a serious scenario of strategic decision-making. Still invoking the game analogy, the situation of Game Theory can be explained as follows:

> "So, Game theory is the study of how players should rationally play games. Each player would like the game to end in an outcome that gives him as large a payoff as possible. He has some control over the outcome since his choice of strategy will influence it." (Bhuiyan, 2018, p. 115)

Space exploration activities typically require cooperation and collaboration, with various international law frameworks being mounted over the years to regulate space activities. Hence, the dynamics of co-operation and sometimes conflict and rivalry about space activities have been understood using Game Theory (Galloway, 2004). For instance, at the level of agencies, the space debris removal dilemma can be modeled as a game in which the players are space agencies, their actions consisting of debris removal strategies and the payoffs being those derived from removal costs, as well as collision risks (Klima et al., 2016, p. 6). By applying Game Theory to the problem of space debris cleanup, the question of whether and how self-interested actors might collaborate in the orbital debris dilemma—which bears similarities to a tragedy of the commons scenario—is posed (Klima et al., 2018). Another study invoking game theory presents a model of space congestion and collision risk in which independent spacefaring agencies and firms decide on the number of satellites to launch, with such an interaction defining a non-cooperative game, the outcome of which is a Nash equilibrium, a situation in which no spacefaring agency or firm can become better off by changing unilaterally its launching decision (Béal et al., 2020). Modeling based on game theory has also been contemplated for future research endeavours surrounding managing space debris in low-orbit satellites in a study by Grzelka and Wagner (2019).

COMBATTING SPACE DEBRIS: TALLYING VARIOUS CONSIDERATIONS

Orbital Debris Technical Issues

When satellites were first launched into orbit more than 50 years ago, it was surmised that after they were rendered inoperable, they would be able to reenter Earth's atmosphere and burn up before reaching the planet's surface and when this did not happen, countless inactive space objects remained in orbit for decades (Akers, 2012). On the other hand, activities to venture more satellites into orbit continue to increase thereby escalating the prospect of more space pollution after satellites end their useful lifespan or be rendered non-operational.

The Space Debris challenge, if humankind is to appreciate its complexity, is a wicked problem. A wicked problem has been considered as constituting a particularly challenging problem such that it may involve various actors at different levels, even transcending policy domains and even resisting attempts to solve it (Termeer et al., 2019). As a wicked problem, the orbital debris challenge does not become susceptible to an easy cut-and-dried single solution. It invites a multi-pronged solution. Since it is a global issue, it inevitably invites global stakeholders to reach consensus agreements and courses of action. As the issue of Space Debris presents a major environmental crisis, it behooves various stakeholders to participate in resolving the matter given the globally shared space resource. The requirement would be that all stakeholders should treat it with urgency and have a shared commitment towards its resolve. How to ensure urgent collective action among global stakeholders to arrest the problem becomes one element of the 'wicked' problem. How willing is humankind prepared to cooperate regarding solving this problem even as there are separate interests at stake from various role-players? As Game Theory suggests as discussed earlier, various role-players may sometimes pursue conflicting strategies serving individual interests. So whilst cooperation has been a theme in international space law, conflict and competition have also been evident (Galloway, 2004).

Amidst the broader dilemma of how to cooperatively deal with the problem are a subset of various technical issues. Granted, the necessity of clearing the orbital refuse circulating in the air is recognised. Yet the intricate details remain as to how this mammoth task of clearing the atmosphere is to be parceled out to various role-players and how this can be undertaken, using what methods, resources, and who should fund the effort.

The necessity for active debris removal (ADR) is acknowledged but even something seemingly as simple as requesting designs for ADR technology concepts will be tied up in further technical and political considerations (Tallis, 2015, p. 85). As Tallis further discusses, technical advancements rarely attract private funding when they offer negligible to no short- or medium-term economic benefits which then leaves prospects for government research funds to cover this shortfall, which may further imply that that political factors will influence the adoption of new technologies. The other consideration is then how the task of removing debris gets allocated to various national or regional jurisdictions. Yan (2023) suggests that advanced spacefaring nations should share responsibility for helping to reduce future space debris and provide developing spacefaring nations with technical assistance to develop their capacity in the areas of space object design, manufacture, operation, and disposal, at the very least. Such an approach, as Yan suggests, would be underpinned by a principle of common but differentiated responsibility and respective capabilities (CBDR-RC) in relation to the responsibility among nations to clean up orbital debris.

Yet there is still a need to appreciate the challenging and complex technical task of active orbital debris removal. Key questions that arise include identifying the most critical region for active debris removal; what debris should be removed first and how to carry out the operations (Liou, 2013, p. 3). According to Barry (2022, p. 248), even if all legal issues that pertain to the task were waived, capturing space debris, and particularly interacting with non-compliant careening objects in space, is a very technically challenging task that make ADR services difficult to realise. Among other contemplations have been how to de-orbit or re-orbit around 15 space objects per year, yet such an action becomes technically difficult considering the variety of shapes and sizes of debris in orbit (Slann, 2014). Another challenging undertaking is to classify the various debris as mitigation measures tend to differ with fragmentation debris being among the more difficult to assess (Taylor, 2007, pp. 72–73). Further to the intricate technical issues, military concerns will surface. For instance, technologies intended for active debris removal may be considered as weapons since they also can remove any objects including enemy satellites from space (Bongers & Torres, 2023, p. 2). In a sense, the technical aspects will inevitably also interface with political issues.

Legal Issues

Space debris removal, as alluded to above, also raises non-technical considerations that arise. Among these have been prominently legal matters. The first immediate legal hurdle that confronts orbital debris removal is a lacuna in international legal frameworks. It doesn't seem the legal context has been clear on space debris with no international agreement that has explicitly discussed the definition of space debris or orbital debris (Sheer et al., 2023). Hence, some views have been that the lack of definitions makes it impossible to recognize which space objects can be removed (Emanuelli et al., 2014). The agitation, among other reasons, for the need for an international legal framework defining space debris is considered necessary to distinguish between space debris and space objects, so that the international community may evolve a regime that supports space debris remediation (Haroun et al., 2021, p. 65).

The lacuna in international law not being explicit about orbital debris can be attributed to the fact most of the legal frameworks were developed at a time before space debris was considered to pose any hazards (Taylor, 2007, p. 74). Since the 1960s, there have been five international treaties that constituted the legal framework for outer space activities these being the 1967 Outer Space Treaty; the 1968 Rescue Agreement; the 1972 Liability Convention; the 1975 Registration Convention, and the 1979 Moon Agreement as well as some resolutions of the General Assembly of the United Nations adopted by that organ since 1982 (Popova & Schaus, 2018, p. 4). Most commonly, the term "space law" refers to the norms, guidelines, and standards of international law found in the five international treaties and five sets of guidelines governing space that was created under the United Nations (UN) auspices.

The United Nations General Assembly established five sets of guidelines that are relevant to space exploration and utilization including:

- The 1963 Declaration of Legal Principles that Oversee State Activities Related to Outer Space Exploration and Use
- The 1982 Guidelines for States Regarding the Use of Man-made Satellites for International Direct TV Transmission
- The 1986 Guidelines Concerning Earth Sensing from Space
- The 1992 Guidelines Concerning the Use of Nuclear Power Sources in Outer Space and
- The 1996 Declaration on International Collaboration in the Study and Utilization of Space for the Gain and Interest of Every State, Considering in Particular the Requirements of Developing Nations (Bohlmann & Petrovici, 2019, p. 4).

The UN organ set up to manage space affairs is the UN Committee on the Peaceful Uses of Outer Space (UNCOPUOS) which examines international cooperation in the peaceful uses of space whilst also promoting research initiatives and examining legal issues resulting from space-related behaviour. These various pieces of legal frameworks have not been considered explicit in defining orbital debris. For instance, the Outer Space Treaty of 1967 condemns harmful contamination of space and yet is not clear on what contamination means as well as mechanisms of enforcement (Tallis, 2015, p. 82).

The space laws have been noted to not offer a comprehensive legal framework for the environmentally responsible and sustainable exploration and use of space, especially not in light of the rapidly growing number of satellites and actors that are currently in space, as they were developed as a result of negotiations that started in the Cold War era of Space 2.0 (Bohlmann & Petrovici, 2019, p. 4). The environmental risk posed by space debris is potentially catastrophic as has been alluded to in an earlier discussion. The ultimate catastrophic scenario may culminate in what is often referred to as the 'Kessler Effect' or the 'The Kessler Syndrome,' whereby orbit space becomes very crowded with the result that it becomes economically illogical to launch satellites and eventually even technically impossible (Matton, 2023, p. 2).

Aside from the impending environmental risk posed as more satellites get launched, several legal issues are raised by the nature of orbital space as a globally shared resource or a global common. One factor that arises is who becomes liable or responsible for harm caused by space debris in the context of a globally shared airspace. Attendant to the issue is how compensation is to be enacted. The 1967 Outer Space Treaty does assert the principle of liability for damage caused by space objects and the Convention on International Liability for Damage Caused by Space Objects (1972), also known as the 'Liability Convention,' sets a framework for liability (Roberts, 1992, p. 64). However, the topic of liability is not dealt with comprehensively in the UNCOPUOS (United Nations Committee on the Peaceful Uses of Outer Space) in terms of rules that would address how a nation or business that produced the fragment will pay compensation to the owner of the satellite if it hits one (Wang, 2016, p. 23). Another legal issue that arises pertains to ownership and jurisdiction in terms of who owns and has jurisdiction over debris in orbit. Overall, there would need to be a clear international regulatory framework consisting of standards and rules for debris mitigation and removal. It has been noted there is no internationally agreed mandatory governance framework for managing space debris with national laws also tending to be vague on debris removal(Newman & Cheney, 2023, p. 116).

Political Issues

The conundrum of space debris removal compels international cooperation or at least some engagements among nation-states on how to manage the shared atmospheric resource of space and its clean maintenance. Hence, the various challenges regarding space debris management may be intertwined with politics. For instance, some observers have pointed to the lack of international cooperation as the reason why the legal framework has not evolved to address the debris problem (Lambach & Wesel, 2021). Geopolitical issues inevitably arise in space-related matters as orbital space is key in military and national security agendas. For countries that consider space important for national security, the geopolitical nature of issues arising from space matters may make some countries reluctant to share information on space affairs (Nomura et al., 2024, p. 26). Given that active debris removal does require the sharing of information or data to track or monitor space debris, nation-states would need to cooperate to facilitate such matters. Even when active debris removal (ADR) technology is contemplated, national security concerns arise as in some instances, there may be similarities between space debris removal systems and space weapons (Sannigrahi, 2017, p. 831). The technological innovations discussion may once more also be mired in political insecurities.

According to Newman and Cheney (2023, p. 127), in the contemporary environment trust between major space power nations is low and so introducing new technological methods for space debris remediation may further aggravate the potential for misunderstanding which in turn fuels instability. It must be understood that space activity has been characterized by political rivalry during the Cold War era with a 'space race' being influenced by Cold War motives (Bohlmann & Petrovici, 2019, p. 1). Hence the issue of political mistrust among some major space fairing nations may be a remnant of this past era. However, the fact that international cooperation efforts are being contemplated and some degree of progress has been arrived at in negotiations for a legal framework for outer space activities suggests there are prospects of better international cooperation efforts in what can be considered a Post-Cold War Era or new space age characterized by increasing international cooperation and more participation of non-state actors (Bohlmann & Petrovici, 2019, p. 2).

Economic Issues

The ultimate solution to the space debris problem requires a technological solution as a practical intervention. The debris must be removed using technological means. This would require investments on the part of countries or private stakeholders participating in space fairing activities. Active Debris Removal technologies as they may exist in the current era or as they may be emerging from technological innovation, have huge costs. A key question, again emanating within the stream of international cooperation, would be how the costs of removing debris will be passed out among global stakeholders involved in space activities.

From an economic perspective, the space debris issue becomes an economic challenge to the sustainable management of a common-pool resource (Palmroth et al., 2021, p. 9).

The issue to be contemplated is who is to bear the brunt of investing, whether they be a state or non-state sector, in the active debris removal technologies in orbital space which is globally shared. Discussions have revolved around whether active debris removal, as a method to combat space debris, is a public good to be provided through an international fund or space fairing governments on a per-case basis or as a private good that should come from commercial stakeholders (Migaud, 2020, p. 7). Another perspective contemplates the possibility of it being a market issue, in terms of whether customers would be willing to pay for a space debris service although this is economically inefficient as it entails paying for another party's mess after the fact of its occurrence (Barry, 2022, p. 248).

The methods of funding space debris management have been contemplated by various economic scholars. Macauley (2015, p. 163) considers the possibility of a tax or fee levied upon launch and proportional to the likelihood of causing harm to others as well as a tax or fee applied at the launch phase but rebated upon maneuvering or graveyarding to avoid collision. There could also be a possibility of environmental taxation, a method used by national governments, which would be difficult to apply internationally to govern taxation in a global commons when there is no global government to receive the taxes (Morin & Richard, 2021, p. 3). In fact, in all the various tax approaches for space debris management, the overall question becomes who becomes the receiving entity for such fees in the context of the lack of a world government that would serve as the tax collection authority. According to Weinzierl (2018, p. 187), the space debris problem is a classic example of negative externalities but in a context in which the conventional remedies suggested by economic analysis will have limited traction. One remedy would be whereby the affected parties negotiate to internalize externalities, but this may be difficult in space debris management because it requires delineated property rights, which don't exist in space. An approach to economic incentives has also been pondered. For instance, those who adopt virtuous behaviour can be subsidized (Morin & Richard, 2021, p. 3).

A further question arises as to who the role-players in the governance of space debris management should be. Wienzierl (2018) contemplates a structure of polycentric governance, whereby public and private role-players collectively manage orbital debris similarly to how large-scale irrigation projects are managed. However, such an approach would require cooperation including the requirement for discipline to be enforced, something which has been hard to attain in the space sector. The role of space agencies has also been considered in the mix of suggested remedies. In terms of their role, a space agency will be faced with a dilemma to either take the individually expensive action of debris removal, which benefits all parties, or wait and hope that others will step in but if everyone waits, catastrophic consequences of space debris may be borne (Klima et al., 2016). At the level of governments, there may also be further dilemmas in how urgently the matter of space debris is prioritized. States are ultimately faced with two options when it comes to financing a response to space debris: either they pay the expense now and create a system for collecting debris, or they pay it later on when there is an orbital collision and space needs to be cleaned up (Newman & Cheney, 2023). Irrespective of who the role-players may be in the space debris mitigation and removal, there will be major cost implications to be contemplated and therefore economic viability is a matter to be contemplated by all the stakeholders. Whichever stakeholder is involved, the costs of orbital debris management gravitates to millions of dollars spent. For both public and private sector role-players, priorities to clean up space may still be subservient to other commercial and national interest priorities.

CONTEMPLATING ORBITAL DEBRIS MITIGATION AND ERADICATION INTERVENTIONS

Managing space debris necessitates several political, economic, technical, and legal factors, as the previous discussion has elucidated. Therefore, in what might be regarded as one of the major environmental sustainability concerns in our modern atmospheric geography, a single-tracked solution would be futile. Solutions must be taken into consideration while keeping the same considerations of technical, economic, political, and legal factors. In this discussion, we examine several alternatives and emphasize that a key factor supporting potential solutions for orbital debris control should be underpinned by sensitivity to environmental sustainability.

Technical Interventions

Any technical innovation being developed to contain space debris must first question whether the technological solution's constituent parts promote environmental sustainability. In using technologies to remove space junk, are the technologies we use environmentally sustainable? Stated differently, in as much as there needs to be an activity of space refuse collection using the appropriate technologies, it would be foolhardy for those very same technologies to in turn further contribute to space debris when they may have mechanical failures or become dysfunctional while in orbit.

Sustainable management of waste in general often calls for reducing, reusing, and recycling waste to limit unsustainable disposal in the environment. The intention to reduce space missions is of course not pragmatic as more satellite technologies are increasingly being flighted into space for various commercial or non-commercial reasons. The launch of space missions has been demonstrating an upward increase with every passing decade thereby always increasing the prospect of orbital debris that may be introduced through satellite collisions as the satellite population grows. Yet the effort to reduce or eliminate orbital waste can be carefully considered at the satellite design level invoking greener or environmentally sustainable designs. Some brandished technology ideals in the literature include 'orbit friendly' satellites that can operate for several weeks and then subsequently burn up into the atmosphere thereby avoiding long-term buildup of orbital debris (Damjanov, 2017, p. 178). Another technological innovation that was contemplated in Russia entailed the development of a 'space pod' that would have the capacity to absorb over 600 satellites and de-orbiting them such that they burn up when reentering the earth's atmosphere (Mullick et al., 2019, p. 881). Yet again, it can be asked if burning up these technologies in the atmosphere won't release toxic gases at atmospheric level. Other innovative technologies that have been considered are:

- Tether-based techniques employing electrodynamic tethers, which involve deploying tethered nets around space debris to accelerate their rate of decay and de-orbit;
- Laser-based active debris removal technologies, which target debris with lasers to cause their sublimation with the pull of gravity eventually creating their de-orbit and decay;
- Spraying frozen gas mists into low Earth orbits from specially deployed satellites to gather and bring down smaller pieces of orbital debris;
- Sail-based methods using solar sails that attach themselves to larger space debris in low Earth orbit automatically so that it will decay over time;
- Unconventional methods such as the "CleanSpace" program, use contact-less maneuvering, pushing, and pulling techniques to deorbit space debris (Damjanov, 2017; Mark & Kamath, 2019).

The phase of recycling waste in terms of the inclination as to how space technology materials can be recycled would also be an environmentally progressive undertaking. In terms of recycling, one technology suggested by researchers is the collection of space debris through a recyclable capsule, which separates the debris components at the collection stage, and which when returning to Earth, would allow different parts such as magnetic, metallic, and non-metallic components to be gathered as part of preserving the planet's rare-earth elements (Mullick et al., 2019, p. 882). It is encouraging to note that green design thinking to reduce or avoid the creation of debris is being contemplated such as the deployment of technologies with explosion-safe batteries; deorbit technologies and satellites designed for demise (OECD, 2020). Whatever technological innovations are being modeled, tested, or experimented upon, the ultimate sustainable solution would be the sustainable disposal of space debris without necessarily using technology that again contributes to space debris. As Damjanov (2017, p. 179) notes, there can be an "irony of sending more technologies to clean the waste of other technologies." Hence the pros and cons of technology solutions should always be weighed against the environmental sustainability soundness so that the situation is not further aggravated when the waste collection technologies in turn further the debris problem.

The luxury of time is no longer available even as experimental technological solutions continue to be explored. An economic Kessler Syndrome—in which orbital space becomes unprofitable rather than unusable—is anticipated to emerge before the commencement of a physical variety of the syndrome and is estimated that it will occur in about the year 2235 (Nomura et al., 2024).

Economic Interventions

The economic Kessler Syndrome, as alluded to above, represents almost a death knell of space activity. In such a scenario, the predictions are that orbital space will be rendered physically unusable, even with remediation, with orbital space initially becoming economically unprofitable perhaps well before it becomes physically unusable (Adilov et al., 2018). In a sense, failure to summon technological solutions to solve the space debris problem could in turn have dire economic consequences given the prominence of the space economy entailing a range of space-related activities. The potential economic impact of space debris is being contemplatively tallied by some studies. Some estimates are that debris will cause negative damage of approximately 1.95% of the global Gross Domestic Product (GDP) in the long term if no debris is remediated (Nozawa et al., 2023).

As has been alluded to previously in the discussion in this chapter, the costs of cleaning technology as much as the satellite technologies themselves, are costly. The issue at stake is who should pay for the debit removal costs. Both the private and public sectors may be reluctant given the costs that gravitate towards hundreds of millions in dollar or euro-denominated currencies. Given the financial burden, the private sector may not be able to afford the costs without some government subsidy and likewise government may not be willing to commit the large funds involved (Newman & Cheney, 2023, p. 125). The stalemate could then call for a private-public sector collaborative approach regarding sharing the costs.

Logically space debris will be emanating from a space object launched by a particular entity – a private or public sector role-player, who ought to bear some cost. The question is then, how do entities launching the satellites get to be held accountable for the costs of space debris?

Costs have been considered at various levels of space activity entailing:

- At the *launching phase* whereby there could be launch taxes, an orbital usage fee or a bottle deposit system where users pay for each space object they place in orbit with the proceeds of taxes going towards the purchasing of ADR systems (Emanuelli et al., 2014; Macauley, 2015; Newman & Cheney, 2023);
- At the *occurrence* level when orbital debris occurs whereby there are penalties enforced for the production of debris (Bernhard et al., 2023);
- At the institutional level entailing the setting up of a supranational global authority to administer tax proceeds as well as an international debris removal economic fund that contributes towards the costs of clearing space debris (Emanuelli et al., 2014; Sheer et al., 2023).

Aside from the above proposals, a more incentivizing approach is characterized by offering incentives to satellite operators or companies who undertake measures to prevent or mitigate satellite collisions. Suggestions have entailed rebate of the tax or fee that is initially paid during the launch stage when an operator undertakes measures such as implementing maneuvering (adding maneuvering capabilities to spacecraft) or graveyarding (retiring space vehicles to a relatively unused graveyard orbit) to avoid collisions(Macauley, 2015). Along the same lines and to implore the sustainability imperative, incentives can also be offered to operators that align with green/sustainability designs in their technologies as well as sustainable practices of deorbiting non-functional spacecraft.

In molding the space sector towards the sustainability imperative direction, among the undertakings could be the advancement of sustainable space practices which should be an arena where both governments and companies should invest in research and development efforts that bring to the fore sustainable innovations for managing space debris operations. A key area of research focus would be the life cycle assessments of spacecraft as part of embedding a sustainable circular economy in the space sector. For instance, throughout their life cycle, space launch vehicles impact the environment through them directly emitting combustion gases, and yet the atmospheric consequences of these activities are understudied and insufficiently addressed (Sirieys et al., 2022). Given the necessity for orbital debris removal services as the hazards of orbital debris collapsing the sector approach, there is also a chance for the activation of a debris removal sub-sector of the space economy. Such a sector would once more focus on sustainable technological solutions including design-for-demise approaches to create spacecraft that can be safe and have re-entry capabilities as well as developing technologies for controlled de-orbiting of satellites. The development of technologies for active debris removal, which culminated as soon as the orbital debris arose, represents an ongoing area of activity. If no concrete actionable steps are made to reinvent the sectors' legacy operational practices, the disastrous effects on the space industry await. There have been warnings that the industry's growing environmental impact from spacecraft production and orbital launches is also increasing the risk of a paradox whereby using space to support the Sustainable Development Goals (SDGs) becomes unsustainable from the perspectives of both Earth and space environments (Wilson & Vasile, 2023).

International Cooperation on Legal and Policy Interventions

Whatever the technological and economic solutions that may be contemplated, there is a compelling need for greater international cooperation which will need to be undergirded by political will and buy-in among governments. Global agreements and treaties to effectively regulate space debris management and mitigation sustainably will inevitably have to be agreed upon as part of a collaborative effort from governments, industry, and international organizations. A key organ and platform for global cooperation is the UN Committee on the Peaceful Uses of Outer Spaces (COPUOS) which should be empowered and strengthened in its role of developing and implementing debris mitigation frameworks.

Revisiting legal interventions to manage global orbital debris is now a matter of urgency. One area of intervention that needs to be firmed in global legal frameworks includes orbital debris removal treaties or agreements specifically an international treaty covering debris removal obligations as well as outlining a clear liability regime for debris-related incidents. In the contemporary era, space debris is not covered, in legal terms, in the existing framework of the five treaties which offer little guidance for both mitigation and removal (Lambach & Wesel, 2021). At the foremost level of revisiting the legal framework for space debris is to create a concrete definition of space debris. Legal experts contend that the legal framework for orbital debris should be reopened in a debate on an international forum to firm up and also clarify liability, ownership, causation, and negligence aspects, among others, about space debris (Haroun et al., 2021, p. 66). Issues of liability remain sketchy in the contemporary environment. For instance, launching states are required to declare the objects they send into space, but they are not accountable for their removal, and satellite stakeholders need to obtain permission from the owner of the space debris before collecting it but would then be liable for any harm that results from the removal (Nomura et al., 2024). Hence, suggestions of a private-public sector shared model of responsibility might be worth looking into (Emanuelli et al., 2014, p. 200). Other suggestions proffer a market-share liability system, where, in the issue of unidentified orbital debris, each party is held liable in proportion to its contribution to the overall debris problem, and such a system, in reality, would assign a large amount of responsibility to the nations most likely to produce space debris, which are China, Russia, and the United States (Poonuganti, 2023, p. 508).

Environmental sustainability must also be given a prominent airing in legal frameworks. This has been evidenced at national and local levels with some policies, albeit few, regulating local environmental impacts of space launches such as in France, which compels space vehicle operations to undergo an environmental assessment before receiving authorization to launch in terms of that countries' French Space Operations Act (FSOA) of 2009 (Sirieys et al., 2022). Sustainable space practices can only be advanced through legal and regulatory frameworks as the enabling pillars to drive the sector towards sustainability.

Raising Awareness and Implications for the Sustainable Development Goals

The Sustainable Development Goals, set as the target date of 2030, have been greatly propagated globally. The publicity, education, and awareness initiatives surrounding these goals should incorporate space sector activities as integral to mitigating global environmental threats to our planet. There have certainly been efforts to create awareness of how space technologies can be leveraged to achieve the Sustainable Development Goals (SDGs) (Di Pippo, 2019). There has been a noticeable emergence of a discourse on how space science and technologies can aid in the achievement of the SDGs. For example, some of the literature examines how six space-related technologies—satellite Earth observation, satellite communication, satellite navigation and positioning, human spaceflight and microgravity research, space technology transfer, and basic scientific research—are being used to realize the SDGs' vision (Wood et al., 2024). However, there not only needs to be an awareness of *sustainability through space* but also considerations of *sustainability for space*. The former would entail activities on how space can be used as a platform to address global sustainable development issues. The additional dimension of *sustainability for space* would entail protecting the terrestrial environment from the impacts of space activities (Wilson & Vasile, 2023, p. 3). Hence an education and awareness initiative should be undertaken along all spheres of the interface between space and sustainability to avoid the world eventuating towards 'the space paradox' whereby "the more that the space sector implements space applications and technologies aimed at positively contributing towards sustainable development challenges for the benefit of humankind, the less sustainable the space sector becomes"(Wilson & Vasile, 2023, p. 4). An awareness and education campaign, together with implementation efforts, should in parallel target the 2030 year in which there is an expectation of significant progress made on the Sustainable Development Goals.

Beyond awareness raising in general, the notion to sustainability ought to permeate the space sector industries more vigorously. An important imperative would be how to re-orient the sector to embrace sustainability in its various operational dimensions. The notion of 'space sustainability' is beginning to earn mindshare in the sector. When the UN Committee on the Peaceful Uses of Outer Space (CUPUOS) took an initiative to publish guidelines for the long-term sustainability of space in 2018, space sustainability was defined as "the ability to conduct space activities into the future in a way that accomplishes the aims of fair access to the advantages of space exploration and utilization for peaceful purposes, to meet the requirements of current generations while protecting the environment" (Mane, 2024, p. 2061). However, as Mane further states, even though the guidelines define the term, their primary focus is on maintaining the space environment and when the industry discusses space sustainability, the topics that typically arise in discussions are space data and space garbage. There is a need to broaden discourse to go beyond these two discussion issues. One of the contemplative platforms to discuss space sustainability in contemporary era is regarding the Sustainable Development Goals.

Pertaining to the Sustainable Development Goals, a base scenario would be to consider the extent to which the issue of orbital debris, if left unattended or pursued diligently, might pose a threat to the achievement of some of the Sustainable Development Goals. Commendably space sustainability is beginning to gain some traction in space organisations such as for instance in the European Space Agency Clean Space initiative. The European Space Agency has developed a catalogue of projects and considered the ways in which they contribute to the SDGs such as SDG Goal 12 (Responsible Consumption and Production); SDG Goal 13 (Climate Action) and SDG 17 (Partnership for the Goals) (El-Shawa et al., 2021). The issue of orbital debris has even direct implications for Goal 13 on Climate Action given how debris may contribute to the risk of satellite collisions. The SDG Goal 17 regarding partnerships would represent a call to action by all stakeholders in the space sector as international cooperation has become an urgent imperative to address the escalating space debris matter. Space Debris, among further other SDGs, also has implications for SDG 9 (Industry, Innovation, and Infrastructure) considering how the debris may impact essential satellite infrastructure used for weather forecasting, navigation, and global communication. To heighten the importance of the space sector in the SDGs, proposals have even been contemplated on having an 18th goal dedicated to the space sector in addition to the 17 SDGs (Losch et al., 2024). However space issues are fashioned into sustainability, recognition is emerging that concerns regarding the outer space environment must be included in global sustainable development agendas (Alfathimy, 2019). However, the space sustainability agenda is pursued in global forums such as UN, there is also still a need for sustainability to be an underlying ethos that should be informing space operations starting from design, technical operations through to the disposal phase of unusable or defunct spacecraft. All space activities should be mindful of sustainability and be adopted in the 'cradle-to-grave' approach that embeds sustainability in various phases of space operations. Partnerships, collaborations, and sectoral discussions ought to continue with a sense of urgency to plot interventions that arrest the emerging global threat of space debris polluting outer space.

CONCLUSION

Space sustainability is becoming an inevitable and unavoidable area of contemplation amidst the major sustainability concerns focused on the earth. The drive for sustainability should spread beyond earthly concerns and look at the entire universe beyond the earth, where human ventures are causing pollution as much as the more visibly glaring environmental pollution that is witnessed on earth. This chapter addressed the fundamental concerns for combating space debris, including legal, economic, technical, and political aspects that must be considered. All these factors may need to be weighed against the larger goal to form and improve global alliances that promote global consensus. The space sector can examine how it can build sustainable space technologies. Opportunities further exist for promoting sustainable space operations and improving space debris collection and mitigation through collaborations among governments, regulatory agencies, and industry. In addition, further cooperation between international organizations, the commercial sector, and non-profit interest groups would be required to create partnerships and forge consensus in the legal, technological, and economic domains that may address the worrying issue of orbital debris. In line with the Sustainable Development Goals deadline of 2030, there ought to be a growing sense of urgency as each year and decade goes by and more satellites are placed into orbit. The catastrophic consequences of satellites colliding with various objects in space, akin to a Space Armageddon disaster, is quite ghastly to contemplate. It would in turn spell disaster for developmental goals linked to satellite technology operations.

REFERENCES

Adilov, N., Alexander, P. J., & Cunningham, B. M. (2018). Corrigendum to "An economic 'Kessler Syndrome': A dynamic model of earth orbit debris" [Econom. Lett. 166 (2018) 79–82]. *Economics Letters*, 170, 185. DOI: 10.1016/j.econlet.2018.04.012

Akers, A. (2012). To Infinity and Beyond: Orbital Space Debris and how to clean it up. *University of La Verne Law Review, 33*(2).

Alfathimy, D. H. A. (2019). Application of Sustainability Concept to Near-Earth Space as an Integral Part of Earth-system in the Context of Sustainable Development Goals (SDGs). *A I.*

Atney-Yurdin, I. (1991). Space Debris Legal Research Guide. *Pace International Law Review*, 3(1), 167. DOI: 10.58948/2331-3536.1019

Barry, K. (2022). Space Debris Mitigation and Remediation: Historical Best Practices and Lessons Learned for Economically Preserving and Utilizing Common Areas. *New Space*, 10(3), 246–258. DOI: 10.1089/space.2021.0022

Béal, S., Deschamps, M., & Moulin, H. (2020). Taxing congestion of the space commons. *Acta Astronautica*, 177, 313–319. DOI: 10.1016/j.actaastro.2020.07.036

Bernhard, P., Deschamps, M., & Zaccour, G. (2023). Large satellite constellations and space debris: Exploratory analysis of strategic management of the space commons. *European Journal of Operational Research*, 304(3), 1140–1157. DOI: 10.1016/j.ejor.2022.04.030

Bhuiyan, B. A. (2018). An Overview of Game Theory and Some Applications. *Philosophy and Progress*, 111–128. DOI: 10.3329/pp.v59i1-2.36683

Blake, J. A. (2022). Looking out for a sustainable space. *Astronomy & Geophysics, 63*(2), 2.14-2.20. DOI: 10.1093/astrogeo/atac022

Bohlmann, U. M., & Petrovici, G. (2019). Developing planetary sustainability: Legal challenges of Space 4.0. *Global Sustainability*, 2, e10. DOI: 10.1017/sus.2019.10

Bongers, A., & Torres, J. L. (2023). Star Wars: Anti-Satellite Weapons and Orbital Debris. *Defence and Peace Economics*, •••, 1–20. DOI: 10.1080/10242694.2023.2208020

Clormann, M., & Klimburg-Witjes, N. (2022). Troubled Orbits and Earthly Concerns: Space Debris as a Boundary Infrastructure. *Science, Technology & Human Values*, 47(5), 960–985. DOI: 10.1177/01622439211023554

Damjanov, K. (2017). Of Defunct Satellites and Other Space Debris: Media Waste in the Orbital Commons. *Science, Technology & Human Values*, 42(1), 166–185. DOI: 10.1177/0162243916671005

Di Pippo, S. (2019). The contribution of space for a more sustainable earth: Leveraging space to achieve the sustainable development goals. *Global Sustainability*, 2, e3. DOI: 10.1017/sus.2018.17

El-Shawa, S., Cattani, B. M., Innocenti, L., & Delaval, J. (2021). *From Cradle to Grave: ESA Clean Space's Approach to Space Sustainability.*

Emanuelli, M., Federico, G., Loughman, J., Prasad, D., Chow, T., & Rathnasabapathy, M. (2014). Conceptualizing an economically, legally, and politically viable active debris removal option. *Acta Astronautica*, 104(1), 197–205. DOI: 10.1016/j.actaastro.2014.07.035

Galloway, J. F. (2004). Game theory and the law and policy of outer space. *Space Policy*, 20(2), 87–90. DOI: 10.1016/j.spacepol.2004.02.006

Grzelka, Z., & Wagner, J. (2019). Managing Satellite Debris in Low-Earth Orbit: Incentivizing Ex Ante Satellite Quality and Ex Post Take-Back Programs. *Environmental and Resource Economics*, 74(1), 319–336. DOI: 10.1007/s10640-019-00320-3

Hall, L. (2014). *The History of Space Debris*. Space Traffic Management Conference. 19. https://commons.erau.edu/stm/2014/thursday/19

Haroun, F., Ajibade, S., Oladimeji, P., & Igbozurike, J. K. (2021). Toward the Sustainability of Outer Space: Addressing the Issue of Space Debris. *New Space*, 9(1), 63–71. DOI: 10.1089/space.2020.0047

Hosseini, S. M., Fathpour, F., & Chanda, S. (2021). Space Debris Mitigation: Some Lessons from International Environmental Law. *Environmental Policy and Law*, 51(6), 391–401. DOI: 10.3233/EPL-210015

Hutagalung, J. M., Tobing, C. I., Debastri, J., & Amanda, R. T. (2020). Space debris as environmental threat and the requirement of Indonesia's prevention regulation. *IOP Conference Series. Earth and Environmental Science*, 456(1), 012081. DOI: 10.1088/1755-1315/456/1/012081

Kessler, D. J. (1993). A Partial History of Orbital Debris: A Personal View. *Orbital Debris Monitor*, 6(3), 16–20.

Klima, R., Bloembergen, D., Savani, R., Tuyls, K., Hennes, D., & Izzo, D. (2016). Space Debris Removal: A Game Theoretic Analysis. *Games*, 7(3), 20. DOI: 10.3390/g7030020

Klima, R., Bloembergen, D., Savani, R., Tuyls, K., Wittig, A., Sapera, A., & Izzo, D. (2018). Space Debris Removal: Learning to Cooperate and the Price of Anarchy. *Frontiers in Robotics and AI*, 5, 54. DOI: 10.3389/frobt.2018.00054 PMID: 33500936

Lambach, D., & Wesel, L. (2021). TACKLING THE SPACE DEBRIS PROBLEM: A GLOBAL COMMONS PERSPECTIVE. *8th European Conference on Space Debris,* Darmstadt, Germany, 20-23 April, 2021.

Liou, J.-C. (2013). Engineering and technology challenges for active debris removal. *Progress in Propulsion Physics*, 735–748. DOI: 10.1051/eucass/201304735

Losch, A., Galli, A., & Schildknecht, T. (2024). History, concepts, and challenges of proposing an 18th UN Sustainable Development Goal related to Space. *Space Policy*, 68, 101631. DOI: 10.1016/j.spacepol.2024.101631

Macauley, M. K. (2015). The economics of space debris: Estimating the costs and benefits of debris mitigation. *Acta Astronautica*, 115, 160–164. DOI: 10.1016/j.actaastro.2015.05.006

Mane, S. (2024). Theoretical Overview on Space and Sustainability. *International Journal of All Research Education and Scientific Methods*, 12(8), 2060–2066.

Matton, E. (2023). Avoiding the Kessler Effect: An Evaluation of the International Regulation of Space Debris. *Master's Thesis in International Law, Uppsala Universitet, Sweden.*

Medvedeva, A. (2015). *Space Debris Remediation: An International Relations Approach.* TU Wien Continuiting Education Center, Vienna School of International Studies.

Migaud, M. R. (2020). Protecting Earth's Orbital Environment: Policy Tools for Combating Space Debris. *Space Policy*, 52, 101361. DOI: 10.1016/j.spacepol.2020.101361

Morgan, F., McLeod, G., Nixon, M., Lynch, C., & Hura, M. (2018). *Gaming Space: A Game-Theoretic Methodology for Assessing the Deterrent Value of Space Control Options.* RAND Corporation., DOI: 10.7249/RR694

Morin, J., & Richard, B. (2021). Astro-Environmentalism: Towards a Polycentric Governance of Space Debris. *Global Policy*, 12(4), 568–573. DOI: 10.1111/1758-5899.12950

Mullick, S., Srinivasa, Y., Sahu, A. K., & Sata, J. T. (2019). A Comprehensive Study on Space Debris, Threats Posed by Space Debris, and Removal Techniques. SSRN *Electronic Journal.* DOI: 10.2139/ssrn.3511445

Newman, C., & Cheney, T. (2023). Barriers and Gateways to Cleaning Up Earth Orbit: The Legal, Economic, and Political Dimensions of Debris Remediation. *Air and Space Law*, 48(Special Issue), 113–136. DOI: 10.54648/AILA2023034

Nomura, K., Rella, S., Merritt, H., Baltussen, M., Bird, D., Tjuka, A., & Falk, D. (2024). Tipping Points of Space Debris in Low Earth Orbit. *The International Journal of the Commons*, 18(1). Advance online publication. DOI: 10.5334/ijc.1275

Norozpour, S., & Safaei, M. (2020). An Overview on Game Theory and Its Application. *IOP Conference Series. Materials Science and Engineering*, 993(1), 012114. DOI: 10.1088/1757-899X/993/1/012114

Nozawa, W., Kurita, K., Tamaki, T., & Managi, S. (2023). To What Extent Will Space Debris Impact the Economy? *Space Policy*, 66, 101580. DOI: 10.1016/j.spacepol.2023.101580

OECD. (2020). *Space sustainability: The economics of space debris in perspective* (OECD Science, Technology, and Industry Policy Papers 87; OECD Science, Technology and Industry Policy Papers, Vol. 87). DOI: 10.1787/23074957

Palmroth, M., Tapio, J., Soucek, A., Perrels, A., Jah, M., Lönnqvist, M., Nikulainen, M., Piaulokaite, V., Seppälä, T., & Virtanen, J. (2021). Toward Sustainable Use of Space: Economic, Technological, and Legal Perspectives. *Space Policy*, 57, 101428. DOI: 10.1016/j.spacepol.2021.101428

Poonuganti, S. (2023). It's Raining Rockets: Heightening State Liability for Space Pollution. *Chicago Journal of International Law*, 23(2).

Popova, R., & Schaus, V. (2018). The Legal Framework for Space Debris Remediation as a Tool for Sustainability in Outer Space. *Aerospace (Basel, Switzerland)*, 5(2), 55. DOI: 10.3390/aerospace5020055

Raoof, O., & Al-Raweshidy, H. (2010). Theory of Games: An Introduction. In Huang, Q. (Ed.), *Game Theory*. Sciyo., DOI: 10.5772/46930

Roberts, L. D. (1992). Addressing the Problem of Orbital Space Debris: Combining International Regulatory and Liability Regimes. *Boston College International and Comparative Law Review*, XV(1), 51–73.

Sannigrahi, A. K. (2017). Removal of Active Man-made Orbital Debris – A Great Challenge to Space Scientists. *International Journal of Scientific Research*, 6(2).

Sheer, A., Fatima, S., Owais Farooqui, M., & Ramzan Kasuri, M. (2023). A panacea to address the legal, administrative and economic aspects of space debris. *Advances in Space Research*, 72(7), 2616–2626. DOI: 10.1016/j.asr.2023.02.023

Sirieys, E., Gentgen, C., Milton, J., & De Weck, O. (2022). Space sustainability isn't just about space debris:On the atmospheric impact of space launches. *MIT Science Policy Review*, 3, 143–151. DOI: 10.38105/spr.whfig18hta

Slann, P. A. (2014). Space debris and the need for space traffic control. *Space Policy*, 30(1), 40–42. DOI: 10.1016/j.spacepol.2013.10.007

Tallis, J. (2015). Remediating Space Debris: Legal and Technical Barriers. *ASPJ Africa & Francophonie, First Quarter*, 80–90.

Taylor, M. W. (2007). *Orbital Debris: Technical and Legal Issues and Solutions*. Library and Archives Canada = Bibliothèque et Archives Canada.

Termeer, C. J. A. M., Dewulf, A., & Biesbroek, R. (2019). A critical assessment of the wicked problem concept: Relevance and usefulness for policy science and practice. *Policy and Society*, 38(2), 167–179. DOI: 10.1080/14494035.2019.1617971

Wang, T. (2016). A Liability and Insurance Regime for Space Debris Mitigation. *Science & Global Security*, 24(1), 22–36. DOI: 10.1080/08929882.2016.1127036

Weinzierl, M. (2018). Space, the Final Economic Frontier. *The Journal of Economic Perspectives*, 32(2), 173–192. DOI: 10.1257/jep.32.2.173

Wilson, A. R., & Vasile, M. (2023). The space sustainability paradox. *Journal of Cleaner Production*, 423, 138869. DOI: 10.1016/j.jclepro.2023.138869

Wood, D., Rathnasabapathy, M., Stober, K. J., & Menon, P. (2024). Challenges and progress in applying space technology in support of the sustainable development goals. *Acta Astronautica*, 219, 678–692. DOI: 10.1016/j.actaastro.2024.03.064

Yan, Y. (2023). Application of the principle of common but differentiated responsibility and respective capabilities to the passive mitigation and active removal of space debris. *Acta Astronautica*, 209, 117–131. DOI: 10.1016/j.actaastro.2023.04.033

Zhang, Y. (2022). An Analysis of the Regulations on Environmental Damage to Outer Space Caused by Space Debris. *Science of Law Journal*, 1(1), 19–26.

KEY TERMS AND DEFINITIONS

Active Debris Removal (ADR): The notions of refers the intentional and planes removal of non-functioning space objects as well as any items from the Earth's orbital space to reduce pollution in the atmosphere and prevent the risk of satellite collusions.

Kessler Effect: Also termed the Kessler Syndrome, is named after Donald J. Kessler, who discussed the phenomenon in a conceptual paper published in 1978, refers to a scenario whereby objects in space will collide leading to an increase in space debris that will overwhelm the orbital environment.

Low Earth Orbit: It is an orbital space that is the closest to the earth at an altitude range of 160 km to 2 000 km which suitable for earth observation and imaging as well as satellite communications.

Orbital Debris: Sometimes referred to as Orbital Debris or Space Junk, refers to the defunct non-functioning objects in the Earth's orbit which result from collisions or explosions as well as defunct satellites and other objects left flying in orbit.

Sustainability: Considers the ability of meeting the needs of the present generation without compromising the capacity of future generations to mee their own needs with key concerns around environmental sustainability, entailing the projection of ecosystems, economic sustainability (long-term economic growth) and social sustainability (equitable distribution of social resources).

Spacecraft: A propelled vehicle or moving object designed to travel and operate outside of the Earth's atmosphere.

Sustainable Development Goals: Refers to the 17 global goals for sustainable development adopted by the United Nations in 2015 for achieving a sustainable future by the 2030 deadline.

Chapter 4
Effectiveness of Leadership Qualities on Sustainable Supply Chain Practices:
Achieving Operational Excellence - A Framework

S. Yoganathan
https://orcid.org/0009-0007-5583-0924
Jain University, India

Ravishankar S. Ulle
https://orcid.org/0009-0008-1011-8738
Jain University, India

Balaji Gopalan
https://orcid.org/0000-0002-1082-5597
Faculty of Management Studies, CMS Busines School, Jain University, India

ABSTRACT

Effective leadership focused on sustainability is key to achieving operational excellence in a supply chain. Leaders with the right qualities implement sustainable practices throughout the chain, considering environmental, social, and economic factors. Success is influenced by company culture, stakeholder engagement, regulations, and data use. This approach leads to operational excellence with better efficiency, quality, and resilience, along with a competitive advantage and happy stakeholders. The framework is a roadmap for businesses to leverage leadership for a sustainable and successful supply chain.

1. INTRODUCTION

In a rapidly changing global landscape marked by shifting market dynamics, environmental concerns, and heightened social expectations, organizations are increasingly recognizing the imperative of integrating sustainability into their supply chain practices. The pursuit of operational excellence has become

DOI: 10.4018/979-8-3693-5748-4.ch004

inextricably linked with sustainable supply chain management (Chen, S. L.,et al., 2023). Key to this pursuit is the role of leadership qualities, which serve as the guiding principles that navigate the complex intersection of sustainable practices and operational efficiency. This conceptual framework seeks to offer a comprehensive exploration of the intricate and dynamic relationship between leadership qualities, sustainable supply chain practices, and the aspiration for operational excellence (Bag, S., et al., 2020).

Leadership qualities encompass a multifaceted range of attributes, competencies, and behaviours that define effective leaders (Kim, M. S., & Thapa, B. 2018). These qualities include visionary thinking, strategic acumen, adept communication, adaptability in the face of change, and a steadfast commitment to ethics and social responsibility. Effective leaders embody these qualities, which serve as the driving force behind the advocacy and implementation of sustainable supply chain practices. In a constantly evolving business environment, leaders play a pivotal role in inspiring and guiding their organizations to embrace sustainability principles that encompass environmental stewardship, social accountability, and economic viability, all while fostering collaborative partnerships with suppliers.

The interplay between leadership qualities and sustainable supply chain practices culminates in the achievement of operational excellence. Operational excellence encompasses a spectrum of attributes, including enhanced efficiency, superior product quality, a culture of innovation, adept risk management, and the establishment of supply chain resilience (Bastas, A., & Liyanage, K. 2018). Attaining these goals hinges on the seamless integration of sustainability practices into supply chain operations, a task that necessitates leadership qualities as a driving force.

However, the complex landscape in which these processes unfold is shaped by several mediating factors. These factors include the organizational culture, stakeholder engagement, the regulatory environment, and the strategic use of technology and data analytics. Organizational culture and values play a fundamental role in shaping how leadership qualities are embraced and manifested throughout the organization (Ahsan, M. J. 2024). The degree of stakeholder engagement, including employees, customers, suppliers, and the broader community, can significantly influence the success of sustainable practices and, consequently, the pursuit of operational excellence.

Furthermore, the regulatory environment can either enable or constrain the adoption of sustainable supply chain practices. Leaders must navigate complex compliance requirements while still striving for sustainability goals. The strategic deployment of technology and data analytics plays a pivotal role in measuring and optimizing the impacts of sustainable practices, providing insights that inform decision-making and enhance performance.

The outcomes of this multifaceted interrelationship ripple throughout the organization. They extend to the attainment of operational excellence (Naik, S.,et al., 2024), the cultivation of a competitive advantage, and the satisfaction of diverse stakeholders, such as customers, employees, and investors by illustrating the following possible research questions.

RQ1: How do leadership qualities, such as vision, impact on the adoption and implementation of sustainable supply chain practices and their contribution to achieving operational excellence?

RQ2: To what extent does leadership adaptability and change management skills affect the successful integration of sustainability principles into supply chain operations, and how does this correlate with operational excellence outcomes?

RQ3: What is the role of ethical and social responsibility leadership in driving the development and execution of sustainable supply chain practices, and how does this influence the attainment of operational excellence?

Organizations that invest in leadership qualities aligned with sustainability and operational excellence are not only better equipped to optimize their supply chain operations in terms of cost-efficiency and product quality but are also strategically positioned to attract a discerning clientele and investors who prioritize environmental and social responsibility (Govindan, K. 2018).

In this conceptual framework lies a roadmap for organizations aiming to harness the power of leadership qualities as a promoter for sustainable supply chain practices. It is within this framework that we embark on an exploration of the dynamic synergy between leadership, sustainability, and operational excellence (Hahn, T, Figge, F.; Pinkse, J.; Preuss, L. 2018). By unravelling the complex connections and pathways that underpin this relationship, we aim to empower organizations to thrive in a dynamic and ever-evolving business landscape, all while promoting a more sustainable and responsible future.

2. THEORETICAL FRAMEWORK FOR SUSTAINABLE SUPPLY CHAIN TO ACHIEVE OPERATIONAL EXCELLENCE

The effectiveness of leadership qualities in the context of sustainable supply chain practices to achieve operational excellence is a critical and complex relationship (Hahn, T. et al., 2015). A conceptual framework for understanding this relationship can help organizations identify the key factors and processes involved. Here is a conceptual framework that outlines the relationship between leadership qualities, sustainable supply chain practices, and operational excellence:

The proposed theoretical framework for sustainable supply chain to achieve operational excellence:

Figure 1. The proposed conceptual framework

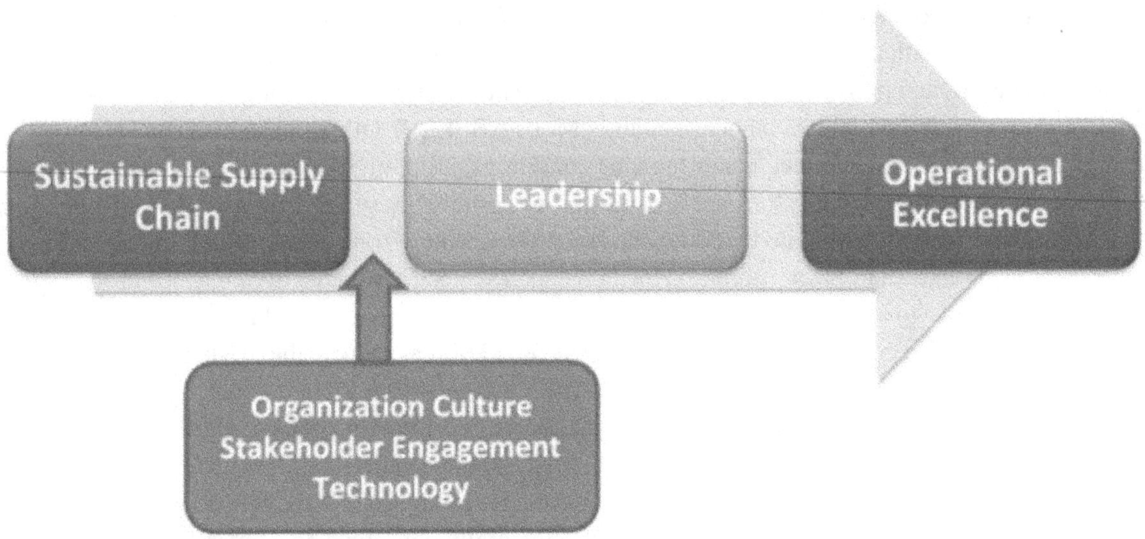

2.1 Sustainable Supply Chain

A sustainable supply chain represents a comprehensive approach to managing the flow of goods and services from their source to their ultimate destination, while considering and minimizing the environmental, social, and economic impacts throughout the entire process (Paulraj, A., et al., 2017). This model takes into account the three key pillars of sustainability - people, planet, and profit.

One of the fundamental principles of a sustainable supply chain is environmental responsibility. Companies operating in such supply chains seek to reduce their ecological footprint by implementing practices that minimize harm to the environment (Singh, R. K. 2024). This includes efforts to lower greenhouse gas emissions, conserve natural resources, and reduce waste generation. For example, companies may adopt energy-efficient transportation methods, eco-friendly packaging, and sustainable sourcing of raw materials. These measures not only contribute to the preservation of the environment but can also lead to cost savings in the long run.

In addition to environmental concerns, social responsibility is a core element of sustainable supply chains. Martins, C.L.; Pato, M.V.(2019). This aspect focuses on fair labour practices, human rights protection, and support for local communities. Businesses in sustainable supply chains ensure fair wages, safe working conditions, and human rights protection for labourers throughout the supply chain. By prioritizing social responsibility, companies contribute to ethical and humane labour practices, which is increasingly important to consumers and stakeholders.

Economic responsibility is the third fundamental component of a sustainable supply chain. It is essential for the overall viability of the operation. Sustainability doesn't mean sacrificing profitability. Brandenburg M, Gruchmann T, Oelze N. (2019) Instead, it encourages companies to optimize their processes, reduce costs, and increase overall profitability. Businesses are driven to find innovative cost-saving methods and adopt lean operational practices to maintain economic sustainability while adhering to environmental and social responsibilities (Dubey, R., Gunasekaran, A., & Ali, S. S. 2015).Ethical sourcing is another critical aspect of sustainable supply chains. Companies in these chains make an effort to source materials and products from ethical suppliers. This often involves using suppliers who adhere to fair trade practices and ethical labour standards Mousa, S. K., & Othman, M. (2020). This practice ensures that products are produced under humane conditions, aligning with social responsibility goals.

Green transportation and logistics are also central to sustainable supply chains. (Gong, Ruifeng, et al., 2019). These supply chains aim to reduce the carbon footprint of transportation and logistics. This is achieved by using eco-friendly transportation methods, such as electric vehicles or biofuels, and optimizing transportation routes to reduce emissions. By doing so, businesses not only reduce their environmental impact but also realize cost savings and enhance their reputation as responsible organizations (Ahmad, A. F., & Karadas, G. 2021).

Sustainable supply chains also emphasize eco-friendly packaging. This involves the use of sustainable materials that reduce waste and are recyclable or biodegradable. (Manurung, Hakim, et al., 2023). Sustainable packaging practices help minimize the environmental impact of products, improve the supply chain's ecological footprint, and resonate with environmentally conscious consumers.

The adoption of a circular economy approach is another hallmark of sustainable supply chains. This principle encourages the minimization of waste through practices like reusing, recycling, or repurposing materials and products. By reducing waste and the need for new raw materials, this approach lessens the strain on natural resources and decreases landfill waste.

Collaboration and transparency play pivotal roles in sustainable supply chains. Organizations work closely with suppliers, manufacturers, distributors, and customers to ensure shared sustainability goals. Transparency is critical in supply chain management as it fosters trust among stakeholders, enables better understanding of social and environmental impacts, and facilitates data-driven decision-making for continuous improvement.

The commitment to continuous improvement is a fundamental characteristic of sustainable supply chains. Companies regularly assess their operations, measure their environmental and social impacts, and strive to reduce these impacts. This data-driven approach ensures that sustainability is not just a one-time initiative but a continual process of improvement and adaptation.

Sustainable supply chains are a holistic approach to managing the flow of goods and services with a focus on minimizing environmental, social, and economic impacts. They align with ethical practices, drive economic viability, and are environmentally responsible Nagariya, R., Kumar, D. and Kumar, I. (2022). Such supply chains not only contribute to the well-being of our planet and society but also offer long-term benefits, including cost savings, enhanced reputation, and resilience to environmental and social disruptions. As consumer and investor preferences increasingly favour sustainable practices, more organizations are integrating sustainability into their supply chain strategies as an essential component of their business model.

2.2 Effective Leadership

Several mediating factors influence the effectiveness of leadership in achieving sustainability and operational excellence. Organizational culture, which encompasses the values and behaviours within an organization, plays a significant role in shaping the implementation of sustainability practices (Kafetzopoulos, D., & Gotzamani, K. 2022). A culture that prioritizes sustainability can drive successful supply chain initiatives. Stakeholder engagement involves collaborating with customers, suppliers, and communities to align with sustainability goals, ensuring that all parties are invested in achieving shared objectives. Regulatory environments influence supply chain practices by setting legal and compliance standards that organizations must meet. Technology adoption, including the integration of advanced technologies such as blockchain, IoT, and AI, enhances transparency, efficiency, and innovation within supply chains.

Transformational leadership is characterized by its ability to inspire and motivate employees to transcend personal interests for the greater good of the organization (Burawat, P. 2019). In supply chain contexts, transformational leaders are pivotal in driving sustainability initiatives by cultivating a shared vision and engaging employees at all organizational levels. Ethical leadership, with its emphasis on integrity and fairness, plays a critical role in fostering ethical sourcing and social accountability within supply chains. Adaptive leadership, which focuses on navigating and adapting to evolving environments, is essential for responding to regulatory changes and market demands related to sustainability. Meanwhile, visionary leadership is instrumental in setting long-term sustainability goals and steering organizations towards innovative practices that align with future-oriented sustainability objectives.

Effective leadership is a cornerstone of operational excellence. Leadership qualities such as vision, strategic thinking, decision-making, and communication skills are essential for setting clear goals and driving performance improvement Teoman, S. and Ulengin, F. (2018). Strong leaders inspire their teams, provide direction, and foster a culture of continuous improvement. They also play a critical role in aligning the workforce with the organization's strategic objectives, ensuring that everyone is working towards a common goal. Additionally, effective leadership promotes accountability and responsibility at all levels, which is crucial for achieving and sustaining operational excellence (Naik, S., et al., 2024)

2.3 Organizational Culture

It defines the values, beliefs, and norms that shape how employees behave and interact within an organization. A positive culture that values continuous improvement, innovation, and efficiency is essential for achieving operational excellence. A culture of excellence encourages employees to take ownership of their work, seek out opportunities for improvement, and collaborate effectively Mondal, S. and Samaddar, K. (2023). It also fosters a culture of learning and adaptability, where mistakes are seen as opportunities for growth rather than failures. An operational excellence culture is underpinned by transparency, trust, and open communication, enabling organizations to identify and address operational challenges more effectively.

2.4 Stakeholder Engagement

It includes customers, suppliers, employees, and other relevant parties, is integral to operational excellence. Engaging stakeholders provides valuable insights into their expectation and needs. For instance, listening to customer feedback can lead to product or service improvements. Engaging suppliers can lead to more efficient and reliable supply chain operations (Nagariya, R., et al., 2022). Engaging employees ensures they feel valued and motivated, which directly impacts productivity and quality. Stakeholder engagement also enhances transparency and builds trust, which is essential for maintaining a positive reputation. In summary, stakeholders play a critical role in identifying opportunities for operational excellence and sustaining it.

2.5 Technology

It is a powerful enabler of operational excellence. It allows organizations to streamline processes, improve efficiency, reduce errors, and enhance decision-making. Technology solutions like data analytics, process automation, and the Internet of Things (IoT) provide real-time insights into operations, enabling organizations to make data-driven decisions and proactively address issues (Bag, Surajit. et al., 2019). Technology can also support the implementation of best practices and standardized processes. Furthermore, it facilitates collaboration, communication, and knowledge-sharing within organizations, which are essential for continuous improvement. Embracing emerging technologies is often a key differentiator for organizations striving for operational excellence.

2.6 Operational Excellence

Operational excellence in the context of sustainability involves achieving resilience, efficiency, and innovation. Resilience refers to the ability of supply chains to recover from disruptions and adapt to environmental, social, and economic challenges. Efficiency entails optimizing processes to reduce waste and enhance resource utilization, which contributes to both sustainability and cost-effectiveness. Innovation is crucial for advancing sustainability and operational performance, driving the adoption of new technologies and practices that support sustainable supply chain management (Bastas, A., & Liyanage, K. 2018).

The impact of leadership qualities, organizational culture, stakeholder engagement, and technology on operational excellence is profound and interrelated. Effective leadership provides direction, motivates teams, and aligns efforts toward operational excellence Ulle, R. S., & Kumar, A. S. (2014). A positive organizational culture fosters a climate of continuous improvement and adaptability. Stakeholder engagement ensures that operational excellence is driven by the needs and expectations of all relevant parties. Technology serves as an enabler, providing the tools and insights necessary to streamline operations and drive performance improvements. Together, these factors create a robust framework for achieving and sustaining operational excellence, allowing organizations to thrive in an ever-evolving business landscape.

Figure 2. Proposed conceptual framework (detailed)

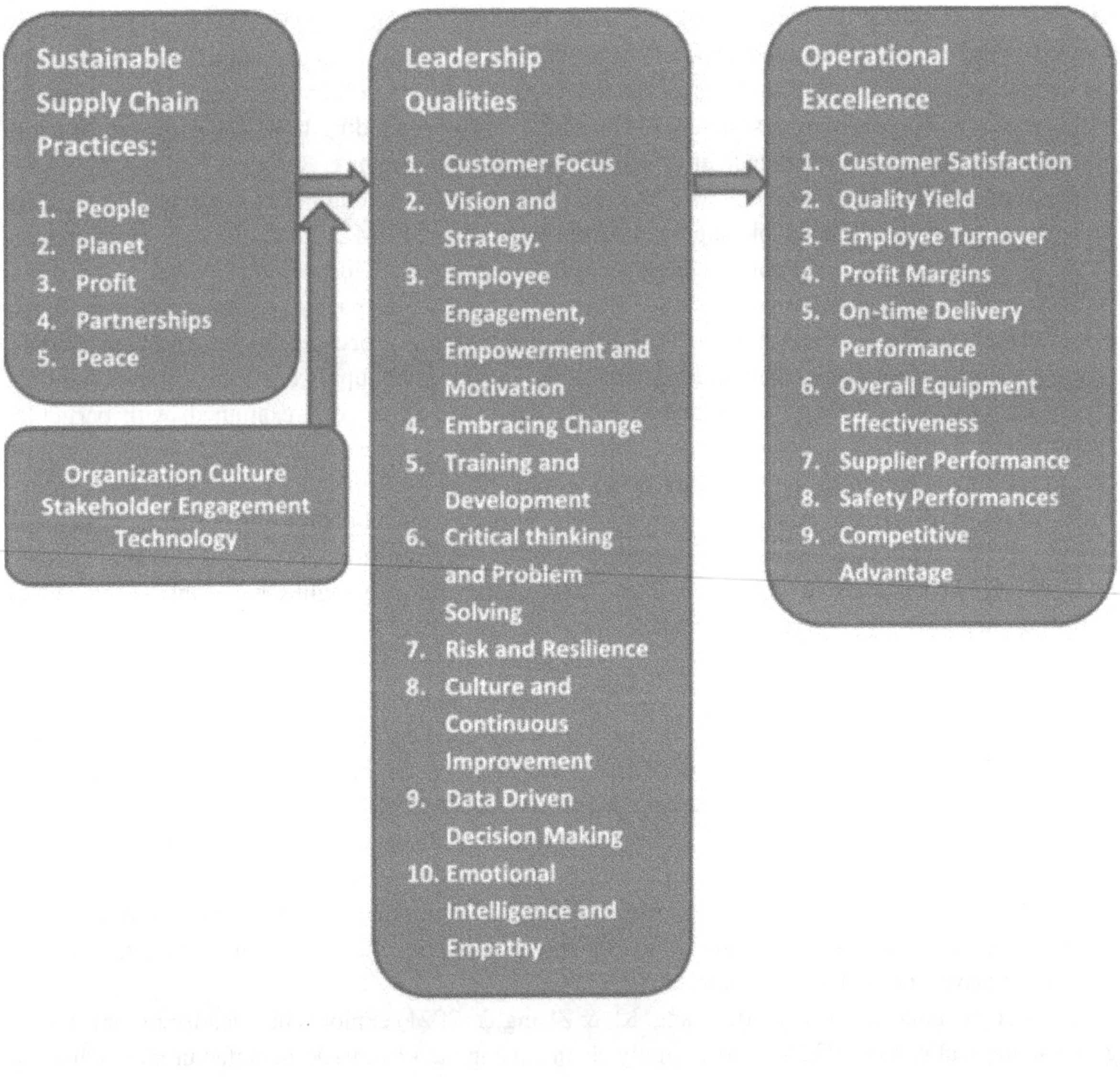

3. LITERATURE REVIEW

The literature review for this research paper will comprehensively examine several key areas of interest. First, it will explore various leadership theories, including transformational, ethical, adaptive, and visionary leadership, to understand their influence on sustainable supply chain practices and organizational performance. (Burawat, P. 2019) Transformational leadership's role in motivating employees towards sustainability, ethical leadership's impact on ethical sourcing and social responsibility, adaptive leadership's ability to manage change in response to regulatory demands, and visionary leadership's contribution to setting long-term sustainability goals will be explored. The review will then focus on sustainable supply chain practices, encompassing environmental responsibility, social accountability, ethical sourcing, and economic sustainability, to assess their effectiveness in enhancing sustainability performance. Additionally, the review will address operational excellence by investigating resilience, efficiency, and innovation within supply chains and their roles in achieving sustainability goals. Mediating factors such as organizational culture, stakeholder engagement, regulatory environments, and technology adoption will be examined to understand their impact on sustainability and operational success. Finally, the review will integrate these findings to identify gaps in the current literature, highlighting areas for future research and providing a structured framework for understanding how leadership, sustainable practices, and operational excellence intersect in modern supply chains.

The purpose of the literature review is to give a detailed and thorough examination of the existing research on leadership, sustainable supply chain management (SSCM), and operational excellence. It serves as a foundation to highlight how these areas have been studied, identifying key themes, theories, and findings from previous work. The review also aims to uncover gaps in the current research and demonstrate how the paper contributes to the field by either building on or diverging from previous studies.

In this study conducted by (Singh, R. K. 2024), the influence of supply chain capabilities and total quality management (TQM) on sustainable supply chain performance is examined, with particular attention to the roles of leadership and institutional pressures. The research validates that both supply chain capabilities and TQM significantly boost sustainability performance, especially under institutional pressures. The findings, supported by evidence from the cement manufacturing sector, affirm the importance of institutional theory and illustrate how leadership can strengthen TQM and supply chain capabilities to enhance sustainability outcomes. This research uniquely highlights the moderating effect of institutional pressures on these relationships.

The study conducted by (Jia, F., Gong, Y., & Brown, S. 2019) addresses the gap in understanding how multinational corporations (MNCs) lead their supply chains in promoting sustainability practices within emerging economies. By employing a multiple-case study approach, the research investigates the proactive sustainability initiatives of three MNCs operating in China. The proposed framework explores the interplay between supply chain leadership, multi-tier governance, supply chain structure, and supply chain learning. Findings reveal that effective supply chain leadership and governance mechanisms significantly influence both the structure and learning processes within supply chains. MNCs adjust their supply chain structures to enhance sustainability learning among their members. The study advances three propositions and discusses implications for future research, offering valuable insights into how MNCs can drive sustainability in complex supply chains.

The study conducted by (Huo, B., Wang, K., & Zhang, Y. 2021) explores how leadership influences green strategy alignment (GSA) across supply chains, using data from 206 manufacturers in China. It finds that internal GSA drives alignment with suppliers and customers. Transformational leadership positively impacts all dimensions of GSA, while transactional leadership influences only internal GSA. Additionally, both internal and customer GSA improve operational performance. The research contributes to green supply chain management (GSCM) by presenting GSA as a multi-dimensional construct and linking leadership styles to better operational outcomes through green strategy alignment.

The study conducted by (Dubey, R., Gunasekaran, A., & Ali, S. S. 2015) explores the impact of supplier relationship management (SRM) and total quality management (TQM) on environmental performance, examining the influence of leadership and the moderating role of institutional pressures (IP). Using a structured, pre-tested questionnaire and a split survey approach, the study collected 187 and 174 valid responses. Rigorous checks for non-response bias, constant variance, and normality were conducted, alongside confirmatory factor analysis and hierarchical regression to ensure reliability, validity, and hypothesis testing. The results validate the constructs within the theoretical framework, showing a strong fit with the data and supporting all hypotheses. This empirical evidence reinforces existing literature and contributes to supply chain management (SCM) and operations management (OM) by integrating TQM, SRM, and green supply chain practices. The study uniquely extends institutional theory to green supply chain networks, providing both practical and theoretical insights into how leadership and institutional pressures influence environmental performance.

The study conducted by (Pham, H., & Kim, S. Y. 2019) examines the impact of sustainable practices on environmental, social, and economic dimensions of sustainability in construction, highlighting the role of leadership competencies in these processes. Using a survey and Structural Equation Modeling (SEM), the research found that environmental, economic, and social practices positively influence sustainability performance. Leadership competencies were shown to strengthen the relationship between environmental practices and sustainability performance, although no evidence was found for their moderating effect on the relationships between economic and social practices and sustainability performance. The study addresses gaps in the literature by providing empirical insights into how leadership can enhance sustainable construction practices, offering valuable guidance for researchers and construction managers.

The research conducted by (Burawat, P. 2019) explores the structural relationships among transformational leadership, sustainable leadership (SL), lean manufacturing practices, and sustainability performance within the Thai manufacturing sector. Data from 598 managers and 40 qualitative participants revealed that lean manufacturing partially mediates the relationship between both transformational leadership and SL with sustainability performance. Findings indicate that lean practices are particularly suited to the automotive industry and emphasize the importance of managerial understanding and buy-in for successful lean implementation. The study highlights a positive relationship between transformational leadership and lean manufacturing, while SL shows no significant effect on lean practices. This research provides new insights into the impact of leadership styles on lean manufacturing and sustainability performance, offering valuable contributions to both academic literature and practical management in manufacturing industries.

(Pantouvakis, A., & Vlachos, I. 2020) investigates the influence of talent and leadership on sustainable performance within the maritime industry, considering the moderating role of organizational culture. Grounded in Service Dominant Logic (SDL) and based on a large-scale survey of 308 shipping companies conducted in 2018, the analysis identifies four leadership styles (authoritarian, stakeholder, formal, growth), four types of talent (innate/developable, exclusive, inclusive, visible), and two organizational culture types (formal, learning). It finds that talent has a greater impact on sustainability performance compared to leadership. While organizational culture moderates the relationship between talent and sustainability, it does not affect the leadership-sustainability relationship. This research highlights the critical role of talent and organizational culture in enhancing sustainability performance in the maritime sector.

(Svensson, G., & Wood, G. 2006) the research explores the sustainability components of leadership effectiveness and their impact on organizational performance through a conceptual discussion. It develops descriptive models that address how leadership influences organizational performance, highlighting both supportive and critical perspectives from the literature. The study acknowledges that leadership effectiveness can vary over time and across contexts, sometimes leading to successful outcomes and other times to failures. It emphasizes the need for organizations to integrate sustainability into their management practices to ensure long-term success. The paper contributes to the field by offering models of timely and contextual leadership effectiveness, along with a typology of leadership effectiveness in corporate decision-making. These insights provide valuable theoretical and practical guidance for understanding and enhancing leadership effectiveness in relation to organizational performance.

The study conducted by (Ahsan, M. J. 2024) examines the impact of organizational culture (OC) and transformational leadership (TL) on corporate social responsibility (CSR) performance—specifically environmental and social performance—and financial performance (FP) within the Italian manufacturing sector. Utilizing a resource-based view theory and analyzing data from 260 employees, the research reveals that recognizing the role of TL and OC is crucial for achieving both sustainable and financial performance. The study establishes a significant positive relationship between CSR performance and FP, highlighting the strategic value of integrating CSR initiatives into core business practices. This research provides novel insights into the interplay between TL, OC, and performance, offering both theoretical contributions and practical guidance for enhancing firm performance through robust organizational culture and effective leadership.

The study develops a framework to analyze how various leadership styles impact talent management (TM) and, in turn, how these styles and TM influence firms' sustainable performance, with environmental dynamism as a moderating factor conducted by (Kafetzopoulos, D., & Gotzamani, K. 2022). An empirical survey of 480 firms in Greece, analyzed using hierarchical regression, reveals that authoritative and transformational leadership significantly affect TM, while talent development positively impacts sustainability. Entrepreneurial leadership is particularly effective for economic sustainability, whereas transformational and transactional leadership styles are more influential for social and environmental sustainability. The study highlights the need for appropriate leadership styles to address different sustainability dimensions and underscores the value of talent development for enhancing sustainable performance. By contributing to upper echelon theory and advancing human resource development literature, this research provides insights into how leadership styles and TM interact to influence sustainability, considering the role of environmental dynamism.

The study conducted by (Suriyankietkaew, S. 2023) addresses the challenge faced by small enterprises in emerging economies, such as Thailand, to enhance long-term financial performance and stakeholder satisfaction while transitioning towards corporate sustainability. By surveying 280 business leaders and entrepreneurs and employing factor analysis and multiple regression, the research identifies seven key leadership factors that impact business sustainability. Notably, it finds that a trusting, innovative team orientation and a strong, shared vision are significant drivers of financial performance and stakeholder satisfaction. The study contributes to the field by offering a contextualized perspective on leadership factors that support sustainability in small businesses and suggests that adopting these leadership practices can enhance business competitiveness and resilience. The findings provide valuable insights for entrepreneurs, policymakers, and practitioners and set the stage for further research on entrepreneurial growth and sustainability in emerging economies.

The paper explores the relationship between leadership behaviors and occupational safety in frontline operations, focusing on how authentic leadership impacts safety outcomes through psychological capital and organizational citizenship behaviour, research done by (Cavazotte, F., Mansur, J., & Moreno, V. 2021). Grounded in authentic leadership theory, the study hypothesizes that leader morality and selflessness influence safety performance by enhancing employees' psychological capital and organizational citizenship. Data from 307 employees at a large Brazilian fuel company were analyzed using structural equation modeling. The results reveal that while leader morality and selflessness positively affect employees' psychological capital, only selflessness influences organizational citizenship behavior. Both psychological capital and organizational citizenship behavior mediate the relationship between authentic leader attributes and frontline safety participation and compliance. These findings underscore the importance of virtuous leadership in improving safety performance and suggest that developing such leadership qualities can significantly enhance sustainable operations in high-risk industries.

(Ahmad, A. F., & Karadas, G. 2021) research tests a model examining how effective leadership impacts organizational performance through green supply chain management (GSCM) practices. Using a survey that assessed dimensions of effective leadership, GSCM, and organizational performance, and analyzing the data with partial least squares–structural equation modeling (PLS-SEM), the study reveals several key findings. It identifies significant positive relationships between the dimensions of effective leadership—leading organization, leading people, and leading self—and GSCM practices. Additionally, GSCM practices are shown to positively influence organizational performance. The research confirms that GSCM practices fully mediate the relationship between leading organization and leading people with organizational performance, and partially mediate the relationship between leading self and organizational performance. These findings highlight that effective leadership indirectly enhances organizational performance through GSCM practices, which also contribute to improved social and environmental outcomes.

The study conducted by (Nan, S., & Chaiprasit, K. 2023)investigates the impact of resilient leadership on sustainable business performance in the manufacturing sector of Jiangxi, China, a region previously unexplored in this context. Using structural equation modeling, the research examines both the direct effects of resilient leadership on sustainable performance and the indirect effects mediated by enterprise innovation. Additionally, hierarchical regression analysis is employed to explore the moderating role of executive incentives. The findings reveal that resilient leadership significantly enhances sustainable business performance, with enterprise innovation serving as a key mediator in this relationship. Furthermore, executive incentives are confirmed to moderate this effect. These insights provide valuable implications for practitioners, highlighting how resilient leadership and strategic incentives can drive sustainable competitive advantages and improve business performance.

The study done by (Awan, U., & Khan, S. A. R. 2021) examines how sustainable leadership moderates the effect of supplier relationship management on social performance. Analyzing data from 224 manufacturers using partial least squares structural equation modeling (PLS-SEM), the research finds that sustainable leadership traits—such as selflessness, enthusiasm, coaching, and innovative thinking—positively influence social performance. The study highlights the critical role of sustainable leadership in enhancing social sustainability and suggests that managers should focus on refining their leadership skills and practices to improve sustainability outcomes.

The research by (Esangbedo, C. O., Zhang, J., Pérez, P. B., & Skitmore, M. 2024) explores the impact of supply chain leadership, digital supply chain practices, and corporate sustainability strategies on the sustainability performance of logistics firms in Nigeria. Using data from 468 firms, the study reveals that proactive sustainability strategies substantially enhance performance, while reactive strategies can diminish the effectiveness of supply chain leadership. The study also identifies a significant mediating role for digital supply practices in this context. The findings advise managers to focus on proactive strategies and digital innovations to improve sustainability outcomes. However, the study's cross-sectional design and reliance on self-reported data highlight the need for longitudinal research and diverse contextual analysis.

The study done by (Sriyakul, T., Singsa, A., Sutduean, J., & Jermsittiparsert, K. 2019) investigates how commitment to change influences operational excellence in Indonesian supply chain companies, addressing a gap in the existing literature. By analyzing data from 190 respondents out of 350 distributed questionnaires, the research employs PLS 3 to explore the impact of cultural traits, transformational leadership, and commitment to change on supply chain performance. The findings indicate that these factors positively contribute to operational excellence, emphasizing the need for effective organizational change strategies to enhance overall company performance. The study suggests that Indonesian supply chain firms should focus on adopting robust change management practices to improve their operational outcomes.

The research done by (Muralidharan, E., & Pathak, S. 2018) explores how culturally endorsed transformational leadership theories (CLTs) and societal sustainability influence social entrepreneurship. Analyzing data from 107,738 individuals across 27 countries in the 2009 Global Entrepreneurship Monitor survey, along with country-level data from GLOBE and the Sustainability Society Foundation, the study reveals that both transformational CLTs and societal sustainability positively impact social entrepreneurship. Notably, the effectiveness of transformational CLTs becomes more significant when societal sustainability is low, highlighting an interaction between cultural leadership styles and societal conditions. This research enriches comparative entrepreneurship studies by identifying key cultural and societal factors that drive social entrepreneurship and suggests areas for further investigation.

4. FINDINGS AND SYNTHESIS OF THE LITERATURE

Leadership qualities such as vision, communication, and adaptability are critical in driving sustainable practices across various contexts. Visionary leaders, who articulate clear and compelling sustainability goals, have been shown to foster long-term commitment and alignment within organizations (Ahmad, A. F., & Karadas, G. 2021). Effective communication enables leaders to effectively engage stakeholders, align objectives, and drive sustainability initiatives. Adaptability is crucial for navigating the evolving landscape of sustainability regulations and practices. Research indicates that visionary leadership is particularly effective in industries where strategic foresight and long-term planning are essential, while strong communication skills are vital in complex, multi-stakeholder environments.

In the ecosystem of supply chains, sustainable practices like green logistics and ethical sourcing have been extensively studied for their impact on overall performance and stakeholder engagement. Literature shows that integrating sustainability into supply chain management not only enhances environmental performance but also improves operational efficiency and stakeholder trust. Sustainable practices contribute to reduced operational costs through energy savings and waste reduction while strengthening relationships with socially conscious consumers and partners.

Operational excellence is increasingly linked to leadership-driven sustainable practices. Evidence suggests that leaders who prioritize sustainability can drive significant improvements in risk management and operational efficiency. By embedding sustainability into core business processes, leaders can enhance operational resilience, reduce risks associated with environmental regulations, and optimize resource use.

Despite these insights, gaps remain in the literature. There is a notable scarcity of empirical data directly linking specific leadership styles to quantifiable improvements in operational excellence. More research is needed to establish concrete connections between different leadership approaches and measurable outcomes in sustainability and performance. Addressing these gaps will provide a clearer understanding of how leadership can be leveraged to achieve both sustainable and operational goals effectively.

In the literature, several conceptual frameworks have been developed to link leadership with sustainability and operational performance. One prominent model is the Transformational Leadership Theory, which emphasizes the role of leaders in inspiring and motivating employees to embrace sustainable practices and drive organizational change. This model highlights how transformational leaders can influence sustainability outcomes by articulating a compelling vision and fostering an environment conducive to innovation and commitment.

While these frameworks offer valuable insights, there are notable limitations. For instance, the Transformational Leadership Theory often focuses on the overall impact of leadership on organizational culture and change but may not sufficiently address specific leadership qualities such as adaptability and communication in the context of sustainability. Additionally, this theory can sometimes overlook how leadership interacts with other factors like institutional pressures and stakeholder expectations (Burawat, P. 2019).

This paper offers a unique perspective by integrating leadership qualities more deeply into the sustainability-operational excellence equation. Unlike traditional models that may emphasize general leadership impacts or resource utilization, our framework specifically explores how distinct leadership qualities—such as vision, communication, and adaptability—directly influence sustainability practices and operational performance.

5. DISCUSSIONS AND MANAGERIAL IMPLICATIONS

The complex relationship between sustainable supply chain practices and effective leadership qualities brings forth a wealth of managerial implications, illuminating a roadmap towards achieving operational excellence. These implications, rooted in research-backed insights, shed light on the pivotal role of leadership and its symbiotic relationship with sustainability (Cavazotte, F., Mansur, J., & Moreno, V. 2021). As organizations strive to navigate the complex landscape of sustainability, managers and business leaders are presented with a compelling opportunity to harness this synergy for tangible benefits (Ahmad, A. F., & Karadas, G. 2021).

Sustainable supply chain practices, when harmonized with visionary leadership, emerge as a source of substantial cost-saving prospects. Managers should proactively align sustainability initiatives with financial objectives (Awan, U., & Khan, S. A. R. 2021), effectively utilizing leadership qualities to optimize resource allocation, reduce waste, and enhance energy efficiency. The resultant cost savings can be channelled into innovation or other strategic endeavours, thus fortifying the foundation of operational excellence.

The implications of research underscore the direct correlation between sustainability practices and improved product and service quality. Managers need to grasp the significance of this relationship, recognizing that leadership's role in facilitating change and adaptation is instrumental (Mondal, S. and Samaddar, K. 2023). This underscores the need for investment in training and development programs that nurture adaptive leadership qualities. The resultant enhancement in product quality is not only a hallmark of operational excellence but also a potent catalyst for bolstering customer satisfaction and loyalty.

In the face of supply chain disruptions that can potentially derail operational excellence, managers and leaders are urged to prioritize ethical and socially responsible leadership attributes. These qualities have been demonstrated to be instrumental in risk mitigation. As part of managerial responsibilities, the development of resilience strategies, such as supplier diversification and contingency planning, is paramount. (Nan, S., & Chaiprasit, K. 2023) Effective leadership, with its embodiment of ethical values, becomes the linchpin in managing disruptions, ensuring business continuity, and minimizing operational setbacks.

Operational excellence is intrinsically linked to innovation, and leaders who foster innovative thinking within a sustainability framework stand to gain a competitive edge (Sriyakul, T., et al., 2019). Managers should take an active role in promoting a culture of innovation. Recognizing that innovation can propel the organization ahead of competitors, managers must actively foster this culture within their teams. Not only does this position the organization as a market leader, but it also amplifies its journey toward operational excellence.

The inherent alignment of sustainable supply chain practices with customer and stakeholder expectations forms a cornerstone of this relationship (Bastas, A., & Liyanage, K. 2018). Effective communication and engagement underpin this alignment. Managers and leaders should leverage this convergence by effectively conveying their unwavering commitment to sustainability. This communication has the potential to elevate customer and stakeholder satisfaction, a vital facet of operational excellence. Nurturing this alignment is an on-going process that necessitates vigilant management.

Adherence to environmental and social regulations is not merely a legal obligation; it also carries profound reputational implications. (Suriyankietkaew, S. 2023) Managers must ensure that compliance is seamlessly integrated with the proactive embrace of sustainability practices. It becomes imperative for leaders to understand the intricate regulatory landscape and fuse it harmoniously into the organization's sustainability strategy. This harmonization between compliance and sustainability falls within the purview of managerial responsibilities.

The reduction of an organization's environmental footprint is a dual imperative, both ethical and operational. Managers play a central role in delineating the strategic objectives and tactics for environmental stewardship (Colin, B. et al., 2023) Effective leadership must guide the execution of these strategies, underscoring the organization's commitment to resource conservation and waste reduction (Çop, S.,et al., 2021). Sustainability objectives should be seamlessly woven into the fabric of overall management strategies.Managers must recognize that the organization's reputation and brand value are deeply influenced by its sustainability initiatives. (Kim, M. S., & Thapa, B. 2018) Ethical leadership stands at the forefront in this context, shaping the perception of the organization in the public eye. Effective reputation management necessitates proactive leadership and consistent communication of the organization's sustainability endeavours. Managers should actively collaborate with leaders to ensure the effective communication of sustainability commitments, attracting stakeholders who share the organization's values.

In the broader context, discussions surrounding the relationship between sustainable supply chain practices and leadership qualities should revolve around strategy development, cultural transformation, talent development, and continuous improvement (Martin B. Osei, Thanos Papadopoulos, Adolf Acquaye, Teta Stamati, 2023). This encompasses the imperative of nurturing an organizational culture that places a premium on sustainability, innovation, and social responsibility, with leadership qualities at the helm of driving this transformation (Ahsan, M. J. 2024). It is the realm of successful managers and leaders to not only acknowledge the innate connection between sustainability and operational excellence but also to proactively leverage this dynamic for the benefit of their organizations and stakeholders Yu, W., Jacobs, M.A., Chavez, R. and Yang, J. (2019). In this confluence of sustainability and leadership, the path to operational excellence becomes illuminated, serving as a beacon for forward-thinking organizations.

6. SCOPE AND FUTURE RESEARCH DIRECTIONS

The scope of research on the effectiveness of leadership qualities in sustainable supply chain practices and their impact on achieving operational excellence is multifaceted (Brandenburg M, Gruchmann T, Oelze N. 2019). This area offers extensive opportunities for exploration, encompassing various dimensions of leadership, sustainability, and operational performance. The scope includes but is not limited to leadership attributes, industry-specific studies, cultural and geographical variances, small and medium-sized enterprises (SMEs), technology integration, long-term sustainability.

Building upon the existing body of research, several potential future research directions can further enrich our understanding of the relationship between leadership qualities and sustainable supply chain practices in achieving operational excellence in various dimensions like firstly quantitative analysis to establish empirical evidence of the impact of leadership attributes on specific key performance indicators (KPIs) related to operational excellence. Carter, C.R.; Washispack, S. (2016). Secondly, case based analysis, where in-depth case studies of organizations that have successfully integrated sustainability and leadership, offering insights into best practices and challenges faced in different industries and contexts. Thirdly behavioural analysis which investigates leadership practices in driving sustainability initiatives, focusing on how leaders influence organizational culture and employee engagement (Ahsan, M. J. 2024). Qualitative research can be conducted to know the underlying role of leadership and sustainability in fostering collaboration and partnerships across the supply chain, involving suppliers, customers, and other stakeholders. Furthermore, research on how multinational corporations can effectively implement sustainability practices and leadership strategies in global supply chains, addressing cross-cultural challenges and regulatory variations.

7. CONCLUSION

In conclusion, the exploration of the effectiveness of leadership qualities within the realm of sustainable supply chain practices and their consequential impact on achieving operational excellence stands as a research domain of paramount significance. The accrued insights gleaned from extant research underscore the dynamic interplay that underscores the indispensable role of visionary leadership, adaptability, ethical responsibility, and innovative thinking in navigating organizations towards the dual objectives of sustainability and operational excellence. The manifold implications that arise from this synthesis of leadership and sustainability practices are far-reaching, encompassing cost efficiencies, quality enhancements, risk management, innovation, stakeholder engagement, regulatory adherence, environmental stewardship, and reputation management.

Notwithstanding the pivotal contributions of extant research, this domain bears certain limitations that necessitate acknowledgement to facilitate a comprehensive understanding of the subject matter. Future inquiries should earnestly endeavour to encompass a more diverse spectrum of organizational typologies and global geographies, ensuring broader relevance. Prospective longitudinal investigations and controlled experiments are mandated to provide more rigorous causal substantiation. Leadership attributes, sustainability, and operational excellence are inherently subjective constructs, vulnerable to varying interpretations. The development and standardization of measurement instruments and objective evaluative metrics serve as an imperative in elevating the scientific rigor of forthcoming research endeavours.

8. LIMITATIONS

In light of these recognized limitations, it is evident that the research domain pertaining to leadership attributes in the context of sustainable supply chain practices and operational excellence persists as a dynamic and unfolding avenue of inquiry. Subsequent research endeavours must confront and navigate these limitations with diligence, while continuing to explore the intricacies of the relationship between leadership, sustainability, and operational excellence within an array of diverse organizational contexts and industries. These research undertakings not only make a substantial contribution to the academic landscape but also impart valuable guidance to managers and leaders striving to propel their organizations towards a future characterized by enhanced sustainability and operational excellence.

REFERENCES

Ahmad, A. F., & Karadas, G. (2021). Managers' perceptions regarding the effect of leadership on organizational performance: Mediating role of green supply chain management practices. *SAGE Open*, 11(2), 21582440211018686. DOI: 10.1177/21582440211018686

Ahsan, M. J. (2024). Unlocking sustainable success: Exploring the impact of transformational leadership, organizational culture, and CSR performance on financial performance in the Italian manufacturing sector. *Social Responsibility Journal*, 20(4), 783–803. DOI: 10.1108/SRJ-06-2023-0332

Amoako, G.K., Obuobisa-Darko, T., Dartey-Baah, K. and Sedalo, G. (2023), "A conceptual framework on finding the nexus between sustainability and desired outcomes for smart cities – the moderating role of green leadership", *International Journal of Tourism Cities*, Vol. ahead-of-print No. ahead-of-print. .DOI: 10.1108/IJTC-12-2021-0238

Ansari, Z. N., & Kant, R. (2017). A state-of-art literature review reflecting 15 years of focus on sustainable supply chain management. *Journal of Cleaner Production*, 142, 2524–2543. DOI: 10.1016/j.jclepro.2016.11.023

Antony, J., Swarnakar, V., Gupta, N., Kaur, J., Jayaraman, R., Tortorella, G. L., & Cudney, E. (2023). Critical success factors for operational excellence initiatives in manufacturing: A meta-analysis. *Total Quality Management & Business Excellence*, 34(9-10), 1152–1172. DOI: 10.1080/14783363.2022.2157714

Awan, U., & Khan, S. A. R. (2021). Mediating role of sustainable leadership in buyer-supplier relationships: an supply chain performance: an empirical study. *LogForum, 17(1).*

Azadegan, K., & Dooley, K. (2021). Dooley, A typology of supply network resilience strategies: Complex collaborations in a complex world. *The Journal of Supply Chain Management*, 57(1), 17–26. DOI: 10.1111/jscm.12256

Bag, S., Wood, L., Xu, L., Dhamija, P., & Kayikci, Y. (2019). Big data analytics as an operational excellence approach to enhance sustainable supply chain performance. *Resources, Conservation and Recycling*. Advance online publication. *10.1016*/j.resconrec.2019.104559

Bag, S., Wood, L. C., Xu, L., Dhamija, P., & Kayikci, Y. (2020). Big data analytics as an operational excellence approach to enhance sustainable supply chain performance. *Resources, Conservation and Recycling*, 153, 104559. DOI: 10.1016/j.resconrec.2019.104559

Bastas, A., & Liyanage, K. (2018). Sustainable supply chain quality management: A systematic review. *Journal of Cleaner Production*, 181, 726–744. DOI: 10.1016/j.jclepro.2018.01.110

Brandenburg, M., Gruchmann, T., & Oelze, N. (2019). Sustainable Supply Chain Management—A Conceptual Framework and Future Research Perspectives. *Sustainability (Basel)*, 11(24), 7239. DOI: 10.3390/su11247239

Burawat, P. (2019). The relationships among transformational leadership, sustainable leadership, lean manufacturing and sustainability performance in Thai SMEs manufacturing industry. *International Journal of Quality & Reliability Management*, 36(6), 1014–1036. DOI: 10.1108/IJQRM-09-2017-0178

Carter, C. R., & Washispack, S. (2018). Mapping the Path Forward for Sustainable Supply Chain Management: A Review of Reviews. *Journal of Business Logistics*, 2018(39), 242–247. DOI: 10.1111/jbl.12196

Cavazotte, F., Mansur, J., & Moreno, V. (2021). Authentic leadership and sustainable operations: How leader morality and selflessness can foster frontline safety performance. *Journal of Cleaner Production*, 313, 127819. DOI: 10.1016/j.jclepro.2021.127819

Chams, N., & García-Blandón, J. On the importance of sustainable human resource management for the adoption of sustainable development goals, *Resources, Conservation and Recycling*, Volume 141, 2019, Pages 109-122, ISSN 0921-3449, DOI: 10.1016/j.resconrec.2018.10.006

Chen, S. L., Su, Y. S., Tufail, B., Lam, V. T., Phan, T. T. H., & Ngo, T. Q. (2023). The moderating role of leadership on the relationship between green supply chain management, technological advancement, and knowledge management in sustainable performance. *Environmental Science and Pollution Research International*, 30(19), 56654–56669. DOI: 10.1007/s11356-023-26304-x PMID: 36920606

Çop, S., Olorunsola, V. O., & Alola, U. V. (2021). Achieving environmental sustainability through green transformational leadership policy: Can green team resilience help? *Business Strategy and the Environment*, 30(1), 671–682. DOI: 10.1002/bse.2646

Dubey, R., Gunasekaran, A., & Ali, S. S. (2015). Exploring the relationship between leadership, operational practices, institutional pressures and environmental performance: A framework for green supply chain. *International Journal of Production Economics*, 160, 120–132. DOI: 10.1016/j.ijpe.2014.10.001

Esangbedo, C. O., Zhang, J., Pérez, P. B., & Skitmore, M. (2024). Sustainable performance and supply chain leadership in logistic firms: The role of corporate sustainability strategies and digital supply chain. *Supply Chain Management*, 29(6), 963–977. DOI: 10.1108/SCM-02-2024-0131

Gabler, C. B., Landers, V. M., Agnihotri, R., & Morgan, T. R. (2023). Environmental orientation on the frontline: A boundary-spanning perspective for supply chain management. *Journal of Business Logistics*, 44(3), 369–386.

Gligor, D. M., & Holcomb, M. C. (2012). Understanding the role of logistics capabilities in achieving supply chain agility: A systematic literature review. *Supply Chain Management*, 17(4), 438–453. DOI: 10.1108/13598541211246594

Gong, R., Xue, J., Zhao, L., Zolotova, O., Ji, X., & Xu, Y. (2019). A Bibliometric Analysis of Green Supply Chain Management Based on the Web of Science (WOS) Platform. *Sustainability (Basel)*, 11(12), 3459. DOI: 10.3390/su11123459

Govindan, K. (2018). Sustainable consumption and production in the food supply chain: A conceptual framework. *International Journal of Production Economics*, 195, 419–431. DOI: 10.1016/j.ijpe.2017.03.003

Hahn, T., Figge, F., Pinkse, J., & Preuss, L. (2018). A Paradox Perspective on Corporate *Sustainability: Descriptive, Instrumental, and Normative Aspects.Journal of Business Ethics*, 148(2), 235–248. DOI: 10.1007/s10551-017-3587-2

Hahn, T., Pinkse, J., Preuss, L., & Figge, F. (2015). Tensions in Corporate Sustainability: Towards an Integrative Framework. *Journal of Business Ethics*, 127(2), 297–316. DOI: 10.1007/s10551-014-2047-5

Huo, B., Wang, K., & Zhang, Y. (2021). The impact of leadership on supply chain green strategy alignment and operational performance. *Operations Management Research : Advancing Practice Through Research*, 14(1-2), 152–165. DOI: 10.1007/s12063-020-00175-8

Jia, F., Gong, Y., & Brown, S. (2019). Multi-tier sustainable supply chain management: The role of supply chain leadership. *International Journal of Production Economics*, 217, 44–63. DOI: 10.1016/j.ijpe.2018.07.022

Kafetzopoulos, D., & Gotzamani, K. (2022). The effect of talent management and leadership styles on firms' sustainable performance. *European Business Review*, 34(6), 837–857. DOI: 10.1108/EBR-07-2021-0148

Khana, M. P., Talibb, N. A., & Kowangc, T. O. (2020). The development of a sustainability framework via lean green six sigma practices in SMEs based upon RBV theory. *Development*, 12(5), 135–156.

Kim, M. S., & Thapa, B. (2018). Relationship of ethical leadership, corporate social responsibility and organizational performance. *Sustainability (Basel)*, 10(2), 447. DOI: 10.3390/su10020447

Manurung, H., Yudoko, G., & Okdinawati, L. (2023). A conceptual framework of supply chain resilience towards sustainability through a service-dominant logic perspective. *Heliyon*, 9(3), e13901. DOI: 10.1016/j.heliyon.2023.e13901 PMID: 36915553

Martins, C. L., & Pato, M. V. (2019). Supply chain sustainability: *A tertiary literature review.Journal of Cleaner Production*, 225, 995–1016. DOI: 10.1016/j.jclepro.2019.03.250

Mokhtar, A. R. M., Genovese, A., Brint, A., & Kumar, N. (2019). Supply chain leadership: A systematic literature review and a research agenda. *International Journal of Production Economics*, 216, 255–273. DOI: 10.1016/j.ijpe.2019.04.001

Mondal, S., & Samaddar, K. (2023). Reinforcing the significance of human factor in achieving quality performance in data-driven supply chain management. *The TQM Journal*, 35(1), 183–209. DOI: 10.1108/TQM-12-2020-0303

Mousa, S. K., & Othman, M. (2020). The impact of green human resource management practices on sustainable performance in healthcare organisations: *A conceptual framework.Journal of Cleaner Production*, 243, 118595. DOI: 10.1016/j.jclepro.2019.118595

Muralidharan, E., & Pathak, S. (2018). Sustainability, transformational leadership, and social entrepreneurship. *Sustainability (Basel)*, 10(2), 567. DOI: 10.3390/su10020567

Nagariya, R., Kumar, D., & Kumar, I. (2022). Sustainable service supply chain management: From a systematic literature review to a conceptual framework for performance evaluation of service only supply chain. *Benchmarking*, 29(4), 1332–1361. DOI: 10.1108/BIJ-01-2021-0040

Naik, S., Sony, M., Antony, J., McDermott, O., Tortorella, G. L., & Jayaraman, R. (2024). Operational excellence framework for sustainability in the organisation: A design science approach. *Production Planning and Control*, 35(11), 1215–1231. DOI: 10.1080/09537287.2023.2165188

Nan, S., & Chaiprasit, K. (2023). Resilient leadership, innovation, executive incentives, and sustainable business performance: An empirical study. *Journal of Logistics. Informatics and Service Science*, 10(4), 1–19.

Osei, M. B., Papadopoulos, T., Acquaye, A., & Stamati, T. (2023). Improving sustainable supply chain performance through organisational culture: A competing values framework approach. *Journal of Purchasing and Supply Management*, 29(2), 100821.

Pantouvakis, A., & Vlachos, I. (2020). Talent and leadership effects on sustainable performance in the maritime industry. *Transportation Research Part D, Transport and Environment*, 86, 102440. DOI: 10.1016/j.trd.2020.102440

Paulraj, A., Chen, I. J., & Blome, C. (2017). Motives and performance outcomes of sustainable supply chain management practices: A multi-theoretical perspective. *Journal of Business Ethics*, 145(2), 239–258. DOI: 10.1007/s10551-015-2857-0

Pham, H., & Kim, S. Y. (2019). The effects of sustainable practices and managers' leadership competences on sustainability performance of construction firms. *Sustainable Production and Consumption*, 20, 1–14. DOI: 10.1016/j.spc.2019.05.003

Singh, R. K. (2024). Impact of leadership, TQM and supply chain capabilities on sustainable supply chain performance: Moderating role of institutional pressure. *The TQM Journal*. Advance online publication. DOI: 10.1108/TQM-09-2023-0275

Sriyakul, T., Singsa, A., Sutduean, J., & Jermsittiparsert, K. (2019). Effect of cultural traits, leadership styles and commitment to change on supply chain operational excellence. *Journal of Computational and Theoretical Nanoscience*, 16(7), 2967–2974. DOI: 10.1166/jctn.2019.8203

Suriyankietkaew, S. (2023). Effects of key leadership determinants on business sustainability in entrepreneurial enterprises. *Journal of Entrepreneurship in Emerging Economies*, 15(5), 885–909. DOI: 10.1108/JEEE-05-2021-0187

Svensson, G., & Wood, G. (2006). Sustainable components of leadership effectiveness in organizational performance. *Journal of Management Development*, 25(6), 522–534. DOI: 10.1108/02621710610670100

Teoman, S., & Ulengin, F. (2018). The impact of management leadership on quality performance throughout a supply chain: An empirical study. *Total Quality Management & Business Excellence*, 29(11-12), 1427–1451. DOI: 10.1080/14783363.2016.1266244

Ulle, R. S., & Kumar, A. S. (2014). A review on total quality leadership in TQM practices-industrial management and organizations. *International Journal of Emerging Research in Management &Technology*, 3(5), 152–155.

Yu, W., Jacobs, M. A., Chavez, R., & Yang, J. (2019). Dynamism, disruption orientation, and resilience in the supply chain and the impacts on financial performance: A dynamic capabilities perspective. *International Journal of Production Economics*, 218, 352–362. DOI: 10.1016/j.ijpe.2019.07.013

Zhou, H., & Li, L. (2020). The impact of supply chain practices and quality management on firm performance: Evidence from China's small and medium manufacturing enterprises. *International Journal of Production Economics*, 230, 107816. DOI: 10.1016/j.ijpe.2020.107816

KEY TERMS AND DEFINITIONS

Leadership Qualities: Core attributes that define effective leaders, such as visionary thinking, strategic acumen, communication, adaptability, and ethical responsibility.

Sustainable Supply Chain: A holistic approach to managing the flow of goods and services that integrates environmental, social, and economic sustainability principles, aiming to minimize negative impacts throughout the supply chain.

Operational Excellence: The pursuit of enhanced efficiency, quality, innovation, and resilience within an organization.

Visionary Thinking: The ability of leaders to anticipate future trends and challenges, which helps in setting a long-term direction for sustainability efforts within the supply chain.

Adaptability: A leader's capacity to respond effectively to changes in the supply chain environment, such as regulatory shifts or market dynamics, which supports sustainable practices and resilience.

Risk Mitigation: Strategies aimed at reducing potential disruptions or negative impacts on the supply chain.

Chapter 5
Epistemological Dilemma: Evaluating Conflicting Sources in AI Plagiarism Detection of Students' Work

Abdalellah Omer Mohmmed
https://orcid.org/0000-0003-2270-1273
De Montfort University, Dubai, UAE

Abdelmoniem Sahal Elmardi
Dubai Medical College for Girls, UAE

Muhammad Wajid Saleem
https://orcid.org/0000-0002-3163-6410
De Montfort University, Dubai, UAE

Eltaib Ganawa
University of Khartoum, Sudan

ABSTRACT

The recent advancement in sophisticated AI technologies in numerous sectors, including education and research, poses several challenges. The utilization of AI chatbot systems, particularly ChatGPT, has become increasingly prevalent and exhibits heightened capabilities in various aspects. However, ChatGPT poses various challenges to the conventional educational and research framework such as the potential for academic dishonesty. Consequently, Plagiarism has emerged as a pressing issue in academic writing and publishing, prompting the creation of numerous plagiarism-detection mechanisms. Thus, this chapter explores the challenges of plagiarism in the AI-empowered era by examining several written essays from the fields of engineering and medicine. The plagiarism in these essays was assessed using various free AI plagiarism-detection tools. The outcomes demonstrated different similarity levels for the same content, which may confuse educators during the assessment process. It delves into the potential measures that academic institutions can implement to preserve academic integrity.

DOI: 10.4018/979-8-3693-5748-4.ch005

1. INTRODUCTION

The advancement of Generative AI has considerably transformed the information generation process in various industries. By leveraging copious datasets of images, music, and text, Generative AI engender novel forms of text, images, and predictable data (Euchner, 2023). This innovative feature of Generative AI has sparked widespread excitement, as it opens numerous potential benefits that could enhance processes, administration, written content, musical creations, and artistic endeavours, and serve as a beneficial asset in various sectors. Generative AI has been utilized in the engineering field to produce diverse written documents and develop numerous programming codes. On the contrary, this technology is being utilized in the medical sector for disease forecasting and image processing analysis. It is demonstrating immense promise in revolutionizing the healthcare industry through improving the patient care standard. Furthermore, the emergence of technology has provided a plethora of resources to enhance education. These resources consist of fundamental document-sharing instruments as well as intricate systems, such as Virtual Learning Environments (VLEs). Recently, there has been a notable increase in modern intelligent systems those can learn from previous embedded data and accordingly make decisions using Artificial Intelligence algorithms (Steponenaite and Barakat, 2023). A chatbot is a modern innovation in the field of AI that facilitate anthropoid conversations through text or verbal communications (Wollny et al., 2021). It has the ability to engage with people and provide instant responses in a natural language, mimicking the way a human partner would communicate (Følstad and Brandtzæg, 2017). ELIZA, a groundbreaking AI chatbot program aimed at simulating human conversation, was created during the infancy of artificial intelligence in 1966 (Weizenbaum, 1966). The evolution of artificial intelligence over the past few decades has led to the development of sophisticated chatbots that are capable of comprehending and addressing intricate inquiries (Liu et al., 2023). The staggering proliferation of technologies such as Dall-E 2, GPT-4, and Copilot is profoundly altering conventional approaches to work and communication between individuals. Currently, conversations have arisen around ChatGPT, the novel AI chatbot developed by OpenAI (Brown et al., 2020). ChatGPT, a revolutionary concept, is a deep learning-based AI chatbot designed to engage in conversation with human users online. Stands for Chat Generative Pre-Trained Transformer. Its transformer architecture was introduced by Vaswani et al. (2017), and has since gained extensive application in natural language processing tasks. According to recent online blogs, this chatbot has generated excitement among certain groups of people, with some even suggesting that it may eventually replace the broadly used Google search engine (Khalil and Er, 2023). ChatGPT can be used for artistic purposes, such as producing images that resemble the illustrations of particular artists or generating text that imitates the writing style of specific authors. However, this language model, boasting an impressive 175 billion parameters, is one of the most expansive language models currently accessible. One of its inconceivable features that its adaptability in handling widespread language tasks, comprising translation, summarization, and text generation for tasks such as writing emails, essays, and computer code, without requiring extensive task-specific training. Therefore, ChatGPT will operate as an intelligent system that assists people in addressing broad form of inquiries (Feuerriegel et al., 2024).

As a mean of Large Language Model (LLM), ChatGPT has garnered significant attention from the media (Dale, 2021). For instance, there have been several reports about the publication of a journal paper of Alarie and Cockfield (2021) that was entirely generated by GPT-3. Due to its advanced language proficiency and ability to deliver high-quality outputs, ChatGPT and other similar AI tools are perceived as a potential threat to the education sector and the integrity of online assessments (Surahman and Wang,

2022, Abd-Elaal et al., 2022). Recently, the intensive use of this advanced and user-friendly technology has sparked concerns about plagiarism in academic sector. In a recent blog post by Stephen Marche titled "The College Essay Is Dead," the issue of using ChatGPT to generate lengthy, high-quality written content using natural language processing and chatbots is discussed (Marche, 2022). According Stokel-Walker (2022), ChatGPT possesses considerable potential in assisting college students with various tasks, including essay writing, homework completion, code creation, and exam preparation. However, some educational institutions, such as in Queensland university Australia, Tasmania schools, New York City school, and Seattle school, have taken measures to prohibit the use of ChatGPT on students' devices and networks. A number of universities, colleges, and schools are currently considering implementing similar restrictions (Wilcox, 2023). Therefore, the utilization of ChatGPT to generate academic essays for assignments has gained significant traction among students, consequently leading to heightened concerns surrounding the issue of plagiarism within the academic realm.

On the other hand, the Sustainable Development Goals (SDGs) are a set of global objectives that stated by the United Nations member, which aims to achieve by 2030, with the ultimate goal of creating an equitable, prosperous, and peaceful world (Sdg, 2019). ChatGPT can play a major role in promoting SDG-4 (Quality Education) if this tool is used carefully considering the academic integrity and ethics of education. On the contrary, the heavy usage of ChatGPT can cause severe environmental impact which violates the recent trend of promoting sustainability in education. In fact, ChatGPT consumes approximately 2.9 Watt per hour (W/h) per inquiry which is nearly 10 times the consumption of the traditional search engine such as Google search engine (Arbiv, 2024). Therefore, the extensive usage of ChatGPT by the students to address their queries will increase the energy consumption and consequently increase the environmental footprint in higher education (Bitter and Loney, 2015, de Vries, 2023). This usage of ChatGPT is against the primary focus of SDG-12 (sustainable consumption and production).

Helgesson and Eriksson (2014) defined plagiarism as the act of presenting or passing off someone else's intellectual property as one's own, while Perkins et al. (2018) defined it as the act of misrepresenting the authorship by providing information that originates from another source or providing unreferenced information. There is a heightened interest in higher education regarding plagiarism and overall academic integrity, particularly due to the prevalence of online resources, comprising articles and solution manuals, which arc readily accessible. Consequently, many students turn to rely on unauthorized websites and publications without confirming their credibility or legitimacy to complete their assessments. As discussed by Sun and Hoelscher (2023), certain academic tasks assigned to students can sometimes be extremely difficult and demanding, with strict submission deadlines and a broad range of topics to cover, which can make it challenging for the students to produce original content. As a result, when students encounter difficulties in fulfilling these deadlines due to inadequate knowledge, experience, or skills, they may opt to seek assistance to complete their work. The hype surrounding ChatGPT has been extraordinary since its release, especially in light of its remarkable capability in producing high-quality content. However, this has also caused apprehension among educators worldwide, compelling the implementation of plagiarism detection software more critical than ever. A plagiarism detection tool is a technological software that compares submitted work to a database of sources to identify potential instances of plagiarism (Boggaram et al., 2023). However, the use of new digital writing tools, encompassing Machine Translators (MTs), Digital Writing Assistants (DWAs), and Automated Paraphrasing Tools (APTs), has the potential to circumvent AI plagiarism detection systems, as indicated by Perkins (2023). These tools generate text based on a comprehensive analysis of LLMs, revealing that even highly skilled educators are impotent to differentiate between written content generated by humans and that produced by these LLMs chatbots. Consequently, the text generated by these LLMs tools poses an existential challenge impacting academic integrity. According to Topinka (March, 2024), a senior lecturer in media and cultural studies at the University of London, he was perplexed when the AI plagiarism detector indicated a 100% match for a brilliant student's essay. This student was previously known as one of the top students in their cohort, prior to the emergence of ChatGPT. Topinka was unsure whom to believe, the student or the AI plagiarism detector. Furthermore, according to Furze (April, 2024) article "AI Detection in Education is a Dead End", after two years of working with generative artificial intelligence software, he concluded that there were no reliable and accurate AI detection tools. Recent instances have shed light on the precision of AI plagiarism detectors and their suitability to serve as a means of judgment in academia. Several issues have come to the fore regarding the results generated by these systems, given the severe repercussions of violating academic integrity in Higher Education Institutes (HEIs). In light of these concerns, Montclair University, after a year of the launch of ChatGPT, declared that academicians should not employ any AI-detector feature apart from Turnitin. This was followed by similar decisions from other institutions, including Vanderbilt University, the University of Texas at Austin, and Northwestern University (Coffey, February, 2024).

Digital tools for writing have been widely adopted by students since the 1980s as a means of providing assistance in their assessments (Palmquist, 2003). The incorporation of style and grammar analysis software into word processor technology has led to significant advancements. Over time, these tools have undergone considerable transformation, and students widely believe that they can enhance their writing skills through the use of these resources (Nobles and Paganucci, 2015). Although digital writing has been prevalent in HEIs for a considerable time (Kozma, 1991), the emergence of new technologies necessitates continuous evaluation of the tools that are accessible to students and the provision of explicit guidelines for their utilization. Moreover, ensuring an impartial and consistent assessment process for student submissions is of utmost importance in maintaining academic integrity within higher educational institutions. Therefore, this chapter assesses the precision and authenticity of AI plagiarism detection software through a comprehensive comparison of essays written by ChatGPT-4. These essays from the fields of engineering and medicine were employed to evaluate the impact of data accessibility on the accuracy of the AI detectors. In addition, the generated essays were paraphrased using existing AI paraphrasing tools to assess the ability of these tools in circumventing AI plagiarism detectors.

2. METHOD

This chapter aims to evaluate the performance of AI plagiarism detectors in capturing the generated essays utilizing ChatGPT-4. The study's focus is limited to the engineering and medicine fields, given the scarcity of research in the engineering field and the abundance of studies in the medicine field regarding generative AI applications. Besides, this study also aims to gauge the efficiency of AI paraphrasing tools in circumventing AI plagiarism detectors by comparing their AI similarity indices against non-paraphrased generated essays using ChatGPT-4. To accomplish this, the available AI detectors are examined, and the generated essays are paraphrased using AI paraphrasing tools. The similarity indices of the paraphrased essays are then compared to assess the paraphrasing tools' ability to bypass AI plagiarism detectors. Moreover, a request was made to distinguish between essays that were initially generated by ChatGPT or not. A total of eight essays were created, four from each field, and their similarity was evaluated using various software programs listed in Table 1.

Table 1. List of AI Plagiarism Detectors used in this study

AI Detector Name	Website	Comment
Dupli Checker	https://www.duplichecker.com/	Subscription with free use up to 25000 words
Small Seo Tools	https://smallseotools.com/plagiarism-checker/	Subscription with free use up to 1000 words
ZeroGPT	https://www.zerogpt.com/	Subscription with free use up to 15000 characters
Plagiarism checker	https://www.plagiarismchecker.co/	Subscription with free use up to 800 words
Plagiarism detector	https://plagiarismdetector.net/	Subscription with free use up to 1000 words
Contentdetector.ai	https://contentdetector.ai/	Free
Editpad	https://www.editpad.org/tool/plagiarism-checker	Free
Seo.ai	https://seo.ai/tools/ai-content-detector	Free
Rewriteguru	https://rewriteguru.com/	Subscription with free use up to 800 words

In addition, different AI paraphrasing tools from those used to check the similarity index were utilized to perform the paraphrasing process for the entire generated essays. Table 2 lists the AI paraphrasing tools along with their subscription terms. Finally, the similarity index of these essays was determined using similar AI detection tools.

Table 2. List of AI paraphrasing tools used in this study

AI Detector Name	Website	Comment
Copy.ai	https://www.copy.ai/tools/	Free
Language Tool	https://languagetool.org/paraphrasing-tool	Free
Ahrefs	https://ahrefs.com/writing-tools/paraphrasing	Free
Paperpal	https://paperpal.com/	Subscription with five use per day for AI features

ChatGPT is a leading AI language model specifically designed for academic writing, and it has showcased remarkable proficiency across various application sectors. For instance, it has exhibited excellence in creating cohesive content and essays, formulating chatbot responses, rendering translations, addressing inquiries, and producing programming code (Kasneci et al., 2023, Qadir, 2023). Consequently, it was employed to produce diverse essays in both engineering and medical domains. The test cases comprising the command texts written to ChatGPT-4 are showcased in Table 3.

Table 3. Command text used in ChatGPT-4 to generate different essays

Field	Command Text
Engineering	1. Write a comprehensive essay of 1000 - 1200 words about the development of the car engine. including its history, and common functioning features, and critically discuss the current issue raised due to the heavy usage of these engines. Besides, discuss the future directives in terms of sustainability. support your writing with scientific journals and scholarly articles where applicable
	2. Write a comprehensive essay of 1000 - 1200 words about vibrational one degree of freedom. including its equation, and fundamental benefit, and critically discuss the equations along with solution procedures of problems that entail this type of vibration. support your writing with scientific journals and scholarly articles where applicable.
	3. write a comprehensive essay of 1000 - 1200 words about non-polymer material as an alternative material for plastic, and critically discuss its future benefits in terms of sustainability and cost. support your writing with scientific journals and scholarly articles where applicable.
	4. Critically evaluate the development of the finite element method. highlight its history and the future directive of this analysis in terms of supporting the engineering design and failure analysis. Use scientific papers to write your essay and present your results in 1000-1200 words. support your writing with scientific journals and scholarly articles where applicable.
Medical	1. Critically evaluate the mycetoma disease and available treatments. Consider disease cause, progression, visible and diagnostically detectable symptoms, and possible prognosis. Use scientific papers to write your essay and present your results in 1000-1200 words.
	2. Critically evaluate the TB disease and available treatments. Consider disease cause, progression, visible and diagnostically detectable symptoms, and possible prognosis. Use scientific papers to write your essay and present your results in 1000-1200 words.
	3. Write a comprehensive essay of 1000 - 1200 words about the kidney as organs in the back of the abdomen. highlight its function, and possible disease, and critically discuss the reasons behind having two kidneys. support your writing with scientific journals and scholarly articles where applicable.
	4. Write a comprehensive essay of 1000 - 1200 words about the Covid-19 pandemic. highlight the history of its appearance, and possible causes of the disease, and critically discuss the post-recovery symptoms. besides, discuss the scientist's opinions about the COVID-19 vaccines. support your writing with scientific journals and scholarly articles where applicable.

3. RESULTS AND OBSERVATIONS

3.1 Comparison between the Plagiarism Checkers

The evaluation of essays generated by ChatGPT-4 involved determining their plagiarism and AI originality. To accomplish this, AI plagiarism detectors listed in Table 2 were employed, providing a similarity index. The majority of the detectors displayed minimal AI plagiarism index, with less than 10% similarity, with the exception of ZeroGPT, Contentdetector.ai, and Seo.ai, as depicted in Figure 1. However, from Figure 1a, the similarity index of the engineering essays is inconsistant, particularly for essay-2, which encompasses many theoritical equations imported to the word file as images. Figure 1b, illustrates the similarity index of medical essays that constatnly detected by AI plagiarism detectors. In general, the similarity index for medical essays is typically higher than that of engineering essays, a phenomenon that can be ascribed to the abundance of medical data and its extensive utilization in the training of AI plagiarism detection tools.

Figure 1. Similarity index of the generated (a) engineering essays (b) medical essays

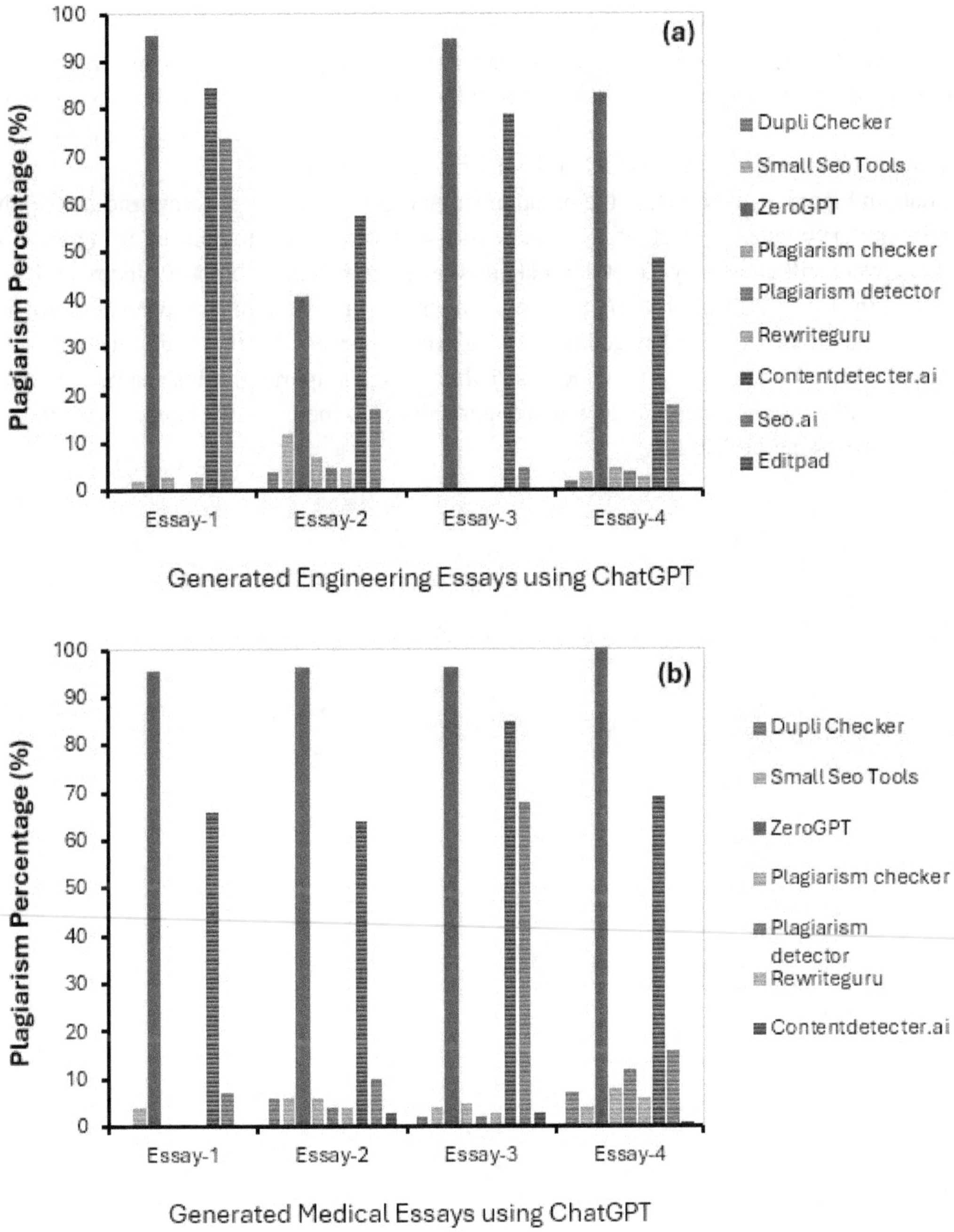

Figure 1: Similarity index of the generated (a) engineering essays (b) medical essays

Given the fact that the entire essays were produced using ChatGPT-4, the plagiarism checker that yields a higher similarity percentage approaching 100% is generally regarded as reliable.

3.2 Similarity Index of Paraphrased Essay

The essays were rewritten using AI paraphrasing tools, which are shown in Table 2, to evaluate their ability to evade plagiarism detectors. It should be mentioned that the paraphrased essays were not thoroughly examined, and the process was executed automatically. Figure 2 presents the similarity index of the paraphrased engineering essays. With regard to the Copi.ai and Ahref paraphrasing tools, four and three detectors, namely Small Seo Tools, ZeroGPT, Contentdetector.ai, and Seo.ai, identified that the engineering essays were generated by AI and not written by a human.

The data in Figure 3 suggests that the paraphrasing process for medical essays did not significantly alter their similarity index. In fact, essay-3 and essay-4, which were paraphrased using the full range of AI tools, were still caught by ZeroGPT plagiarism detector with a similarity index of 100%. This suggests that the use of AI paraphrasing tools without proper understanding or awareness might lead to an increased likelihood of being flagged by AI plagiarism checkers. Furthermore, the similarity index of the medical essays was found to have increased after being paraphrased. This can be attributed to the summary algorithm used by these tools, which either reduces or increases the language content, thereby affecting the bypassing probability.

Figure 2. Similarity index of the paraphrased engineering essays using (a) Copi.ai (b) Ahref (c) Language tool (d) Paperpal

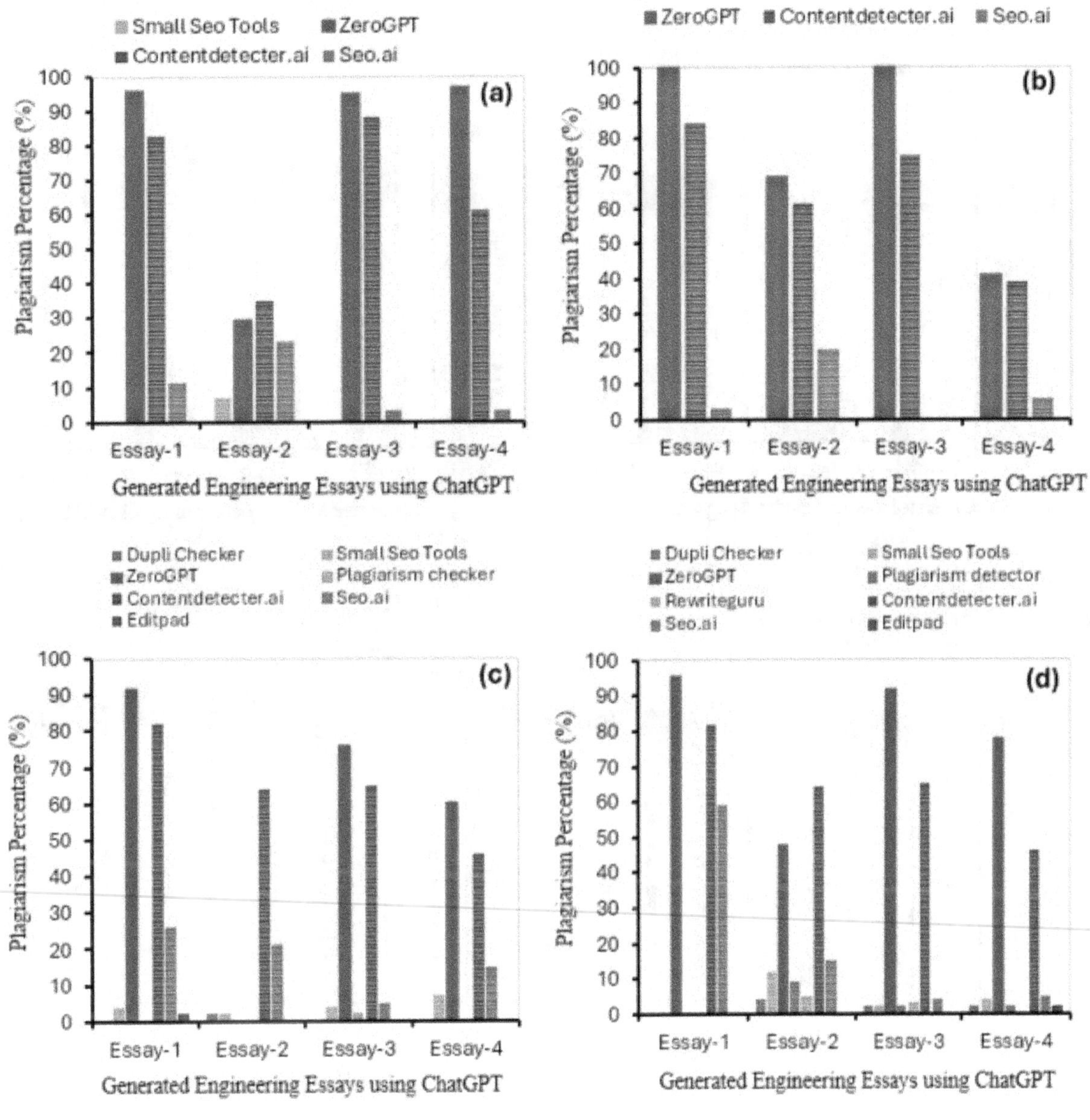

Figure 2: Similarity index of the paraphrased engineering essays using (a) Copi.ai (b) Ahref (c) Language tool (d) Paperpal

Figure 3. Similarity index of the paraphrased medical essays using (a) Copi.ai (b) Ahref (c) Language tool (d) Paperpal

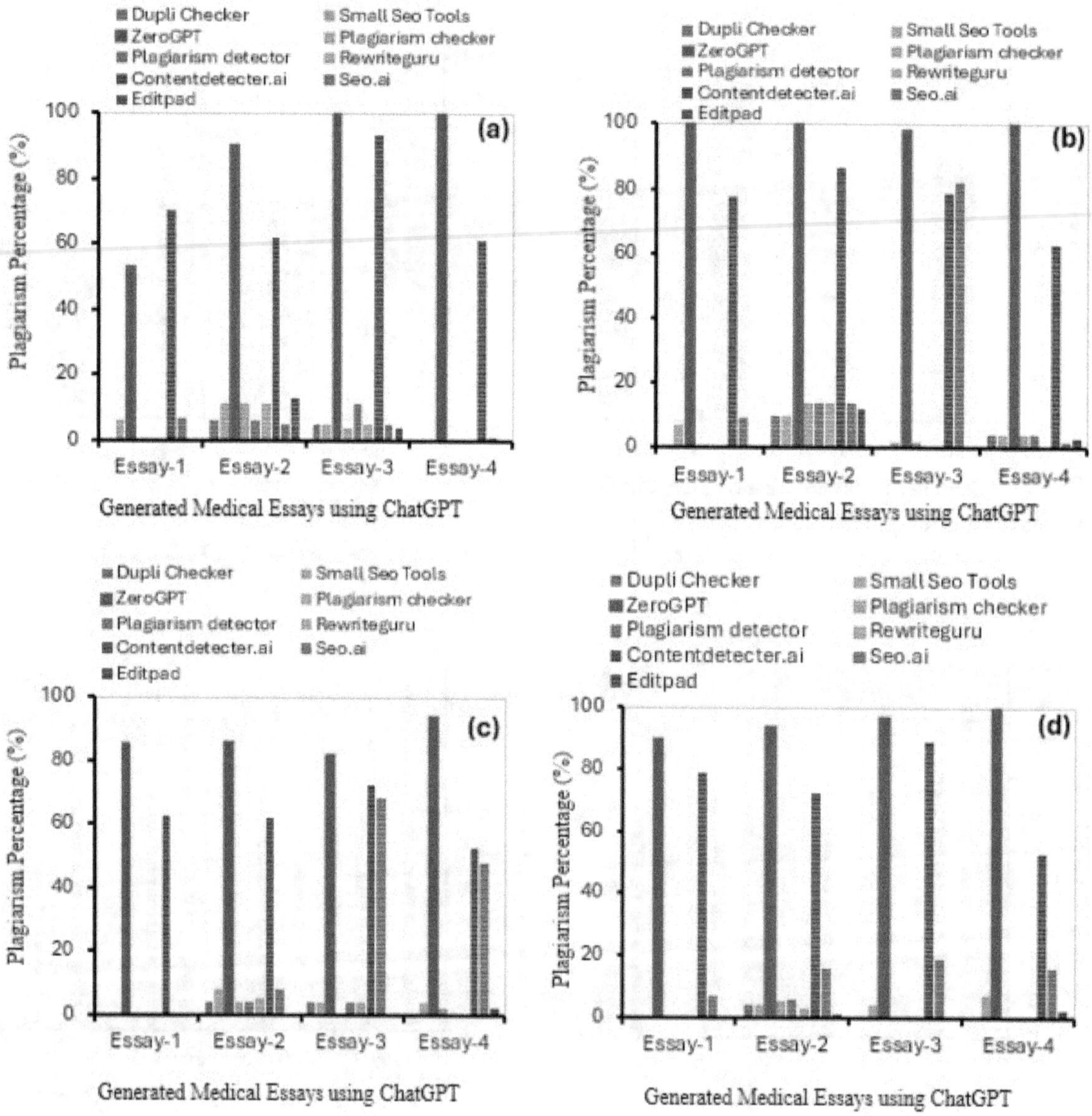

Figure 3: Similarity index of the paraphrased medical essays using (a) Copi.ai (b) Ahref (c) Language tool (d) Paperpal

The likelihood of being detected by plagiarism checkers was found to be highly dependent on the topic category. For instance, in engineering, topics that contain detailed fundamental theories are expected to have a lower probability of being flagged, unlike other topics that discuss historical and future directives, as illustrated in Figure 4. On the other hand, for the medical field, all topics have a similar probability of 25% for the entire essay, as shown in Figure 4.

Figure 4. Likelihood of plagiarism detection by ZeroGPT for engineering and medical essays

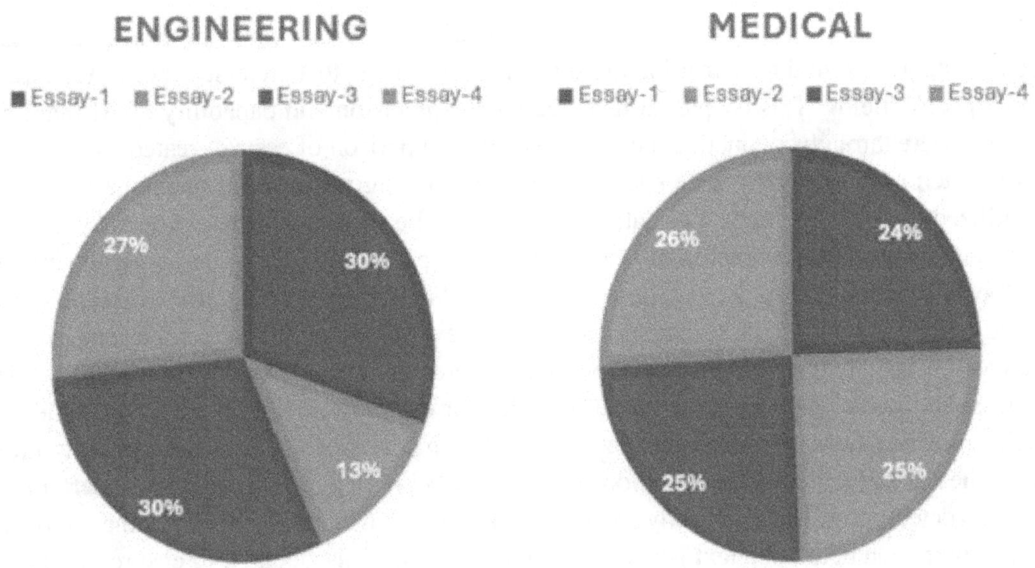

Figure 4: Likelihood of plagiarism detection by ZeroGPT for engineering and medical essays

3.3 Potential Measures of Plagiarism by HEIs

It is evident that academic institutions must establish and enforce policies that promote academic integrity and utilize plagiarism detection software, while also providing staff members with adequate training on ChatGPT and its limitations and advancements in plagiarism software. This is crucial in preventing improper use of ChatGPT (Archibald and Clark, 2023). A consensus must be reached among higher education institutions regarding the acceptable detection percentage, which may be subjective and differ based on the field of study. Furthermore, evaluations ought to be redirected from typical essay writing, which can be completed using AI assistants, towards problem-solving that involves critical thinking and practical application. The demonstrated results revealed that ChatGPT-4 has substantial proficiency in open discussions essays, involving those related to the futural prospects and forecasting. Therefore, such topics should not be considered as critical thinking problems. Moreover, writing a continuous essay can pose a challenge for ChatGPT-4, as it typically prefers to demonstrate information in bullet points. To overcome this, it is advisable to emphasize the importance of avoiding the usage of bullet points in essay writing. In addition, it should be noted that depending solely on AI plagiarism detectors to identify potential instances of plagiarism is not permanently reliable, as these tools may produce inaccurate results in certain circumstances. Therefore, it is strongly advised to thoroughly review the essay's structure, referencing, citation style, language style, and sentence coherence.

4. CONCLUSION

The importance of identifying instances of plagiarism in HEIs cannot be exaggerated. Ensuring academic integrity is vital for nurturing the skills and knowledge that are essential for students' success in their specific fields. This chapter aims to assess the precision and capability of AI plagiarism detection software through conducting a comprehensive comparison of essays created using ChatGPT-4. The generated essays covered wide spectrum areas in the engineering and medicine fields, and they were utilized to assess the impact of data availability on the accuracy of the AI detectors. In addition, the generated essays were paraphrased using existing AI paraphrasing tools to assess the efficiency of these tools in circumventing AI plagiarism detectors. Among the AI plagiarism detectors utilized in this study, ZeroGPT, Contendector.ai, and Seo.ai have exhibited impressive competence in identifying text created using ChatGPT-4, whereas the other AI detectors exhibited a similarity index of less than 10%. It is vital to recognize that the similarity index for the medical field is typically higher than that of the engineering field. While AI paraphrasing tools can sometimes reduce the similarity index, they also run the risk of increasing the likelihood of the essay being flagged as plagiarized. Certain AI-based plagiarism detection tools are currently employed in academic institutions; however, these tools were not able to capture and detect the text generated by ChatGPT-4. Consequently, to ensure that future work meets academic standards, it is recommended that the outputs of ChatGPT be evaluated using the same plagiarism detection tools commonly used in academia, such as Turnitin.

ACKNOWLEDGMENT

The authors hereby attest that this chapter has not been supported by any funding source.

REFERENCES

Abd-Elaal, E.-S., Gamage, S. H., & Mills, J. E. (2022). Assisting academics to identify computer generated writing. *European Journal of Engineering Education*, 47(5), 725–745. DOI: 10.1080/03043797.2022.2046709

Alarie, B., & Cockfield, A. (2021). Will machines replace us?: Machine-authored texts and the future of scholarship. *Law. Technology and Humans*, 3(1), 5–11. DOI: 10.5204/lthj.2089

Arbiv, O. E. (2024). *ChatGPT and Sustainability in Universities: Exploring Environmental and Educational Impacts*. University of Twente.

Archibald, M. M., & Clark, A. M. (2023). *ChatGTP: What is it and how can nursing and health science education use it?* Wiley Online Library.

Bitter, C. & Loney, E. (2015). Deeper Learning: Improving Student Outcomes for College, Career, and Civic Life. Policy Brief. *Education Policy Center at American Institutes for Research.*

Boggaram, C., Omotosho, O. & Tenghjiao, Z. (2023). Plagiarism Checker Based on Machine Learning and OpenAI.

Brown, T., Mann, B., Ryder, N., Subbiah, M., Kaplan, J., Dhariwal, P., Neelakantan, A., Shyam, P., Sastry, G., & Askell, A. (2020). Language Models are Few-Shot Learners. *Advances in Neural Information Processing Systems*, 33.

Coffey, L. (2024). Professors Cautious of Tools to Detect AI-Generated Writing. *Inside Higher Ed (IHE).*

Dale, R. (2021). GPT-3: What's it good for? *Natural Language Engineering*, 27(1), 113–118. DOI: 10.1017/S1351324920000601

De Vries, A. (2023). The growing energy footprint of artificial intelligence. *Joule*, 7(10), 2191–2194. DOI: 10.1016/j.joule.2023.09.004

Euchner, J. (2023). Generative AI. *Research Technology Management*, 66(3), 71–74. DOI: 10.1080/08956308.2023.2188861

Feuerriegel, S., Hartmann, J., Janiesch, C., & Zschech, P. (2024). Generative ai. *Business & Information Systems Engineering*, 66(1), 111–126. DOI: 10.1007/s12599-023-00834-7

Følstad, A. & Brandtzæg, P. B. (2017). Chatbots and the new world of HCI. *interactions,* 24, 38-42.

Furze, L. (2024). *AI Detection in Education is a Dead End* [Online]. Available: https://leonfurze.com/2024/04/09/ai-detection-in-education-is-a-dead-end/ [Accessed].

Helgesson, G. & Eriksson, S. (2014). Plagiarism in research. Medicine Health Care and Philosophy. doi.

Kasneci, E., Sessler, K., Küchemann, S., Bannert, M., Dementieva, D., Fischer, F., Gasser, U., Groh, G., Günnemann, S., Hüllermeier, E., Krusche, S., Kutyniok, G., Michaeli, T., Nerdel, C., Pfeffer, J., Poquet, O., Sailer, M., Schmidt, A., Seidel, T., & Kasneci, G. (2023). ChatGPT for good? On opportunities and challenges of large language models for education. *Learning and Individual Differences*, 103, 102274. DOI: 10.1016/j.lindif.2023.102274

Khalil, M., & Er, E. (2023). Will ChatGPT G et You Caught? Rethinking of Plagiarism Detection. *International Conference on Human-Computer Interaction*, 475-487.

Kozma, R. B. (1991). Computer-based writing tools and the cognitive needs of novice writers.

Liu, H., Azam, M., Bin Naeem, S., & Faiola, A. (2023). An overview of the capabilities of ChatGPT for medical writing and its implications for academic integrity. *Health Information and Libraries Journal*, 40(4), 440–446. DOI: 10.1111/hir.12509 PMID: 37806782

Marche, S. (2022). The college essay is dead. Nobody is prepared for how AI will transform academia. *Atlantic (Boston, Mass.)*, •••, 6.

Nobles, S., & Paganucci, L. (2015). Do digital writing tools deliver? Student perceptions of writing quality using digital tools and online writing environments. *Computers and Composition*, 38, 16–31. DOI: 10.1016/j.compcom.2015.09.001

Palmquist, M. (2003). A brief history of computer support for writing centers and writing-across-the-curriculum programs. *Computers and Composition*, 20(4), 395–413. DOI: 10.1016/j.compcom.2003.08.013

Perkins, M. (2023). Academic Integrity considerations of AI Large Language Models in the post-pandemic era: ChatGPT and beyond. *Journal of university teaching & learning practice,* 20, 07.

Perkins, M., Gezgin, U. B., & Roe, J. (2018). Understanding the relationship between language ability and plagiarism in non-native English speaking business students. *Journal of Academic Ethics*, 16(4), 317–328. DOI: 10.1007/s10805-018-9311-8

Qadir, J. Engineering education in the era of ChatGPT: Promise and pitfalls of generative AI for education. *2023 IEEE Global Engineering Education Conference (EDUCON)*, 2023. IEEE, 1-9. DOI: 10.1109/EDUCON54358.2023.10125121

Sdg, U. (2019). Sustainable development goals. *The energy progress report.Tracking SDG*, 7, 805–814.

Steponenaite, A., & Barakat, B. Plagiarism in AI empowered world. International Conference on Human-Computer Interaction, 2023. Springer, 434-442.

Stokel-Walker, C. (2022). AI bot ChatGPT writes smart essays-should academics worry? *Nature*. Advance online publication. DOI: 10.1038/d41586-022-04397-7

Sun, G. H., & Hoelscher, S. H. (2023). The ChatGPT storm and what faculty can do. *Nurse Educator*, 48, 119–124. DOI: 10.1097/NNE.0000000000001390 PMID: 37043716

Surahman, E., & Wang, T. H. (2022). Academic dishonesty and trustworthy assessment in online learning: A systematic literature review. *Journal of Computer Assisted Learning*, 38(6), 1535–1553. DOI: 10.1111/jcal.12708

Topinka, R. March, 2024. The software says my student cheated using AI. They say they're innocent. Who do I believe? *The Guardian*.

Vaswani, A., Shazeer, N., Parmar, N., Uszkoreit, J., Jones, L., Gomez, A. N., Kaiser, Ł., & Polosukhin, I. (2017). Attention is all you need. *Advances in Neural Information Processing Systems*, 30.

Weizenbaum, J. (1966). ELIZA—A computer program for the study of natural language communication between man and machine. *Communications of the ACM*, 9(1), 36–45. DOI: 10.1145/365153.365168

Wilcox, H. (2023). 2023. Cheating Aussie student fails uni exam after being caught using artificial intelligence chatbot to write essay-now Australia's top universities are considering a bizarre solution to stop it happening again. DailyMail., Retrieved February 2, 2023, from.

Wollny, S., Schneider, J., Di Mitri, D., Weidlich, J., Rittberger, M., & Drachsler, H. (2021). Are we there yet?-a systematic literature review on chatbots in education. *Frontiers in Artificial Intelligence*, 4, 654924. DOI: 10.3389/frai.2021.654924 PMID: 34337392

Chapter 6
Education for Sustainable Awareness With Integrating Eco Awareness Into Educational Curricula: Strategies and Challenges

Seema Yadav
https://orcid.org/0000-0002-8477-5895
Bhopal School of Social Sciences, India

ABSTRACT

The integration of sustainability into education is essential for addressing pressing global challenges and achieving sustainable development goals. Education for sustainable development (ESD) equips individuals with the knowledge, skills, and values needed to promote sustainable behaviors and address issues such as climate change, resource scarcity, and environmental degradation. This highlights the importance of collaborative efforts from educators, policymakers, civil society organizations, and businesses to overcome challenges such as rigid curricula, resistance to change, and disparities in funding and resources. Effective training programs for teachers in sustainable teaching and eco-awareness are crucial for nurturing environmentally conscious citizens. By integrating sustainability into teaching strategies, providing access to educational resources, and fostering continuous professional development, educators can inspire future generations to embrace sustainability and contribute to building a balanced and sustainable world for all.

INTRODUCTION

One of the most significant objectives for humanity today is sustainable development. Students pursuing higher education, particularly those preparing to become teachers, have the potential to actively contribute to the achievement of the Sustainable Development Goals (SDGs) for global citizenship (Akçay et al., 2024). Our economies and communities are under threat from pressing sustainability issues. The availability of freshwater, agricultural production, animal extinction, human health, and

DOI: 10.4018/979-8-3693-5748-4.ch006

ecological systems are all negatively impacted by problems like climate change and ecological system degradation (Georgallis & Bruijn, 2022). Global issues that affect people's lives all throughout the world include climate change, resource scarcity, and an ever-increasing global population. The United Nations (UN) has established 17 interconnected sustainable development goals to address the pressing concerns and promote peace and prosperity going forward. Education is taken into consideration as a key enabler to accomplish these goals. In the pursuit of an environmentally sustainable civilization, political and educational contexts play a critical role in promoting environmental sustainability (Winter et al., 2022). As such, institutions of higher learning have begun to include sustainable development in their curricula (Buerkle et al., 2023). The biggest problem currently facing humanity is sustainable development or SD. For many years, academic institutions worldwide have been attempting to implement the ideas. Higher education institutions (HEIs) must teach the next generation to recognize the difficulties that lie ahead and to provide them with the skills and information necessary to overcome them, which makes integrating SD essential (Cornet et al., 2023).

Education is essential for promoting sustainable development, guiding society towards a future in which people care more about the environment and society, and shaping the attitudes and behaviors of both the current and next generations (Veckalne & Tambovceva, 2022). The evolution of sustainable development is the process of meeting human needs without jeopardizing the ability of future generations to meet their own. The goals of the Millennium Development Goals were expanded upon with the establishment of the Education for Sustainable Development (ESD) and Education for Sustainable Development Goals (ESDG) (Kopnina, 2020). A crucial instrument for accomplishing the Sustainable Development Goals (SDGs) is education. The relevance of global civic education, lifelong learning, and high-quality education is emphasized in the 2030 Agenda for Sustainable Development. In recent times, academics and policymakers have been paying increasing attention to education as a means of transforming communities and fostering sustainable and responsible behaviors (Veckalne & Tambovceva, 2022). Education is a great means of increasing public understanding of sustainable development. Over the last few decades, an increasing number of individuals have enrolled in all levels of schooling. According to UN estimates for 2021, education will boost social and economic metrics, end poverty, and lower inequality by doing so. To encourage tolerance and create a more peaceful society, education is essential. Many educational institutions now incorporate education for sustainable development into their curricula to reap the benefits of education outlined above (Veckalne & Tambovceva, 2022). One of the most crucial objectives for humanity today is sustainable development. The achievement of the Sustainable Development Goals (SDGs) for global citizenship might involve higher education students, particularly teacher candidates (Akçay et al., 2024). Enhancing environmental understanding and influencing young people's eco-friendly behaviours are major goals of environmental education. The daily deeds and habits of youth will define the course of Earth's future as a planet (Ma et al., 2023).

To promote sustainability, educators should aim to incorporate students' lived experiences and personal concerns by asking them to consider how sustainability issues directly impact them and their communities (Micalay-Hurtado & Poole, 2022). Sustainability may be the lens through which we view all other subjects, including food, agriculture, water, forests, industry, livelihoods, and perhaps civilization itself. The prerequisite for this universal application is an integrative but pluralistic perspective (Purushothaman, 2019). Discussions in the classroom address the notion in both its tangible and abstract forms, as well as factual real-world issues. In the first course, structured debates centered on questions arising from discussions and the lived experiences of the students are employed as pedagogical methods (Purushothaman, 2019).

It is anticipated that education for sustainable development (ESD) will provide individuals and groups with the principles, information, and abilities necessary to address challenging socio-ecological issues. By raising students' understanding of sustainability issues, ESD seeks to actively involve them in supporting sustainable development (Veckalne & Tambovceva, 2022). In order to achieve sustainable development, it is crucial to raise awareness. from education to ongoing training on the job, in every stage of personal development. People will change their daily behaviors to lead more sustainable lifestyles as they become more conscious of their role in contributing to environmental and social degradation (Veckalne & Tambovceva, 2022). Higher education is linked to a greater understanding of environmental issues and a propensity to adopt sustainable habits, which makes it a significant factor in determining sustainable purchasing behavior (Hasibuan & Judijanto, 2023).

In higher education, including sustainability into teaching strategies is crucial. Giving students the information, comprehension, and abilities they need to succeed as well as cultivating the next wave of leaders and innovators may bring about the necessary change and have a significant influence on the transition to a sustainable future (Gamage et al., 2022). Higher education now serves a wider range of purposes, including addressing poverty, promoting health and well-being, and enabling sustainable livelihoods, rather than just honing professional skills and competencies (Buerkle et al., 2023). Higher education places a high premium on teaching techniques that incorporate sustainability. The next generation of innovators and leaders may be developed, and the necessary change can potentially be brought about and have a significant impact on the transition to a sustainable future by providing students with the information, comprehension, and skills they need. To successfully meet most of the sustainable development goals, advancements in the STEM fields and related industries are seen to be essential. For this reason, methods used to integrate sustainability into teaching and learning in STEM higher education programs are noteworthy in a variety of ways (Gamage et al., 2022b). A society that can accomplish the Sustainable Development Goals (SDGs) must prioritize education. To lessen environmental harm and stop global warming, one SDG is to switch energy use from fossil fuels to renewable energy (RE) sources. Citizens must be made aware of RE concepts and technologies from an early age in order to inspire students to seek employment in the field and higher education (Hoque et al., 2022).

Because the ideas it encompasses are vague and multidisciplinary, sustainability is a difficult and paradoxical topic. For students to tackle this challenge and get the necessary information, they must be able to critically examine challenges involving various stakeholders (Despeisse, 2018).

Sustainable development is a critical challenge facing humanity, with pressing issues such as climate change, resource scarcity, and environmental degradation threatening our economies and communities. Education is a key enabler for achieving the United Nations' Sustainable Development Goals (SDGs), and higher education institutions are increasingly incorporating sustainability into their curricula. This integration aims to equip students with the knowledge and skills needed to address complex socio-ecological issues, fostering sustainable and responsible behaviors. Higher education not only enhances professional skills but also promotes health, well-being, and sustainable livelihoods. Teaching methods that incorporate sustainability, particularly in STEM fields, are essential for developing future innovators and leaders. By raising awareness and encouraging critical examination of multidisciplinary challenges, education for sustainable development seeks to transform communities and guide society towards a more sustainable future.

THE ROLE OF ECO-AWARENESS AND EDUCATION IN ACHIEVING SUSTAINABLE DEVELOPMENT

A major challenge to human-nature coexistence and balance has come from the effects of rapid economic expansion brought on by scientific and technological advancements. As a result, many people, including scientists, environmentalists, world leaders, and others, are extremely concerned about the notion of sustainability as well as sustainable development and its goals (Gamage et al., 2022b). In the decades that followed, awareness of sustainability-related issues gained momentum, becoming the subject of conference themes and scholarly articles as well as the catchphrase for several projects, development plans, and environmental initiatives. Sustainability has been characterized from a variety of angles since its reasons are multifaceted and intricate. Sustainability is, broadly speaking, the capacity to maintain a particular degree of stable, balanced development without jeopardizing the balance between the natural world and the physical world created by humans (Gamage et al., 2022b). The development of an environmentally friendly lifestyle for a clean environment that satisfies the needs of present and future generations can be greatly aided by environmental knowledge and education (Ma et al., 2023).

The idea of education for sustainable development (ESD), a transformative learning approach intended to give people and communities the information, abilities, and moral principles required to promote moral behavior, is examined. For ESD to have a complete and comprehensive effect, it must be incorporated into a variety of learning contexts. This emphasizes the necessity of interdisciplinary and holistic approaches that take into account various learning styles, backgrounds, and situations (Veckalne & Tambovceva, 2022). There is a greater likelihood of pro-environmental attitudes and sustainable behavior among students who obtain environmental education. In this decade, higher education institutions have become increasingly essential as tools for promoting sustainability through teaching, training, research, and development. Education for sustainability is one of these tools. Given that education raises environmental knowledge, awareness of climate change, and attitudes toward the environment all of which contribute to pro-environmental intentions and sustainable practices it is a critical factor in determining how people consume sustainably (Hasibuan & Judijanto, 2023). Education takes up a significant portion of the lives of both adults and younger generations. Any academic institution plays a critical role in achieving sustainable development goals because it fosters the development of fundamental skills, values, behaviours, and lifestyles that are beneficial to society and a sustainable future, develops future professionals as change agents, and helps create spaces where ideas can be freely expressed (Gamage et al., 2022b). Many attempts have been made to incorporate ESD into various learning environments as a result of the realization that educational institutions must be changed to accommodate a range of learning styles, backgrounds, and situations (Veckalne & Tambovceva, 2022).

Eco-awareness and sustainable development are closely interconnected concepts that play a critical role in shaping a future that balances environmental health, economic growth, and social well-being. Here's how they are connected:

1. Definition and Purpose

Eco-Awareness: This refers to the understanding and knowledge of environmental issues and the impact of human activities on the planet. It entails realizing the value of conserving ecosystems, cutting pollution, and maintaining natural resources.

A comprehensive strategy for growth known as "sustainable development" seeks to satisfy current demands without jeopardizing the capacity of future generations to satisfy their own. Economic development, social inclusion, and environmental conservation are its three primary foundations.

2. Raising Awareness to Drive Action

Understanding the environment is essential to putting sustainable development into practice. People, groups, and organizations are more likely to behave in ways that promote sustainability when they are aware of environmental challenges. This entails implementing techniques like trash reduction, energy conservation, and endorsing eco-friendly laws and goods.

3. Educating for Sustainable Practices

Promoting eco-awareness requires education above everything else. People can learn about the value of sustainability and how they can help by incorporating environmental education into programs offered in their communities, companies, and schools. With this information, people may make better decisions and act in ways that support the objectives of sustainable development.

4. Policy and Governance

Sustainable development policies are usually developed and implemented by governments and other organizations that place a high priority on environmental awareness. A high level of eco-awareness among the populace increases the likelihood that policies like renewable energy subsidies, environmental controls, and conservation activities will be supported and successfully implemented.

5. Business and Innovation

Customer demand for sustainable goods and services is driven by eco-consciousness. In order to meet this need, businesses innovate and implement sustainable methods. This entails utilizing renewable energy sources, cutting carbon emissions, and putting circular economy ideas into practice. Thus, market-driven improvements motivated by eco-awareness advance sustainable development.

6. Community Engagement

Sustainable development efforts are more actively pursued by eco-aware communities. Collective efforts of knowledgeable and driven residents are beneficial to local initiatives like community gardens, recycling programs, and conservation projects. The accomplishment of more general sustainable development goals depends on these grassroots initiatives.

7. Addressing Global Challenges

Eco-awareness makes it easier to address global environmental concerns such as climate change, biodiversity loss, and water scarcity. An informed public is necessary for international cooperation and the implementation of international agreements like the Paris Agreement, which are crucial for sustainable development.

8. Long-Term Vision

Eco-awareness naturally supports long-term visions, which are necessary for sustainable growth. Practices that guarantee the longevity and health of our planet are encouraged when people are aware of the long-term effects of their current actions on future generations.

Eco-awareness is the cornerstone that supports sustainable development. It provides knowledge, inspiration, and mobilization for people, organizations, communities, and governments to embrace policies that guarantee the concordance of social progress, economic expansion, and environmental preservation. By encouraging eco-awareness, we pave the road for everyone to live in a sustainable future. Sustainable development and eco-awareness are closely related, with eco-awareness acting as the cornerstone of sustainable behaviors. Understanding environmental issues and the effects of human activity on the environment are two components of eco-awareness. This information motivates people to embrace sustainable practices that strike a balance between social inclusion, economic growth, and environmental protection in their communities, enterprises, and governments. Eco-awareness fosters educated decision-making and behaviors that support long-term sustainability in education, policy-making, innovation, and community engagement. Developing eco-awareness is ultimately crucial to solving world issues and building a sustainable future.

Scientific and technical breakthroughs have fuelled rapid economic growth, but they have also created serious environmental problems that have raised concerns about sustainability and sustainable development around the world. In scholarly and policy discourse, sustainability—broadly construed as preserving balanced development without endangering the environment—has gained significance. For people to be able to encourage sustainable behaviors, they must be given the values, knowledge, and skills of education for sustainable development (ESD). By encouraging students to adopt sustainable habits and an understanding of the environment, higher education institutions play a crucial role in this. Enhancing pro-environmental attitudes and behaviors through the integration of ESD into varied learning contexts promotes sustainable consumption and lifestyle choices. Sustainable development is predicated on an awareness of environmental challenges and the effects of human activity. It motivates citizens, companies, and governments to take action in favor of laws, creative solutions, and community-based projects that pursue sustainable objectives. Through education, we can raise awareness of environmental issues and work towards ensuring that everyone has a sustainable, balanced future.

PEDAGOGICAL APPROACHES FOR INTEGRATING ECO-AWARENESS

There has been much discussion about the goal of education, with many opinions and theories being put up. The primary goal of education can be the one-way transfer of knowledge, beliefs, skills, and facts from instructor to student (Misseyanni et al., 2020). As prospective future leaders and decision-makers, kids benefit from effective teaching and learning about sustainability when it comes to understanding their environment and the intricacy and interdependence of its issues (Abeydeera, 2021). Active and experiential learning strategies have been employed by EE and EFS/ESD to accomplish their long-standing aims of producing critical and engaged thinkers as well as responsible and active citizens (Misseyanni et al., 2020). In order to promote quality education and improve each student's quality of life, it is imperative that teacher education institutions integrate education for sustainable development. HEIs should also encourage their students to participate in sustainability-related activities on campus (Flores, 2023). For the past 20 years, forest pedagogy has been making its way into classrooms (Kollarova, 2024). In addition to the fundamentals of forest pedagogy as a viable teaching approach, they place a strong focus on environmental education and raising awareness of sustainable development (Kollarova, 2024). The Sustainable Development Goals were adequately brought to people's attention through the presentation and implementation of the Ecological Projects. Gender disparities were found to be statistically significant in three questions. In conclusion, based on the responses to the questionnaire, pairs of questions exhibiting a moderate to high monotonic connection are also included (Manolis & Manoli, 2021). The development of resilience as a manifestation of self-reliance, self-confidence, a strong desire to rediscover meanings and a sense of purpose, and the ability to have broad perspectives and propose audacious interpretations and ambitious solutions are all made possible by innovative pedagogies, such as those that address sustainability issues (Rogalska-Marasińska, 2022).

ESD is influenced by the global movement towards competency-based education, which places an emphasis on the acquisition of specific information, abilities, and attitudes required for tackling challenging sustainability issues. ESD incorporates concepts from global citizenship education, environmental education, and development education to show the connections between social, environmental, and economic sustainability challenges (Veckalne & Tambovceva, 2022). Higher education institutions have a responsibility to train, nurture, and develop the next generation of business leaders, given the power and responsibility that business has to address sustainability concerns (Georgallis & Bruijn, 2022). It is imperative that schools prioritize environmental teaching and that green behaviors are supported by political incentives. Campaigns in the media can also be a useful instrument for increasing societal awareness (Lopes et al., 2024). It is possible to increase students' awareness of environmental issues, help them comprehend the psychological effects of climate change, foster environmental citizenship, motivate climate action, and contribute to the development of more sustainable communities by implementing an integrated approach that combines traditional science, ecological education, affective education, and community psychology. The Future Generation program is transdisciplinary in that it integrates participatory techniques, emotional/psychological components, and ecological theoretical features (Mebane et al., 2023).

Teachers are leading the way in implementing ESD in classrooms and courses. They are responsible for developing and carrying out engaging and fruitful learning opportunities that improve students' understanding of sustainability concepts, hone their critical-thinking skills, and promote environmentally conscious behavior (Veckalne & Tambovceva, 2022). Students are vital to ESD because they are active learners and future change agents who participate in the curriculum and apply what they learn to solve practical sustainability issues (Veckalne & Tambovceva, 2022). A variety of stakeholders must work together and actively participate for ESD to be implemented successfully. By combining their unique skills and perspectives, educators, students, governments, communities, and corporate partners can create more sustainable societies and progress ESD (Veckalne & Tambovceva, 2022). It is imperative to incorporate strategies that propel or facilitate the development of students' systemic vision, enabling them to contextualise the causes and solutions of social and environmental issues. It is important to remember that the strategies employed will impact the students' understanding of this concept (García-González et al., 2020).

Universities focus on vast volumes of underutilized human resources. A few short computations provide some astonishing numbers (Evans et al., 2015). To choose from a variety of facilities within a set budget while taking economic, social, and environmental considerations into account, students were put in the position of a planning team for an expansion of a university campus. This exercise is based on the pedagogical theory that more students will engage substantively with ideas and problems related to sustainability if they are allowed to approach and think about sustainability issues in an open-ended, team-based exercise (Evans et al., 2015). In addition to raising awareness of the sustainable consumer options currently accessible to UK children and their parents or guardians, the workshops are meant to be a part of a continuum of experiences needed to assist the next generation in adapting to and responding to the challenges associated with biodiversity (Counsell et al., 2024). Analysis of the teaching and learning enterprise and a suitable examination of the crucial aspects of the tutorship program are made possible in the setting of higher education (Sithole & Gumede, 2022). It is advantageous to use the DBL approach to teach sustainability competency in engineering education since it helps students become more adept at overcoming obstacles related to sustainability practice (Huang et al., 2020). Given the increasing severity of environmental issues, special consideration should be given to the role that environmental groups play in schools. The goal of these clubs is to significantly alter students' perceptions and knowledge of the environment. They are intended to be an embodiment of environmental education (Houmam & Aomar, 2023). Environmental clubs help students become more environmentally conscious and help them embrace eco-citizen behaviors. In order to develop a generation that is conscious of and concerned about environmental issues, more has to be done to enhance and expand the effectiveness of these environmental clubs in educational institutions (Houmam & Aomar, 2023).

Several pedagogical strategies have been proposed to support transdisciplinary and comprehensive approaches to ESD. These strategies emphasize authentic, collaborative, and experiential learning events that involve students in real-world sustainability concerns. Examples of these strategies include problem-based learning, project-based learning, service learning, and place-based education (Veckalne & Tambovceva, 2022). To advance sustainable development and impact the beliefs and behaviors of present and future generations, education for sustainable development, or ESD, is crucial. Adopting transformative learning strategies is necessary for a major impact. These methodologies include ESD in a variety of learning situations, emphasizing the need for multidisciplinary and integrative approaches that include different learning styles, backgrounds, and circumstances (Veckalne & Tambovceva, 2022). It might be challenging to integrate practical community-based projects with traditional classroom instruction. This necessitates investigating alternative pedagogical strategies for teaching SD in field-based and community-oriented programs. Given that the community members are the ultimate beneficiaries of these initiatives, their engagement in community-oriented programming ought to come as second nature (Cornet et al., 2023). Though there is little proof that field-based teaching techniques like experiential learning are effective in teaching sustainability, they are known to be effective in other domains (Cornet et al., 2023). Field engagement that is oriented on solving the actual problems facing the communities is regarded as a more advanced kind of HEI engagement with SD (Cornet et al., 2023). With HEI's main goal being the education of future generations, the pedagogical approach of these programs is crucial to their success (Cornet et al., 2023). In order to influence the opinions and attitudes of future citizens, leaders, and politicians, HEIs should work with students to cultivate lasting knowledge, values, and understanding. By including more students, we hope to generate a new generation of ICT professionals who will graduate from college equipped with the knowledge and abilities needed to tackle sustainability-related problems (Angelaki et al., 2024).

Participatory approaches are acknowledged as more inclusive, therefore it stands to reason that HEI committing to SD would use them in SI initiatives with communities (Cornet et al., 2023). Pupils practice reasoning and making decisions as they are exposed to thorough and well-researched material derived from actual circumstances. Similar goals can be achieved via debates because of the active participation and critical thinking that they foster in pupils (Georgallis & Bruijn, 2022). Especially during the COVID-19 epidemic, politicians, business executives, and national leaders have been emphasizing the phrase sustainability more and more. Hence, the pandemic's unparalleled disruption presents a chance and a turning point for leaders and administrators, both domestically and internationally, to reconsider and realign their strategies in order to improve sustainability (Abeydeera, 2021). There needs to be a shift from an instructor-centered to a learner-centered approach, from individual to group work, from low cognitive to higher cognitive, and from merely acquiring knowledge to experiential (creative problem-solving) and lifelong learning (Abeydeera, 2021).

The goal of education has been widely debated, with the transfer of knowledge, beliefs, skills, and facts often cited as its primary objective. However, the growing emphasis on sustainability has reshaped this perspective, highlighting the importance of teaching students to understand the complexity and interdependence of environmental issues. Effective education for sustainable development (ESD) incorporates active and experiential learning strategies, fostering critical thinking and responsible citizenship. In this endeavor, higher education institutions (HEIs) play a crucial role by incorporating ESD into courses and promoting student involvement in sustainable projects. To solve socioeconomic and environmental concerns, ESD, which is influenced by competency-based education, integrates ideas from global citizenship, environmental education, and development education. Instructors are essential to the implementation of ESD because they design stimulating lessons that encourage environmentally friendly behaviour. As the next generation of leaders, students actively participate in these programs and apply what they learn to actual environmental problems. In order to build sustainable societies, effective ESD necessitates cooperation between educators, students, governments, communities, and corporations. With an emphasis on real-world, collaborative, and experiential learning, pedagogical techniques like problem-based, project-based, service-learning, and place-based education assist comprehensive ESD approaches. Despite challenges in integrating community-based projects with traditional instruction, these methods are crucial for advancing sustainable development and transforming the attitudes and behaviors of current and future generations. The COVID-19 pandemic has further underscored the importance of sustainability, presenting an opportunity for leaders to realign strategies towards a more sustainable future. This shift calls for a move from instructor-centered to learner-centered approaches, fostering lifelong learning and creative problem-solving.

CURRICULAR FRAMEWORKS FOR ECO-AWARENESS

Our educational system is focused on the past and worships it exclusively. It is necessary for the educational system to redefine itself in new terms. Our objective should be to create an education system that is focused on the future, with activities and issue-based learning taking precedence. In order to provide students with the skills to handle problems in real-world settings, the government should make ESD a required topic in schools. This would help to raise awareness among head teachers and ESD teachers

(Sharar et al., 2022). Important stakeholders, such as educators, legislators, and civil society organizations, must be involved in the development and implementation of transformative learning experiences that encourage behavioral change, critical thinking, and social involvement. Even though there can be challenges in putting ESD into practice, awareness-building is still important. The call for a renewed commitment to ESD as a vital instrument in addressing the global sustainable development agenda is necessary for a better future for everybody (Veckalne & Tambovceva, 2022). Universities have had difficulty incorporating sustainability into their business plans and organizational frameworks (Evans et al., 2015). It is becoming unacceptable for the curriculum to ignore climate change and sustainability, and English language schools need to acknowledge how these issues affect students' real-world experiences (Micalay-Hurtado & Poole, 2022). It is impossible to exaggerate the significance of including ESD in teacher education. Teachers have a pivotal role in molding the brains of children and impacting the attitudes and actions of society. A multiplier effect can be created by providing teachers with the resources they need to integrate sustainability ideas into their pedagogical techniques, as these educators help students become knowledgeable, accountable, and environmentally conscientious citizens (Flores, 2023).

The establishment of roles in the academic and administrative hierarchy has ingrained social responsibility into university governance, while environmental sustainability has been positioned as an enabling strategy that supports the institution's capacity to carry out its objectives in research, teaching, and social responsibility (Evans et al., 2015). The potential of sustainability to teach science and promote universal values like respect for human rights and other forms of life, the holistic development of the individual, and engagement in democratic processes has led to its recent adoption as a common context for science education. Teachers and researchers have a responsibility to address the Sustainable Development Goals in higher education, by means of deep didactic interventions that promote sustainable attitudes and increase environmental awareness (Míriam Andrea et al., 2022).

(Flores, 2023) described how the landscape of ESD integration is changing and emphasized the value of faculty development, student participation, information sharing, and all-encompassing sustainability activities in teacher education programs. These observations support comprehensive approaches to education, producing graduates who are socially conscious, economically savvy, and mindful of the environment, making them ready for difficulties that will arise in the future.

Teachers frequently employ a "Doom and Gloom" approach when teaching pupils about difficult subjects like prejudice or climate change, which can be depressing and lead to resignation. Future studies should examine efficient communication techniques that engage students while conveying the seriousness of the problems (Buerkle et al., 2023). Teachers can design curricula to raise consumer awareness and encourage environmentally friendly behavior (Hasibuan & Judijanto, 2023). Instead of assuming that student engagement will happen on its own, teaching and learning activities need to be intentionally designed to foster it. Students will be able to make connections between material from disparate disciplines thanks to the thoughtful design of interdisciplinary teaching and learning activities, which will improve their higher-order cognitive skills such as problem-solving, critical thinking, and the ability to employ multiple perspectives (Abeydeera, 2021). As fostering pro-environmental attitudes is the first step towards creating sustainable societies, it is critical to shift instruction to one that is more transformative (Misseyanni et al., 2020). Instead of sticking to boring, rigid, and authoritarian teaching methods, educators will be able to embrace more problem-based, experiential, reflective, critical, and radical ways of teaching and learning with an increased understanding of sustainability (Abeydeera, 2021). Academics who teach and study sustainability will need to comprehend that for businesses to succeed in the new normal, more fundamental adjustments will be required. A crucial step in such a shift will be to place more emphasis on the integration of business with the community, society, and environment, as well as values like diversity, equity, and inclusion, as opposed to just business growth (Abeydeera, 2021). To support attempts to foster pro-environmental behaviors, teacher training in active learning and technology-enhanced learning is regarded as a crucial prerequisite. Administration training is particularly crucial since administrators are supposed to give staff, students, and teachers the proper frameworks and support. To achieve active learning, flexible learning environments and infrastructure appear to be crucial. Furthermore, cooperation must be encouraged (Misseyanni et al., 2020). Analyzing legal underpinnings, policy, and the ESD framework reveals varying degrees of knowledge about ESD legalities and policies, highlighting the need for thorough communication. Social, environmental, and economic components are all integrated into the ESD framework; practices are still in the early stages of adoption.

There is an opportunity to improve inclusivity and shared governance given the variation in student participation in institutional decision-making (Flores, 2023). Energy efficiency is a component of environmental sustainability initiatives, however, problems with environmentally friendly procedures and visually appealing learning spaces still exist. Students who participate in commerce-focused learning activities demonstrate how economic empowerment through ESD is achieved. The importance of faculty development in enhancing educational practices and facilitating effective integration of ESD is evident (Flores, 2023). Meticulous preparation, including a precise description of the goals to be met and the abilities that the students are expected to acquire. Even when students work in teams, there should be an emphasis on group, cooperative, open, and shared learning while still taking individual assessments into account to address asymmetries. Encouraging students to experiment with volumetric materials when creating mockups and models, even if they do it in their homes, and to take pictures of the process as they go (De Sampaio, 2022). Designing curriculum and programming that allow the integration of sustainability material and competencies in a way that is adaptable to many subject areas and professional sectors is crucial for effective integration, as supported by Education for Sustainable Development (ESD) (Garzón-García et al., 2023).

The current educational system needs a transformative shift from its past-focused orientation to a future-oriented one, emphasizing activities and issue-based learning. Making Education for Sustainable Development (ESD) a mandatory part of the curriculum is crucial for equipping students with the skills to address real-world problems. This requires the involvement of educators, policymakers, and civil society organizations to create transformative learning experiences that foster critical thinking, social involvement, and behavioral change. Integrating sustainability into higher education's business models and curricula is essential, as is promoting interdisciplinary and experiential learning to prepare students for future challenges. Teacher education must focus on sustainable practices, encouraging active learning and the development of pro-environmental attitudes. Effective communication strategies, flexible learning environments, and faculty development are critical to successfully incorporating ESD and fostering environmentally conscious, socially responsible graduates.

CHALLENGES IN EMBEDDING ECO-AWARENESS INTO EDUCATIONAL CURRICULA

It is essential in higher education to incorporate sustainability into instructional practices. Encouraging the next generation of leaders and innovators while equipping students with the knowledge, skills, and understanding they need to thrive may effect the change that is required and play a big role in the shift to a sustainable future (Gamage et al., 2022b). Because of their dedication to sustainability, business schools face challenges to their legitimacy. A major part of how business schools prove their validity is through instruction. The separation of theory and practice in management education, the propensity for academic silos, and the overuse of normative theoretical models in research and instruction are three factors that are linked to legitimacy (Williamsson et al., 2022). There are various obstacles to overcome when incorporating eco-awareness into school curricula. One major barrier is the curricula's current rigidity, which frequently provides little opportunity for new subjects or the multidisciplinary approaches needed for effective teaching of sustainability. Educators may also struggle to confidently incorporate complicated environmental ideas into their lessons due to a lack of resources and expertise. Another factor impeding development is resistance to change among educators and administrators who are used to old teaching practices. Moreover, unequal eco-awareness program implementation can result from differences in funding and resources throughout schools. Another problem is keeping the curriculum relevant in the face of fast-evolving environmental science while engaging students in meaningful ways and completing required topics. To overcome these obstacles, systemic support, sufficient financing, ongoing professional development, and a cooperative, sustainability-focused curriculum design approach are needed. Many educators struggle with the interdisciplinary elements of sustainability, feel underprepared for the content and student-centered pedagogy of ESD, and receive insufficient resources, materials, time for collaboration, and opportunities for ongoing professional development. The fact that these barriers continue to exist in spite of the fact that it is widely acknowledged that teachers play a crucial role in supporting learning for sustainability emphasises the necessity of tying education for sustainability policy more closely to the viewpoints, experiences, and learning requirements of teachers (Parry & Metzger, 2023).

For students to be equipped with the knowledge, abilities, and comprehension needed to spearhead the shift to a sustainable future, sustainability must be incorporated into higher education teaching methodologies. Significant obstacles stand in the way of this integration, though, including insufficient resources and training for instructors as well as strict curricula that restrict the addition of fresh, interdisciplinary subjects. Further complicating the implementation of eco-awareness programs include differences in funding and resources among schools, educators accustomed to traditional ways of teaching who are resistant to change, and the challenge of engaging pupils while covering required information. To tackle these obstacles, systemic support, sufficient financing, ongoing professional development, and a cooperative, sustainability-focused curriculum design approach are needed.

STRATEGIES FOR OVERCOMING CHALLENGES

Given that sustainability is a complex, multidisciplinary topic, teaching it can be challenging. We must adopt a student-centered and participative strategy in order to inspire pupils to think about societal challenges and their personal responsibilities in fixing them. Thus, we need the appropriate educational resources that encourage creativity, an open mind, and broad thinking in order to raise awareness and teach about sustainability (Despeisse, 2018). One approach to researching personalized learning is looking at tactics that are tailored to an individual's needs, preferences, and learning style. The effects of personalized learning on student development and engagement, adaptive tutoring systems, personalized assessment, and customized content could all be research topics (Buerkle et al., 2023). The best places to study sustainability issues are universities (Evans et al., 2015). While colleges are increasingly turning to co-production and living labs as strategies to address sustainability concerns, there is still a lack of discussion on the connections between the two (Evans et al., 2015). Agroecological approaches should be prioritised in reforms, but contentious 2030 Agenda concerns and technical deficiencies in official targets and indicators should also be addressed through critical multidisciplinary research, education, community outreach, evaluation, and evidence-based policy discussion.

To maximize their efforts in addressing sustainability concerns, universities need to create new frameworks that incorporate teaching, research, and social responsibility into their core activities. Living laboratories provide creative ways to work that can carefully structure co-production possibilities in order to achieve precisely that (Evans et al., 2015). Gamification and serious games have been popularised in recent years as useful tools for involving students in decision-making and exposing them to a range of challenging situations. Gamification works especially effectively for teaching sustainability because it provides students with an enjoyable and safe environment in which to explore, make challenging decisions, and think through the implications of those actions. It can impart the abilities needed to address global concerns, like brainstorming, critical thinking and reflection, teamwork, and making decisions under uncertain circumstances (Despeisse, 2018). When the program is implemented, prospective teachers who experience good and stimulating emotions believe that the experience will change the way they teach science when they become teachers. With this intervention, aspiring teachers encounter actual issues, grow in their ability to perceive and analyze troubling global situations, and enhance their scientific knowledge, environmental attitudes, and emotional connection to science education (Míriam Andrea et al., 2022). When it comes to teaching sustainability, gamification works especially well because it gives students a fun and secure environment in which to experiment, make difficult decisions, and consider the consequences of their choices (Despeisse, 2018). Games have become a potentially effective teaching tool that may immerse students in a variety of difficult situations and allow them to actively participate in decision-making (Despeisse, 2018). To encourage environmentally conscious behaviour and mould eco-friendly lives for long-term environmental sustainability, environment-related courses should be required in all degree programs (Ma et al., 2023).

Continuing to invest time and effort in the design of stimulating scientific activities in teacher training that generate attitudinal and awareness changes will lead to competent citizens who are scientifically literate and with the capacity to make decisions. Generating positive emotions during science learning is therefore decisive. The conclusion is that the time and effort dedicated to research in didactics for integral sustainability are indeed worthwhile (Míriam Andrea et al., 2022). It is imperative to consider the ethical implications that arise when integrating technology into teaching. Concerns about data privacy, various types of discrimination, and the advantages and disadvantages of using technology over in-person instruction could all be discussed (Buerkle et al., 2023). Academic research may focus on strategies for developing students' capacity for self-regulated thought. This may mean setting goals, adjusting their approaches, and monitoring their own academic development. Therefore, self-regulated learning may increase motivation, autonomy, and well-being (Buerkle et al., 2023). It could be investigated what barriers students encounter when attempting to get educational resources and how to overcome them. This may include, among other things, considering the perspectives of students from underrepresented groups, low-income families, and international students (Buerkle et al., 2023). These factors work together to support sustainable behaviors, as demonstrated by the positive correlations between government legislation, consumer awareness, education, and sustainable purchasing (Hasibuan & Judijanto, 2023). There may be educational advantages to both experiential learning and interactive methods. By transitioning from participatory to hands-on participation, the participatory approach improves students' experiential learning (Cornet et al., 2023). The use of case-based discussion techniques in teaching sustainability in business schools or other educational contexts can be advantageous in several ways. It can build on each strategy by combining the best features of the discussion and case approaches (Georgallis & Bruijn, 2022).

The complexity and transdisciplinary nature of sustainability make it difficult to teach. Educational materials that encourage creativity, open-mindedness, and broad thinking are necessary for a student-centered and participative strategy that effectively engages students and fosters awareness. Personalized learning techniques that are catered to each learner's requirements and preferences might improve participation and advancement. Through cutting-edge techniques like co-production and living labs, universities play a critical role in researching and tackling sustainability challenges. Engaging students in intricate scenarios and decision-making processes through gamification and serious games can effectively develop soft skills such as collaboration and critical thinking. However, incorporating technology into the classroom necessitates giving ethical consideration to issues like bias and data privacy. Educational outcomes can be further improved by scholarly investigation into strategies for promoting self-regulated learning and removing barriers to instructional materials access. The significance of comprehensive methods to sustainability education is highlighted by the interrelationships among education, consumer awareness, government regulation, and sustainable behavior. In the context of business schools and other educational settings, case-based discussions provide a useful method for teaching sustainability, while participatory tactics and experiential learning can improve student knowledge and involvement.

TEACHER TRAINING AND CAPACITY-BUILDING INITIATIVES

Fostering a generation that appreciates and practices sustainability requires the implementation of effective teacher training programs in eco-awareness and sustainable education. A full grasp of sustainability concepts, including global issues like resource depletion, biodiversity loss, and climate change, should be covered in such training programs, with a focus on the value of ecosystems and biodiversity. It is imperative that educators possess an understanding of sustainable practices in the classroom, including ways to lessen schools' environmental impact and integrate recycling initiatives. Embracing immersive and outdoor learning approaches is essential, as is integrating sustainability into a variety of disciplines through project-based and interdisciplinary lessons. Access to current scientific data, digital technologies, and instructional materials will improve the efficacy of instruction. Learning can be enhanced even further by interacting with neighborhood environmental organizations and promoting community participation in sustainability initiatives. Opportunities for professional development, such as conferences, workshops, and online courses that offer continual learning, are essential for educating educators about the most recent developments in sustainability. A supportive environment for teachers can be created by establishing collaborative learning communities, providing assistance with curriculum development, and putting mentorship programs into place. The curriculum will also be improved with the use of feedback mechanisms and assessment tools to gauge the success of the instruction. Sustainable teaching practices will be further promoted by forming alliances with environmental organizations, using technology for resource sharing and simulation, and rewarding excellent teachers and schools with incentives. Teacher training programs can effectively prepare educators to implant sustainable beliefs and practices in their students, fostering a future generation of environmentally conscious and responsible individuals, by utilizing these comprehensive components and tactics. HEIs can improve their students' comprehension of SD by a variety of parallel approaches, but primarily by including into the current curriculum fundamental ideas that are closely related to SDs, such as concerns pertaining to poverty, diversity, and gender equality (Abowardah et al., 2024).

In the rapidly changing, internationally interconnected world of the twenty-first century, with growing environmental restrictions and concerns, environmental education and sustainability are priorities from elementary school to higher education. Education will create people who are socially and environmentally sensitive, as well as capable of handling present and future environmental challenges (Misseyanni et al., 2020). Teacher preparation programs ought to include an exercise in which participants construct and optimize a filter of this type, in order to enable PTs to really understand the new Integral Sustainability (IS) paradigm. This will lead to a change in mindset, promote creativity, and raise awareness (Míriam Andrea et al., 2022). The aim of a proposed cognitive and emotional intervention is to generate positive emotions when learning about sustainability physics and chemistry, as well as to foster a higher sense of self-efficacy when performing laboratory tasks. This intervention may result in a structural shift in the way science is taught to students (Míriam Andrea et al., 2022).

The financial capacity to cover tuition costs or the requirement for part-time work throughout a university degree program can have a big influence on a student's performance in advanced education as well as their future professional prospects. Thus, in order to combat poverty in education, teachers' understanding and pedagogical competencies may be able to empower students from low-income families (Buerkle et al., 2023). Future studies might concentrate on assisting educators in recognizing discriminatory practices, microaggressions, and unconscious prejudice in the curriculum they teach. Non-biased instructional materials can also support culturally sensitive instruction, a universal design for learning, and the development of an all-around more inclusive learning environment (Buerkle et al., 2023). Studying the effects of expanding the representation of varied voices and viewpoints within curricula across disciplines is one way to achieve curriculum diversity. Subsequent studies could investigate the effects of this inclusive design on the performance and engagement of students (Buerkle et al., 2023).

Universities have unquestionably played a crucial role in establishing the groundwork for sustainable development. The United Nations Sustainable Development Goals (SDGs) can be greatly aided by these institutions, which serve as the primary instructors of tomorrow's leaders (Abowardah et al., 2024). Rethinking how universities, which are frequently seen as role models for professional managers, approach sustainability education is another important task for educational administrators (Abeydeera, 2021). As collaborators in education, students need to shift their perspective from seeing businesses as part of the issue to seeing them as a solution. This necessitates engaging in active, experiential, and reflective learning as well as a radical shift in our understanding of what sustainability actually is (Abeydeera, 2021). For head teachers to become more knowledgeable about ESD, they must receive professional orientation and in-service training on the subject. Headteachers at each school will benefit from the training as they implement the ESD dimension (Sharar et al., 2022). Effective training programs for teachers in sustainable teaching and eco-awareness are crucial for fostering a generation that values and practices sustainability. With an emphasis on ecosystems, biodiversity, and sustainable practices in education, these programs ought to offer a comprehensive grasp of sustainability concepts and global concerns. It is imperative to integrate experiential and outdoor learning approaches with project-based, multidisciplinary lessons. The effectiveness of instruction will be improved by having access to instructional materials, digital tools, and current scientific data. Additionally, participation in sustainability projects by the community and local environmental organizations will boost the learning process. Mentorship programs, collaborative learning communities, and ongoing professional development opportunities are essential for educating and sustaining educators. The curriculum will be improved with the use of assessment tools and feedback mechanisms, and sustainable teaching practices will be encouraged through collaborations with environmental organizations and recognized rewards. To further equip teachers to establish enduring values in their pupils, teacher training programs should incorporate activities and interventions that promote creativity, pleasant emotions, and inclusivity.

CONCLUSION

Achieving sustainable development and tackling the many issues confronting humanity requires the integration of sustainability into education, especially in higher education. The goal of education for sustainable development (ESD) is to provide people with the values, information, and abilities needed to address urgent problems including resource shortages, climate change, and environmental degradation. Education has a key role in altering communities and guiding society toward a more sustainable future through the promotion of environmental awareness, critical evaluation of transdisciplinary concerns, and active engagement in sustainability projects. But in order to overcome obstacles like inflexible curricula, opposition to change, and discrepancies in funding and resources, this revolutionary transformation necessitates cooperation from corporations, civil society organizations, legislators, and educators. To raise ecologically conscious and responsible citizens, effective teacher training programs in sustainable education and Eco-awareness are crucial. By incorporating sustainability into their lesson plans, making educational resources accessible, and encouraging ongoing professional growth, educators can encourage the next generation of sustainability advocates and help create a more equitable and sustainable society for all.

REFERENCES

Abeydeera, S. (2021). Teaching and Learning Sustainability during and beyond COVID-19: A Review of Business and Management Studies. *University of Colombo Review*, 2(1), 138–154. DOI: 10.4038/ucr.v2i1.43

Abowardah, E. S., Labib, W., Aboelnagah, H., & Nurunnabi, M. (2024). Students' Perception of Sustainable Development in Higher Education in Saudi Arabia. *Sustainability, 16*(4, 1483), 1–23. DOI: 10.3390/su16041483

Akçay, K., Altinay, F., Altınay, Z., Daglı, G., Shadiev, R., Altinay, M., Adedoyin, O. B., & Okur, Z. G. (2024). Global Citizenship for the Students of Higher Education in the Realization of Sustainable Development Goals. *Sustainability (Basel)*, 16(4), 1–13. DOI: 10.3390/su16041604

Angelaki, M. E., Bersimis, F., Karvounidis, T., & Douligeris, C. (2024). Towards more sustainable higher education institutions: Implementing the sustainable development goals and embedding sustainability into the information and computer technology curricula. *Education and Information Technologies*, 29(4), 5079–5113. DOI: 10.1007/s10639-023-12025-8

Buerkle, A., O'Dell, A., Matharu, H., Buerkle, L., & Ferreira, P. (2023). Recommendations to align higher education teaching with the UN sustainability goals – A scoping survey. *International Journal of Educational Research Open, 5*(April, 100280), 1–19. DOI: 10.1016/j.ijedro.2023.100280

Cornet, S., Barpanda, S., Guidi, M. A. D., & Viswanathan, P. K. (2023). Sustainability education and community development in higher education using participatory and case based approaches in India. *International Journal of Sustainability in Higher Education*, 25(9), 78–93. DOI: 10.1108/IJSHE-07-2022-0242

Counsell, G., Edney, G., & Dick, S. (2024). Improving awareness on sustainable palm oil: Measuring the effectiveness of a repeat-engagement zoo outreach programme. *Environmental Education Research*, 30(2), 153–169. DOI: 10.1080/13504622.2023.2225805

De Sampaio, C. P. (2022). THE CHALLENGE OF TEACHING DESIGN FOR SUSTAINABILITY IN A PANDEMIC CONTEXT: A DIDACTICAL-PEDAGOGICAL EXPERIENCE. *MIX Sustentável*, 8(2), 121–132. DOI: 10.29183/2447-3073.MIX2022.v8.n2.121-132

Despeisse, M. (2018). Teaching Sustainability Leadership in Manufacturing: A Reflection on the Educational Benefits of the Board Game Factory Heroes. *Procedia CIRP*, 69, 621–626. DOI: 10.1016/j.procir.2017.11.130

Evans, J., Jones, R., Karvonen, A., Millard, L., & Wendler, J. (2015). Living labs and co-production: University campuses as platforms for sustainability science. *Current Opinion in Environmental Sustainability*, 16, 1–6. DOI: 10.1016/j.cosust.2015.06.005

Flores, G. T. (2023). Fostering Awareness of Integrating Education for Sustainable Development in Teacher Education Institutions across the Caraga Region, Philippines. *International Journal of Membrane Science and Technology*, 10(3), 1277–1284. DOI: 10.15379/ijmst.v10i3.1700

Gamage, K. A. A., Ekanayake, S. Y., & Dehideniya, S. C. P. (2022a). Embedding Sustainability in Learning and Teaching: Lessons Learned and Moving Forward-Approaches in STEM Higher Education Programmes. *Education Sciences*, 12(3), 225. Advance online publication. DOI: 10.3390/educsci12030225

Gamage, K. A. A., Ekanayake, S. Y., & Dehideniya, S. C. P. (2022b). Embedding Sustainability in Learning and Teaching: Lessons Learned and Moving Forward—Approaches in STEM Higher Education Programmes. *Education Sciences*, 12(225), 1–20. DOI: 10.3390/educsci12030225

García-González, E., Jiménez-Fontana, R., & Goded, P. A. (2020). Approaches to teaching and learning for sustainability: Characterizing students' perceptions. *Journal of Cleaner Production*, 274, 1–9. DOI: 10.1016/j.jclepro.2020.122928

Garzón-García, R., Florido-Trujillo, G., & Ramírez-López, M. L. (2023). The integration of sustainability and the Sustainability Development Goals (SDGs) in the geographical teaching of tourism: A didactic and methodological proposal for University. *Investigaciones Turísticas*, 26(26), 86–113. DOI: 10.14198/INTURI.22411

Georgallis, P., & Bruijn, K. (2022). Sustainability teaching using case-based debates. *Journal of International Education in Business*, 15(1), 147–163. DOI: 10.1108/JIEB-03-2021-0039

Hasibuan, F. U., & Judijanto, L. (2023). The Influence of Education, Consumer Awareness, and Government Regulation on Sustainable Consumption Behavior: A Case Study on Ciawitali Market in Garut Regency. *West Science Journal Economic and Entrepreneurship*, 1(03), 109–119. DOI: 10.58812/wsjee.v1i03.406

Hoque, F., Yasin, R. M., & Sopian, K. (2022). Revisiting Education for Sustainable Development: Methods to Inspire Secondary School Students toward Renewable Energy. *Sustainability (Basel)*, 14(14), 1–18. DOI: 10.3390/su14148296

Houmam, L., & Aomar, I. (2023). Effects of School-based Environmental Clubs: Fostering Environmental Awareness and Strengthening Eco-citizen Behaviors Among Students. *Journal of Law and Sustainable Development*, 11(12), e1775. Advance online publication. DOI: 10.55908/sdgs.v11i12.1775

Huang, Z., Peng, A., Yang, T., Deng, S., & He, Y. (2020). A design-based learning approach for fostering sustainability competency in engineering education. *Sustainability (Basel)*, 12(7), 1–17. DOI: 10.3390/su12072958

Kollarova, D. (2024). Forest Pedagogy and Its Didactic Approach in Pre-Primary and Primary Education. *Slavonic Pedagogical Studies Journal*, 13(1), 61–74. DOI: 10.18355/PG.2024.13.1.6

Kopnina, H. (2020). Education for the future? Critical evaluation of education for sustainable development goals. *The Journal of Environmental Education*, 51(4), 280–291. DOI: 10.1080/00958964.2019.1710444

Lopes, J. M. M., Gomes, S., & Trancoso, T. (2024). Navigating the green maze: Insights for businesses on consumer decision-making and the mediating role of their environmental concerns. *Sustainability Accounting. Management and Policy Journal*, 15(4), 861–883. DOI: 10.1108/SAMPJ-07-2023-0492

Ma, L., Shahbaz, P., Haq, S., & Boz, I. (2023). Exploring the Moderating Role of Environmental Education in Promoting a Clean Environment. *Sustainability (Basel)*, 15(10), 1–19. DOI: 10.3390/su15108127

Manolis, E. N., & Manoli, E. N. (2021). Raising awareness of the Sustainable Development Goals through Ecological Projects in Higher Education. *Journal of Cleaner Production*, 279, 1–31. DOI: 10.1016/j.jclepro.2020.123614

Mebane, M. E., Benedetti, M., Barni, D., & Francescato, D. (2023). Promoting Climate Change Awareness with High School Students for a Sustainable Community. *Sustainability (Basel)*, 15(14), 1–15. DOI: 10.3390/su151411260

Micalay-Hurtado, M. A., & Poole, R. (2022). Eco-critical language awareness for English language teaching (ELT): Promoting justice, wellbeing, and sustainability in the classroom. *Journal of World Languages*, 8(2), 371–390. DOI: 10.1515/jwl-2022-0023

Míriam Andrea, H. B., Jesús, S. M., Isaac, C. C., & Florentina, C. C. (2022). Is sustainability an exhausted concept? Bridging the gap from environmental awareness to emotional proficiency in science education through integral sustainability. *Heliyon*, 8(12), e12546. Advance online publication. DOI: 10.1016/j.heliyon.2022.e12546 PMID: 36593825

Misseyanni, A., Marouli, C., & Papadopoulou, P. (2020). How teaching affects student attitudes towards the environment and sustainability in higher education: An instructors' perspective. *European Journal of Sustainable Development*, 9(2), 172–182. DOI: 10.14207/ejsd.2020.v9n2p172

Parry, S., & Metzger, E. (2023). Barriers to learning for sustainability: A teacher perspective. *Sustainable Earth Reviews*, 6(2), 1–11. DOI: 10.1186/s42055-022-00050-3

Purushothaman, S. (2019). Sustainability: Challenges in Teaching. *Ecology. Economy and Society*, 2(2), 33–36. DOI: 10.37773/ees.v2i2.73

Rogalska-Marasińska, A. (2022). Reimagining Education Towards Sustainability. "Project X" – an Innovative Learning Strategy from Poland. *Studia Edukacyjne*, 66(66), 17–34. Advance online publication. DOI: 10.14746/se.2022.66.2

Sharar, T., Qutoshi, S. B., & Sharar, T. (2022). The level of school leadership awareness about education for sustainable development. *Journal for Educators. Teachers and Trainers*, 13(5), 358–370. DOI: 10.47750/jett.2022.13.05.032

Sithole, M. P., & Gumede, P. R. (2022). Sustaining a tutorship programme at a university of technology: A systems approach. *Perspectives in Education*, 40(3), 224–240. DOI: 10.18820/2519593X/pie.v40.i3.15

Veckalne, R., & Tambovceva, T. (2022). The Role of Digital Transformation in Education in Promoting Sustainable Development. *Virtual Economics*, 5(4), 65–86. DOI: 10.34021/ve.2022.05.04(4)

Williamsson, J., Schaad, G., & Sandoff, A. (2022). Building legitimacy for sustainable business schools: Using the business model concept when teaching corporate sustainability. *Journal of Cleaner Production*, 367(13311), 1–8. DOI: 10.1016/j.jclepro.2022.133116

Winter, J., Zhai, J., & Cotton, D. R. E. (2022). Teaching environmental sustainability in China: Opportunities and challenges for business and economics faculty in higher education. *Environmental Education Research*, 28(2), 318–332. DOI: 10.1080/13504622.2021.2012560

KEY TERMS AND DEFINITIONS

Sustainability Education: To promote awareness and action towards sustainable development, refers to the integration of sustainability-related knowledge, skills, and values into educational curricula and learning experiences.

Higher education: Post-secondary educational institutions, including universities, colleges, and technical schools, are included in this phrase. These are the places where students go after advanced degrees in their fields of study.

Education for Sustainable Development (ESD): To enable people to address global concerns and contribute to sustainable solutions, ESD incorporates teaching and learning strategies that support environmental stewardship, social responsibility, and sustainability ideals.

Sustainable Teaching: Using instructional strategies and approaches that integrate sustainability principles, attitudes, and practices into learning activities, sustainable teaching fosters critical thinking, problem-solving, and action-oriented learning.

Eco-Awareness: Understanding ecological systems, environmental challenges, and how human activity affects the environment is referred to as eco-awareness. It includes understanding environmental preservation, sustainability, and the interdependence of social, economic, and ecological systems.

Teacher Training: In order to improve teachers' knowledge, abilities, and competencies in instructing particular subjects or pedagogical approaches, teacher training programs, courses, and workshops are offered. Teacher training in the context of sustainability education is centered on providing teachers with the instruments and materials required to incorporate sustainability ideas into their instructional strategies.

Collaborative Efforts: In order to solve shared objectives or issues, collaborative activities entail cooperative initiatives and collaborations among stakeholders, including corporations, educators, legislators, civil society organizations, and community people. Collaborative activities in the context of sustainability education seek to advance sustainability projects and promote favorable social and environmental outcomes by utilizing a variety of viewpoints, resources, and skills.

Chapter 7
Pixelated Preservation:
Digital Art for Sustainable Architecture, Cultural Preservation, and Interactive Experiences

Maryam Saleem
https://orcid.org/0009-0004-6207-0124
Universiti Sains Malaysia, Malaysia

Nur Zaidi Bin Azraai
Universiti Sains Malaysia, Malaysia

Amna Saleem
National College of Arts, Pakistan

ABSTRACT

Digital arts, guided by New Media Design principles, are transforming architectural heritage conservation. This chapter explores how digital tools, including 3D mapping, VR/AR, and simulations, enhance sustainability in preservation practices. By creating immersive experiences and fostering community engagement, digital art not only documents but reimagines historical sites, ensuring their accessibility and relevance for future generations. This multidisciplinary approach tackles challenges and embraces opportunities, showcasing the potential of digital technology to reshape heritage conservation towards a more sustainable and interactive future.

1. INTRODUCTION

The emergence of 'Digital Art' as a vital tool for preserving cultural heritage items has been a prominent focus in the late 20th and early 21st centuries. The rapid advancement of multimedia technology and its widespread dissemination across the globe has underscored the significance of digital art. This phenomenon surpasses other forms of innovation. Utilizing new media design technologies, it is possible to optimize multimedia systems in a way that holds meaningful and symbolic value. A distinct digital

DOI: 10.4018/979-8-3693-5748-4.ch007

"language" is created, seamlessly integrating technical, artistic, and value-oriented elements, thus playing a crucial role in this context.

Lev Manovich's five "New Media Design" principles are transcoding, numerical representation, modularity, automation, and variability (Fig. 1) (Jeknic, 2014). The digitization of historical assets has become more accessible due to significant advancements in automation, processing speeds, accuracy, and precision. Scanning and image-based modeling techniques are now at the forefront of large-scale applications, including analyzing architectural historical sites and features. This is due to their rapid acquisition times, cost-effectiveness, efficient processing, and repeatable workflows.

Figure 1. Lev Manovich' proposed five principles of 'New Media'

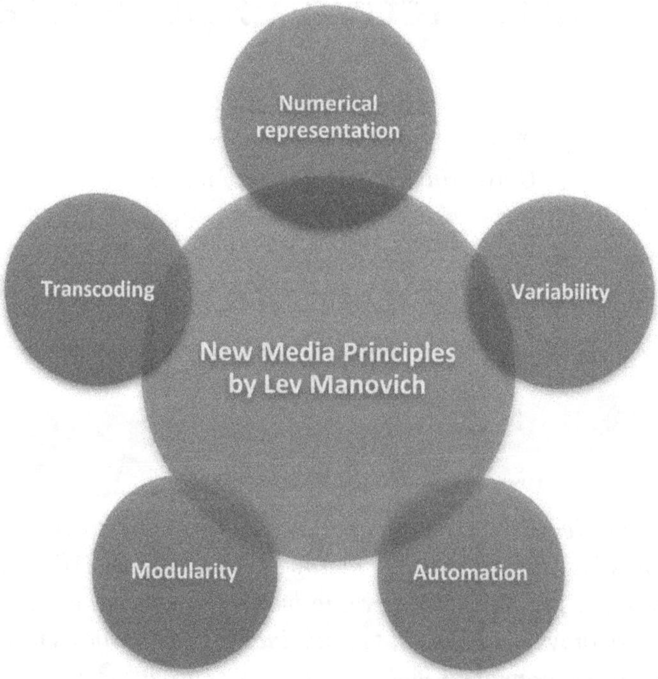

(Lev Manovich, 2001)

Preserving artistic and cultural heritage presents significant challenges due to its unique characteristics and exposure to severe environmental threats. There are various methods for assessing the condition of cultural heritage and conserving it from potential harm. Additionally, leveraging advanced digital technology to create digital replicas of relevant artifacts can reduce uncertainty in the analysis process. The development of products and experiences related to heritage tourism has the potential to impact the social and economic well-being of local communities positively. However, heritage tourism initiatives face obstacles in maintaining cultural integrity while transforming them into attractive tourist destinations.

In this chapter, the submersion method is employed to solve scenario-design difficulties involving the application of video projection and 3D mapping in exhibition space design and for projections onto architectural landmark facades to preserve components of cultural heritage, explains in further detail how state-of-the-art 3D modeling and scanning can help with the virtual presentation of history. The functional goal of the context in which the interactive work is displayed determines the content of these digital simulations featuring digital arts and the particular qualities of the artistic pictures they use. The objective is to gather data regarding the definition of Digital media art, its developmental trends, and research techniques for digital art to promote sustainable transmission and preservation in architectural contexts.

This study investigates both active and passive methods of digitizing heritage in three dimensions that are constantly being improved. More profound scientific knowledge of interactive art is still needed to improve professional practices in preserving cultural heritage works and sustaining heritage data for future generations. With a multidisciplinary strategy that uses art and technology for legacy conservation and dissemination, the chapter offers a new paradigm called "*Intelligent Architectural Heritage Preservation for Future Sustainability*." Moreover, it aims for interdisciplinary collaboration and innovative solutions with sustainable approaches to heritage preservation that can be applied to create interactive experiences for the next generation at heritage sites by incorporating real-world case studies to illustrate theoretical concepts with practical examples, enhancing their applicability.

2. OVERVIEW OF DIGITAL ARTS AND NEW MEDIA

Background

The digital arts can be traced historically to the Dada movement (or Dadaism), founded in New York in 1915 and in Zurich, Switzerland, in 1916. Moreover, it was active internationally during the early twentieth century. Dadaists responded to the social unrest of World War I, the spread of conservative politics throughout Europe and beyond, and the continuation of established artistic methods. Dadaism employs techniques including collage, photomontage, assemblage, appropriation, and audience engagement to blur the lines between various art forms and the lives of ordinary people. Artistic activities are impacted by technology. "We must expect great innovations to transform the entire technique of the arts, thereby affecting artistic invention itself and perhaps even bringing about an amazing change in our very notion of art," said French poet Paul Valéry in the early 20th century, foreshadowing the rise of the digital arts (Charles Ryan, 2014)Lovejoy states that public art—defined as art with a social use value conceptualized, made for, and existing in the public domain—is another essential art-historical effect on digital art creation. (Julie Wosk, 2015). The Russian Constructivists, who fused art and technology in anticipation of a future social order, are the source of contemporary public art.

New Media Digital Arts: Here, it is crucial to consider how the terms "medium" and "media" are utilized. A medium denotes the substance used to create the artwork within the art context. There are different ways in which an artwork can be considered digital. In the past, combinations of text, audio (music or voice), static visuals (scanned photos or computer graphics), and moving images (video, film, or computer animation) have been used in works referred to as "multimedia," as described by Wilson (1993). More recently, an interactive, non-linear structure that allows viewers to choose their paths through information has been termed "hypermedia." (Wilson, 1993)

Media art theorists distinguish between a "tool" and a "medium" when describing how artists such as Lovejoy, Paul, and Vesna 2011 employ digital technologies. As an instrument, technology makes it easier to create conventional works of art like sculptures, paintings, or dances, or it can convert analog works into digital formats to improve their distribution, storage, or long-term sustainable preservation.

(Lovejoy, 2011; Paul, 2003.) Digital imaging and photography are two examples of digital technology used as instruments in the creation of art. Artists utilize digital technologies in many media, such as painting, drawing, sculpture, photography, and video.

Three dimensions are particularly relevant to the digital arts, despite the term's wide range of applications: (a) the blending of the real and virtual worlds, which produces novel "hybrid" realities in which a work is experienced, performed, and preserved; (b) the blurring of lines between artistic practice categories, which challenges conventional notions of what constitutes a creative work of art; and (c) the merging of various media, such as graphic art, animation, and video, to create new digital artworks. (Charles Ryan, 2014)

Lev Manovich's New Media Approach

In the study of digital arts, the Dada combination of order and chance is analogous to the algorithms, or the codes, instructions, or rules, that enable artists and other computer software users. Many of today's digital arts initiatives take shape thanks to the fundamental ideas of coding as the rules regulating online environments and programming as a set of instructions—the connection between technology and the arts. Lev Manovich splits the digital arts into two categories in his 1996 article "The Death of Computer Art": "Duchamp-land" and "Turing-land," the latter named for the British mathematician and "father of computer science" Alan Turing (1912–1954) (Manovich, 1996) . Manovich considers the just-outlined movements he believed would eventually unite art and technology. These movements peaked in intensity throughout the 1990s. Duchamp-land is, on the one hand, the conventional art world of esteemed publications, galleries, and museums, along with the creative ideas and methods that are supported by them. Conversely, Turing-land describes the emerging field of computer art that was linked to the International Symposium on Electronic Art (ISEA), Ars Electronica, and other prominent gatherings around 1995. (Jeknic 2014).

Advancements in computer software, defined as programs that control a computer's CPU and enable the execution of specific functions, are also responsible for developments in the digital arts. Lev Manovich contends that "data searchable structures" and "algorithms" are the two types of software artists use. In the first, the artist creates "hypernarratives" with no beginning or conclusion, navigating a path via a database, a software program with non-hierarchically presented things. (Manovich, 2001) Later, in 2001, 'Lev Manovich' proposed five New Media principles: Numerical representation, Modularity, Automation, Variability, and Transcoding in his book 'The Language of New Media' (Fig. 1). (Madeleine Sorapure, 2003)

Lev Manovich's approach to new media art bridges this historical development with the rise of digital arts. His classification of digital arts into "Duchamp-land" and "Turing-land" represents the divide between traditional, institution-based art and the emerging field of computer-driven art. Manovich's new media principles further explain how art creation has changed due to digital technologies, distribution, and experience. By highlighting the importance of algorithms, coding, and the database as creative tools, Manovich extends the legacy of Dadaism into the digital realm, showing how both chance and order, as seen in the Dada movement, shape the world of digital art. The integration of these principles emphasizes the convergence of art and technology, leading to innovative and interactive artistic practices in the digital age.

Developments in Digital Arts

The evolution of digital arts, from Dadaism to contemporary public art, shows how technology has progressively intertwined with artistic practices. The Dadaists' embrace of techniques like collage, photomontage, and audience interaction reflected a fusion of art and life, laying the groundwork for integrating technology into creative processes. This historical shift anticipated how digital arts blend real and virtual worlds, challenge traditional artistic boundaries, and merge various media forms like graphic art, animation, and video. Just as Dada blurred the lines between mediums and engaged with society, modern digital art continues this tradition by using digital tools to create hybrid realities and reimagine what art can be. Technology, particularly digital imaging and photography, plays a dual role as both a tool for traditional art forms and a medium for new forms of expression.

To be more precise, Dada, Fluxus, Conceptualism, and public art are all sources of inspiration for new installations and art forms, which in turn influence the digital arts. Rather than focusing on art's conventional, tangible objects, these early to mid-20th-century art movements investigate ideas, procedures, events, and audience participation. (Jeknic 2014). Digital artists like Stelarc explore embodiment and question how people relate to emerging technologies and virtual worlds. Advancements in computer software, defined as programs that control a computer's CPU and enable the execution of specific functions, are also responsible for developments in the digital arts. Many ways to change an artwork are made feasible by digital technologies, leading to the synthesis of art forms and the integration of media. Installation, music, sound, cinema, video, animation, internet art, software art, and virtual reality-based works are examples of digital technology used as mediums. (Stelarc & Smith, 2005)

Figure 2. Digital arts initiatives are shaped by coding and programming, connecting technology with the arts

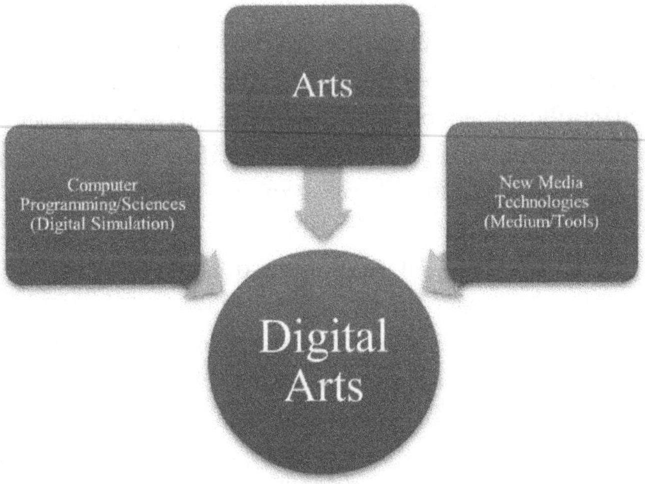

(Jeknic 2014);– Illustrated By Author

The rise of digital arts, influenced by movements, introduced a shift from traditional, static art forms to more dynamic, process-driven, and participatory works. These early movements laid the groundwork for contemporary digital artists who utilize emerging technologies to push the boundaries of art and interaction. With software and digital media advancements, artists blend various forms—installation, video, and virtual reality—into immersive experiences that redefine the artistic medium and engage with broader cultural and social issues. This evolution in digital media parallels the growing need for sustainable approaches to cultural heritage preservation.

Just as digital arts synthesize multiple media forms, digital technologies can be employed in heritage conservation to create virtual reconstructions, interactive databases, and immersive experiences that preserve cultural assets for future generations. By integrating digital tools into conservation efforts, we can document and actively engage the public in maintaining endangered heritage, fostering a deeper understanding and appreciation of these cultural legacies. (Kryvuts et al., 2021). This convergence of art, technology, and heritage conservation reflects a broader movement toward sustainability, ensuring that cultural heritage remains relevant and protected in an increasingly digital world.

3. SUSTAINABLE APPROACHES TO DIGITAL CULTURAL HERITAGE PRESERVATION

The term "cultural heritage" was first used in the middle of the 20th century, primarily by organizations concerned with protecting it. One such organization is the United Nations Educational, Scientific, and Cultural Organization (UNESCO), which defines it as "all tangible and intangible cultural expressions" in a document that was produced as a result of the 1972 Convention for the Protection of the World Cultural and Natural Heritage, which was held in Paris. (UNESCO, 1972) . The activities, representations, expressions, and knowledge that a nation or region acknowledges as a component of its cultural legacy are defined by UNESCO in the same document as intangible cultural heritage. (Unesco, 2023)

The need to protect cultural heritage from threats imposed by human activity and the natural world results in an increased focus on its conservation and condition assessment. With significant cultural assets that must be conserved and located in seismically vulnerable areas, protecting cultural heritage becomes challenging for many Mediterranean countries. Several stakeholders have focused on the relationship between cultural heritage and sustainable development for several years. Consequently, cultural heritage was formally incorporated into the sustainable development agenda in September 2015, highlighted within the Sustainable Development Goals (SDGs) established by the United Nations (UN). (McDermott et al., 2017). Grounded in this perspective, the chapter explores the strengths and limitations of different disciplines in contributing to attaining the Sustainable Development Goals (SDGs) through cultural heritage conservation and management (CHC&M).(Khalid, 2021)

The devastation caused by significant seismic disasters has highlighted the necessity of reducing hazards and preserving artistic and architectural heritage. However, these objectives are challenging to accomplish because of the characteristics of this cultural evidence and the relationships between structural and non-structural elements—that is, between creative assets and structures. (Unesco, 2023) .National and international recommendations have been approved to address these problems and provide appropriate approaches for the knowledge and analyses of cultural heritage for future sustainability.(ICOMOS, 2003)This heritage includes magnificent architectural creations, museum-displayed artwork, and priceless items like sculptures, frescoes, and stuccoes, as well as other elements essential to architectural legacy. (Marra et al., 2021)

Digital Cultural Heritage and Disaster Cycle Implication

Disasters pose significant threats to both tangible and intangible natural and architectural heritage. Whether man-made (such as war, armed conflict, terrorism, arson) or natural (like floods, earthquakes, and volcanic eruptions), they threaten the survival of local populations and cultural heritage while causing irreversible material and financial loss as well as a loss of memory that coexists with the heritage. (Khan, 2008). Catastrophic disasters, both natural and man-made, can affect World Heritage properties and heritage sites, putting their integrity at risk and perhaps reducing their value. These exceptional properties and assets have an enormous detrimental effect on local and national populations when they become unavailable or damaged because of their socioeconomic and cultural value. A community's identity is derived from its heritage, which also serves as a gateway to teachings, abilities, and information. (Wali Shah et al., 2024). Therefore, it is essential to preserve these locations and make sure that future generations can take advantage of them. Heritage properties enhance social cohesiveness and pride in the nation and community.

Figure 3. Architectural Disaster cycle flow sheet diagram and application of digital technologies

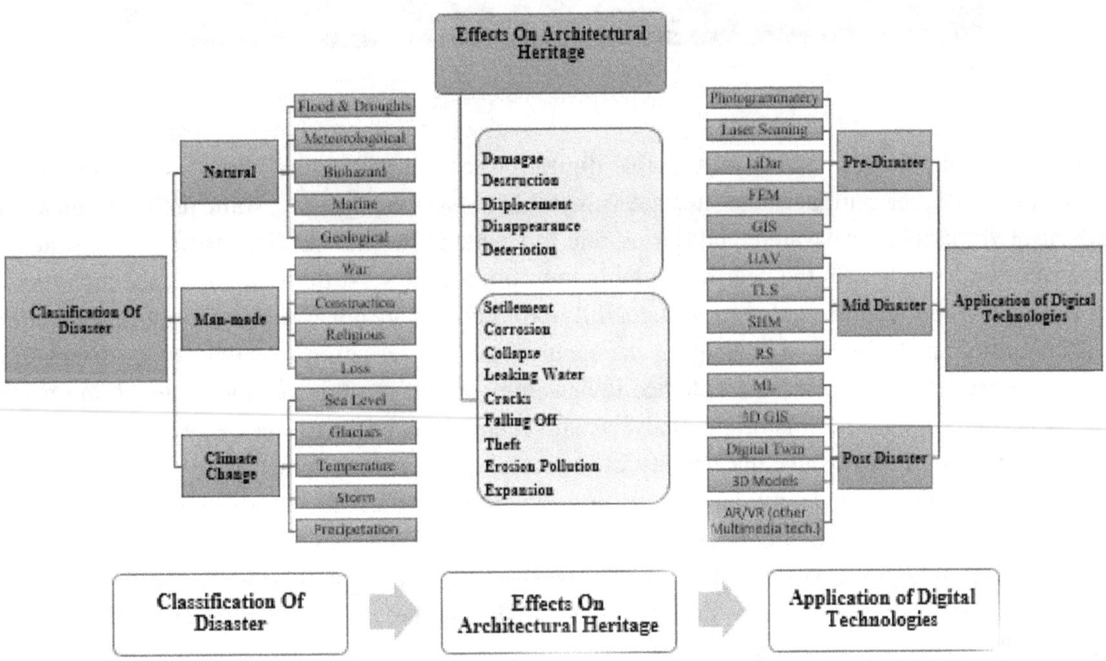

(Li, 2023)

"Digital cultural heritage—DHT" is accepted as standard terminology for identifying digital heritage techniques within a historical context (Baucom, 2019). Architectural heritage digital preservation is a relatively multidisciplinary field that spans multiple disciplines. Three stages need to be addressed to protect architectural heritage from disasters: pre-disaster, mid-disaster, and post-disaster. While corrective repairs during disasters and restoration afterward are crucial, pre-disaster activities concentrate on preventive maintenance.(Wali Shah et al., 2024). A study by Li, Du, et al. (2023) presents a flow sheet diagram (Fig.3) classifying disaster causes into three categories—natural, man-made, and climate-related. (Li, 2023) The graphic also illustrates how they affect architectural heritage, highlighting the importance of digital technologies for efficient management and mitigation throughout crisis cycles.

Figure 4. Cultural Heritage Conservation Complexity Data Acquisition Model Key Factors

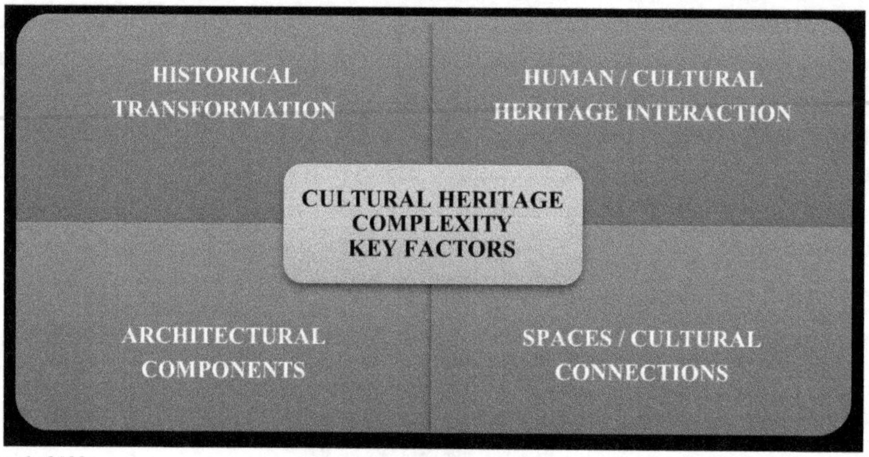

(Bevilacqua et al., 2022)

This discussion focuses on the sustainable digital protection of cultural heritage. In selected geographical sections, the built heritage sites are vulnerable to various disasters; some archaeological sites are excavated, partially excavated, and buried due to financial limitations. The artifacts excavated are transferred to the shelves of the nearby available museum displays. At the same time, the architectural features and the implicit knowledge are still not fully explored and are not available for future generations.

This study aims to assist in evaluating the security and preservation of cultural assets and determining interventions to protect and enhance their value. To achieve this, a strategy involving various skills is needed, integrating historical analysis with meticulous examination of geometry, materials, and deterioration state to reduce uncertainty in structural and seismic evaluation. For this, the cultural heritage conservation complexity data acquisition model is a critical factor that needs to be followed while working on conserving architectural heritage. (Fig. 4)(Bevilacqua et al., 2022). Digital arts and new media technologies are crucial for sharing, preserving, and educating the next generation about sustainable cultural heritage preservation worldwide. (Mendoza et al., 2023).

4. HERITAGE PROTECTION USING ADVANCED VISUALIZATION TECHNOLOGIES

Digital Arts and Interactive Preservation of Cultural Heritage

The architectural heritage's wisdom, principles, and significance depend on the natural world to endure and contend with various factors. A multidisciplinary approach is still required to address the three crucial avenues of heritage protection and technological involvement in this endeavor. The first is an accurate disaster prediction methodology; the second is an early warning system for heritage damage; and the third is an intelligent monitoring system that integrates human-computer interactions (HCI). (Mendoza et al., 2023). Many state-of-the-art digital technologies are used to gather information from these settlements and buildings and to disseminate knowledge.

Rapid advancements in multimedia technology have fueled the emergence of digital art in the late 20th and early 21st centuries, leading to its widespread dissemination globally. This unique art form integrates technical, artistic, computer sciences, and value-oriented elements and has proven exceptionally impactful (Fig. 2). A focus on developing a distinct multimedia "language" has garnered significant attention. (Charles Ryan 2014; Jeknic 2014). This study explores the potential of utilizing immersive methods, such as videos, 3D mapping, and digital simulations, to address challenges related to preserving cultural heritage items. It delves into the use of these techniques in designing exhibition spaces and projecting visuals onto prominent buildings. The nature of the exhibition space determines the content and artistic style of the visual experiences. A deeper scientific understanding of interactive art is essential to enhance professional practices in cultural heritage preservation. (Kryvuts et al., 2021)

Around the world, various technical methodologies have been employed globally to preserve cultural assets and provide insights into technological strategies for advancing heritage promotion from a sustainable approach, have some direct relationship with the promotion, dissemination, and appropriation of heritage in general, to make available to those interested, support to adopt good practices in future research work on the subject with the help of visual interactive representations to make it more interactive (Fig. 4) (Bec et al., 2019; Little et al., 2020) The interplay between architectural elements and historical changes makes it challenging to comprehend the built heritage, which may not always be readily accessible or still in existence. Over the past two decades, the widespread integration of Information and Communication Technologies (ICT) in heritage has significantly transformed access to knowledge, enjoyment, and enrichment of cultural heritage. (Bevilacqua et al., 2022). These tools assist interested parties in implementing best practices for upcoming research on the subject.

Visual Representation / Interactive Art (Human-Computer Interaction HCI)

Interactive technologies have significantly impacted the dissemination of information, the transmission of events through artistic and computer-based mediums, the replication of cultural values using digital simulation techniques, and the development of new spectacular scenarios that mirror historical and cultural events since the turn of the twenty-first century. Many scientists have been preoccupied with finding a solution to this issue from the perspective of how innovative approaches to disclosing exhibition content by various techniques have evolved. (Kryvuts et al., 2021)

The result of a continuous conversation between the artist, the work, and the audience—whether in person at an exhibition or virtually in an online setting—*is the 'interactive quality of digital arts.'* (Bec et al., 2019).Interactive installations, like Landscape One, require viewers to physically interact with the work of art and transition from passive observers to active participants who make decisions; another term, *'Human-Computer Interaction' (HCI), is used for interactive art.* Thus, one goal of interactive projects is to create environments that encourage innovative conversations and allow projects to proceed unexpectedly.(Little et al., 2020).An essential interactivity component is the possibility of unexpected or distant audience participation. (Charles Ryan, 2014)

Figure 5. A model with four stages for preserving heritage and Transforming it into digital interactive experiences

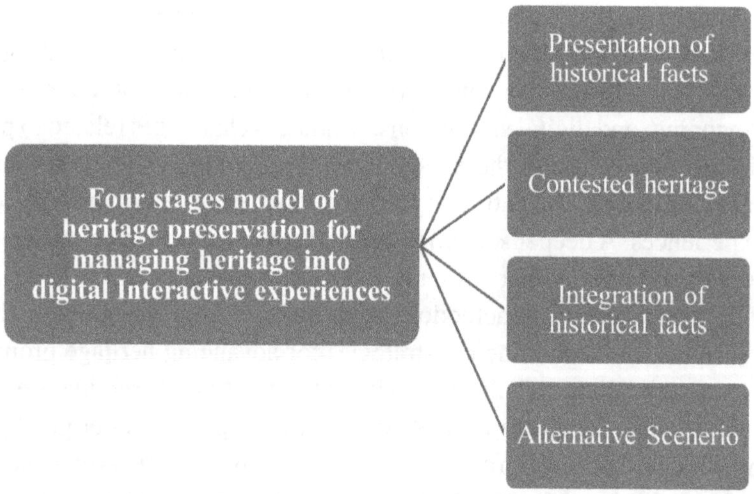

(Bec et al., 2019; Little et al., 2020)

As technology advances, there is a growing focus on integrating architectural and natural elements through virtual representation to shape the emotional atmosphere of public spaces. (Bec et al., 2019). This involves a complex blend of architectural-historical facts and digital arts (Fig.5). When creating interactive works, 3D displays aesthetically combine personal and social spiritual space, influencing individual preferences and promoting an active social presence. The transformation function, popular among modern audiences, transcends boundaries of knowledge and cultural information, enabling the portrayal of narratives rooted in historical, cultural, or mythical events.(Little et al., 2020)

Today's primary objectives include converting cultural heritage into digital formats, employing art for virtual portrayal, constructing cyberspace, and generating 2D and 3D visualizations. These ideas influence the advancement of fresh and sustainable pathways in the humanities, especially in art education, and contribute to conserving global cultural heritage via virtual tours. (Dean, 2002) And information transfers to future generations.

5. PRESERVING, SIMPLIFYING, AND DISSEMINATING KNOWLEDGE FOR FUTURE GENERATIONS; DIGITIZING CULTURAL HERITAGE AND FORMING A COLLECTION OF TRADITIONAL INFORMATION

Submersive Methods For Sustainable Architectural Heritage Preservation

The immersive method, also known as submersion, revolves around the impact of being fully immersed in the environment of the depicted era or event. This approach produces an artistic representation that sparks the audience's imagination and invites their involvement. Involving spectators in the overall exhibition space can convey the significant narrative of the display. Through interactive displays, the concept aims to stimulate viewers' interest and evoke their own experiences. What sets the submersive method apart from other approaches is its emphasis on drawing the viewer into the exhibition's storyline and fostering complete immersion. (Kryvuts et al., 2021). A wide-ranging, queryable, and interoperable 3D data system can be accessed using active or passive survey approaches, providing a multi-scale reality-based context observation. A virtual replica of cultural heritage assets can be created, and data can be transmitted efficiently, contributing to integrating iconographic and bibliographic data to generate interpretive 3D models and comprehend architectural modifications (Digital Twin, DT). Within the last ten years, there has been a dramatic advancement in the interaction and immersion of digital models in virtual worlds. (W. Kritzinger, 2018).

Knowledge Visualization Theory Application in Heritage Protection

This section describes the approach taken in this study for sustainable digital architectural heritage protection and usage based on knowledge visualization. Digital preservation is achieved by utilizing digital art technologies. According to the study by Zhang (2022), knowledge acquisition, knowledge reconstruction, knowledge innovation, and knowledge management were all incorporated into this approach, which used digital media technologies and knowledge visualization for interactive art theory (Fig. 6). Innovation and knowledge reconstruction are applied to digital use.(Zhang et al., 2022)

Figure 6. The general technical framework of this study is for digital preservation and the virtual presentation of architectural heritage conservation

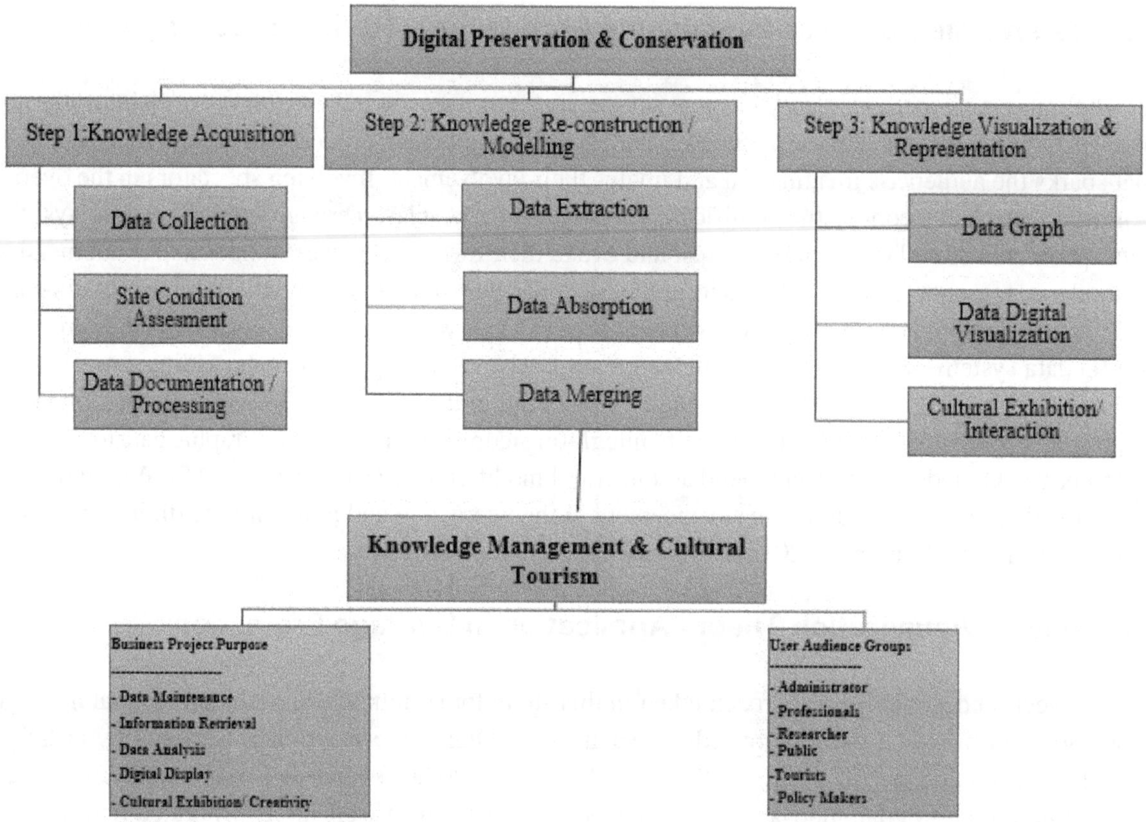

(Zhang et al., 2022)

The scientific community links the necessity of heritage object documentation to the need for preservation, valuation, and protection. The international literature lists augmented or virtual reality, digital archiving, conservation, 3D printing of replicas, real-time documentation of archaeological excavations, monitoring, and digital restoration as some of its areas of applicability.(Hong et al., 2023). Methods for digitizing tangible heritage in three dimensions are constantly being updated, including computer systems, active and passive sensors, data-gathering techniques, and applied algorithms. These advancements make it possible to digitize historical materials with greater precision and accuracy at faster automation and processing speeds, following new media principles. (Adamopoulos et al., 2021)

Combining qualitative and quantitative data, this knowledge model based on 3D digital archives offers significant advantages for managing, assessing, and protecting architectural history. Within this context, which precise virtual models characterize, the primary areas of interest for research are data representation and management. (Yang et al., 2015). The relationship between qualitative and quantitative data, specifically between conventional methods and creative approaches established in digital technologies and the arts, promotes the creation of Integrated Informative Systems (IISs) that aid in conserving and assessing cultural heritage.(Marra et al. 2021).

Figure 7. Digital Preservation to the virtual presentation of Architectural heritage Conservation

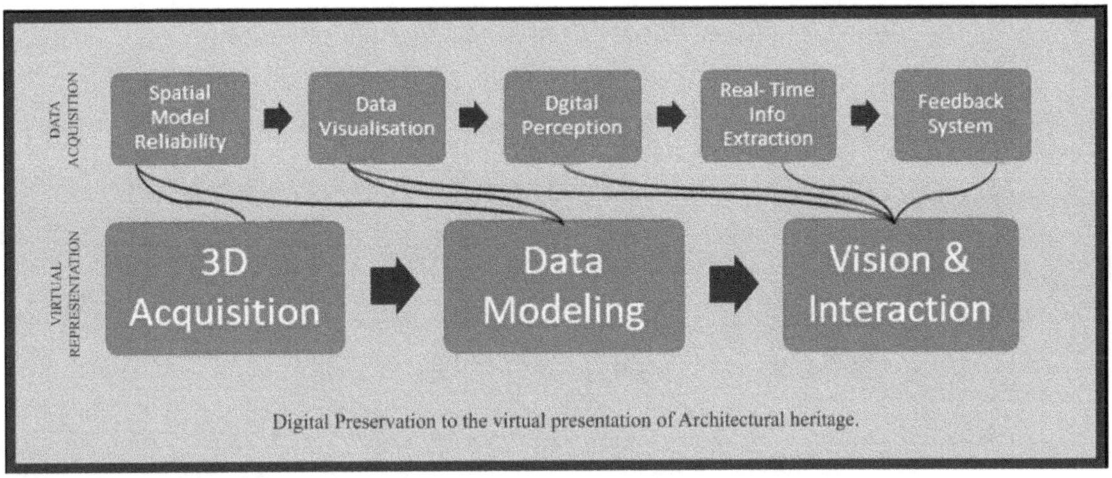

(Bevilacqua et al., 2022)

A.–V. Kerlow covers the fundamental aspects of producing three-dimensional animation, including modeling, visualization, and digital work execution, demonstrating practical design implementation with various techniques. (GERE 2006). Interactive visual representations are artistic images used on contoured surfaces, like architectural landmarks, to give the impression of multidimensional movement. This innovative technique has opened up new possibilities for representing reality and establishing a shared visual language for design solutions (Fig. 7)(Bevilacqua et al., 2022). Scanning and image-based modeling techniques have taken the lead for large-scale applications, including examinations of archaeological sites, artifacts, or architectural elements, since these methods are fast to gather, reduce costs, require minimal time and resources to process, and have repetitive workflows. (Adamopoulos et al., 2021). Media design is growing in modern culture, and rapid changes drive its optimization.

Technical Review of the Literature

The technical factors to consider while evaluating various investigations are the technologies used to maintain tangible and intangible cultural property and identify the types of heritage most entwined with technological processes. It should be noted that, given the distinction between tangible and intangible cultural heritage, the majority of interventions are focused on tangible cultural heritage, which includes things like churches, museums, buildings, sculptures, and paintings, among other things. Of these, 131 articles (92%) are related to tangible cultural heritage (Fig. 8) (Ricca, et al., 2020)

Multiple studies show several countries have committed to preserving tangible and intangible cultural heritage. A particular study, which was presented in a systematic literature review on 3D technologies used for the preservation of tangible and intangible cultural heritage, analyzed international articles and found that the most popular technologies for preserving tangible cultural heritage are motion capture systems, augmented reality, virtual reality, 3D modeling, 3D scanning, and 3D visualization. (Maffei, Canevese, De Gottardo, & Pizzol, 2019) . It has been implemented across multiple buildings and structures, including castles, churches, sculptures, and archeological sites, primarily to conserve cultural assets.

Figure 8. Type Of Cultural Heritage Entwined with Digital Preservation

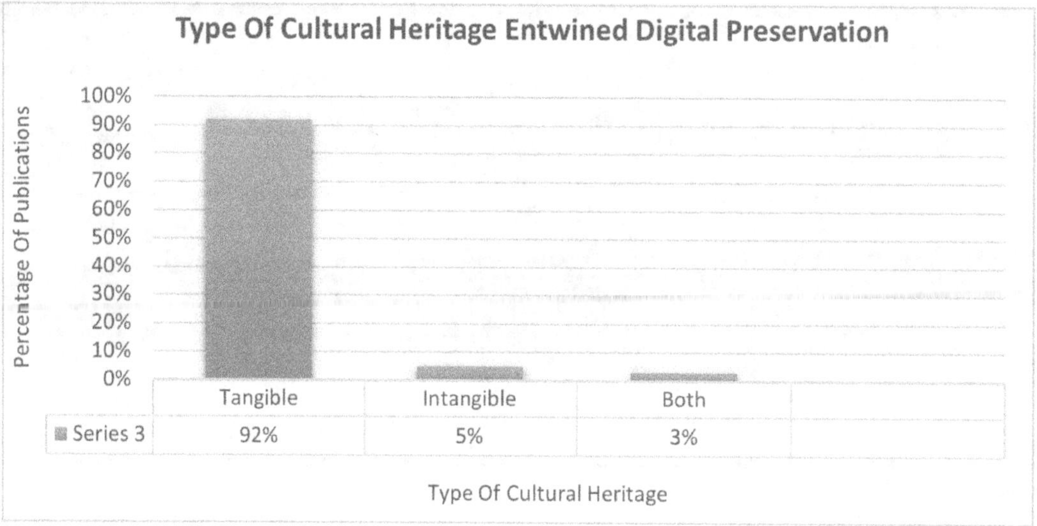

(Ricca, et al., 2020)

Table 1 illustrates that the most popular technologies for preserving cultural heritage are 3D modeling (44%), followed by virtual and augmented reality (15%) and AR/VR (15%), according to Mendoza 2023, a review of 101 articles on technological intervention. Information systems, social networks, gamification, digital restoration, and various web technologies are also used worldwide to conserve cultural property. (Mendoza et al., 2023)

Table 1. Type Technologies Used for Digital Preservation of Cultural Heritage (Mendoza et al., 2023)

Type Of Technologies	Number Of Papers	Percentage
3D Digital Technologies	45	45%
Other Technologies	34	31%
3D AR/VR	17	17%
IOT	04	5%
Sensors	02.	2%.

Visualizations illustrating abstract connections contribute to speeding up processing and lessening cognitive stress. They facilitate the communication of knowledge, its sharing with others, and the development of a thorough comprehension of the topic. Digital media arts visualizations encourage creativity by providing methods that utilize the creative potential of imagery, such as visual metaphors and symbolic representations.

7. DIGITAL ARTS IN ARCHITECTURAL CULTURAL HERITAGE PRESERVATION: CASE STUDIES OF SUCCESSFUL IMPLEMENTATION

Use of 3D Mapping and Video Projection

Video projections broaden the possibilities for transforming the exhibition space selected for the design by producing stunning optical illusions with the play of light. Thus, we can identify the self-regulation mechanisms within the medium's exposure system by observing the systems-forming features incorporated in media design. These include content that reveals the idea of an artistic multimedia work, elements that determine which parts create the system itself, the form that reveals the degree of harmony of images, a function that allows one to identify the nature of the system as a whole; and the submersion method, which controls the level of immersion in the interactive world of the video projections.

Screen City Festival 2013 and 2012 Exhibition, Bordos

The artists use professional-grade artistic tools to blend static and dynamic data. Exhibition designers influence the narrative rhythm in a multimedia digital setting by working within the constraints of time and space. Interactive audio-visual technologies maximize impact, appealing to a wide range of audiences. The design integrates diverse elements into a cohesive whole, influenced by modern technology. The responsibility of maintaining the creations of great masters from the past and present, as well as the processes of translating cultural values, are impacted by digital art. Using transformation, modularity, and cyclicity methodologies, this indicator enables the designer to investigate the salient features of the natural or architectural object using 3D video mapping (Fig. 9-10). This case study shows how a multimedia video sequence intended to be projected onto famous buildings' façade is made up of multiple distinct images to produce a stereo effect. (Kryvuts et al., 2021)

Figure 9. 3D video mapping design transformation techniques by BORDOS.ARTWORKS at Screen City Festival 2013, Stavanger, Norway.

(Kryvuts et al., 2021)

Figure 10. 3D video mapping design transformation techniques and stereo 3D mapping at Mapping Festival 2012, Geneva (Museum of Art and History)

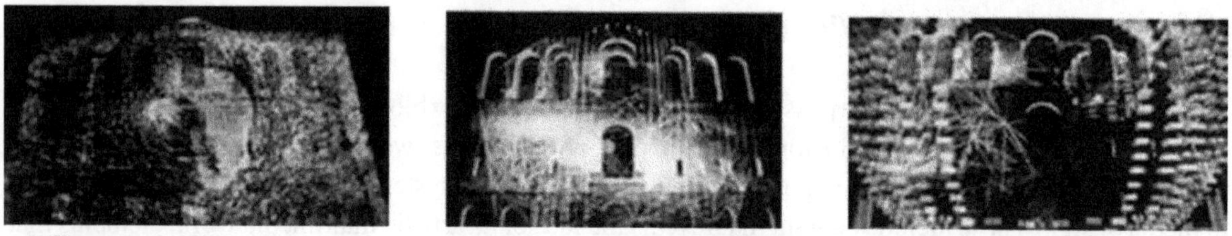

(Kryvuts et al., 2021)

Leading artists today are creating design solutions for visual representation of realistic, fantasy-based, or abstract artistic concepts in the context of globalization. Exhibition spaces with creative media designs feature interactive multimedia complexes and informational films. They immerse viewers in the realm of digital art, enabling them to journey across historical and contemporary dimensions. 3D video mapping is an important technique for creating both simple abstract and complex-modulated visuals from real architecture.

BYBLOS 3D Project Mapping on the Water

The yearly free summer lighting and technology event on this UNESCO World Heritage site, the Byblos-Jbeil projection mapping show on the old port castle, is becoming increasingly attractive. With the Municipality of Byblos' initiatives, there is a new and viable model for open public events in Lebanon that should inspire similar ones around the nation. Significant historical events influence not only the character of the design concept for 3D video projection but also current developments. Every historical era of the city was represented, effectively capturing the atmosphere of that era. (Fig. 11). The 6-minute show's flow was thoughtfully crafted to lead viewers into a succession of surprising visual cliffhangers and maintain their attention. The 3D image's composition has unique elements related to the geographic, cultural, and other traits of the area in which it was made. (Fig. 12)

Figure 11. 3D Project Mapping on the Water in BYBLOS showcased every historical era of the city

(Kryvuts et al., 2021)

Figure 12. Original exhibition space: 3D Project Mapping on the Water in BYBLOS

(Kryvuts et al., 2021)

The BYBLOS 3D Projection Mapping event honoring the 8000-year history of Byblos-Jibail, a UNESCO World Heritage Site and one of the most ancient cities in the world, is a spectacular example of water video. The 3D initiative aimed to promote modern technology advancements while also positively affecting tourism. The design by Studio Mr. White and Minus5 Architects prioritized Byblos' architectural legacy, selecting the old port's medieval stronghold as the backdrop for a specially created 3D film that was projected above the ocean (Fig. 13). The design's narrative concept aimed to convey each era of the city's history by illustrating significant events in a way that captured the spirit of the eras on screen. The way the plot developed kept viewers interested throughout the show by showcasing several surprising visual multimedia compositions. (Kryvuts et al., 2021)

Figure 13. BYBLOS 3D Project Mapping on the Water, Byblos-Jibail, Lebanon

(Kryvuts et al., 2021)

The architectural environment is profoundly shaped by socio-cultural reality, as seen through interactive artistic images. 3D video images play a crucial role in enhancing exhibitions' overall aesthetic while contributing to the sustainable, interactive preservation and elevation of irreplaceable cultural heritage. This, in turn, allows for updating the cultural and historical context of events.

Digital Twin with VR and AR Technologies

The purpose of virtual and augmented reality is growing, and it is quickly becoming a recognized and crucial component of the domain of cultural computing (Greengard., 2019). At different stages of design, knowledge, and learning, virtual representation of cultural heritage can have an impact on asset understanding, time, cost, and quality management (M.K. Bekele, 2018). The user acts as a dynamic participant, effortlessly engaging with cultural heritage and offering newfound perspectives on culture to a wide audience. This fosters interdisciplinary study and promotes the spread of knowledge. The digital representation of cultural heritage through VR/AR technology involves spatial model reliability, data visualization, digital perception, real-time information extraction, and feedback systems on virtual experiences. It encompasses building 3D models, data processing, vision systems, and interactivity, linking cultural elements to the fields of architecture and digital/computer science.(Bevilacqua et al., 2022)

Content generation in cultural heritage is central and critical as it involves experiential and cultural aspects. Several applications of digital twins for virtual representation in the field of cultural heritage emphasize the potential of VR and AR to assist in understanding and interpreting architectural legacy. These scenarios encourage reflection on how digital arts, interactivity, and immersive tools relate to one another, which is advantageous for future growth and cultural legacy sustainability.

Qinglian Temple in Shanxi, China

The Qinglian Temple in Shanxi, China, was selected as the case study for this investigation; the experiment was initiated in 2019 to demonstrate the methodology better. Situated in the southeast of Shanxi Province, China, on Xiashi Mountain, is Qinglian Temple; the temple is separated into the Upper Temple and Lower Temple, which are 500 meters apart and have a combined space of 16,000 square meters (Fig. 14). Known as the "Capital of Buddhism," the temple was one of the first Buddhist temples in China, having been constructed between 550 and 559 A.D. Along with 25 historic structures, there are cultural artifacts such as paintings, stone tablets, painted sculptures, and buildings. Specifically, many mixed-use buildings were constructed at various times, and their blended styles offer helpful information for architectural studies. (Li & Gao, 2003)

Figure 14. The Qinglian Temple location UAV image

(Zhang et al., 2022)

Qinglian Temple has great artistic, technical, and historical significance overall. Several buildings date back more than 1,000 years. Building styles from many eras were expertly blended to create distinctive designs with a refined appearance. Additionally, the building's combination of colored sculptures, brickwork, and stone carvings creates a harmonious and vibrant structure with unique artistic features. The temple received recognition as a Chinese national historic protection unit in 1988. It is now known as National Scenic Spot 4A.(Shen et al., 2022). However, because of their remote mountain position, adverse environmental circumstances, and delicate building designs, these architectural legacies continue to be in danger of vanishing, and the general public cannot comprehend their underlying culture. (Zhang et al., 2022)

Figure 15. Point cloud processed data using laser scanning

(Zhang et al., 2022)

The Qinglian Temple in Shanxi, China, is a case study demonstrating the Digital Protection and Utilization based on Knowledge Visualization (DPUKV) approach developed in this study. Knowledge was gathered using sophisticated data collection methods like terrestrial laser scanning (TLS), UAV photogrammetry, and ground photography. All 25 of the temple complex's buildings' point cloud data, UAV photos, and high-definition ground photos were gathered using these methods (Fig. 15). To ensure precise capture of geometric and textural information, architectural elements—including ornamentation and structural components—were carefully documented. The acquired data in this case study was transformed into CAD drawings and 3D models, enabling a detailed and accurate representation of the intricate architectural features and structures of the Qinglian Temple. Cross-platform applications thoroughly present architectural heritage through various platforms, including text, photographs, videos, animations, and augmented reality. Additionally, a cross-platform application was created using Vuforia and Unity3D. This application can be installed in museums and other locations and is extensively used as an app on smartphones, tablets, and PCs. (Fig.16). (Zhang et al., 2022)

Figure 16. Virtual representation of Cross-platform applications and AR/VR interaction

(Zhang et al., 2022)

This promotes the cultural transmission of the Qinglian Temple's historical, technical, and artistic significance to a broader audience and assists in preserving its physical buildings. Thus, these digital tools fulfill two purposes, they increase public awareness and appreciation of architecture while offering a way to preserve it.

Monumental Complex of the Charterhouse of Pisa in Calci (VR)

One of the most important monasteries in Tuscany, the Charterhouse of Pisa near Calci (Fig.17), is the subject of this case study. It is an enormous complex. The charterhouse was established around 1360 and underwent several changes known to have occurred between then and the end of the 18th century. (M. A. Giusti, 1993). The National Museum of the Monumental Charterhouse of Calci and the University of Pisa Natural History Museum are two significant museums housed in the Charterhouse, which lost its original purpose in 1962. The work discussed here aims to improve the National Museum of the Charterhouse visits using immersive virtual reality experiences. The University of Pisa is funding a more extensive, multidisciplinary research project that includes this investigation to conserve and improve the Charterhouse. The development of the immersive experiences detailed here began with a thorough historical investigation and a 3D digital survey of the entire complex, which was conducted using LIDAR and 3D photogrammetric methods, among other activities made possible by the research project. (Bevilacqua et al., 2022)

Figure 17. Monumental complex of the Charterhouse of Pisa in Calci, Tuscany

(Bevilacqua et al., 2022)

The first preliminary phase of the study, which focused on the evolution of the monastery, its most valuable architectural and decorative elements, and the individuals who have historically played a significant role in the history of the Charterhouse, helped define the goals and contents of the new visit tour. A new tour trail was then developed to enhance the immersive interactive experience and increase the number of spaces that could be visited to promote tourism while heritage preservation.

Figure 18. 3D Graphic animation of 3D models that replicate an exterior view of the complex's different construction phases

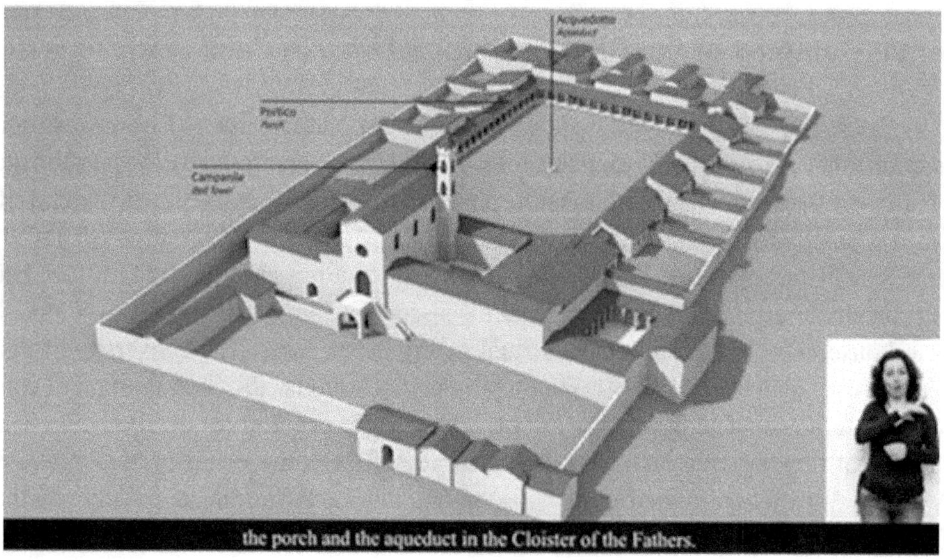

(Bevilacqua et al., 2022)

Tourists navigate the path with audio-visual guidance on tablets and smartphones. Based on the material that will be presented, all of the areas have been divided into three categories: mixed information, art and architecture, and monastic life. As a result, the research project concentrated on creating the three VR applications listed in the comprehensive plan:

- The opening animated video, which appears at the museum's entrance in the Chapel of St. Sebastian, features graphic animation of 3D models that replicate an exterior view of the complex's different construction phases. The video is narrated in Italian and has multiple subtitles (Fig. 18).
- The virtual reality experience focused on the 18th-century Pasquale Cioffo quadratures frescoes in the Chapel of St. Antony. It attempted to understand better the geometric principles employed in the painted composition and offered a three-dimensional reconstruction of the imaginary space portrayed that visitors could access and experience through the use of a virtual reality (VR) helmet;
- The Chapter in the Cloister offers a virtual reality (VR) activity that requires a VR helmet. It immerses users in virtual environments that depict the cloister's layout during some of its most significant historical periods (Fig. 19).

Figure 19. The chapter in the cloister offers a virtual reality (VR) activity that requires a VR helmet

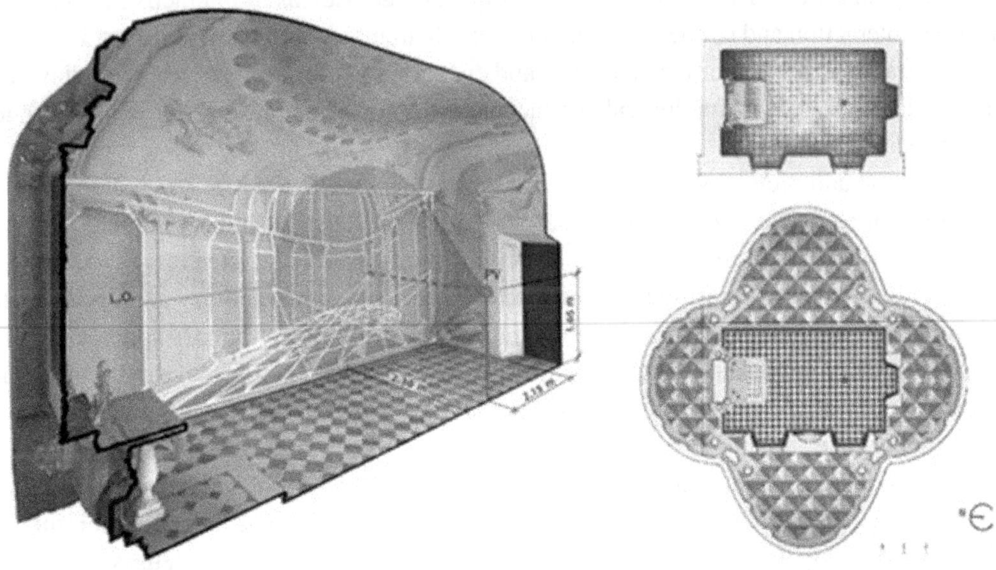

(Bevilacqua et al., 2022)

Moreover, it immerses users in virtual environments.(Bevilacqua et al., 2022). Blender, an open-source program, was used to create 3D modeling. Audacity was used to process informative texts; Adobe After Effects was used for the introductory video in the Chapel of St. Sebastian. Unreal Engine was utilized to create the immersive experiences in the Chapel of St. Anthony and the Cloister of the Chapter. A larger multidisciplinary research project for the preservation and improvement of the Charterhouse of

Calci and its Museums, led by M.G. Bevilacqua and supported by the University of Pisa from 2018 to 2020, includes the study of virtual reality experiences in the National Museum of the Monumental Charterhouse of Calci. The University of Pisa's ASTRO Laboratory produced LIDAR and digital photogrammetric surveys. Prof. E. Karwacka of the University of Pisa developed the historical analysis of the Charterhouse. (Bevilacqua et al., 2022)

8. CONCLUSION

Digital art significantly impacts how modern culture is informed about values. Kryvuts and Gonchar state that creating an image for HD graphics begins with synthesizing sound and images and progresses to creating a new structural model of environmental reflection. Multimedia technologies are then used to combine these models into a single image, document, and game. (SALGADO, 2009).

Exploring and preserving architectural heritage is increasingly being facilitated by immersive visualization and interactive experiences. These methods expand the audience for architectural preservation and enhance our understanding of architectural environments and their evolution. The development of new evaluation criteria and the rise of digital arts culture are propelling the creation of innovative approaches to optimize media design projects. Furthermore, contemporary interactive communication platforms promote cross-cultural understanding, while digital arts influence various disciplines.

In recent years, three-dimensional art created by artists from diverse cultural backgrounds has gained significance. Digital art enables viewers to deeply immerse themselves in interactive experiences, fostering personal interaction and connecting creators with their audience. This integration is instrumental in passing cultural legacies to future generations and encouraging tourism. The use of audiovisual tools in arts and media, alongside technological advancements, holds great promise for conserving and disseminating cultural heritage.

The preservation and protection of architectural heritage are facing increasing challenges due to the passage of time. In response, three-dimensional digital art and other technologies play a vital role in conserving and disseminating cultural heritage. By creating immersive and interactive experiences, these technologies preserve historical artifacts and educate and engage contemporary audiences, fostering a deeper appreciation and understanding of cultural heritage. This multidisciplinary approach ensures that cultural material remains accessible and relevant, fostering a dynamic and inclusive relationship between the past and the present and setting the stage for further advancements in the sustainable preservation of architectural heritage.

The chapter concludes that although digital art technologies present encouraging potential for safeguarding and passing down architectural heritage, their present uses still need to realize their full potential fully. This gap is filled by the suggested knowledge visualization approach, which uses digital advances like VR and AR and tools like mind maps to translate implicit architectural information into transparent, understandable formats. The approach's success, as shown in the case study of Qinglian Temple, highlights how adaptable it is for more extensive cultural heritage protection. To increase the reach and influence of digital heritage conservation, future studies should concentrate on improving knowledge translation methods and incorporating cutting-edge technologies like the metaverse.

9. FUTURE RECOMMENDATION

Based on the literature review, several recommendations for future research and practice have been identified. The first recommendation pertains to using technology to conserve and disseminate tangible and intangible cultural heritage. Enhancing educational processes is suggested to safeguard cultural heritage. Currently, 92% of technological solutions primarily cater to tangible heritage. Thus, it is imperative to extend these technologies to intangible heritage. Creating immersive experiences using VR, AR, and 3D mapping can help engage broader audiences and foster an appreciation for cultural heritage.

Heritage tourism has the potential to enhance local communities' social and economic conditions significantly. However, sites must maintain their cultural legacy while becoming viable tourist attractions. This can be achieved by fostering public-private partnerships and securing increased funding to facilitate the integration of digital technologies in heritage conservation for sustainable tourism development. Adhering to these recommendations, digital arts and new media technologies can play a pivotal role in ensuring sustainable preservation and transmission of cultural heritage for future generations.

REFERENCES

Adamopoulos, E., Rinaudo, F., & Ardissono, L. (2021). Geo-Information A Critical Comparison of 3D Digitization Techniques for Heritage Objects †. *ISPRS International Journal of Geo-Information*, 10(1), 10. DOI: 10.3390/ijgi10010010

Bec, A., Moyle, B., Timms, K., Schaffer, V., Skavronskaya, L., & Little, C. (2019). Management of immersive heritage tourism experiencs: A conceptual model. *Tourism Management*, 72, 117–120. DOI: 10.1016/j.tourman.2018.10.033

Bekele, M. K., Pierdicca, R., Frontoni, E., Malinverni, E. S., & Gain, J. (2018). A survey of augmented, virtual, and mixed reality for cultural heritage. [JOCCH]. *Journal on Computing and Cultural Heritage*, 11(2), 1–36.

Bevilacqua, M. G., Russo, M., Giordano, A., & Spallone, R. (2022). 3D Reconstruction, Digital Twinning, and Virtual Reality: Architectural Heritage Applications. *Proceedings - 2022 IEEE Conference on Virtual Reality and 3D User Interfaces Abstracts and Workshops, VRW 2022*, 92–96. DOI: 10.1109/VRW55335.2022.00031

Charles Ryan, J. (2014). Analogue Angels and Digital Diamonds: Tracing the Origins of New Media Art. In *Philosophy Study* (Vol. 4, Issue 6).

Dean, D. (2002). Museum Exhibition. *Theory into Practice.*

Dore, C., a. M. (2012). Integration of Historic Building Information Modeling (HBIM) and 3D GIS for recording and. *18th International Conference on Virtual Systems and Multimedia*. DOI: 10.1109/VSMM.2012.6365947

GERE. C. (2006). Digital Culture (second edition). *London: Reaction books.*

M. A. Giusti, a. M. (1993). La Certosa di Pisa a Calci. . *Pacini Editore.*

Greengard, S. (2019). *Virtual reality*. MIT Press. DOI: 10.7551/mitpress/11836.001.0001

Haydn, T. (2013). *Using New Technologies to Enhance teaching and learning in History*. Routledge. DOI: 10.4324/9780203075593

Hong, L., Xinyan, K., Sheng, L., & Yangmengliu, Z. (2023). Studying Tian Hock Keng's architectural heritage through the lens of interaction through digital design. *Usability and User Experience*, 110. Advance online publication. DOI: 10.54941/ahfe1003222

ICOMOS. (2003). *ICOMOS PRINCIPLES FOR THE PRESERVATION AND CONSERVATION-RESTORATION OF WALL PAINTINGS (2003) Introduction and Definition.*

Jeknic, O. (2014). New media or new interfaces: The concept of interface in Lev Manovich's' theory of new media. *Kultura*, 145(145), 188–197. DOI: 10.5937/kultura1445188j

Khalid, A. (2021). Conservation Challenges and Emerging Trends of Digital Preservation for UNESCO Architectural Heritage, Pakistan. *Conservation*, 2(1), 26–37. DOI: 10.3390/conservation2010003

Kritzinger, W., Karner, M., Traar, G., Henjes, J., & Sihn, W. (2018). Digital Twin in manufacturing: A categorical literature review and classification. *IFAC-PapersOnLine*, 51(11), 1016–1022.

Kryvuts, S., Gonchar, O., Skorokhodova, A., & Radomskyi, M. (2021). The phenomenon of digital art as a means of preservation of cultural heritage works. *Muzeologia a Kulturne Dedicstvo*, 9(1), 145–156. DOI: 10.46284/mkd.2021.9.1.9

Little, C., Bec, A., Moyle, B. D., & Patterson, D. (2020). Innovative methods for heritage tourism experiences: Creating windows into the past. *Journal of Heritage Tourism*, 15(1), 1–13. DOI: 10.1080/1743873X.2018.1536709

Lovejoy, M. P. (2011). *"Introduction." Context Providers: Conditions of Meaning in Media Arts.*

Manovich, L. (2002). *The language of new media*. MIT press.

Marra, A., Gerbino, S., Greco, A., & Fabbrocino, G. (2021). Combining integrated informative system and historical digital twin to maintain and preservation of artistic assets. *Sensors (Basel)*, 21(17), 5956. Advance online publication. DOI: 10.3390/s21175956 PMID: 34502848

McDermott, D., Binan, D., Bumbaru, D., Jungeblodt, G., Odiaua, I., Soule, J., Kuban, N., Murray, C., Avrami, E., & Wilson, H. Jackson for Australia NSCES, R., Turner, M., Motta, P., Burke, S., & Colletta, T. (2017). *ICOMOS Action Plan: Cultural Heritage and Localizing the UN Sustainable Development Goals (SDGs) Prepared by Ege Yıldırım, ICOMOS Focal Point for the SDGs with contributions from Sofia Avgerinou-Kolonias, ICOMOS Sustainable Development Task Force Chair.* https://www.icomos.org/en/what-we-do/involvement-in-international-conventions/un-sustainable-development-goals-2

Mendoza, M. A. D., De La Hoz Franco, E., & Gómez, J. E. G. (2023). Technologies for the Preservation of Cultural Heritage—A Systematic Review of the Literature. In *Sustainability (Switzerland)* (Vol. 15, Issue 2). MDPI. DOI: 10.3390/su15021059

Organización de las Naciones Unidas para la Educación, l. C. (1972). "Convención sobre la protección del patrimonio mundial, cultural y natural.". *Actas de 17a reunión.*

Recommendations, P. C. (2011). "Directive of the Prime Minister, 09/11/2011. Guidelines for the assessment and the mitigation of seismic risk of cultural heritage with reference to Italian NTC 2008.".

E. Rossi, E. S. (2016). Costruire il Tempio. Alla ricerca del progetto di Baldassarre Peruzzi per il Duomo di Carpi. *APM edizioni, Carpi, .*

SALGADO., M. a. (2009). Re-thinking an Annotation Tool for the Museum Community Generated Content. *Nordic Design Research Conference (Nordes). Oslo: Nordic Design Research.*

Shen, J., Li, L., Zhang, D., Dong, S., Xiang, J., & Xu, N. (2022). A Multi-Analytical Approach to Investigate the Polychrome Clay Sculpture in Qinglian Temple of Jincheng, China. *Materials (Basel)*, 15(16), 5470. Advance online publication. DOI: 10.3390/ma15165470 PMID: 36013607

Sorapure, M. (2003). Five principles of new media: Or, playing Lev Manovich. *Kairos*, 8(2), 26–48.

UNESCO. (1972). *Convention Concerning The Protection Of The World Cultural And Natural Heritage.* 1–58.

Unesco. (2023). *Operational Guidelines for the Implementation of the World Heritage Convention UNITED NATIONS EDUCATIONAL, SCIENTIFIC AND CULTURAL ORGANIZATION INTERGOVERNMENTAL COMMITTEE FOR THE PROTECTION OF THE WORLD CULTURAL AND NATURAL HERITAGE WORLD HERITAGE CENTRE.* https://whc.unesco.org/fr/orientations

Wali Shah, M., Ali, I., Mahmood, A., Ashar Awan, M., Ahmad, N., Syed Hamid Arshad, H., Pakhtunkhwa, K., & Scholar, P. (2024). *Safeguarding Endangered Built Heritage In Khyber Pakhtunkhwa-Pakistan: A Digital Restoration Approach. 21*(S7), 437–449. www.migrationletters.com

Wilson, S. (1993). *The Aesthetics and Practice of Designing Interactive Computer Events Introduction : Design Challenges of Interactive Multimedia.*

Wosk, J. (1991). Postmodern Currents: Art and Artists in the Age of Electronic Media by Margot Lovejoy. *Technology and Culture*, 32(1), 138–140.

Yang, W. B., Yen, Y. N., & Cheng, H. M. (2015). Digitalized preservation and presentation of historical building-taking traditional temples and dougong as examples. *ISPRS Annals of the Photogrammetry, Remote Sensing and Spatial Information Sciences, 2*(5W3), 371–376. DOI: 10.5194/isprsannals-II-5-W3-371-2015

Zhang, X., Zhi, Y., Xu, J., & Han, L. (2022). Digital Protection and Utilization of Architectural Heritage Using Knowledge Visualization. *Buildings*, 12(10), 1604. Advance online publication. DOI: 10.3390/buildings12101604

KEY TERMS AND DEFINITIONS

Interactive Art (IA): A form of art that involves active participation from the audience to influence or modify the artwork, making the viewer a co-creator.

Submersive Method (SM): Techniques used to engage users fully by deeply integrating them into an experience, often through sensory or emotional immersion.

Immersive Art: Art that surrounds and involves the viewer in a complete environment, often using digital technology.

Exhibition Space: The physical or virtual environment where art, artifacts, or installations are displayed to the public, designed to enhance viewer interaction and experience.

3D Mapping: Creating three-dimensional representations of physical objects or environments using data collected from various sources, often for visualization, analysis, or conservation purposes.

Cultural Heritage (CH): The legacy of physical artifacts (buildings, monuments, etc.) and intangible attributes (traditions, languages) inherited from past generations, maintained in the present for future generations.

Digital Twin (DT): A digital replica of a physical object or system used in various fields to simulate real-time performance, monitor conditions, or predict future behavior.

Virtual Representation (VR): A simulated experience created by computer technology that immerses users in a virtual environment

Human-Computer Interaction (HCI): The study and design of how people interact with computers and digital systems, aiming to improve usability and user experience.

Geographic Information System (GIS): A system that captures, stores, analyzes, and visualizes spatial or geographic data.

Chapter 8
Passive Design:
Sustainable Solutions for High-Performance Architecture

Faina Saleem
https://orcid.org/0009-0002-3886-8203
University of Malaya, Malaysia

Mohd Firrdhaus Mohd-Sahabuddin Sahabuddin
University of Malaya, Malaysia

Kamran Shabbir
Bahaudin Zakariya University, Pakistan

ABSTRACT

Passive design is a sustainable approach to architecture that prioritizes natural elements and energy efficiency. It encompasses strategies such as heat recovery ventilation, spatial openness for natural light and airflow, energy-efficient HVAC systems, and heat control mechanisms. Effective insulation, high-performance windows, proper solar orientation, and an airtight envelope are crucial components. By minimizing environmental impact and maximizing occupant comfort, passive design offers a holistic solution for high-performance architecture. This chapter will provide detailed insights into these strategies and their real-world applications.

BACKGROUND

Passive design is an architectonical process that looks to improve the energy performance of a building by using either natural resources or resources with minimal consumption, the idea of this process is to take advantage of the inflows of Energy by a system to offset the Effluents provided by the same system. Passed on principles that predate modern mechanical systems, Passive Design is inspired by traditional building practices that seek to use natural climate conditions to control thermal comfort (Elshafei, Vilcekova, Zelenakova, & Negm, 2021). Passive design has been receiving more attention this last year again, considering the global concern about the increase in carbon emissions and the need for more energy efficiency in the built environment. This philosophy is in harmony with the wider sustainable

DOI: 10.4018/979-8-3693-5748-4.ch008

practice, working with natural ventilation, daylighting, thermal mass, insulation, and solar orientation to produce sustainable yet economically competitive high-performance buildings.

Passive design is a key strategy for modern and future architectural practice since it incorporates these ideas and lowers buildings' operational energy demands while also enhancing indoor air quality and occupant well-being (Hasim Altan, 2016).

INTRODUCTION

In achieving the goal of sustainability in the built environment, passive design is an essential strategy that offers creative solutions for high-performance structures. As the globe fights to reduce carbon emissions and avoid the harmful effects of climate change, the building sector is leading the charge in this issue. Since buildings use a lot of energy and emit greenhouse gases, it's critical to implement environmentally responsible and energy-efficient design principles (Hafe, et al., 2023).

Historical Context of Passive Design

The historical context describes how old architectural traditions have influenced the development of passive design solutions (Salameh, 2022). Many ancient architectural methods, like the Roman hypocaust systems or the wind catchers used in Persia (Ahmadkhani, 2011), were created to control climate without artificial equipment naturally. The basis of passive design was not realized by these technologies; instead, the indoor climate was controlled by natural resources. The relationship to contemporary issues discusses the applicability of these antiquated methods in the present. Passive design has become increasingly important in light of the contemporary concerns of urban heat islands, rising energy use, and climate change (Sheng Liu, 2020).

Passive design strategies decrease energy use and promote the use of natural materials, which enhances occupant well-being and encourages environmental responsibility. In contrast to active design, which primarily relies on mechanical systems and energy-intensive technologies, passive design controls indoor lighting and climate by utilizing the natural environment and the basic qualities of building materials (Jiyoung LEE). This technique reduces the amount of energy needed for buildings to operate while also enhancing occupant comfort and indoor air quality.

The fundamental ideas of passive design come from tried-and-true architectural methods that have been updated to meet contemporary environmental requirements. Using daylight as much as possible, utilizing thermal mass, increasing natural ventilation, and ensuring that insulation and airtightness are sufficient are important tactics. When combined, these strategies result in buildings that are long-lasting, energy-efficient, and resistant to shifting weather patterns.

The application of statistics and data emphasizes the importance and effectiveness of passive design in contemporary buildings. Studies have shown that, for example, passive solar architecture can reduce the amount of energy used for heating and cooling new homes by 50% (Toroxel & Silva, 2024)or more, indicating that it has the potential to significantly lower energy consumption and associated costs. Similarly, the use of natural ventilation in buildings can result in energy savings of up to 30%, highlighting the importance of good design for total energy efficiency (Schulze & Eicker, 2013).

This chapter aims to provide a comprehensive introduction to the principles of passive design and their application in high-performance architecture. Using cases from the real world, we will illustrate how these techniques could be applied to develop sustainable and energy-efficient structures. In addition, we will look into various methods and approaches, assess their effectiveness, and identify potential areas for further research and discipline development. We hope that this inquiry will inspire designers, architects, and scholars to incorporate passive design concepts into their work, thus helping to create a more sustainable built environment. Passive design emphasizes natural solutions and reduces reliance on artificial systems, offering a means to attain net-zero energy buildings and enhance the overall sustainability of our built environment (Lin Chen, 2024) discussed in Figure 1.

Figure 1. Heating and cooling energy use

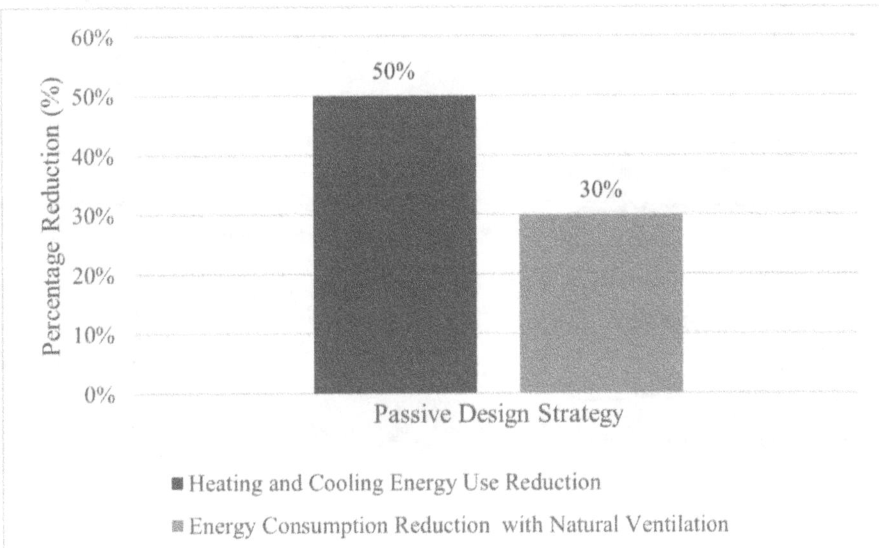

(Toroxel & Silva, 2024)

Heating and Cooling Energy Use can reduce Passive solar design can reduce heating and cooling energy use by up to 50%. Energy Consumption with Natural Ventilation can reduce Buildings utilizing natural ventilation can reduce energy consumption by up to 30%.

1. EMPHASIZE CROSS OR HEAT RECOVERY VENTILATION

Design that improves indoor air quality, thermal comfort, and energy economy. To promote the natural movement of air throughout a building, cross ventilation involves the thoughtful positioning of openings like windows, vents, and doors (Wahab, Ismail, & Kadir, 2019). By forcing fresh air through the structure and removing stale air, this technique uses pressure differences brought on by wind and temperature changes to lessen the need for mechanical cooling mentioned in Figure 2.

Figure 2. Cross ventilation

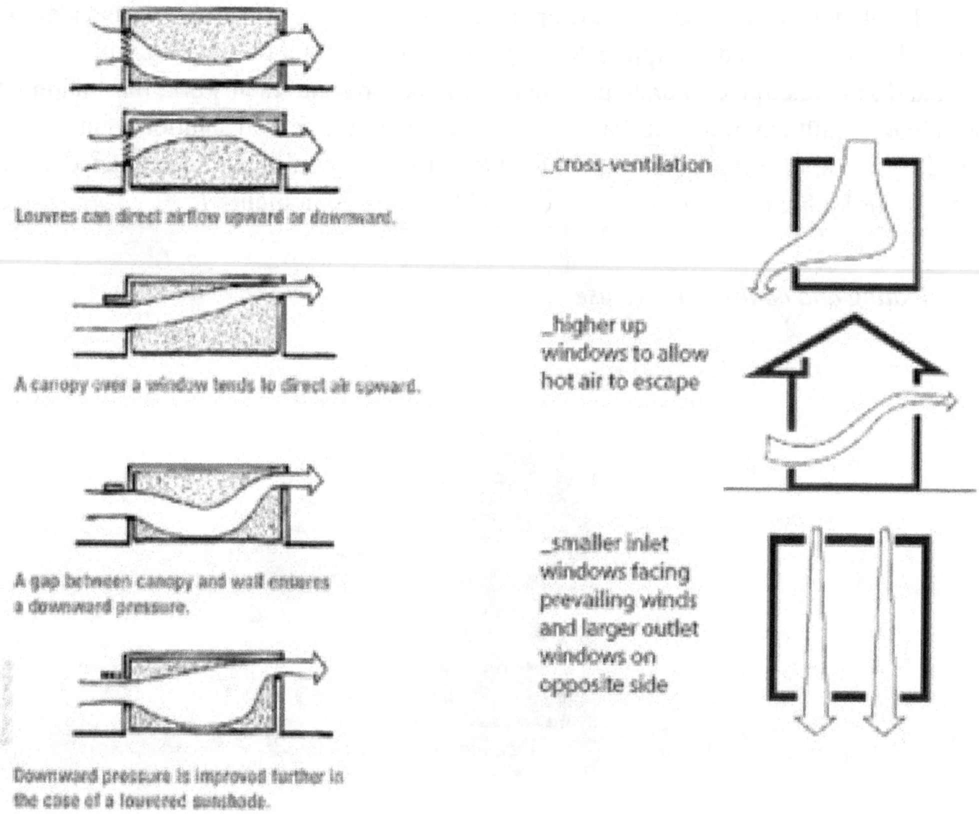

(Hasim Altan, 2016)

Systems for heat recovery ventilation (HRV), on the other hand, are made to absorb heat from exhaust air and move it into incoming fresh air (Świercz, 2021). This procedure greatly lowers the amount of energy needed for heating and cooling while simultaneously guaranteeing a steady flow of fresh air, which enhances indoor air quality. HRV systems improve a building's overall energy efficiency by recovering heat that would otherwise be lost as illustrated in Figure 3.

Figure 3. Heat recovery ventilation

(Dewsbury, 2015)

1. **Energy Savings:** Cross ventilation can lead to energy savings of up to 30% by reducing reliance on mechanical cooling systems (Roberto Stasi, 2024).
2. **Heat Recovery Efficiency:** HRV systems can achieve heat recovery efficiency of up to 70%, significantly improving heating and cooling energy use (Bas Knoll, 2018).
3. **Reduction in Heating Demand:** Both systems contribute to reducing heating demand, with HRV systems showing a significant impact due to heat recovery from exhaust air. Table 1 shows the key benefits of cross-ventilation and heat recovery systems.

Table 1. Key benefits of cross-ventilation and heat recovery ventilation (HRV) systems

Metric	Cross Ventilation	Heat Recovery Ventilation (HRV)
Energy Savings	Up to 30%	Significant, due to heat recovery
Heat Recovery Efficiency	N/A	Up to 70%
Reduction in Heating Demand	Up to 20%	Significant reduction
Indoor Air Quality Improvement	Enhanced natural ventilation	Continuous fresh air supply
Operational Cost Savings	Lower cooling costs	Lower heating and cooling costs

Case Study: The Bullitt Center, Seattle, USA

A complex heat recovery ventilation system is used by the Bullitt Centre, a well-known example of sustainable architecture in Figure 4, to achieve remarkable energy performance. The Bullitt Centre, which is situated in Seattle, Washington, is intended to function as a net-zero energy structure, generating as much energy as it uses each year. Its sophisticated HRV system is one of the main elements that has helped it achieve this (Peña, 2020).

Figure 4. The Bullitt Center's 242 KW PV (photovoltaic) array (Benjamin Benschneider).

Up to 70% of the heat from the building's exhaust air is captured by the HRV system at the Bullitt Centre and transferred to the entering fresh air. This procedure guarantees a steady supply of clean, pre-heated air and drastically lowers the building's heating requirements, particularly during the winter (Peña, 2020) discussed in Figure 5. Additionally, during the warmer months, natural airflow can cool and ventilate the interior areas because of moveable windows that are deliberately placed in the building's architecture to maximize cross ventilation. This dual strategy, which combines mechanical heat recovery with natural ventilation, is an excellent example of how passive design concepts may be applied to improve occupant comfort and energy efficiency.

Figure 5. Bullitt Center integrated systems for power production, daylighting, heating, cooling, and ventilation.

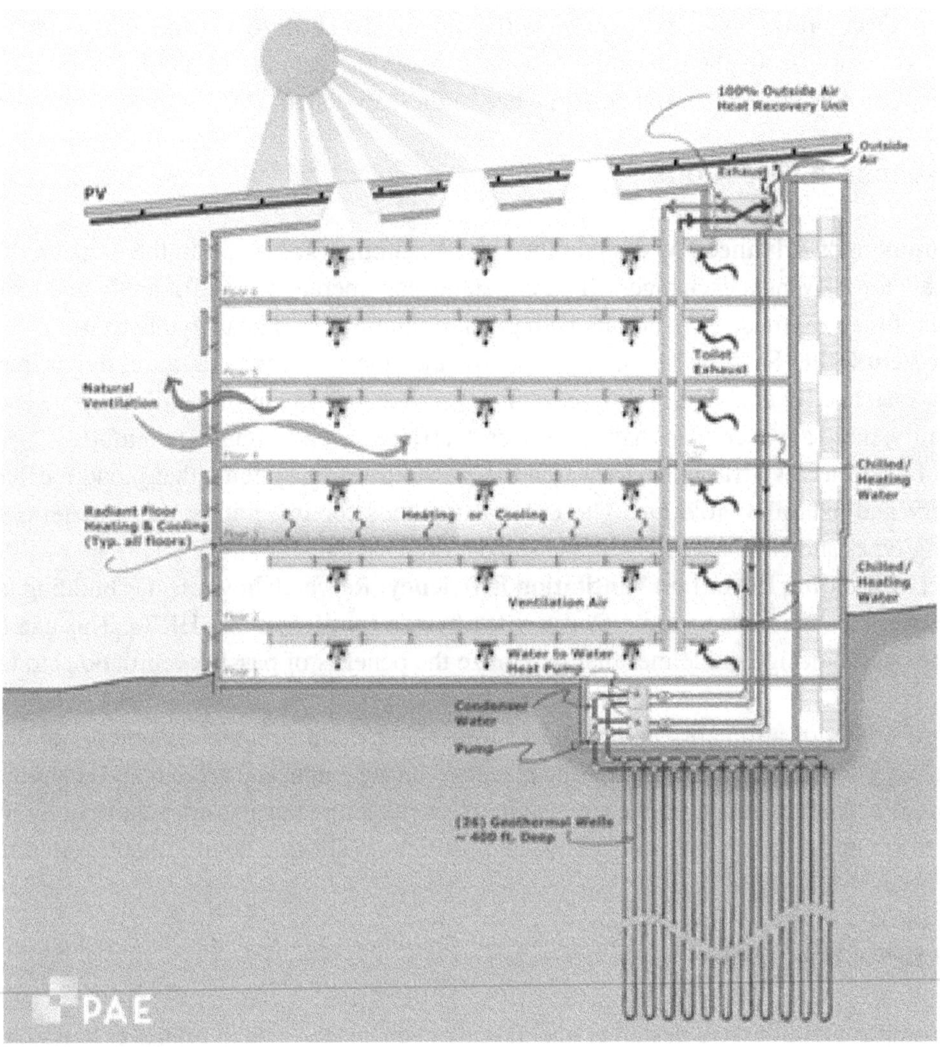

(Peña, 2020).

Solutions and Recommendations

1. **Incorporate Heat Recovery Ventilators (HRVs):** Integrate HRVs into the building design to improve energy efficiency. These systems should be selected based on the building's size, occupancy, and local climate conditions to ensure optimal performance (Pinar Mert Cuce, 2014).
2. **Strategically Place Operable Windows:** Design buildings with operable windows placed to maximize cross ventilation. Consider factors such as prevailing wind direction, building orientation, and internal layout to facilitate effective natural airflow (Hasim Altan, 2016).

3. **Utilize Automated Controls:** Implement automated control systems that adjust ventilation rates based on indoor air quality and occupancy levels. These systems can optimize the performance of HRVs and natural ventilation, ensuring a healthy and comfortable indoor environment.
4. **Educate Occupants:** Provide occupants with guidelines on how to use operable windows and ventilation systems effectively. Educating occupants about the benefits of natural ventilation and heat recovery can enhance the overall performance of these systems.

Future Research Directions

1. **Development of Advanced Materials for Heat Exchangers:** Look into the creation of advanced materials for HRV heat exchangers. These systems can operate more efficiently and have a longer lifespan if they are made of materials with greater durability and heat conductivity.
2. **Smart Ventilation Systems:** Examine how IoT and smart sensor integration might improve ventilation systems. By enabling real-time monitoring and adjustment of ventilation rates and interior air quality, these advances can maximize energy efficiency and occupant comfort.
3. **Hybrid Ventilation Strategies**: Research the possibilities of systems that blend mechanical heat recovery and natural ventilation. The optimal methods for integrating these systems in various building types and climates might be the subject of research.
4. **Impact of Building Layout on Ventilation Efficiency:** Research how various building layouts and architectural features impact the effectiveness of cross ventilation and HRVs. This can lead to the development of design guidelines that maximize the benefits of passive ventilation strategies.

To improve interior air quality and occupant well-being, architects and designers can dramatically lower the energy consumption of buildings by placing a strong emphasis on cross and heat recovery ventilation. These techniques not only advance sustainable architecture but also open the door for the creation of high-performing structures that are flexible and resilient to shifting environmental circumstances.

2. KEEP SPACES OPEN

A key component of passive design is maintaining open spaces, which improves natural ventilation and daylighting, lowers energy use, and improves indoor environmental quality. Better air circulation is created possible by open areas, which is essential for preserving appropriate interior temperatures and lowering the demand for artificial ventilation and cooling systems (V. Faye McNeill, 2022). Furthermore, open areas allow more natural light to enter, which might lessen the need for artificial lighting during the day as mentioned in Figure 6.

Figure 6. Courtyard

(Dai, 2021)

An open layout makes a room more functional and flexible by facilitating a variety of applications and simple adaptability to shifting requirements. This can be especially helpful in residential and commercial structures, as being able to rearrange spaces can increase occupant contentment and usability (Kooi, 2022) Table 2 discusses the benefits of keeping spaces in Passive design.

Table 2. Benefits of keeping spaces open in passive design

Metric	Benefits
Enhanced Air Flow	Better Indoor Air Quality
Increased Natural Light	Lower Reliance on Artificial Lighting
Improved Flexibility	Adaptable Space Use
Reduced Energy Use	Lower Cooling Costs

Case Study: Khoo Teck Puat Hospital, Singapore

Singapore's Khoo Teck Puat Hospital (figure 7 & 8) is a prime example of how open spaces may be used efficiently in passive architecture. Natural daylighting and ventilation are given priority in the hospital's architecture, which takes advantage of spacious, open courtyards and expansive hallways. These open areas improve light and air penetration while fostering a therapeutic atmosphere for patients and staff (Tan, 2012).

Figure 7. Khoo Teck Puat Hospital

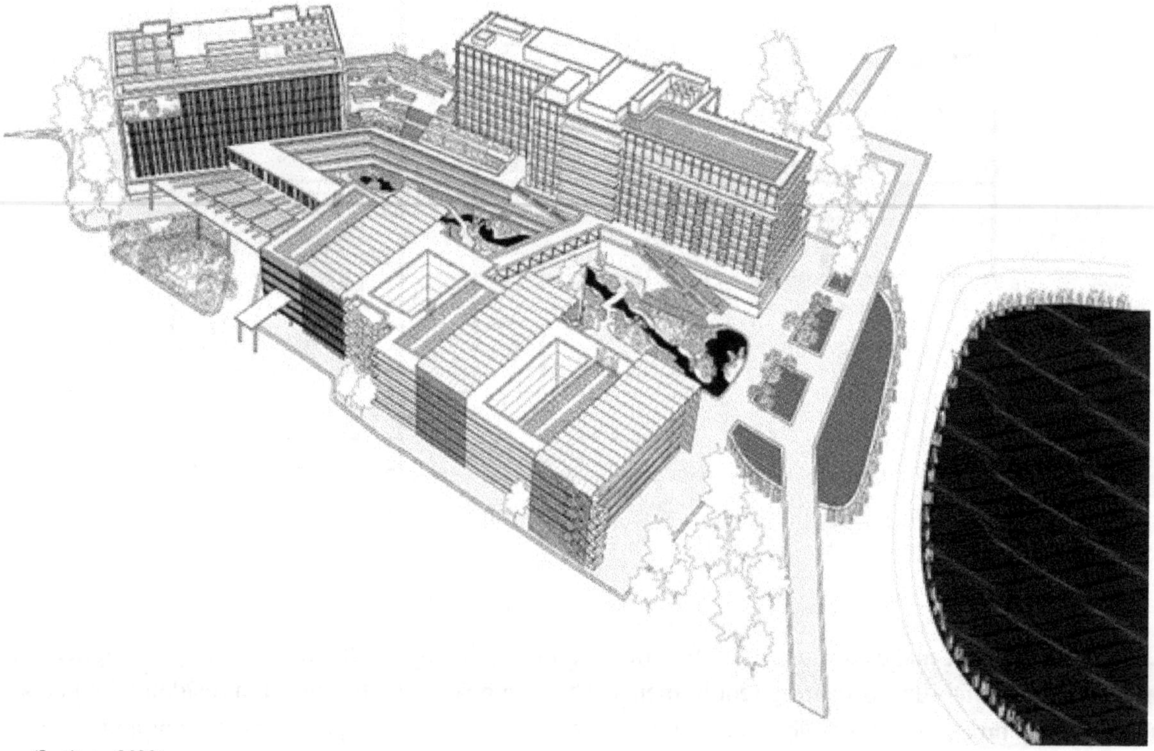

(Institute, 2022)

Many atriums and gardens in the hospital's architecture let natural light into the space, which lessens the demand for artificial lighting. By allowing cross circulation through open corridors and well-placed apertures, it is possible to maintain a suitable indoor climate without using a lot of air conditioning. This design strategy has produced a pleasant, healthful environment for all residents as well as significant energy savings.

Figure 8. Khoo Teck Puat Hospital Atriums

(Institute, 2022)

Solutions and Recommendations

1. **Design Open Floor Plans:** Include open floor plans that bring in natural light and unhindered airflow. When necessary, employ flexible partitions and refrain from over-partitioning.
2. **Make Use of Atriums and Courtyards**: To promote natural ventilation and daylighting, incorporate atriums and courtyards into your structures. These areas can act as major centers for the distribution of light and air.
3. **Place Windows for Maximum Natural Light and Cross Ventilation:** Arrange windows for maximum natural light. Make use of larger windows and pay attention to where you place them so that they can efficiently capture daylight and prevailing winds.
4. **Use Modular Design Components:** Make use of flexible rooms and adjustable partitions that may be rearranged as needed. This adaptability makes the area more functional and facilitates improved light and air distribution.

Future Research Directions

1. **The Effect of Open Spaces on Indoor Air Quality:** This research examines the effects of open spaces on indoor air quality in various building types and climates. It can help with design optimization for a range of environmental circumstances.
2. **Thermal Comfort in Open Spaces:** Research how thermal comfort and open spaces are related. It is possible to develop more effective passive cooling techniques by comprehending how open layouts impact temperature management.

3. **Integration with Smart Building Technologies:** Examine how open space design can be integrated with cutting-edge systems that monitor and regulate lighting and ventilation in real-time.
4. **User Behaviour and Space Utilisation:** Look at how people use open areas and how their actions affect how well daylighting and natural ventilation work. This study can help with designing tactics that raise occupant satisfaction.

3. RETHINK MECHANICAL DESIGN

Incorporating passive architecture into mechanical design entails rethinking HVAC (heating, ventilation, and air conditioning) systems and energy management in novel ways. To maintain indoor environmental quality, the old method frequently depends primarily on mechanical systems, which raises energy consumption and operating expenses. We can attain increased energy efficiency, enhanced indoor comfort, and notable decreases in the environmental effect of buildings by reconsidering these designs.

Advanced Control Systems

By improving both energy efficiency and occupant comfort, the use of smart technologies has revolutionized passive design in recent years. Passive design techniques can be optimized in real time by smart systems, including automated control systems, sophisticated sensors, and Internet of Things (IoT) devices (Chin-Chi Cheng, 2019). To cut down on wasteful energy use, smart thermostats, for example, can modify the interior temperature in response to occupancy levels and meteorological circumstances. In addition to providing automatic lighting and ventilation controls and real-time monitoring, these technologies guarantee that buildings remain comfortable while using the least amount of energy (Alain Starke, 2021). Automatic shade systems are a significant development in passive solar control. These systems, which are frequently linked to sensors, can modify blinds, louvers, or other shading devices in response to variations in temperature and light throughout the day (A.R. Hassani, 2022). By enabling the construction of structures that adapt dynamically to changes in their surroundings, smart technologies present the potential to expand on the advantages of passive architecture. By combining modern technology with age-old passive techniques, buildings are ensured to be both energy-efficient and able to function well under a range of climatic conditions. These developments are essential to help us head towards a time when innovative design and sustainability will inevitably converge (Dewsbury, 2015).

Energy-Efficient HVAC Systems

High-Efficiency Systems: Modern HVAC systems are made to operate with less energy while still offering comparable or superior heating and cooling capacities. Modern heat exchangers, high-efficiency compressors, and variable-speed fans are just a few of the features these systems frequently have (DeQuante Rashon Mckoy, 2023).
Geothermal Heat Pumps: These systems use the steady temperatures found below the surface to offer both heating and cooling, using a great deal less energy than traditional systems (Mamdouh El Haj Assad, 2022).

Zoning Systems: Based on occupancy and usage patterns, energy utilization can be optimized by creating distinct zones within a building and managing each zone's temperature independently (Aitichya Chandra, 2020) discussed in Figure 9.

Figure 9. Mechanical design

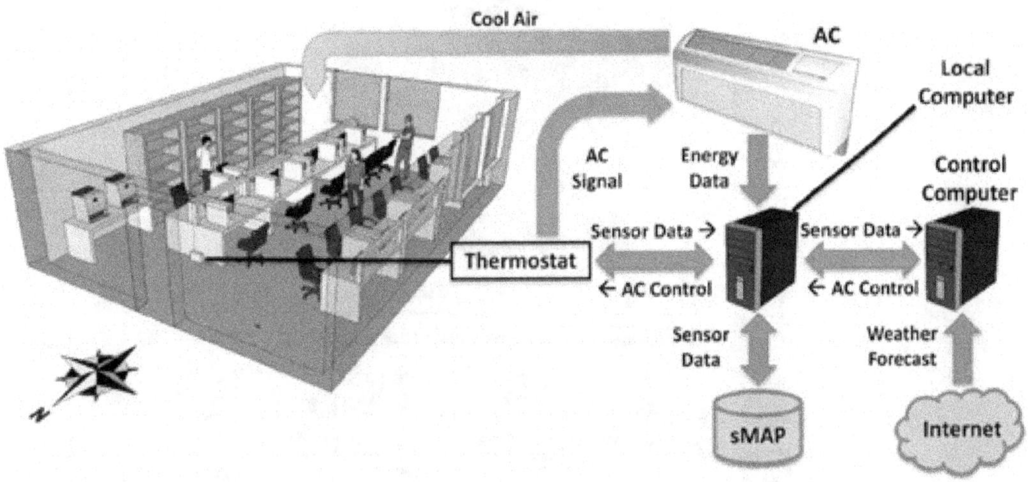

(Anil Aswani, 2012)

Advanced Control Systems

Smart Thermostats: These gadgets automatically alter temperatures to save energy when the building is empty based on the schedule and preferences of the residents (Alain Starke, 2021). Integrated systems that monitor and manage building services including lighting, security, and HVAC are called building management systems, or BMSs. Real-time data can be utilized by them to maximize energy efficiency and performance (Khairy Sayed, 2018).

Occupancy Sensors: By detecting whether a room is inhabited, these sensors prevent energy from being spent on empty rooms by modifying HVAC and lighting systems. Figure 10 discusses the advanced control systems.

Figure 10. Advanced control systems

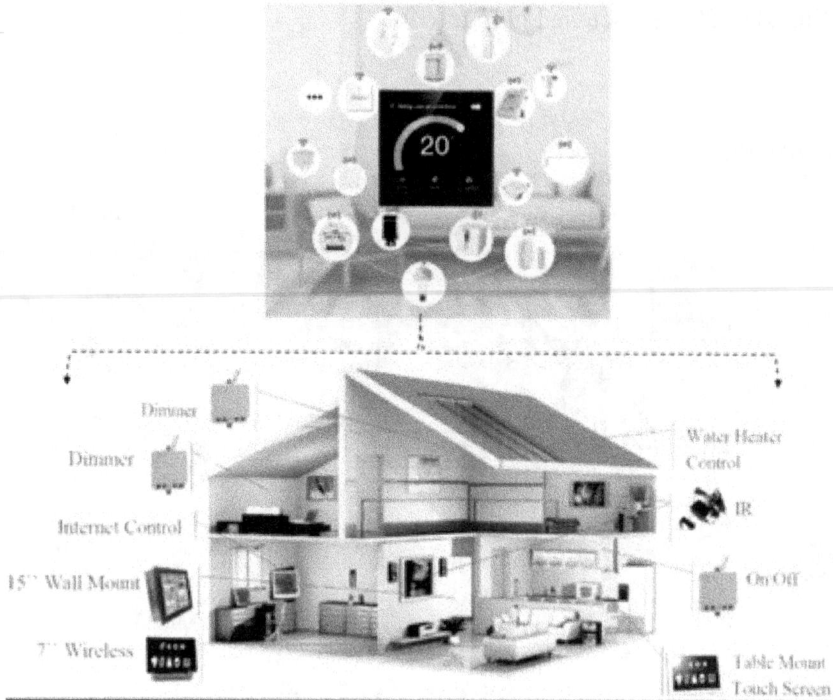

(Shweta Singh, 2019)

Integration of Renewable Energy

1. **Solar Panels:** By installing solar panels to produce electricity locally, you may lessen your carbon impact and dependency on the grid. Batteries can be used to store extra energy for later consumption or to feed it back into the grid (Hamza, 2023).
2. **Solar Thermal Systems:** These systems eliminate the need for traditional water heating systems by heating water using the energy of the sun.
3. **Wind Turbines:** Small-scale wind turbines can serve as an additional renewable energy source in appropriate regions. Renewable energy resources are discussed in Figure 11.

Figure 11. Energy usage

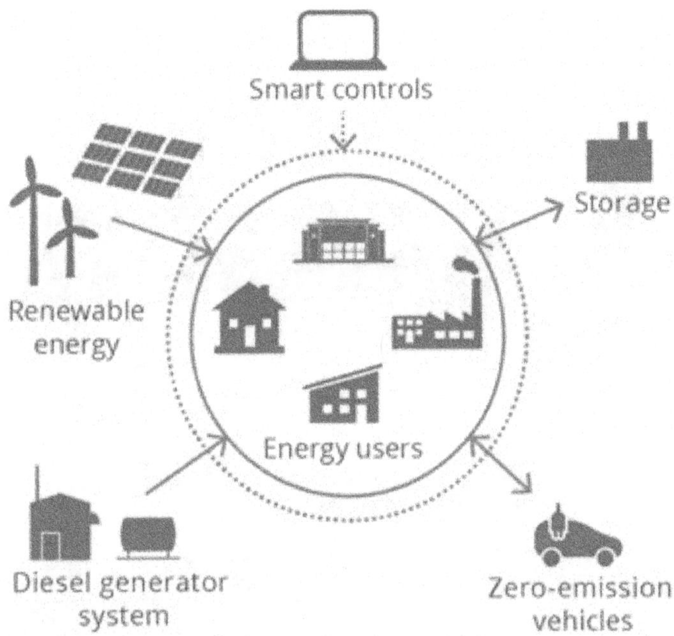

(Mohamed Azeroual, 2019)

Improved Insulation and Sealing

1. **High-Performance Insulation:** By more successfully regulating interior temperatures, improved insulation materials and methods can drastically lower the requirement for heating and cooling (Figure 12).

Figure 12. Insulation and sealing

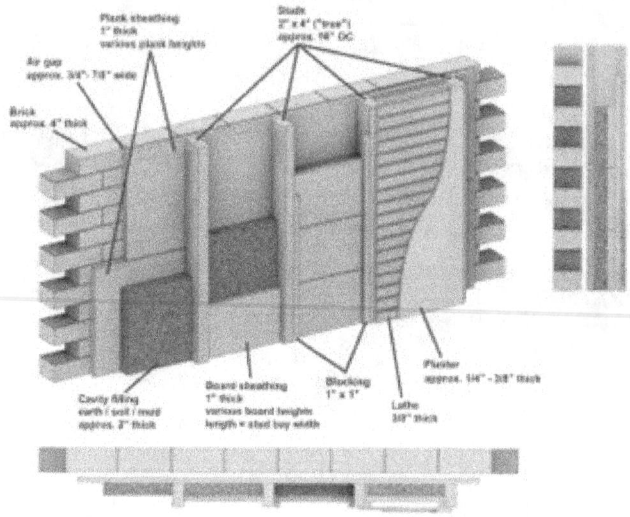

(Mohamed Azeroual T. L., 2019)

2. **Airtight Building Envelope:** Maintaining an airtight building envelope increases the effectiveness of HVAC systems by preventing unintentional heat gain and loss.

Case Study: Edge Building, Amsterdam, Netherlands

A noteworthy example of a mechanical design concept applied with consideration is the Edge Building in Amsterdam (Figure 13). Modern technology is used by the Edge, which is renowned for its sustainability, to reduce energy use and increase comfort and effectiveness (Ibrahim, 2020)

Figure 13. Edge Building, Amsterdam, Netherlands

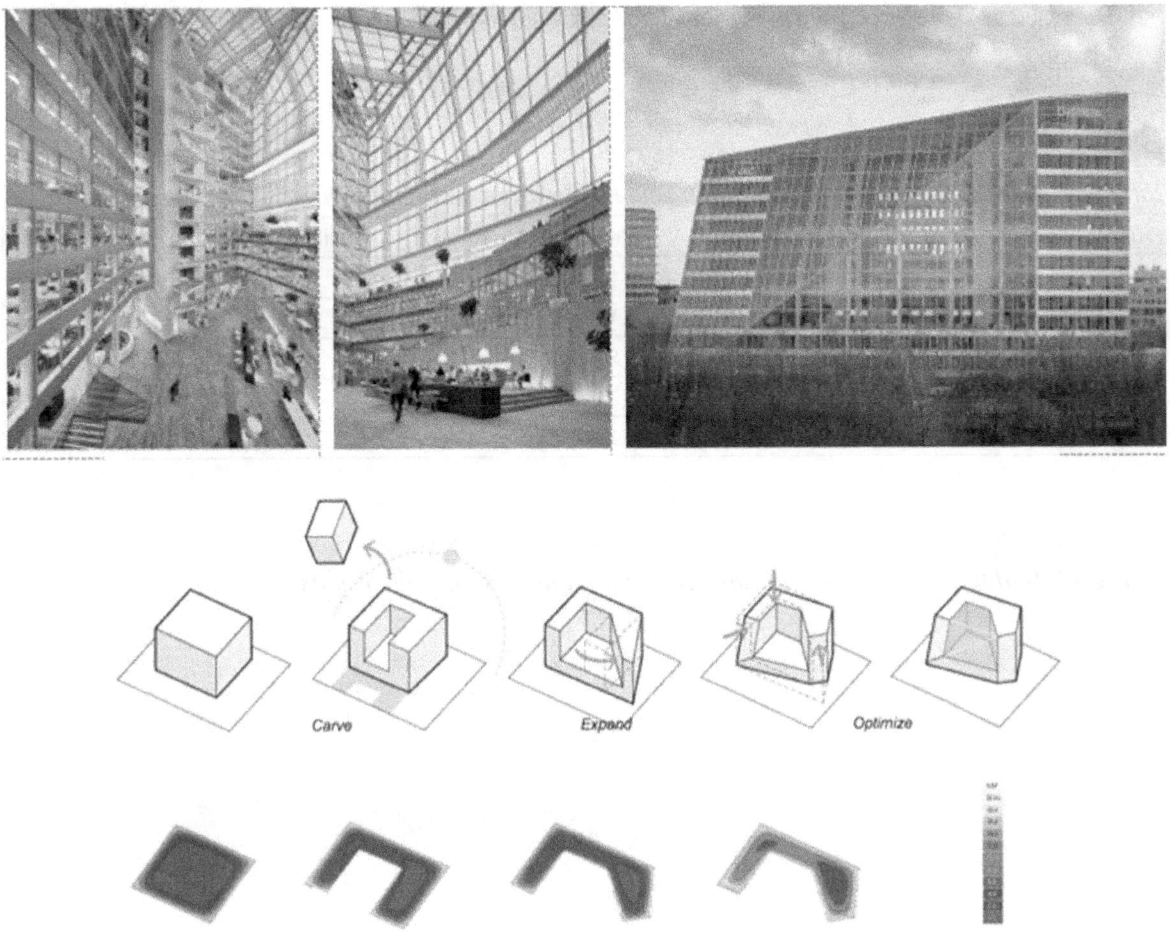

(Ibrahim, 2020)

1. **Energy-Efficient HVAC:** Geothermal energy and cutting-edge heat pumps are two components of the building's extremely efficient HVAC system. This technology improves overall efficiency and lessens dependency on conventional energy sources.
2. **Advanced Controls:** Thousands of sensors are integrated into the building management system to monitor and manage the temperature, lighting, and occupancy of the building, among other aspects of its operation. By enabling dynamic modifications, this real-time data optimizes energy utilization.
3. **Renewable Electricity:** The Edge produces a sizable amount of its electricity thanks to a sizable array of solar panels mounted on its roof and exterior. By utilizing renewable energy, the building's carbon footprint is decreased.
4. **Natural Ventilation:** By providing ample open areas and atriums for natural ventilation, the design lessens the requirement for artificial cooling. This enhances the quality of the air indoors while also saving electricity.

5. **Heat Recovery:** The ventilation system has HRV units that warm incoming fresh air by recovering heat from exhaust air, which drastically lowers the amount of heating energy needed. A comparison is discussed in Table 3 and Figure 14, between traditional mechanical design and rethought mechanical design.

Table 3. Comparison between traditional mechanical design and rethought mechanical design

Metric	Traditional Design	Rethought Mechanical Design
Energy Consumption	High	Reduced
Indoor Environmental Quality	Variable	Enhanced
System Efficiency	Standard	High
Operational Costs	Higher	Lower
Renewable Energy Integration	Limited	Extensive

Figure 14. Building management of Edge Building, Amsterdam, Netherlands

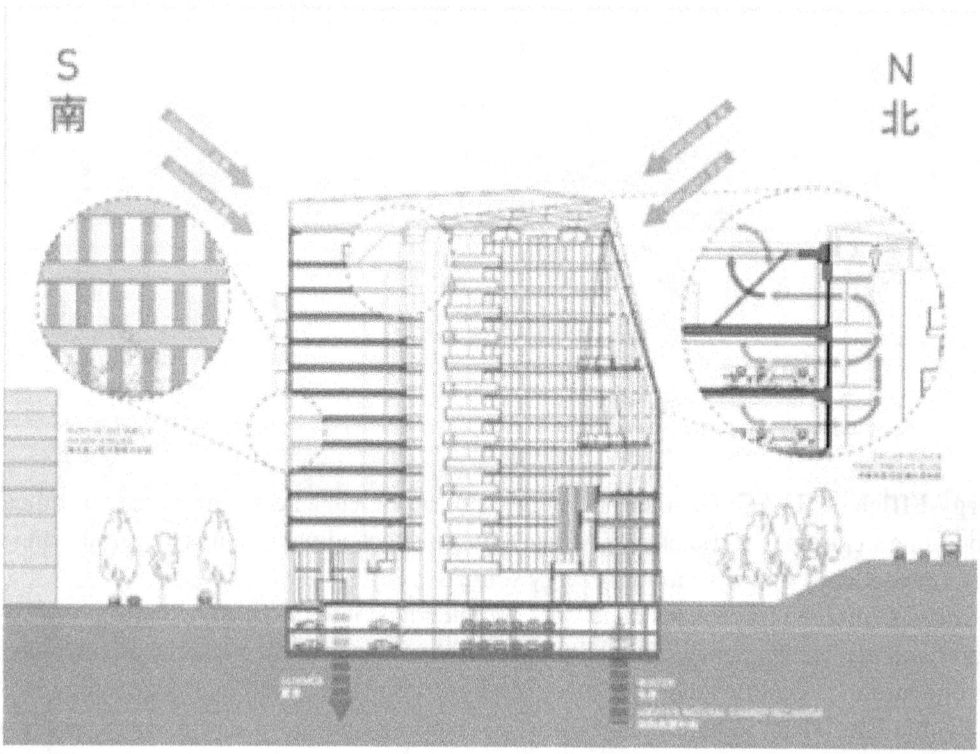

(Ibrahim, 2020)

Solutions and Recommendations

1. **Install High-Efficiency HVAC Systems:** Make sure the HVAC systems are the right size for the needs of the building and select those with high energy efficiency ratings.
2. **Make Use of Smart Controls:** Install cutting-edge control systems that can maximize HVAC performance in response to occupancy trends and real-time data.
3. **Incorporate Renewable Energy:** To power mechanical systems and lessen dependency on non-renewable energy sources, install renewable energy systems like solar panels.
4. **Regular Maintenance and Upgrades:** Make sure mechanical systems receive regular maintenance, and when newer, more effective technologies become available, consider them.
5. **Make Use of Natural Ventilation:** To lessen reliance on mechanical systems, include design features that improve natural ventilation.
6. **Install Heat Recovery Systems:** By catching and reusing waste heat, heat recovery systems help ventilation systems operate more energy efficiently.

Future Research Directions

1. **Development of Advanced HVAC Technologies:** Investigate novel HVAC solutions with reduced environmental impact and increased efficiency.
2. **AI Integration with Building Controls:** Investigate how AI can be used to forecast and enhance HVAC operations using predictive analytics and historical data.
3. **Impact of Renewable Energy Integration:** Research the long-term effects on building sustainability and the overall performance of integrating renewable energy sources with mechanical systems.
4. **User Behavior and System Performance:** Examine how occupant behavior impacts mechanical system performance and devise plans to promote energy-conserving habits.

4. CONTROL HEAT

One of the main principles of passive design is the management of heat in buildings. Efficient heat control techniques provide year-round comfort in buildings while reducing energy usage. This entails controlling heat gains and losses using a range of technology and design interventions, including thermal mass utilization, strategic shading, and passive solar methods (Michailidis, 2015). Table 4 discusses heat control strategies in passive design.

Table 4. Heat control strategies in passive design

Strategy	Description	Benefits
Strategic Shading	Use of overhangs, louvers, and dynamic systems to block direct sunlight	Reduces cooling loads, prevents glare
Thermal Mass Utilization	Incorporation of high thermal mass materials to absorb and release heat	Moderate indoor temperatures reduce energy use
Passive Solar Techniques	South-facing windows and appropriate SHGC window selection	Maximizes winter solar gain, minimizes summer overheating

Strategic Shading

Fixed Shading Devices: During the hottest hours of the day, architectural features such as overhangs, louvers, and brise-soleil are intended to hide direct sunlight while still letting natural light in (Heidari, 2021). Types of fixed shading devices are mentioned in figure 15.

Figure 15. Fixed shading devices

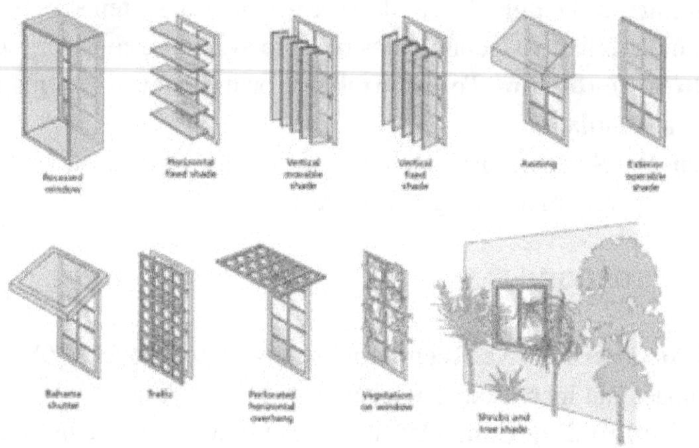

(Heidari, 2021)

Dynamic Shading Systems: To maximize solar exposure and minimize glare, movable shading systems, like motorized blinds or external shades, can be operated manually or automatically (Salwa M. Al-Masrani, 2019) shown in Figure 16.

Figure 16. Dynamic shading systems

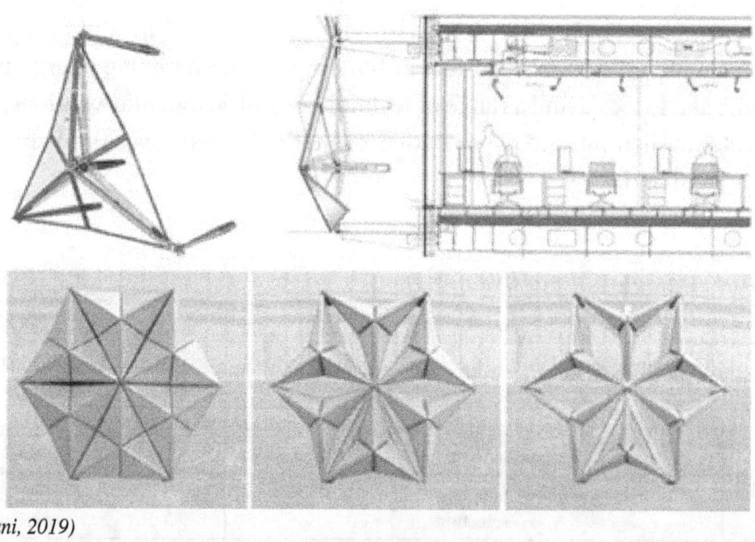

(Salwa M. Al-Masrani, 2019)

Thermal Mass Utilization

Using materials with a high thermal mass, such as stone, brick, or concrete, helps regulate indoor temperatures by absorbing and storing heat during the day and releasing it at night. To maximize the efficiency of thermal mass, place it in areas where it may absorb direct sunlight, such as walls, floors, and ceilings **(Kuczynski, 2020)** shown in Figure 17.

Figure 17. Thermal mass utilization

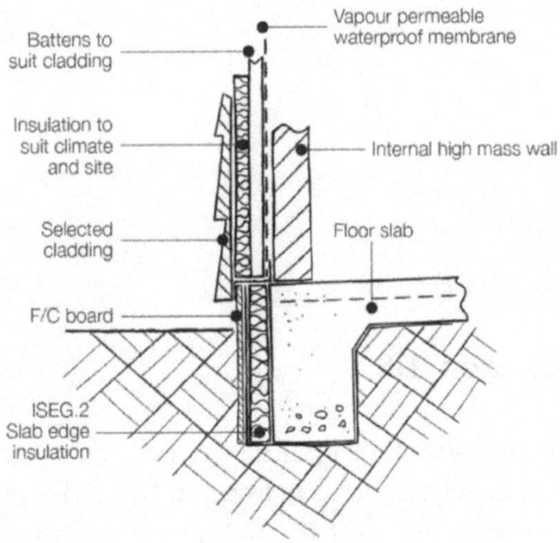

(Kuczynski, 2020)

Passive Solar Techniques

South-Facing Windows: In the northern hemisphere, placing windows on the south side of a structure maximizes winter solar gain and reduces summer heat gain. Solar Heat Gain Coefficient (SHGC): Choosing windows with the right SHGC rating guarantees that, depending on seasonal requirements, solar radiation is either maximized or minimized shown in Figure 18 (A.R. Hassani, 2022).

Figure 18. Solar Heat Gain Coefficient (SHGC)

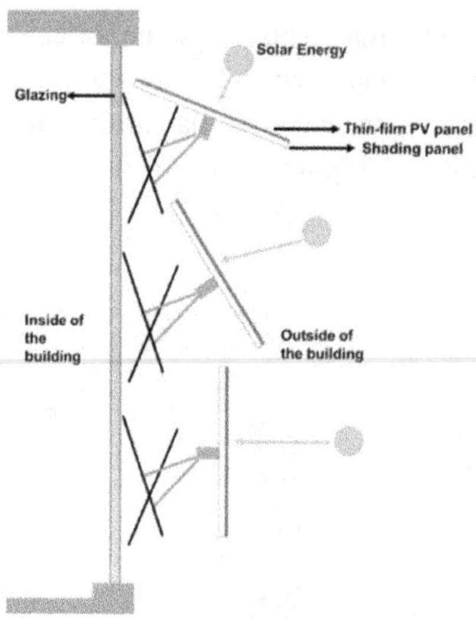

(A.R. Hassani, 2022)

Case Study: BedZED, London, UK

Beddington Zero Energy Development, or BedZED (figure 19), is a prime example of passive design's ability to regulate heat effectively. Situated in London, UK, BedZED is the country's biggest and pioneering carbon-neutral eco-community, created to reduce energy usage by employing creative passive design techniques (Dan Zhu, 2015).

1. **Strategic Shading:** The building has a lot of wide windows facing south, with overhangs that shade the windows in the summer to keep them from overheating while letting the low winter sun enter and warm the interior areas.
2. **Thermal Mass Utilization:** To minimize temperature swings, the structures are made of thick, thermally massive materials that absorb heat during the day and release it gradually at night.
3. **Passive Solar Design:** BedZED maximizes solar gain through orientation and layout. The layout of the buildings guarantees that every unit receives solar heating, and high-performance windows with low SHGCs are used to control heat gain effectively.
4. **Airtight Envelope and High-Performance Insulation:** BedZED's structures have an airtight envelope and high-performance insulation, which prevents heat loss in the winter and keeps interiors cool in the summer (Young, 2015).

Figure 19. Beddington zero energy development

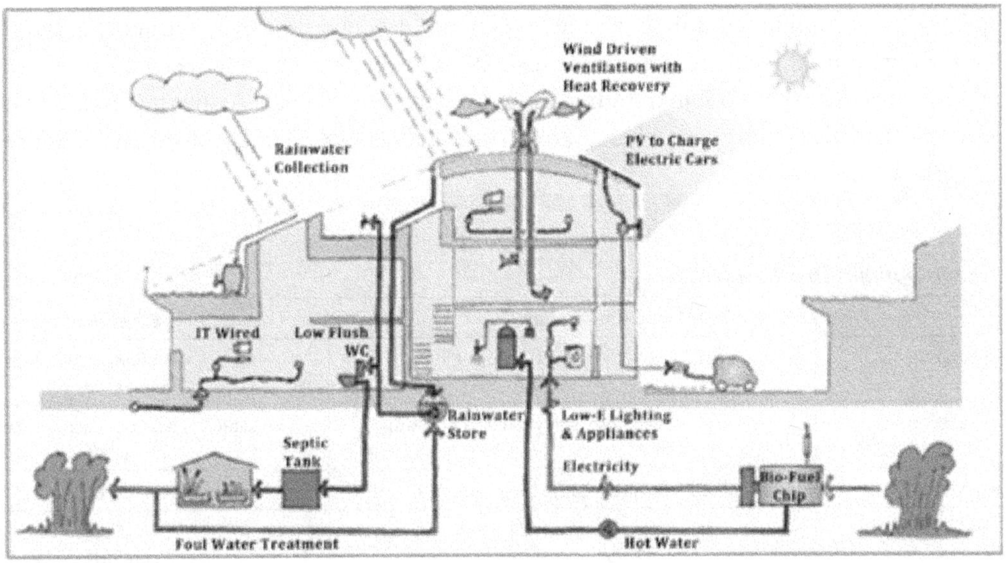

(Dan Zhu, 2015).

Solutions and Recommendations

1. **Utilize Fixed and Dynamic Shading:** To maximize sunlight exposure and minimize glare, combine fixed shading devices with dynamic systems.
2. **Make Use of Thermal Mass:** To help regulate interior temperatures, strategically place items with a high thermal mass.
3. **Optimize Window Placement and Selection:** Choose windows with the right SHGC rating and position them to receive the most sunlight possible in the winter.
4. **Enhance Airtightness and Insulation:** To reduce heat transfer and increase energy efficiency, use high-performance insulation and make sure the building envelope is airtight.
5. **Automate Shading Systems:** Put in place automatic shading systems that change according to occupancy and the current weather.

Future Research Directions

Research on novel materials and technologies for dynamic shading systems that provide better performance and aesthetically pleasing integration is known as advanced shading technologies. The creation of materials with improved thermal storage properties that may be applied to a larger range of situations is known as "smart thermal mass materials."

Innovative Passive Solar Design Techniques: Examine cutting-edge passive solar design strategies that work in a variety of weather conditions and building styles. Integration with Renewable Energy: For optimal efficiency, look into how heat control techniques can be combined with renewable energy systems.

5. HAVE OPTIMAL INSULATION

Maintaining a constant indoor climate, reducing energy use, and improving comfort in buildings all depend on having optimal insulation. Superior insulation lowers heat transfer across floors, roofs, and walls, keeping interior spaces cool in the summer and warm in the winter. Achieving high-performance architecture and boosting energy efficiency are closely related to effective insulation (Adam, 2019) shown in Table 5.

Table 5. Have optimal insulation

Aspect	Description	Benefits
Types of Insulation	Fiberglass, cellulose, spray foam, rigid foam	Versatile applications enhance thermal resistance
Types of Insulation	Fiberglass, cellulose, spray foam, rigid foam	Versatile applications enhance thermal resistance
Installation Methods	Batt and roll, blown-in, sprayed/injected, foam board	Effective coverage, reduced thermal bridging
R-value and Thermal Resistance	Measures resistance to heat flow; higher R-values mean better insulation	Improved energy efficiency, increased comfort
Air Sealing and Moisture Control	Airtight construction, moisture barriers/vapor retarders	Prevents air leakage, protects against condensation and mold

Types of Insulation

1. **Fiberglass Insulation:** One of the most popular forms of insulation, it is composed of fine glass fibers. It reduces heat transfer well and is non-flammable.
2. **Cellulose Insulation:** This insulation provides good thermal performance and is treated with fire retardants. It is made from recycled paper products.
3. **Spray Foam Insulation:** This kind of insulation fills spaces and forms an airtight seal when it is applied. It has superior air sealing and heat-resistant qualities.
4. **Rigid Foam Insulation:** Rigid foam boards have a high heat resistance and can strengthen walls structurally. They are frequently used as exterior wall sheathing.

Installation Methods

1. **Batt and Roll Insulation:** Pre-cut portions that fit in between studs, joists, and rafters are commonly utilized in walls, floors, and ceilings.
2. **Blown-In Insulation:** Installed using specialized equipment that pushes loose fibers into the space, blown-in insulation is frequently utilized in attics and hard-to-reach regions.
3. **Sprayed or Injected Insulation:** This technique, which includes spraying or injecting foam insulation that expands to fill the space, is perfect for retrofits and areas with uneven shapes.
4. **Foam Board Insulation:** To offer continuous insulation and lessen thermal bridging, these rigid panels are utilized on foundation walls, roofs, and external walls.

R-Value and Thermal Resistance

1. **Identifying the R-value:** The resistance of an insulating substance to heat flow is measured by its R-value. Better insulating qualities are indicated by higher R-values. The necessary R-value varies according to the type of building, the environment, and the particular applications.
2. **Ensuring Appropriate Installation:** Appropriate installation is necessary to obtain the required R-value. Misalignment, compression, and gaps can lessen the efficiency of insulation (Ismaiel, 2021).

Air Sealing and Moisture Control

1. **Identifying the R-value:** The resistance of an insulating substance to heat flow is measured by its R-value. Better insulating qualities are indicated by higher R-values. The necessary R-value varies according to the type of building, the environment, and the particular applications.
2. **Ensuring Appropriate Installation:** Appropriate installation is necessary to obtain the required R-value. Misalignment, compression, and gaps can lessen the efficiency of insulation.

Case Study: Manitoba Hydro Place

The Manitoba Hydro Place in Canada (figure 20) achieves remarkable thermal performance via the application of cutting-edge insulating technology. The building's insulated façade helps it maintain net-zero energy status by reducing heat leakage (Potienko, 2017).

Figure 20. Building Performance of Manitoba Hydro Place

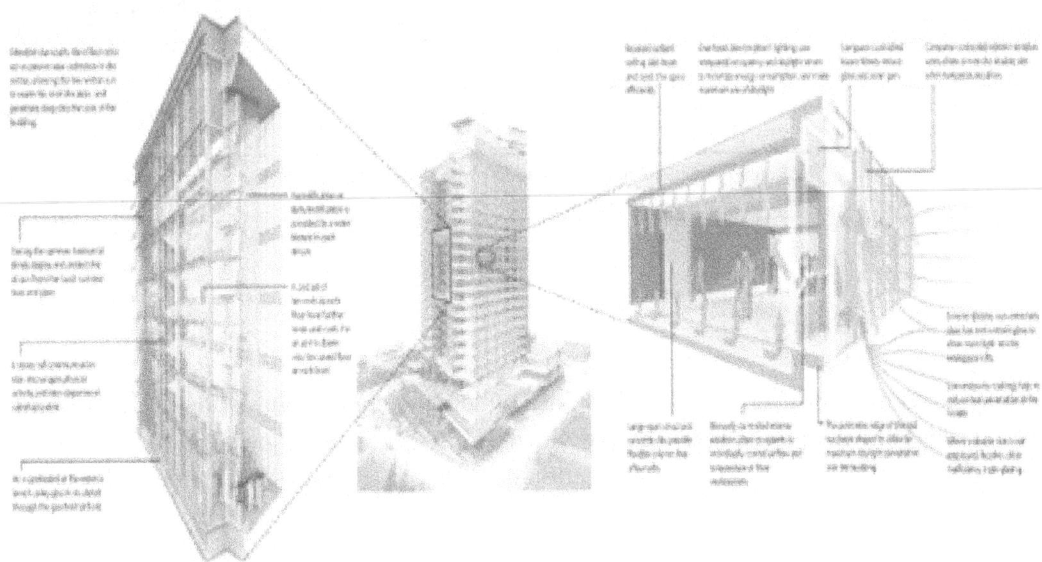

(Potienko, 2017)

Solutions and Recommendations

In building envelopes, use high-performance insulation materials. To prevent thermal bridging, pay attention to the continuity of the insulation. In warmer climates, think about reducing heat gain with reflective insulation.

Future Research Directions

Investigate innovative insulating materials that have reduced environmental effects and higher R-values. Examine the long-term effectiveness of different insulating techniques in a range of climates.

6. USE HIGH-PERFORMANCE WINDOWS AND DOORS

Energy-efficient, comfortable, and well-performing buildings can be greatly improved with high-performance windows and doors, which are crucial elements of passive design. These elements maximize natural light and ventilation while minimizing heat transfer and air leakage. High-performance windows and doors have several glazing layers, gas fills, insulated frames, low-emissivity (Low-E) coatings, and sophisticated sealing methods (Feng, 2021). Table 6 discusses the high-performance windows and doors.

Table 6. High-Performance Windows and Doors

Aspect	Description	Benefits
Multiple Glazing Layers	Double or triple-glazing with air or gas-filled spaces	Improved thermal insulation, reduced heat transfer
Low-Emissivity Coatings	Reflective coatings that reduce heat gain and loss	Enhanced energy efficiency, better climate control
Insulated Frames	Frames made from insulating materials with thermal breaks	Reduced thermal bridging, improved performance
Advanced Sealing Techniques	High-quality weather-stripping and proper installation to prevent air infiltration	Increased airtightness, reduced energy loss

Multiple Glazing Layers

1. **Double and Triple Glazing:** High-performance windows frequently have double or triple glazing, which consists of two or three glass layers divided by gas- or air-filled gaps. With this arrangement, heat transfer is decreased and thermal insulation is improved (Rana, 2018) mentioned in Figure 21.

Figure 21. Double and triple glazing

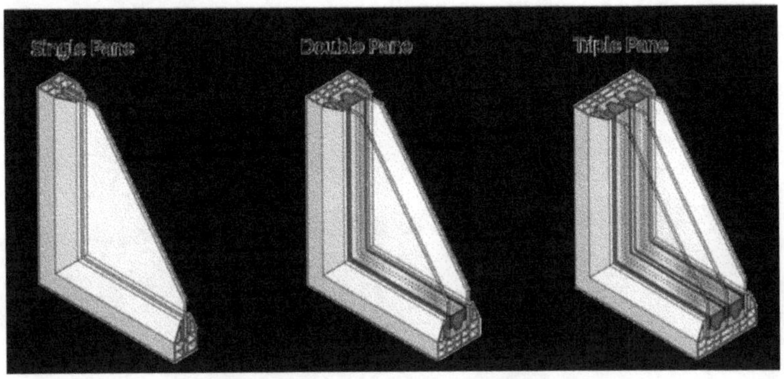

(Rana, 2018)

Figure 22. Gas fills

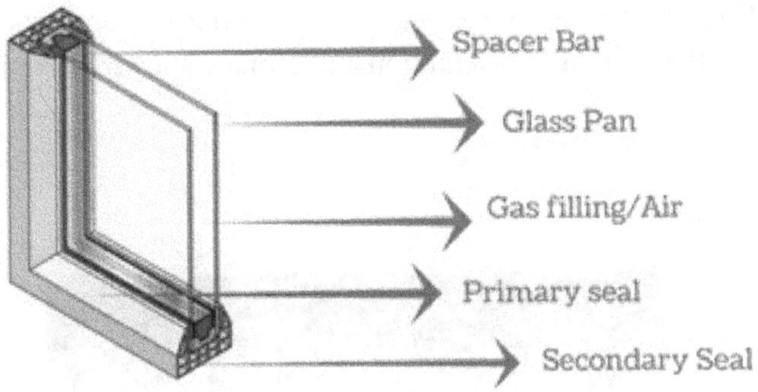

(Aguilar-Santana, 2020)

Low-Emissivity (Low-E) Coatings

1. **Reflective Coatings:** Thin metallic coatings known as "Low-E coatings" are applied to the surface of glass. By letting visible light through while reflecting infrared radiation, they lessen heat gain and loss in the summer and winter in Figure 23. (Raut, 2011).

Figure 23. Low E glass

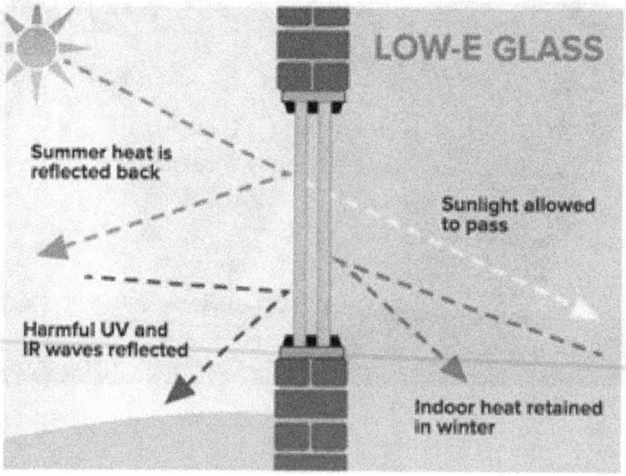

(Raut, 2011)

2. **Customized Coatings:** Various Low-E coatings can be utilized to optimize performance for different conditions, based on the building's orientation and climate shown in Figure 24.

Figure 24. Customized coatings

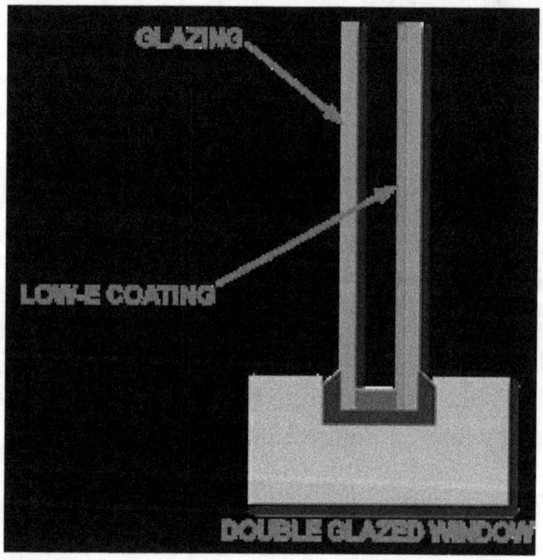

(Aguilar-Santana, 2020)

Insulated Frames

1. **Materials:** Thermal bridging is lessened and overall window efficiency is enhanced by frames composed of materials with good insulating qualities, such as wood, fiberglass, or insulated vinyl.
2. **Thermal Breaks:** To further reduce heat transfer, frames with thermal breaks include insulating materials between the outer and inner portions of the frame (Ijjada & Nayaka, 2022). Insulated frames are discussed in Figure 25.

Figure 25. Insulated frames

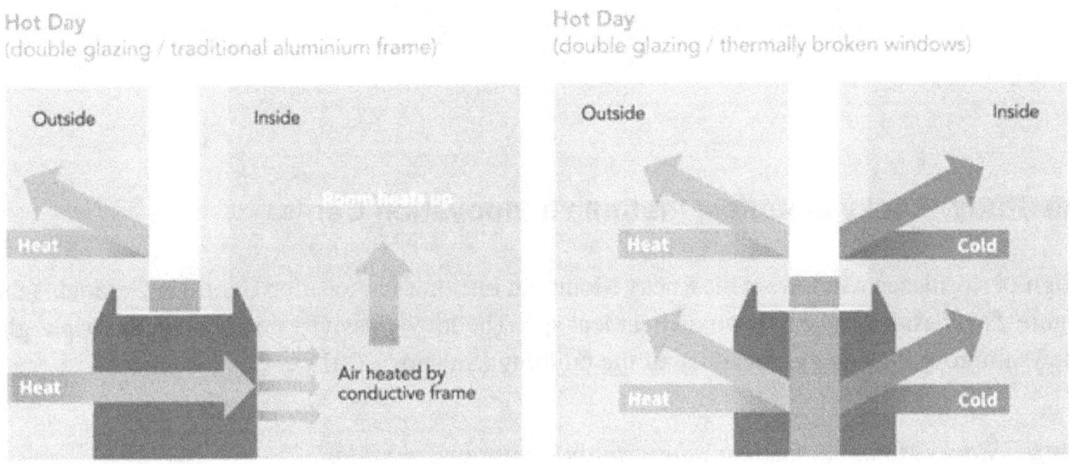

(Ijjada & Nayaka, 2022)

Advanced Sealing Techniques

1. **Weather-Stripping:** To improve airtightness, high-quality weather-stripping materials stop air from leaking around windows and doors.
2. **Appropriate Installation:** To avoid drafts and air leaks, windows, and doors must be installed properly and sealed using the right methods. These techniques are also shown in Figure 26.

Figure 26. Sealing techniques

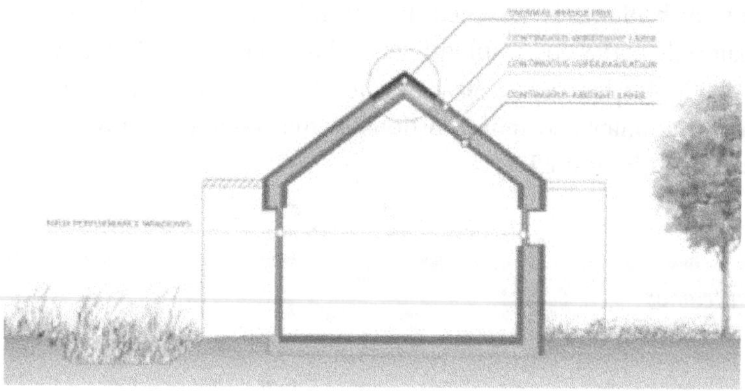

(Feng, 2021)

Case Study: Rocky Mountain Institute's Innovation Centre

High-performance windows at the Rocky Mountain Institute's Innovation Centre in Colorado (shown in Figure 27) considerably cut down on heat leakage. The low-emissivity coatings on the triple-glazed windows improve the energy efficiency of the building (Šujanová, 2019).

Figure 27. Rocky Mountain Institute's Innovation Centre in Colorado

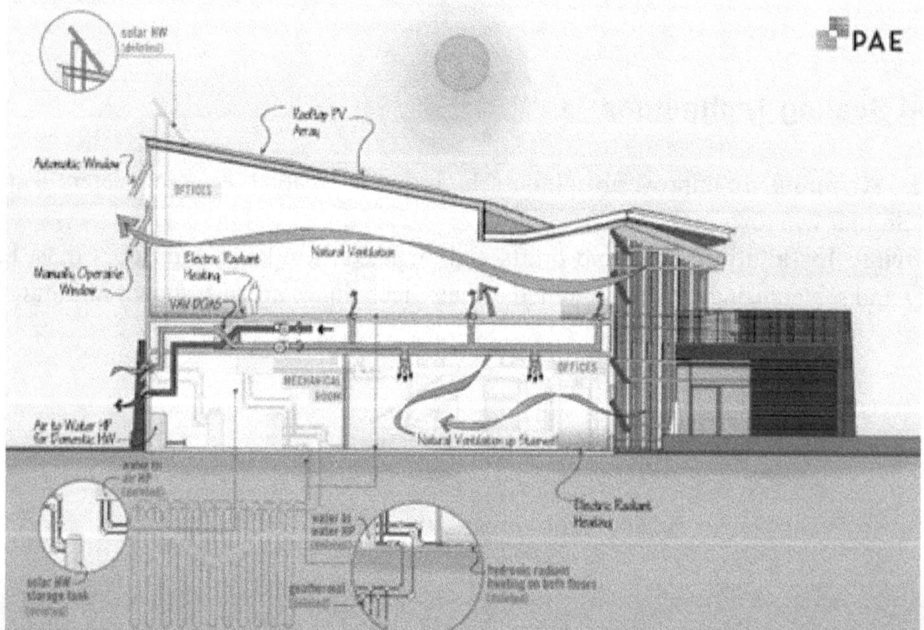

(Šujanová, 2019)

Solutions and Recommendations

Install low-emissivity coatings and low-U-valued high-performance windows. Make use of very thermally resistant airtight doors. Think of utilizing smart glass technologies that modify transparency according to the amount of sunshine.

Future Research Directions

Install low-emissivity coatings and low-U-valued high-performance windows. Make use of very thermally resistant airtight doors. Think of utilizing smart glass technologies that modify transparency according to the amount of sunshine.

7. HAVE PROPER SOLAR ORIENTATION

An essential component of passive design, proper solar orientation has a big impact on a building's natural illumination, thermal comfort, and energy efficiency. It entails arranging the structure and all of its components to maximize the heating, cooling, and lighting benefits of natural sunlight. In addition to minimizing summertime overheating and optimizing wintertime solar gains, proper solar orientation also guarantees a year-round cozy and energy-efficient interior atmosphere (Abid, 2019). Table 7 discusses the sun's orientation.

Table 7. Solar orientation

Aspect	Description	Benefits
Maximizing Solar Gains	Orienting the building to capture maximum sunlight during winter	Reduces heating costs, enhances thermal comfort
Minimizing Overheating	Using shading devices and minimizing direct sun exposure on east and west walls	Prevents overheating, and reduces cooling costs.
Minimizing Overheating	Using shading devices and minimizing direct sun exposure on east and west walls	Prevents overheating, and reduces cooling costs.
Daylighting	Optimizing window placement for abundant natural light	Decreases reliance on artificial lighting, improves comfort
Shading Devices and Overhangs	Implementing adjustable and fixed shading solutions	Controls solar gain adapts to seasonal changes
Site Considerations	Taking topography, vegetation, and neighborhood context into account	Enhances overall design efficiency and comfort

Maximizing Solar Gains

Winter Heating: Orienting the building to maximize solar exposure on the south-facing side can significantly boost passive solar heating in colder climates. Sunlight can enter through large south-facing windows, warming the interior areas.

Solar Path Tracking: By comprehending the sun's course throughout the year, windows, shading structures, and roof overhangs can be designed to maximize solar gains throughout the winter months.

Minimizing Overheating

1. **Summer Cooling:** Reducing the amount of direct sunlight that strikes east- and west-facing walls can assist prevent overheating in warmer areas. Awnings, louvers, and vegetation are examples of appropriate shade structures that can block too much summer sun.
2. **Cross-Ventilation:** Ensuring appropriate orientation and arrangement to promote cross-ventilation naturally benefits the building's natural cooling process.

Daylighting

1. **Natural Light Utilization:** Orienting windows and openings to the south and north ensures abundant natural light without causing glare or excessive heat gain. This reduces the need for artificial lighting and enhances indoor comfort.
2. **Skylights and Light Tubes:** These elements can be incorporated to bring natural light into deeper parts of the building, further optimizing daylighting (Wong, 2017).

Shading Devices and Overhangs

1. **Adjustable Shading:** Windows and other openings facing north and south allow for an abundance of natural light to enter the space without creating glare or undue heat gain. As a result, less artificial lighting is used, improving indoor comfort.
2. **Fixed Overhangs:** By adding these features, daylighting can be further maximized by allowing natural light to enter the building's deeper spaces (Kirimtat, 2016).

Site Considerations

1. **Topography and Vegetation:** Taking into account the site's topography and existing vegetation can influence the building's orientation and its interaction with the sun.
2. **Neighborhood Context:** Considering the surrounding buildings and their impact on sunlight access and shading is crucial in urban settings.

Case Study: PNC Tower

The PNC Tower in Pittsburgh, Pennsylvania shown in Figure 28, sometimes referred to as The Tower at PNC Plaza, is a symbol of energy efficiency and environmental design. Gensler's design for the skyscraper integrates cutting-edge passive design techniques and cutting-edge technologies, with a focus on natural ventilation and solar shading (Kaushal, 2022).

Key Features

1. **Innovative Solar Shading:**
   1. **Double-Skin Façade:** The building has an operable window layer inside and an exterior layer of glass making up its double-skin façade. In addition to providing natural ventilation, the space between the layers serves as a temperature buffer.
   2. **Automated Louvers:** Integrated into the exterior façade, these louvers are programmed to shift in response to the sun's position, minimizing heat gain and glare and maximizing natural light penetration.
2. **Natural Ventilation:**
   a) Breathable Building: The tower is designed to “breathe” with its double-skin façade, which facilitates natural ventilation and reduces the need for mechanical cooling. Operable windows allow fresh air to flow through the building.
   b) Solar Chimney: A solar chimney at the top of the building enhances the stack effect, drawing warm air up and out of the building while bringing in cooler air from lower levels.

Figure 28. PNC tower detail

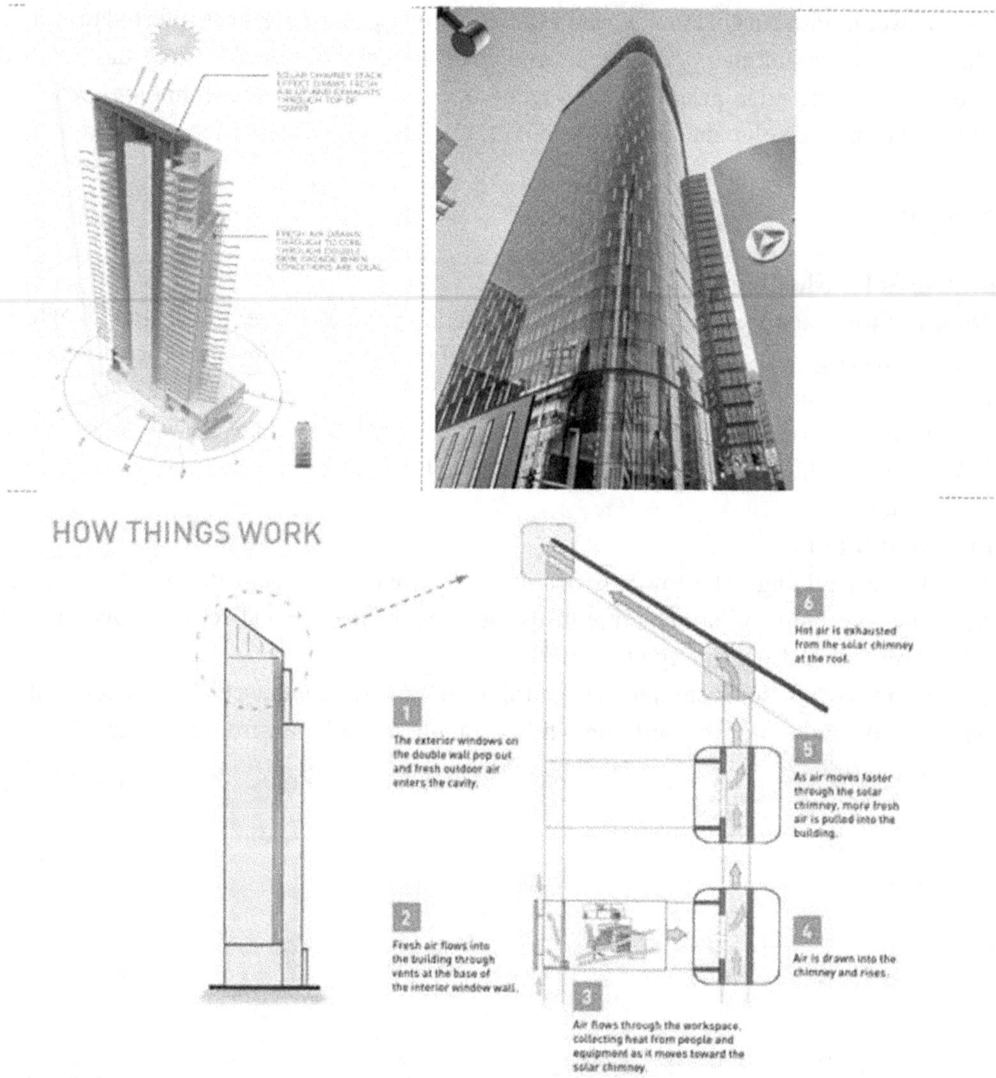

(*Kaushal, 2022*)

Solutions and Recommendations

Consider the local climate while designing buildings to ensure optimal solar orientation. To maximize passive solar heating and natural illumination, use building orientation. Include solar-shading mechanisms to avoid overheating.

Future Research Directions

Provide resources and recommendations for maximizing building orientation in various climates. Examine how building layout and urban density affect sun access and energy efficiency.

8. BUILD HIGH-PERFORMING WALLS

Passive architecture requires high-performing walls since they are essential to attaining thermal comfort, energy efficiency, and overall building performance. These walls are made to reduce heat transmission, stop air leaks, and offer superior insulation, all of which help to lower the amount of energy used for heating and cooling. It is discussed in table 8.

Table 8 Build high-performing walls

Feature	Description	Benefits
High Insulation Value	Thick layers of insulation providing high R-value	Reduces heat transfer, enhances thermal comfort, lowers energy consumption
Thermal Mass	Use of materials like concrete to absorb and release heat	Stabilizes indoor temperatures, reduces heating and cooling loads
Air Tightness	Meticulous sealing of joints, windows, and penetrations	Prevents air leakage, reduces energy loss, improves indoor air quality
Moisture Management	Incorporation of moisture barriers and vapor retarders	Prevents moisture infiltration, protects building structure, enhances durability
Thermal Bridging Reduction	Continuous insulation and thermal breaks in the wall assembly	Minimizes thermal bridging, improves overall thermal performance, increases energy efficiency

Case Study: Vancouver House

High-performing walls with cutting-edge insulation and moisture management systems can be found in the Vancouver House in Canada shown in Figure 29. The walls add to the building's longevity and energy efficiency (Dorbayani, 2019).

Figure 29. Vancouver House in Canada

(Dorbayani, 2019)

Solutions and Recommendations

Use wall systems that can effectively manage moisture and have strong thermal resistance. For improved performance, think about utilizing insulated concrete forms or double-stud walls. To prevent moisture problems and thermal bridging, pay attention to the wall detailing.

Future Research Directions

Examine cutting-edge wall systems with improved moisture and heat control capabilities. Examine how different wall systems operate over time in varying climatic circumstances.

9. THERMAL BRIDGE-FREE DETAILING

Thermal bridging, which may significantly decrease a building's overall thermal efficiency, is the result of a highly conductive material penetrating or breaking through the insulation layer. Thermal bridge-free detailing is crucial to reducing this. By designing and constructing building features to minimize or eliminate thermal bridges, indoor comfort, energy efficiency, and the likelihood of condensation and mold formation are all improved mentioned in Table 9 (Alhawari, 2018).

Table 9. Thermal bridge-free detailing

Feature	Description	Benefits
Continuous Insulation	The insulation layer is continuous across all building elements	Eliminates gaps in insulation, enhances thermal efficiency
Thermal Breaks	Non-conductive materials used in places where structural elements penetrate insulation	Reduces heat transfer, and prevents thermal bridging.
Insulated Connectors	Thermally insulated connectors for attaching components	Maintains continuity of insulation, reduces conductive pathways
Advanced Window Installation	High-performance windows with thermally broken frames and continuous insulation around openings	Prevents thermal bridging at window openings, and improves overall insulation performance.
Sealed Joints and Penetrations	Meticulous sealing of joints and penetrations	Maintains airtightness, prevents heat loss, improves energy efficiency

Key strategies for thermal bridge-free detailing include:

Continuous Insulation

1. Ensuring that the insulating layer is uninterrupted and gap-free across all structural components, including the floors, walls, and roofs.

Thermal Breaks

1. Non-conductive materials are called thermal breaks when structural components break the insulation layer, such as around window frames, roof-wall intersections, and balconies shown in Figure 30.

Figure 30. Thermal breaks

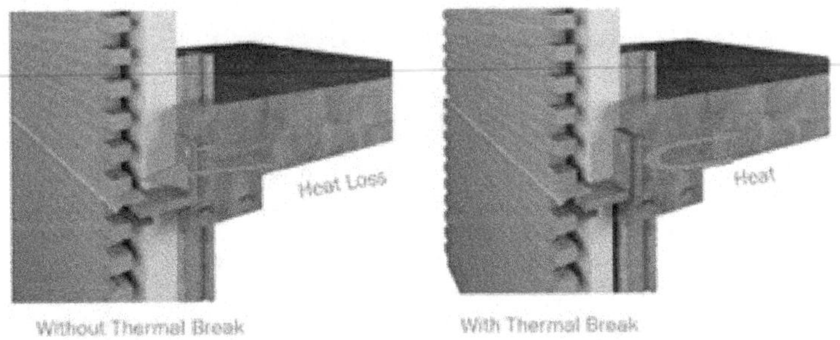

(Alhawari, 2018)

Insulated Connectors and Fasteners

1. Attaching building components with insulated connectors and fasteners minimizes the conductive routes for heat transmission shown in Figure 31 (Aurelija Stonkuvienė, 2021).

Figure 31. Insulated connectors and fasteners

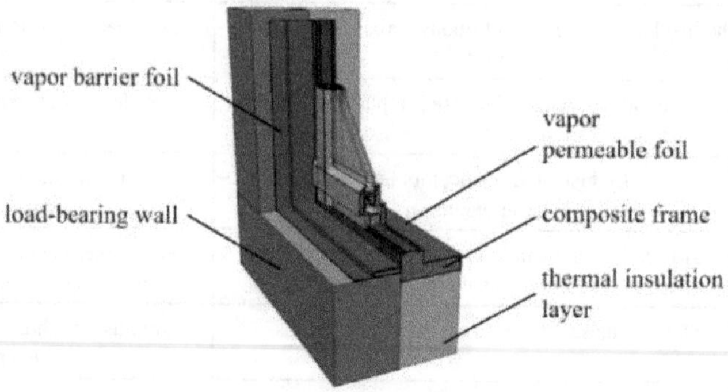

(Aurelija Stonkuvienė, 2021)

Advanced Window Installation

1. Installing windows with frames that are thermally cracked and making sure that the insulation layer is continuous around window openings.

Attention to Joints and Penetrations

- Effectively planning and repairing penetrations and joints to preserve the insulating layer's integrity and avoid thermal bridging shown in Figure 32 (Chaokun Hong, 2020).

Figure 32. Attention to joints and penetrations

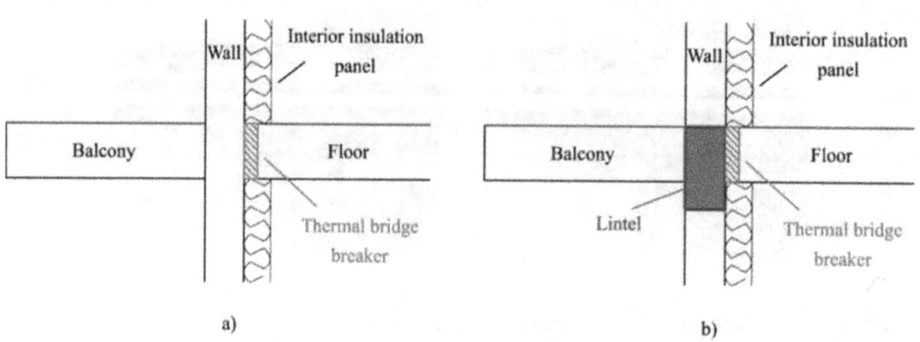

(Chaokun Hong, 2020)

Case Study: LUCIA Living Lab

Thermal bridge-free detailing is used in the LUCIA Living Lab in Norway (shown in Figure 33) to improve energy efficiency. The building's low energy usage can be attributed to the design's elimination of thermal bridges at door and window connections (Dell'Era, 2014).

Figure 33. Section of LUCIA Living Lab in Norway

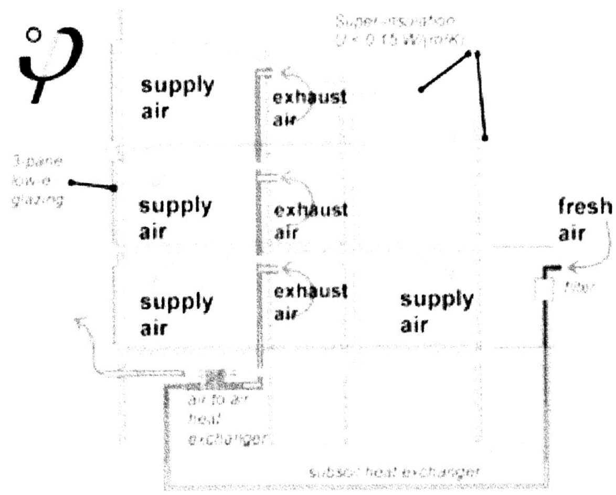

(Dell'Era, 2014)

Design features of the Passive House in Darmstadt-Kranichstein (Bednarova, 2011)

Building component	Description	Photograph of site	U-Value
Roof	Grass roof: Humus, non-woven filter, root protective membrane, 50 mm formaldehyde-free chipboard; Wooden light-weight beam (I-beam of wood, stud link of hardboard), counter lathing, sealing with polyethylene sheeting bonded without jointing, gypsum plasterboard 12.5 mm, wood-chip wallpaper, emulsion paint coating, entire cavity (445 mm) filled with blown-in mineral wool insulation.		0.1
Exterior wall	Fabric-reinforced mineral render; 275 mm of expanded polystyrene insulation (EPS) (installed in two layers at that time, 150+125 mm); 175 mm sand-lime brick masonry; 15 mm continuous interior gypsum plastering; wood-chip wallpaper, emulsion paint coating		0.14
Basement ceiling	Surface finish on fiberglass fabric; 250 mm polystyrene insulation boards; 160 mm concrete; 40 mm polystyrene acoustic insulation; 50 mm cement floor finish; 8-15 mm of parquet, adhesive; sealing solvent-free		0.13
Windows	Triple-pane low-e glazing with Krypton filling: Ug-value 0.7 W/(m^2K). Wooden window with polyurethane foam insulated framework (CO2-foamed, HCFC-free, handcrafted)		0.7
Heat Recovery ventilation	Counterflow air-to-air heat exchanger; Located in the cellar (approx. 9°C in the winter), carefully sealed and thermally insulated, the first one to use electronically commutated DC fans.		Heat recovery rate approx. 80%

Solutions and Recommendations

Consider thermal bridge-free details while designing structures. Prevent localized heat transfer by implementing thermal breaks at intersections and penetrations. To find and fix thermal bridges, perform thermal imaging testing.

Future Research Directions

Provide novel building materials and methods that eliminate thermal bridges. Examine the effects of thermal bridging on occupant comfort and the building's overall energy efficiency

CONCLUSION

The use of passive design techniques is essential to producing high-performance architecture. The passive design prioritizes natural components, minimizes the use of mechanical systems, and ensures precise detailing to increase energy efficiency and occupant quality of life. This chapter's ideas and case examples demonstrate how these tactics can be applied practically and impact real-world applications. The built environment will continue to be shaped by passive design concepts as we move towards a sustainable future.

Critical Review of Existing Studies

Sustainable architectural practices have benefited greatly from research on passive design tactics; yet, the efficiency of these strategies varies dramatically depending on factors such as building type, climate, and technological integration (Sheng Liu, 2020). The potential energy-saving benefits of passive methods including sun orientation, natural ventilation, and thermal mass have been widely acknowledged by existing studies. Passive solar design, for example, has been shown to reduce heating and cooling loads by as much as 50% in temperate climates (Emad Mushtaha, 2021), demonstrating its effectiveness in areas with major seasonal fluctuations. However, there are difficulties when implementing these strategies in extreme weather conditions. Research examining the efficiency of passive techniques in tropical or extremely humid settings reveals constraints. In tropical climates, cross-ventilation can effectively lower the need for cooling; nevertheless, problems with air quality, condensation, and discomfort during periods of high humidity can arise (Mohannad Bayoumi, 2017). This implies that to get the best outcomes, passive techniques must be properly tailored to the local climate. However, there are difficulties when implementing these strategies in extreme weather conditions. Research examining the efficiency of passive techniques in tropical or extremely humid settings reveals constraints (Arun Kumar Shukla, 2023).

In tropical climates, cross-ventilation can effectively lower the need for cooling; nevertheless, problems with air quality, condensation, and discomfort during periods of high humidity can arise. This implies that to get the best outcomes, passive techniques must be properly adjusted to the local climate. Also, there is a growing corpus of research on the combination of smart systems and the Internet of Things (IoT) with passive design (Kılıç, 2017). The performance of conventional passive techniques can be improved by automated shading systems, smart sensors, and AI-driven HVAC systems, according to recent studies (Chin-Chi Cheng, 2019). There is still a lack of study in this field, especially when it comes to long-term performance and large-scale application. There is a deficiency in the practical validation of these integrated systems in various building types because the majority of studies concentrate on simulations or small-scale applications. Significant geographic biases can be observed in the literature Sub-Saharan Africa, South Asia, and South America are under-represented in the study since a large amount of it is based on case studies from these three continents and temperate climates in North America and Europe. The widespread use of passive design would be advantageous and benefit more research conducted in a variety of settings, particularly in areas with more extreme weather (Carolina de los Ríos, 2018). Closing this gap could result in the creation of more locally relevant solutions that take into account the climate, building customs, and available resources.

Future Directions

Ongoing investigation into cutting-edge materials and technology to improve passive design strategies even more. For even greater sustainability, combine renewable energy systems with passive design techniques. Creation of software and design tools to make it easier to apply passive design concepts in architectural practice. Architects and designers can develop buildings that not only satisfy current needs but also contribute to a sustainable and resilient future by using passive design. Further key features are discussed in the Passive design strategy in Table 10.

Table 10. Passive Design Strategy (Rastogi, 2022)

Passive Design Strategy	Key Features	Benefits
Cross or Heat Recovery Ventilation	Recovers heat from exhaust air, improves indoor air quality	Reduces energy consumption, enhances thermal comfort
Open Spaces	Facilitates natural airflow and daylight penetration	Reduces reliance on artificial lighting and mechanical ventilation
Rethinking Mechanical Design	Energy-efficient HVAC systems, advanced controls	Optimizes temperature regulation, reduces energy consumption
Control Heat	Shading devices, thermal mass, passive solar techniques	Manages heat gain/loss, improves thermal comfort
Optimal Insulation	High R-value materials, continuous insulation	Minimizes heat transfer, ensures year-round thermal comfort
High-Performance Windows and Doors	Triple glazing, thermally broken frames, airtight installation	Superior insulation minimizes thermal bridging, maximizes natural light
Proper Solar Orientation	Building alignment to optimize solar gains and reduce overheating	Enhances energy efficiency, improves occupant comfort
Thermal Bridge-Free Detailing	Continuous insulation, thermal breaks, insulated connectors	Prevents heat transfer, enhances overall thermal efficiency
High-Performing Walls	Combination of high insulation value, thermal mass, airtightness, moisture management	Achieves superior thermal performance, reduces energy consumption, enhances durability

Incorporating passive design strategies is crucial to achieving energy efficiency and sustainability in modern architecture. As demonstrated, these techniques not only reduce energy consumption but also enhance indoor comfort and environmental resilience. Moving forward, the integration of these principles will play a vital role in shaping a more sustainable built environment.

REFERENCES

Abid, E. (2019). The optimum tilt angle and orientation for Solar panels. Journal Port Science Research, 259-263.

Adam, A. (2019). Optimum Thermal Insulation Thickness for Building Under Different Climate Regions - A Review. *Thermal Science*, 59, 254–268.

Aguilar-Santana, J., Jarimi, H., Velasco-Carrasco, M., & Riffat, S. (2020). Review on window-glazing technologies and future prospects. International Journal of Low-Carbon Technologies. *The International Journal of Low Carbon Technologies*, 15(1), 112–120. DOI: 10.1093/ijlct/ctz032

Ahmadkhani, B. (2011). Windcatcher: passive and low energy cooling system in Iranian vernacular architecture. Tech Phys Problems Eng., 3.

Aitichya Chandra, A. P. (2020). Designing Zoning Systems for Freight Transportation Planning: A GIS-based approach for Automated Zone Design using Public Data Sources. *Transportation Research Procedia*, 48, 605–619. DOI: 10.1016/j.trpro.2020.08.063

Alain Starke, M. W. (2021). Promoting Energy-Efficient Behavior by Depicting Social Norms in a Recommender Interface. *ACM Transactions on Interactive Intelligent Systems*, 11(3-4), 1–32. DOI: 10.1145/3460005

Alhawari, A. &. (2018). Thermal bridges in building envelopes – An overview of impacts and solutions. International Review of Applied Sciences and Engineering, 31-40.

Anil Aswani, N. M. (2012). Reducing Transient and Steady State Electricity Consumption in HVAC Using Learning-Based Model-Predictive Control. *Proceedings of the IEEE*. DOI: 10.1109/JPROC.2011.2161242

Arun Kumar Shukla, A. K. (2023). Active and passive methods for cooling load reduction in a tropical building: A case study. *Energy Conversion and Management*, 293(117490).

Aurelija Stonkuvienė, R. B. (2021). The impact of connector's thermal and geometrical characteristics on the energy performance of facade systems. *Journal of Building Engineering*.

Bas Knoll, W. B. (2018). Improving the usability and performance of heat recovery ventilation systems in practice. aivc.org.

Bayoumi, M. (2017). Energy saving method for improving thermal comfort and air quality in warm humid climates using isothermal high-velocity ventilation. *Renewable Energy*, 114, 502–512. DOI: 10.1016/j.renene.2017.07.056

Bednarova, P. V. (2011). *An energy-efficient building for the Arctic climate Is a passive house a sensible solution for Greenland?*

Carolina de los Ríos, J. E. (2018). Persistence of methodological, taxonomical, and geographical bias in assessments of species vulnerability to climate change: A review. *Global Ecology and Conservation*, 15, e00412. Advance online publication. DOI: 10.1016/j.gecco.2018.e00412

Chaokun Hong, H. L. (2020). Review of connections for engineered bamboo structures. *Journal of Building Engineering*.

Chin-Chi Cheng, D. L. (2019). Artificial Intelligence-Assisted Heating Ventilation and Air Conditioning Control and the Unmet Demand for Sensors: Part 1. Problem Formulation and the Hypothesis. *Sensors (Basel)*, 19(5), 1131. Advance online publication. DOI: 10.3390/s19051131 PMID: 30845664

Dai, D. (2021). *Future Courtyard 2050*. UNITEC Institute of Technology.

Dan Zhu, M. K. (2015). Analysis of Sustainable Energy Systems in Ecovillages: A Review of Progress in BedZED and Masdar City. *Low Carbon Economy*, 6(1).

Dell'Era, C., & Landoni, P. (2014). Living Lab: A Methodology between User-Centred Design and Participatory Design. *Creativity and Innovation Management*, 23(2), 137–154. DOI: 10.1111/caim.12061

DeQuante Rashon Mckoy, R. T. (2023). Review of HVAC Systems History and Future Applications. *Energies*, 16(17), 6109. Advance online publication. DOI: 10.3390/en16176109

Dewsbury, M. (2015). The Environmental Measurement of Residential Buildings A Technical Guide. Dewsbury, Mark: The Environmental Measurement of Residential Buildings. DOI: 10.13140/RG.2.1.1520.7840

Dorbayani, M. (2019). INVESTIGATING VANCOUVER'S HOUSING CRISIS. Manchester: University of Salford. DOI: 10.17866/rd.Salford.17124347.v1

Elshafei, G., Vilcekova, S., Zelenakova, M., & Negm, A. M. (2021). Towards an Adaptation of Efficient Passive Design for Thermal Comfort Buildings. *Sustainability (Basel)*, 13(17), 9570. DOI: 10.3390/su13179570

Emad Mushtaha, T. S. (2021). The impact of passive design strategies on cooling loads of buildings in a temperate climate. *Case Studies in Thermal Engineering*, 28(101588), 101588. DOI: 10.1016/j.csite.2021.101588

Faye, V., & McNeill, R. C. (2022). Room-level ventilation in schools and universities. *Atmospheric Environment: X*, •••, 13. PMID: 35098105

Feng, F., Kunwar, N., Cetin, K., & O'Neill, Z. (2021). A critical review of fenestration/window system design methods for high-performance buildings. *Energy and Building*, 248, 248. DOI: 10.1016/j.enbuild.2021.111184

Hafez, F. S., Sa'di, B., Safa-Gamal, M., Taufiq-Yap, Y. H., Alrifaey, M., Seyedmahmoudian, M., & Mekhilef, S. (2023). Energy efficiency in sustainable buildings: A systematic review with taxonomy, challenges, motivations, methodological aspects, recommendations, and pathways for future research. *Energy Strategy Reviews*, 45, 101013.

Hamza, A. (2023). Renewable Energy Integration.

Hasim Altan, M. H. (2016). Passive Design. ZEMCH: Toward the Delivery of Zero Energy Mass Custom Homes. DOI: 10.1007/978-3-319-31967-4_8

Hassani, A. R., Domenighini, P., Belloni, E., Ihara, T., & Buratti, C. (2022). Evaluation of the solar heat gain coefficient of innovative aerogel glazing systems: Experimental campaigns and numerical results. *Journal of Building Engineering*, 62, 62. DOI: 10.1016/j.jobe.2022.105354

Heidari, A., Taghipour, M., & Yarmahmoodi, Z. (2021). The Effect of Fixed External Shading Devices on Daylighting and Thermal Comfort in Residential Buildings. *Journal of Daylighting*, 8(2), 165–180. DOI: 10.15627/jd.2021.15

Ibrahim, U. (2020). *Design Strategies and Smart Technologies that Contribute to Well-Being in the Workplace; The Case Studies of the Atlas and the Edge in the Netherlands*. Research in Urbanism and Architecture.

Ijjada & Nayaka, R. (2022). Review on properties of some thermal insulating materials providing more comfort in the building. *Materials Today: Proceedings*.

INSTITUTE. I. L. (2022). KHOO TECK PUAT HOSPITAL. Retrieved from https://living-future.org/: https://living-future.org/case-studies/award-winner-khoo-teck-puat-hospital/

Ismaiel, M. &.-N. (2021). Thermal resistance of masonry walls: A literature review on influence factors, evaluation, and improvement. *Journal of Building Physics*, •••, 528–567.

Jiyoung, L. E. E. K. S. (n.d.). Passive Design Techniques applied To green Buildings as an aesthetic and spatial Design concept. Journal of Green Building, 10(2), 79-109. Retrieved from https://www.researchgate.net/publication/281169946_Passive_design_techniques_applied_to_green_buildings_as_an_aesthetic_and_spatial_design_concept

Kaushal, A. (2022). *The Tower at PNC Plaza by Gensler*. Rethinking the Future.

Khairy Sayed, H. A. (2018). Building Energy Management Systems (BEMS). In H. A. Khairy Sayed, Energy Conservation in Residential, Commercial, and Industrial Facilities (pp. 15-81). DOI: 10.1002/9781119422099.ch2

Kılıç, T. &. (2017). An Investigation on Internet of Things Technology (IoT) In Smart Houses. Uluslararası Muhendislik Arastirma ve Gelistirme Dergisi, 196-207.

Kirimtat, A., Koyunbaba, B. K., Chatzikonstantinou, I., & Sariyildiz, S. (2016). Review of simulation modeling for shading devices in buildings. *Renewable & Sustainable Energy Reviews*, 53, 23–49. DOI: 10.1016/j.rser.2015.08.020

Kooi, Y. K. (2022). Adaptability and Flexibility in Architecture Concepts & Theories Applied in Residential Architecture to Achieve Adaptability.

Kuczynski, T., & Staszczuk, A. (2020). Experimental study of the influence of thermal mass on thermal comfort and cooling energy demand in residential buildings. *Energy*, 195(1), 116984. DOI: 10.1016/j.energy.2020.116984

Lin Chen, Y. H.-S. (2024). Green building practices to integrate renewable energy in the construction sector: A review. *Environmental Chemistry Letters*, 22(2), 751–784. DOI: 10.1007/s10311-023-01675-2

Liu, S., Kwok, Y. T., Lau, K. K. L., Ouyang, W., & Ng, E. (2020). Effectiveness of passive design strategies in responding to future climate change for residential buildings in hot and humid Hong Kong. *Energy and Building*, 228, 110469.

Mamdouh El Haj Assad, M. N. (2022). *Geothermal Heat Pumps: Principles and Applications*. Advances in Science and Engineering Technology International., DOI: 10.1109/ASET53988.2022.9734907

Michailidis, I. (2015). Improving Energy Savings and Thermal Comfort in Large-scale Buildings via Adaptive Optimization. In Miranda, F. (Ed.), *I. &. Michailidis, Control Theory: Perspectives, Applications, and Developments* (pp. 14-1–14-22). Nova Science., DOI: 10.1159/9781634827300.ch14

Mohamed Azeroual, T. L. (2019). Simulation tools for a smart grid and energy management for a microgrid with wind power using a multi-agent system. *Wind Engineering*, 44(6).

Mohamed Azeroual, T. L. (2019). Simulation tools for a smart grid and energy management for a microgrid with wind power using a multi-agent system. *Wind Engineering*.

Peña, R. B. (2020). *Living Proof THE BULLITT CENTER: High Performance Building Case Study*. University of Washington Center for Integrated Design.

Pinar Mert Cuce, S. R. (2014). A comprehensive review of heat recovery systems for building applications. *Renewable & Sustainable Energy Reviews*, •••, 47.

Potienko, N. (2017). Low-grade energy of the ground for civil engineering. MATEC Web of Conferences. DOI: 10.1051/matecconf/201710606026

Rana, A. &. (2018). Thermal performance of double and triple-glazed windows Experimental results from the lab and in-situ measurements. *1st International Conference on New Horizons in Green Civil Engineering*. Victoria, BC, Canada.

Rastogi, M. M. (2022). A Review of Passive Design Strategies for Improving Building Energy Performance. International Journal of Multidisciplinary Innovative Research.

Raut, H., Ganesh, V. A., Nair, A. S., & Ramakrishna, S. (2011). Anti-Reflective Coatings: A Critical, In-Depth Review. *Energy & Environmental Science*, 4(10), 3779–3804. DOI: 10.1039/c1ee01297e

Roberto Stasi, F. R. (2024). Influence of cross-ventilation cooling potential on thermal comfort in high-rise buildings in a hot and humid climate. *Building and Environment*, •••, 248.

Salameh, M., & Touqan, B. (2022). Traditional Passive Design Solutions as a Key Factor for Sustainable Modern Urban Designs in the Hot, Arid Climate of the United Arab Emirates. *Buildings*, 12(11), 1811. Advance online publication. DOI: 10.3390/buildings12111811

Salwa, M., & Al-Masrani, K. M.-O. (2019). Dynamic shading systems: A review of design parameters, platforms, and evaluation strategies. *Automation in Construction*, 102, 195–216. DOI: 10.1016/j.autcon.2019.01.014

Schulze, T., & Eicker, U. (2013). Controlled natural ventilation for energy-efficient buildings. *Energy and Building*, 56, 221–232. DOI: 10.1016/j.enbuild.2012.07.044

Shweta Singh, S. V. (2019). *Home Automation Using Node MCU*. Firebase & IOT.

Šujanová, P. (2019). A Healthy, Energy-Efficient and Comfortable Indoor Environment, a Review. *Energies*.

Świercz, E. Z. (2021). A Review of Heat Recovery in Ventilation. *Energies*, 14(6), 1759. DOI: 10.3390/en14061759

Tan, S. Y. (2012). The Practice of Integrated Design: The Case Study of Khoo Teck Puat Hospital, Singapore. Nottingham: Masters of Science in Sustainable Building Design in BCA Academy.

Toroxel, J. L., & Silva, S. M. (2024). A Review of Passive Solar Heating and Cooling Technologies Based on Bioclimatic and Vernacular Architecture. *Energies*, 17(5), 1006. DOI: 10.3390/en17051006

Wahab, I. A., Ismail, L. H., & Kadir, A. (2019). Opening Design and Position Effect on Building Natural Stack Effect and Cross Ventilation. *International Journal of Engine Research*, 5(1), 13–22.

Wong, I. (2017). A review of daylighting design and implementation in buildings. *Renewable & Sustainable Energy Reviews*, 74, 959–968. DOI: 10.1016/j.rser.2017.03.061

Young, J. (2015). Towards zero energy buildings: lessons learned from the BedZED development.

KEY TERMS AND DEFINITIONS

Passive Design: This is the term for architectural design techniques that regulate a building's temperature and air quality without the need for air conditioners or heaters by utilizing natural resources like direct sunlight and wind. Passive design reduces energy use and produces more comfortable places by cooperating with nature.

Thermal Mass: Materials that can absorb heat during the day and release it at night, such as concrete or brick. Keeping buildings warm at night and cool during the day, contributes to more consistent interior temperatures with limited energy consumption.

Cross Ventilation: A natural cooling method that uses wind pressure to force fresh air through a building from one side to the other. By allowing cool, fresh air to circulate throughout indoor rooms, it reduces the demand for air conditioning.

Heat Recovery Ventilation (HRV): The system heats the fresh air that enters by absorbing heat from the stale indoor air. It's an ideal way to provide a constant flow of fresh air indoors while conserving energy, particularly in colder climates.

Shading Devices: Overhangs, blinds, and louvers are examples of structures used to obscure direct sunlight and keep people from getting too hot. By letting in natural light and decreasing the need for air conditioning, they develop comfortable interior spaces.

Smart Technologies (IoT): Internet-enabled building appliances that change settings in real-time, such as automatic blinds or smart thermostats. These innovations facilitate the creation of more pleasant, energy-efficient dwellings.

Solar Orientation: The technique of designing a structure to best utilize the path of the sun. This involves minimizing direct sunlight in warmer regions to keep the building cool in the winter and optimizing sunshine in cooler climates to warm the building.

R-Value: A measure of how well the extent to which insulation keeps heat from entering or leaving a building. Better insulation has a higher R-value, translating into a more pleasant and energy-efficient building throughout the year.

Airtight Envelope: This is the building's perimeter seal that keeps air from leaking in or out. The indoor temperature of a well-sealed building remains constant, which demands less energy to heat or cool the area.

Return on Investment (ROI): In the context of passive design, it refers to the energy savings and cost reductions to observe over time by using energy-efficient building strategies.

Chapter 9
Navigating the Green Financial Markets:
Opportunities and Challenges

Charu Banga
https://orcid.org/0000-0002-6288-1419
Rochester Institute of Technology, Dubai, UAE

Momina Afzal
https://orcid.org/0009-0002-2247-9043
De Montfort University, UK

ABSTRACT

Green Financial Markets play a significant role in directing capital towards environmentally friendly projects, renewable energy, and other green initiatives. The demand for green financial products such as green bonds and sukuks, green derivatives, as well as green investment funds, is expected to increase phenomenally as consumer awareness of environmental issues grows. However, as the sector expands, it encounters challenges such as greenwashing, identifying and managing green risks, and ensuring data availability and quality. Effective regulations and supportive policies harmonizing global standards and reporting will ensure transparency, reliability, and accountability in financial flows directed towards green investments. The ongoing evolution of the green markets will support moderating the impact of climate change and undoubtedly offer new opportunities for investors and businesses to thrive in a green economy.

1. INTRODUCTION TO GREEN FINANCIAL MARKETS

1.1 Defining Green Financial Markets

According to the United Nations Environment Program, green financing involves increasing financial support from private, public, and non-profit organizations to promote sustainable development. Green finance plays a critical role in addressing environmental damage and climate change threats (Bhatnagar & Sharma, 2022). This type of financing includes banking, insurance, micro-credit, and investment and

DOI: 10.4018/979-8-3693-5748-4.ch009

aims to manage social and environmental costs and risks, while creating opportunities for improved returns and environmental benefits, and ensuring greater accountability. Green finance is commonly used to finance sustainable projects and reduce carbon emissions. Similar to traditional financial markets, green financial markets include bonds, stocks, commodities, cryptocurrencies, and carbon markets, all in support of these sustainable objectives.

1.2 Green Finance and its Growth

The concept of green finance has its origins in the 1970s. However, it wasn't until September 2015, during the adoption of the Sustainable Development Goals (SDGs) and the United Nations Development Programme (UNDP) in Paris, that the issues of sustainability of human activities at a societal and political level were widely debated (Aghion et al., 2022). The sustainability movements started to emerge, and initiatives such as Principles of Responsible Investments (PRI) involved investors to voluntarily adopt and implement six principles of responsible finance. Secondly, the sustainability accounting standards board disclosure standard, the Task Force on Climate-related Financial Disclosure, and recently, the European Union (EU) Taxonomy create a common understanding of what is classified as environmentally sustainable. To shape the behavior of today's markets, through these initiatives, such standards and principles are set to move the needle on green financing.

As green finance emerges, the financial markets are becoming increasingly concerned. For example, the leading financial market Bloomberg estimates a growth of $53 trillion by 2025 on ESG assets, representing more than a third of the total asset representation. However, the basis of this prediction comes from a rather controversial trend of green finance, where global sustainable finance volumes faced a decrease in trend from a $1.8tn peak in 2021 to $1.55tn in 2022 and reached $1.3tn in 2023. On the other hand, green bond issuance saw an 11 percent increase every year until 2023. Figure 1 depicts the increasing growth of green finance within sustainable finance products. The upward trend of green finance can be attributed to various political and economic factors, including the impact of the American elections on the region.

Figure 1. Green finance products

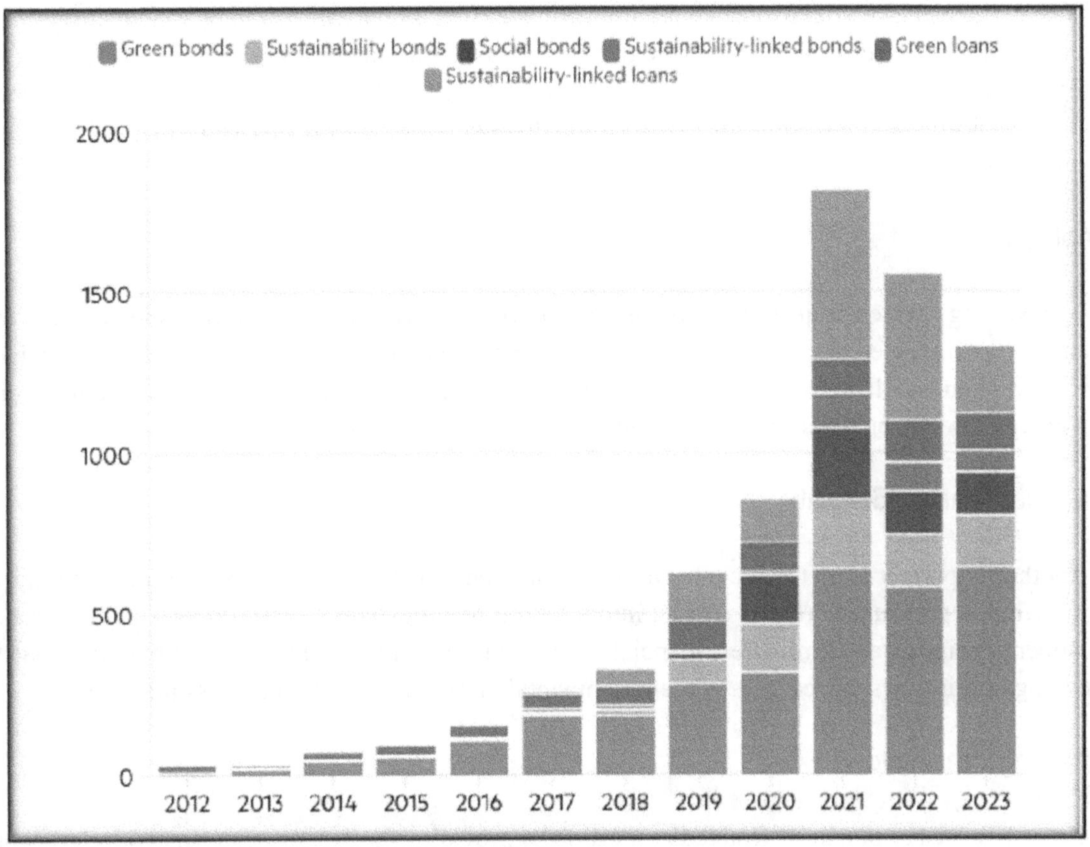

Source: https://www.ingwb.com/binaries/content/assets/insights/themes/sustainable-finance/ing-sustainable-finance-pulse---issue-1.pdf

1.3 Significance of Green Finance for Economic Development

Green finance plays a vital role in driving economic development by channeling financial resources toward environmental and sustainable projects (Allen & Clouth, 2012). Several key aspects underline the significance of this impact are:

Sustainable Growth

Green finance encourages investments that use energy-efficient, renewable energy, and sustainable technology. Transitioning to a carbon-free economy promotes sustainable and environmental growth by omitting the risks of climate change and reducing dependence on fossil fuels. Countries, therefore, try to promote green finance as part of their economic development objectives.

Climate Risk Mitigation

Green finance plays a huge role in mitigating climate risks. Economies do so by funding projects that enhance resilience to climate impacts, reduce greenhouse emissions, and avoid costs associated with extreme weather conditions, long-term environmental damage, and natural disasters. Although countries cannot completely omit the occurrence of a natural disaster, they do try to take steps in advance to avoid the associated risks.

Employment

By investing in green projects, the economy creates more jobs and tries to achieve its second economic objective (the increased employment rate). For example, a renewable energy project such as a wind or solar farm requires a labor force for construction and operation. The project does not have to resonate with energy directly but could be any way involved in promoting sustainability.

Financial Market Stability

Another important aspect that contributes to economic growth is the incorporation of environmental, social, and governance (ESG) criteria into decision-making. This is very important when making investment decisions as it promotes financial market stability. It not only identifies but also assists in managing the risks associated with non-environmental or non-sustainable business practices.

Fostering Innovation

Green finance has popularly fostered innovation by providing funds for research and development on new processes and technologies that are less harmful to society and the environment. While it promotes the innovation culture of the economy, it also assists in enhancing the competitiveness of economies in an attempt to improve their adoption of such innovations. These innovations do not usually provide instant results but are more forward and future-focused.

Regulation and Policy

To promote sustainable finance, governments regulate and implement policies such as carbon pricing, stricter environmental standards, and renewable energy. These regulations and policies assist governments and organizations in helping firms avoid potential risks associated with negative externalities. Therefore, they act as a way to foster and become an environment-friendly country.

2. UNDERSTANDING GREEN INVESTMENTS

An important element of green finance is green investments. These investments prioritize environmentally friendly returns alongside financial returns. Green investments allocate funds and resources to projects that are socially responsible, environmentally friendly and follow good corporate governance.

Below are four different types of green investments: Green Equity/Stocks, Green Bonds and Sukuks, Green Funds and ETFs, and Sustainable & Responsible Investments (SRI).

2.1 Green Equity/Stocks

Green Equity, also known as green stocks, refers to shares in companies that are sustainable and environmentally friendly. These companies are involved in practices such as waste management, CO_2 emissions reduction, energy efficiency, and clean technology. Key characteristics of green stocks include sector focus, adherence to ESG criteria, and transparent reporting. The major benefits of green stocks include potential growth, particularly in renewable energy, due to increased global demand and technological advancement. Additionally, investors who fund such projects contribute to reducing carbon footprint and promoting sustainability. Companies that prioritize sustainability tend to manage regulatory risks better than those that are not environmentally friendly (Roy, 2015). Notable examples of green stocks include Tesla Inc, a leading electric vehicle company, and Beyond Meat Inc, which focuses on reducing environmental damage from meat production and promoting plant-based protein.

2.2 Green Bonds and Sukuks

Green Bonds are fixed-income instruments to raise funds for environmental and climate projects, while Sukuks are Sharia-law-compliant financial instruments similar to bonds, also used in green finance. They are characterized by utilizing funds exclusively for promoting environmental benefits, third-party verification and certification, and providing transparency through reports. These instruments appeal to investors concerned about efficient resource use and preserving resources for future generations, and they also contribute to market growth and provide risk diversification (Chang et al., 2022). Examples include Apple Green Bonds, Green Sukuk (issued in Indonesia and Malaysia), and World Bank Green Bonds.

2.3 Green Funds and ETFs

Exchange-traded funds (ETFs) and Green Funds are investment pools created by multiple investors, comprising a diversified portfolio of green bonds, stocks, and other securities. The main benefit of these funds is diversification, which helps to spread the risk. Additionally, they offer active and passive management by portfolio managers who carefully select investments and track green indices. ETFs, in particular, have low investment thresholds, making them more accessible to all investors. The major benefits include professional management and accessibility, as they encourage individual investors to participate without extensive research (Naqvi, 2022). These funds also offer liquidity as they trade on stock exchanges, providing opportunities for investors. Some common examples of these funds are Green Century Balanced Fund, iShares Global Clean Energy ETF, and Vanguard ESG U.S. Stock ETF.

2.4 Sustainable and Responsible Investments (SRI)

Sustainable and Responsible Investments (SRI) are those investments that integrate the ESG criteria to generate long-term financial and economic returns. The major characteristics include economic, social, and governance factors integration as a criterion for selection. Secondly, engagement through investors and advocacy for sustainable and economical projects. The screening of any negative or positives, such as harmful activities or positive contributions. The benefits are that it goes beyond environmental and includes governance and social. Moreover, it is long-term and short-term focused. It also leads to better shareholder engagement that leads to improved business accountability towards sustainable practices (Allevi et al., 2019). For example, Pax Global Environmental Markets Fund invests in firms that provide solutions to environmental and sustainability challenges at a global level. Another example is Calvert Research and Management, which offers multiple SRI funds to projects focusing on ESG.

3. OTHER GREEN INVESTMENT INSTRUMENTS AND PRODUCTS

3.1 Green Loans and Credit Facilities

Green loans and credit facilities are tailored investment tools intended to support sustainable and environmentally friendly projects. They require that the funds be exclusively allocated to green initiatives, such as waste management, water conservation, energy efficiency, low carbon emission projects, and other sustainable endeavors. Borrowers are expected to meet specific eligibility criteria that demonstrate their dedication to environmentally friendly projects, including compliance with green standards, certification from recognized environmental bodies, and sustainability impact assessments. Similar to traditional loans, green loans may offer competitive interest rates to encourage individuals and businesses to invest in environmentally friendly projects.

The recipients of green loans and credit facilities are often required to submit regular reports detailing how the funds are used and the resulting impact on the environment. This ensures transparency and accountability, allowing lenders to verify that the funds are being used for their intended purpose. Many green loans employ third-party verification to assess whether the projects meet specific environmental, social, and governance (ESG) criteria.

Green loans and credit facilities offer multiple benefits, including fostering sustainable development for the long-term benefit of the environment and society. Additionally, they result in cost savings through energy-efficient upgrades, reducing utility bills and other expenses. Furthermore, green loans enhance a company's reputation and brand image by attracting more investors, customers, and the public who value a commitment to the environment. They also serve to encourage small businesses, particularly those dedicated to serving society and the environment, to enter the market.

3.2 Green Finance Derivatives

The green financial derivatives are derived from underlying indices, assets or rates that particularly focus on environment and sustainability[1]. While the purpose of these derivatives is to promote ESG, they do manage risks pertaining to low carbon emission and climate-related risks. The different types of green financial derivatives are discussed below:

Green Options

The term "green options" refers to the right, but not the obligation, to buy or sell a green asset at a predetermined price before a certain date. These options are commonly used for speculating on the price movements of carbon credits or renewable energy stocks. The most prevalent green options are carbon options, which are based on carbon credits or emissions allowances. By purchasing options, traders can secure a future price for carbon credits, which is essential for budgeting and financial planning in industries subject to emissions regulations. For example, options on the European Union Allowances (EUA) are traded, enabling companies to hedge against future price changes in the cost of carbon emissions. The other environmentally friendly options include renewable energy, disaster preparedness, and weather-related choices, among others[2].

Green Futures

Green futures are agreements to buy or sell a green asset at a predetermined price on a future date. These contracts help businesses mitigate price fluctuations in markets related to environmental goods or compliance obligations. For instance, in the European Union, companies engage in trading Carbon Emission Futures on the Intercontinental Exchange (ICE). These futures enable entities such as utility companies to fix prices for carbon credits, effectively managing their costs of complying with carbon emissions[3].

Carbon Credit Derivatives

These are centered around carbon credits, which are allocated based on the volume of carbon dioxide or greenhouse gases emitted. They encompass carbon futures and options that provide businesses with the means to effectively manage their expenses related to emissions and carbon reductions. Major energy companies in Europe rely heavily on these derivatives to ensure seamless compliance with EU carbon regulations. The trading of these derivatives is facilitated by major exchanges such as ICE Futures Europe, allowing for effective management of carbon-related costs.

Renewable Energy Certificate (REC) Derivatives

Renewable energy certificates certify that the bearer owns one megawatt-hour of electricity. These tradable instruments can allow companies to hedge against price fluctuations in the energy markets. In markets like Texas (ERCOT), renewable energy developers utilize REC derivatives to secure financing for new projects and mitigate the impact of fluctuating market prices of RECs[4].

Weather Derivatives

As the term implies, these derivatives are utilized to protect against adverse weather conditions. They are particularly important for renewable energy projects, which rely heavily on weather patterns (Matsumoto and Yamada, 2023). For example, in the case of wind energy production, options and futures are based on wind speed or temperature. Typically, wind farms purchase wind derivatives that provide payouts when wind speeds are lower than expected, thereby compensating for reduced electricity generation and sales.

Catastrophe Bonds

These bonds are issued to provide insurance against natural disasters, such as hurricanes or earthquakes. Investors in these bonds receive high yields but risk losing their principal if a disaster triggers a payout. The California State Authority utilizes these catastrophe bonds to mitigate the risk associated with large-scale earthquakes, thereby reducing the financial impact on insurance reserves and ensuring liquidity in the aftermath of a major disaster[5].

4. EVALUATING GREEN INVESTMENT OPPORTUNITIES

4.1 Risk and Return Analysis

Every investment requires an evaluation of the risk and return ratio. Generally, higher risk comes with the potential for higher returns. Green investment and finance are particularly intriguing because returns can be both monetary and non-monetary, especially in terms of environmental impact. It is essential for businesses to balance risks with financial and non-financial returns, enabling investors to make well-informed decisions (Purwanti, 2023). Diversification, by spreading investments across different sectors and securities, can help manage risk. Additionally, conducting thorough due diligence provides investors with a comprehensive understanding of associated risks, allowing them to examine environmental and social factors. Lastly, to mitigate short-term volatility, investors can balance risk and return by investing in long-term securities to fully realize the potential and benefits of green investments.

4.2 Environmental, Social, and Governance (ESG) Criteria

Using economic, social, and governance (ESG) criteria is crucial for evaluating green investments. To effectively assess these criteria, investors must dive into ESG reports and conduct thorough research, analyzing data such as sustainability reports and disclosures. This in-depth analysis allows investors to gain a comprehensive understanding of the ESG rating or environmentally friendly score of the company under consideration. Furthermore, it is highly advisable for investors to actively engage with companies, seeking information about their ESG-based projects and the extent of their commitment to such initiatives. This engagement provides valuable insight into the company's future outlook and their dedication to ESG objectives.

In addition to research and engagement, investors should utilize established ESG frameworks and standards, such as the Task Force on Climate-Related Financial Disclosures (TCFD), Sustainability Accounting Standards Board (SASB), and Global Reporting Initiatives (GRI). These frameworks provide a roadmap for evaluating green investments and offer essential information necessary for informed decision-making.

It's important to note that evaluating green investments is an ongoing process. Investors should consistently monitor ESG performance and regularly review their investments to ensure alignment with ESG criteria. This iterative approach allows investors to make well-informed decisions and stay abreast of any changes in the ESG landscape.

4.3 Impact Investing Metrics

Investing Metrics serves as a comprehensive framework for evaluating the economic, social, and governance factors associated with green investments. This framework emphasizes the systematic measurement and reporting of metrics to enable investors to not only achieve financial returns but also make a positive impact on the environment and society. To achieve this, investors must establish clear aims and objectives by defining specific social and environmental goals they intend to accomplish. Furthermore, they need to select relevant metrics that are aligned with their investment objectives. Since each investor has unique interests and perspectives, it is imperative for them to prioritize the most meaningful metrics for their specific investment strategies. Subsequently, investors should collect data as a benchmark for measuring progress toward their goals. Regular monitoring and reporting of progress through these metrics to stakeholders is vital, as it facilitates an informed decision-making process involving all relevant parties. Finally, investors should adjust their strategies based on the progress and insights derived from the metrics, enabling a thorough evaluation of their green investments and the ability to make informed adjustments as needed.

5. REGULATORY FRAMEWORK AND POLICY ENVIRONMENT

The growth of green financial markets relies on effective regulations and supportive policies. These regulations and policies establish the rules and guidelines that ensure transparency, reliability, and accountability in financial flows directed toward environmentally sustainable investments. They can also incentivize investments in green projects through subsidies, tax benefits, and more favorable borrowing terms for green initiatives. The global landscape of green standards is continually evolving and expanding. Some key initiatives include the Green Bond Principles developed by the International Capital Market Association, the Climate Bond Standard issued by the Climate Bond Initiative, the EU Taxonomy, and the Sustainable Finance Disclosure Regulation (SFDR). Other initiatives focus on harmonizing global standards, such as the International Platform on Sustainable Finance, the Network for Greening the Financial System, and the G20 Sustainable Finance Working Group. Table 1 shows how different countries structure their regulatory and reporting mechanisms to boost green finance.

Table 1. Regulatory framework and reporting standards across countries

Country	Regulatory Framework	Reporting Standards and Achievements
European Union	The ***EU Taxonomy*** for Sustainable Activities aims to classify and define environmentally sustainable activities to prevent greenwashing[6]. Additionally, the ***Sustainable Finance Disclosure Regulation (SFDR)*** mandates that financial market participants disclose how they integrate ESG factors into their risk processes[7].	The ***Non-Financial Reporting Directive (NFRD)*** mandates large companies to disclose their environmental and social impacts. It was recently updated to the ***Corporate Sustainability Reporting Directive (CSRD)*** to expand its scope[8].
United States	The ***Climate-Related Financial Risk Executive Order (2021)*** mandates that federal agencies integrate climate-related financial risk into their regulatory and supervisory activities[9].	***Securities and Exchange Commission (SEC) Climate Risk Disclosures:*** Publicly traded companies should prioritize improving climate-related disclosures to provide investors with comprehensive risk information[10].
China	***Green Bond Endorsed Project Catalogue***—Guidelines outlining eligible green projects for bond issuance, aligning with international principles[11]. ***Carbon Trading Market***—Launched to cap-and-trade carbon dioxide emissions among high-emission industries[12].	***Mandatory Environment Information Disclosure*** – Requires key polluters and listed companies to disclose environmental information[13].
United Kingdom	***Green Finance Strategy***: Incorporates green considerations into private sector finance, supporting the transition to a low-carbon economy[14]. ***Green Taxonomy:*** Developed a taxonomy to define environmentally sustainable economic activities by Green Technical Advisory Group[15].	***Green Bond Framework*** – Outlines how the proceeds from green bonds will be used and reported, aligning with the Green Bond principles by ICMA[16].
Japan	***Green Growth Strategy Through Achieving Carbon Neutrality in 2050:*** A strategy that combines industrial policy with green finance[17].	***TCFD-aligned Disclosures***: Encouraging companies to disclose climate-related financial information in accordance with the Task Force on ***Climate-related Financial Disclosures (TCFD)*** guidelines[18].
UAE	***Sustainable Finance Regulation*** – Launched by Dubai Financial Services Authority to promote sustainable finance and investments, particularly in renewable energy and green projects[19]. ***Abu Dhabi Global Market's Sustainable Finance Framework*** – Establishes a platform to facilitate sustainable finance activities[20].	***Dubai Green Fund*** encourages private investment in green projects and supports the city's commitment to sustainability through the Dubai Clean Energy Strategy 2050[21].

Source: Own Contribution

6. CHALLENGES IN GREEN FINANCIAL MARKETS

Green finance stands as a vital component of the global strategy to tackle climate change and achieve sustainable development. It plays a significant role in directing capital towards environment-friendly projects, renewable energy, and other green initiatives. However, as the sector expands, it encounters challenges that could undermine its effectiveness and credibility. These challenges include identifying and managing green risks, combating greenwashing, and ensuring data availability and quality. This section of the chapter will elaborate on these challenges.

6.1 Identifying and Managing Green Risks

One of the significant challenges in green financial markets is the identification and management of green risks. These risks are potential environmental and financial risks associated with green investments that arise from factors such as regulatory changes, technological shifts, market volatility, physical climate-related events, and transition risks from adjusting to a low-carbon economy. Thus, it is crucial to identify and manage these risks for investors and financial institutions to protect their investments and ensure the long-term sustainability of green projects.

When investing in green projects, it is crucial to conduct thorough risk assessments and due diligence in order to effectively identify green risks and develop measures to mitigate them. This involves evaluating potential risks related to management, technical aspects, green materials, and regulatory/economic factors (Hwang et al., 2017). One of the main challenges in managing green risks is the difficulty in quantifying and assessing them. Unlike traditional financial risks, green risks involve complex interactions between environmental, social, and governance (ESG) factors that are hard to predict and quantify. There is no universally accepted framework for measuring and comparing green risks, which complicates the risk assessment process for investors and financial institutions. Furthermore, the environmental impact and associated risks often unfold over long time horizons, making it even more challenging to incorporate them into the short-term focus of financial analysis.

6.2 Issues with Greenwashing

Another challenge in green financial markets is the issue of greenwashing. Greenwashing refers to the misleading or exaggerated claims made by companies or financial institutions about the environmental benefits of their products, services, or investments. This practice undermines the credibility and effectiveness of green financial markets by creating confusion among investors and consumers, which can erode trust and hinder genuine sustainability efforts.

Companies that engage in greenwashing may make false or vague claims about their environmental performance, use misleading labels or certifications, or misrepresent the carbon footprint of their products or projects (Braga Junior et al., 2018; De Freitas Netto, 2020). The lack of uniform regulations governing green claims across different jurisdictions makes it easier for entities to manipulate or exaggerate their environmental performance. Additionally, the opacity of complex global supply chains can obscure the true environmental impact of products and services, facilitating greenwashing.

To address this issue, regulators and industry bodies must establish robust green investment standards and guidelines. They should require companies to provide transparent and verifiable information about the environmental impact of their products, services, and investments. In addition, independent third-party certifications and audits can help verify the green credentials of projects and investments. Table 2 below highlights different types of greenwashing across industries.

Table 2. Different types of greenwashing across industries

Type	Meaning and Example
Vagueness	*Meaning:* Ambiguous claims that are not readily verifiable. *Example:* H&M has been criticized for using vague sustainability claims, such as labeling their clothing line as "conscious," without providing clear details on how the products are more sustainable. This lack of transparency may mislead consumers about the environmental impact of their purchases[22].
No proof	*Meaning:* Claims that are not supported by verifiable evidence or certification. *Example:* TerraCycle and its partners, along with major consumer goods companies, faced accusations of failing to substantiate their claim that their products could be recycled for free. This resulted in a lawsuit compelling them to clearly define the limitations of their recycling program[23].
Irrelevance	*Meaning:* Potentially true claims that are unimportant or unhelpful for consumers seeking better products. *Example:* Ryanair boldly proclaimed to be Europe's "lowest emission airline," a statement that, while technically valid within its specific operational context, conveniently overlooks the larger environmental impact of the aviation sector. This claim was subsequently banned by the Advertising Standards Authority. (ASA)[24].
Lesser of Two Evils	*Meaning:* Claims within the product category may be accurate, yet they have the potential to divert attention from the more substantial environmental impacts associated with the category. *Example:* Toyota must be held accountable for promoting its hybrid vehicles as environmentally friendly. While they may be a greener option compared to traditional cars, their reliance on fossil fuels still poses a significant environmental impact[25].
Fibbing	*Meaning:* Making false claims and spreading outright lies about a product's environmental friendliness is harmful and misleading. *Example:* Volkswagen's "Clean Diesel" campaign was a case where the company falsely advertised its diesel vehicles as low emission, while in reality, they were using software to manipulate emission tests and pass environmental standards. This incident sheds light on the importance of transparent and honest advertising in the automotive industry[26].
Hidden Trade-off	*Meaning:* Claims that imply a product is environmentally friendly based on limited factors without considering other important environmental aspects. *Example:* Starbucks introduced straw-less lids, claiming they were more sustainable. However, these lids contained more plastic than the previous lid and straw combination, representing a trade-off that wasn't clearly communicated[27].
Worshipping False Labels	*Meaning:* Misleading use of labels or badges that give the impression of third-party endorsement where none exists. *Example:* Innocent Drinks, a subsidiary of Coca-Cola, pushed advertisements suggesting that purchasing their smoothies could help save the environment. This was deemed misleading as it exaggerated the environmental benefits of their products, which still involved significant use of single-use plastics[28].

Source: Author's Contribution

6.3 Data Availability and Quality

Accurate and reliable data is absolutely essential for investors to assess the environmental performance and impact of green projects and make well-informed investment decisions. However, the current challenges in obtaining such data must be addressed. Firstly, it is imperative to establish a standardized framework for reporting environmental data to ensure consistency across companies and institutions. Additionally, there must be a concerted effort to improve data availability, particularly for newer or emerging green technologies and industries. Finally, the quality of available data must be rigorously scrutinized to eliminate any inaccuracies or lack of transparency from companies. These steps are crucial in enabling investors to make sound assessments and decisions (El-Sayegh et al., 2018).

In order to overcome the data availability and quality challenge in green financial markets, decisive actions must be taken. Firstly, regulators and industry bodies must collaborate to establish standardized frameworks for reporting environmental data. This will ensure that companies use consistent metrics and methodologies, making it easier for investors to compare and analyze data across projects and entities. Secondly, government and international institutions must play a pivotal role in improving data availability. They must actively support research and data collection efforts to fill the data gaps and provide more accurate and reliable information. Furthermore, they must implement incentives and regulations to compel companies to disclose their environmental impact and ensure the accuracy of the reported data. Additionally, independent third-party verification and audit processes must be implemented to ensure reported data's credibility and ultimately build investors' trust and confidence (Mansour, 2023).

In summary, the growth of green financial markets presents a positive opportunity for progress. However, effectively tackling the challenges of identifying and managing green risks, combating greenwashing, and ensuring high-quality data is crucial for the integrity and success of these markets. Stakeholders can overcome these challenges by proactively implementing standard metrics, enhancing regulatory frameworks, and investing in data infrastructure and actively contribute to a more sustainable and resilient financial system.

7. OPPORTUNITIES IN GREEN FINANCIAL MARKETS

The global imperative for environmental sustainability and the urgent need to combat climate change has unequivocally created a unique opportunity for the financial industry - the green financial markets. This market comprises various financial instruments and strategies with the explicit goal of supporting environmentally friendly projects and businesses. It provides numerous opportunities for investors, governments, and businesses to drive environmental sustainability while actively achieving substantial financial returns. Financial institutions, investors, and policymakers are increasingly realizing and taking advantage of such opportunities. These markets are essential for transitioning to a low-carbon, sustainable global economy. They go beyond what public institutions can provide, opening the door for private investors to play a crucial role. This transition is necessary and presents a promising long-term business opportunity, as the demand for green financial products is expected to rise significantly (Lupu and Criste, 2021).

In the realm of green finance, there lies a remarkable opportunity for the creation and widespread adoption of innovative financial products and services. These encompass a range of instruments such as green bonds, green stock indices, and other financial tools meticulously designed to mitigate climate risks, channel investments into green sectors, and uphold financial stability. Moreover, the financial sector's support for green initiatives goes beyond traditional investments. Venture capital funds, for instance, play a vital role in driving forward innovative projects that offer environmental support or effective mitigation solutions. Furthermore, green financial markets not only foster innovation and research but also drive the development of digital technologies and the application of FinTech services for green financial products and services (Guang-Wen and Siddik, 2023).

Good governance and robust social protections are essential components that underpin the potential of green financial markets. A societal shift is pivotal for the successful transition to a low-carbon economy, enabling the implementation of vital measures to combat climate change. As these financial markets continue to expand, they will progressively assume a more critical role in mitigating climate change and fostering sustainable development (Xue, 2024).

Further in this section of the chapter, we will delve into some sector-specific opportunities, innovation and research opportunities, and contributions to sustainability development goals.

7.1 Sector-Specific Opportunities

i. Renewable Energy

Renewable energy is leading the way in the green financial markets, showing significant growth in solar, wind, hydro, and bioenergy. Investments in renewable energy are driven not just by its potential to reduce global carbon emissions but also by its ability to generate sustainable economic returns. The International Energy Agency (IEA) highlights that this sector is not only reducing carbon emissions but also creating numerous jobs globally (IEA, 2024). The decreasing cost of renewable technology, along with increasing regulatory support such as subsidies and tax incentives, provides a fertile ground for investments in the sector.

Wind farms in Europe, solar parks in Asia, and the expansion of bioenergy capacities in the Americas represent noteworthy examples of investments in renewable energy. Europe has taken a pioneering role in wind energy, making substantial investments in onshore and offshore wind farms. Nations such as Germany, the UK, and Denmark boast thousands of megawatts of wind energy capacity, propelling the region's shift towards green energy[29]. Meanwhile, Asia is experiencing a surge in solar power, led by countries like China and India. These nations are making significant investments in large-scale solar parks to sustainably meet their increasing energy demands[30]. Additionally, there is a rising emphasis on investments in bioenergy, biomass, and waste-to-energy technologies in the Americas.

ii. Sustainable Agriculture

Sustainable agricultural practices are essential for improving food security by creating a reliable food supply that minimizes reliance on synthetic fertilizers and pesticides. Methods such as crop rotation, agroforestry, and natural pest control not only sustain but also enhance crop yields over time. Additionally, the development of green financial markets plays a vital role in driving the transformation of agriculture into a more sustainable industry. It is noteworthy that both the United States and the European Union are increasingly directing government subsidies and private investments towards supporting organic farming, which is a positive step towards a more sustainable agricultural future.

Further, precision agriculture is a rapidly growing industry that utilizes IoT sensors, GPS technology, and big data analytics to optimize the utilization of water, fertilizer, and pesticides. By embracing these technologies, farmers can improve the accuracy of their practices, leading to higher yields and reduced environmental effects (Rehman et al., 2022). Notably, companies like John Deere have heavily invested in the advancement of equipment designed to support precision agriculture[31].

Vertical farming represents an innovative approach to food production, utilizing vertically stacked layers integrated into structures like skyscrapers or repurposed warehouses. Leading companies such as AeroFarms have successfully secured significant funding from private equity sources and grants to further develop their operations. This sustainable farming model is particularly effective in urban areas with limited space and high demand for fresh produce[32].

The aforementioned cases underline the significance of allocating resources towards environmentally sustainable agricultural practices and technologies. In doing so, investors not only contribute to the global effort towards sustainability but also position themselves to capitalize on an expanding market that places a premium on eco-conscious food production.

iii. Green Construction

The construction sector stands poised to reap the benefits of green financial markets, particularly through advancing green buildings. These structures are engineered to optimize energy usage, minimize waste, and decrease greenhouse gas emissions. As environmental awareness grows, many construction stakeholders are adopting sustainable practices to mitigate their environmental impact (Gunduz and Almuajebh, 2020).

Green bonds serve as an effective method for raising capital to support green real estate projects. In 2019, the World Bank issued green bonds to finance sustainable development projects, particularly those related to green real estate, with the goal of enhancing energy efficiency and integrating sustainable materials and technologies[33]. Moreover, in the USA, Freddie Mac introduced the "Energy Efficient Mortgage," enabling borrowers to incorporate the costs of energy-efficient upgrades into their mortgage, thus encouraging higher initial investments for long-term savings[34]. Additionally, various countries worldwide provide incentives and subsidies to encourage and advance green building practices[35].

These financing mechanisms have been instrumental in driving the growth of sustainable building practices, effectively mitigating the impact of global warming, and actively promoting healthier, more sustainable living environments.

iv. Circular Economy and Waste Management

The concept of the circular economy marks a shift away from the conventional economic approach of "make, take, and dispose." It revolves around a production and consumption framework that prioritizes reuse, recycling, waste reduction, and redesign. The objective is to prolong the use of resources, maximize their value during use, and recover and regenerate products and materials at the end of their life cycle. This approach not only supports environmental sustainability but also offers economic advantages, including substantial cost savings and the creation of new business prospects.

The growing shift towards a circular economy is being actively supported by environmentally conscious financial markets. These markets are providing crucial funding for advanced recycling technologies, waste management systems, and sustainable manufacturing methods (Ferraz, 2023). For example, DS Smith, a leading sustainable corrugated packaging provider, has significantly invested in developing fully recyclable solutions. Through the use of funds obtained from green bonds, the company has successfully allocated resources to research and development, creating innovative packaging that strictly follows the principles of the circular economy[36].

Veolia, another firm, is renowned for its expertise in waste management, water, and energy services. The company invests in technology to upgrade waste treatment facilities and develop new solutions for recycling complex materials such as batteries, electronics, and plastics through green financing[37]. As more companies and governments embrace circular economy principles, the role of green financing becomes increasingly important in shaping a sustainable future.

The chapter section firmly emphasizes the critical role of green financing across various sectors. Green financing drives innovations that deliver environmental benefits and substantial economic returns, while also creating jobs. Financial institutions and investors play a crucial role in promoting sustainable development and mitigating the adverse effects of climate change by investing in various green initiatives.

b. Role of Green Finance in Achieving the Sustainable Development Goals (SDGs)

Green financial markets are absolutely necessary for achieving all SDGs, as they provide crucial funding for sustainable development initiatives that aim to combat global challenges. These initiatives encompass the advancement of gender equality, poverty alleviation, the promotion of energy innovations, and the support of climate action. Green finance is a pivotal component in furthering the global sustainable agenda. Table 3 offers a comprehensive perspective on how specific green financing efforts are driving global sustainability advancements.

Table 3. SDGs and Green Financing Initiatives

SDG	Goal Description	Role of Green Finance	Example Initiatives
SDG 1	No Poverty	Commits to supporting initiatives aimed at improving access to essential services and economic opportunities, particularly for the most vulnerable.	Grameen Bank in Bangladesh offers microloans to empower struggling families to launch small businesses[38].
SDG 2	Zero Hunger	Supports sustainable agricultural practices that enhance food security and increase production.	Investments by the International Fund for Agricultural Development (IFAD) in sustainable farming projects in east Africa[39].
SDG 3	Good Health and Well-Being	Invest in projects that reduce pollution and provide clean water to minimize health risks.	Funding for installing the clean water systems in rural India by water,org[40].
SDG 4	Quality Education	Supports the creation of eco-friendly schools and the development of sustainable educational resources.	Solar-powered school projects in Zimbabwe, funded by UNICEF and UK Government[41].
SDG 5	Gender Equality	Empower women by financing businesses and projects led by women, particularly in sustainable industries,	Women's World Banking green funding for women entrepreneurs in Southeast Asia[42],
SDG 6	Clean Water and Sanitation	Support the development of infrastructure for sustainable water resource management.	European Investment Bank provides funding for sustainable water management projects in the Mediterranean through its Water Sector Fund[43].
SDG 7	Affordable and Clean Energy	Investing in renewable energy projects for affordable energy access.	Backing the off-grid solar projects in sub-Saharan Africa by the World Bank through an initiative called DARES[44].
SDG 8	Decent Work and Economic Growth	Invest in green businesses that create jobs and promote local economic growth.	IFC and Bank BTPN issue green bonds to boost inclusive development in Indonesia[45].

continued on following page

Table 3. Continued

SDG	Goal Description	Role of Green Finance	Example Initiatives
SDG 9	Industry, Innovation and Infrastructure	Promote sustainable infrastructure development and encourage innovation in manufacturing and industrial sectors.	Asian Development Bank promotes sustainable transport and green initiatives in Asia-Pacific through the ASEAN Catalytic Green Finance Facility[46].
SDG 10	Reduced Inequalities	Funding for projects aimed at promoting financial inclusion and addressing economic disparities.	HSBC issued social bonds to fund initiatives that address economic inequality in disadvantaged areas[47].
SDG 11	Sustainable cities and communities	Investing in urban projects for resilience and environmental sustainability.	C40 Cities Finance Facility funds sustainable infrastructure projects in developing and emerging economies[48].
SDG 12	Responsible Consumption and Production	Supports initiatives that promote sustainable resource management and energy efficiency.	Initiatives supported by UNDP aim to decrease food waste and encourage recycling in Latin America[49].
SDG 13	Climate Action	Funds endeavors that address climate change through funding mitigation and adaptation projects.	The Green Climate Fund invests climate resilience projects in Small Island Developing States (SIDS)[50].
SDG 14	Life Below Water	Investing in conservation projects to safeguard marine environments and promote the sustainable use of ocean resources.	Seychelles issued Blue Bonds to finance marine conservation projects[51].
SDG 15	Life on Land	Finance the land restoration and anti-deforestation projects to conserve terrestrial ecosystems.	The Amazon Fund, supported by Norway, finances anti-deforestation efforts in Brazil[52].
SDG 16	Peace, Justice, and Strong Institutions	Support the establishment of institutions that are essential for sustainable development and peace.	UNDP initiatives for judicial reforms and anti-corruption projects in post-conflict regions[53].
SDG 17	Partnerships for Goals	Promote collaborations to support sustainable development.	Global Environment Facility (GEF) fosters partnerships to address global environmental challenges through collaborative funding[54].

Source: Author's Contribution

8. CASE STUDIES ON GREEN FINANCE

a. The New York Metropolitan Transport Authority (MTA) Green Bond Issuance

Background: The MTA is among the largest public transit authorities in the United States, overseeing an extensive network of subways, buses, and railroads. Recognizing the crucial role of public transportation in reducing vehicle emissions, the MTA's commitment to expanding and upgrading its services is a significant component of New York's broader sustainability goals. To secure the necessary funding while aligning with broader environmental objectives, the MTA turned to the green bond market.

Structure: In 2016, MTA took a significant step forward by issuing its first green bond, the Transportation Revenue Bonds, Series 2016A, as part of a larger issuance totaling approximately USD500 million. This bond issue received certification from the Climate Bond Initiative, an international, investor-focused non-profit organization dedicated to promoting large-scale investment in a low-carbon economy. Sustainalytics provided verification for this bond issuance. The bonds were thoughtfully offered with maturities ranging from 15 to 30 years and carried a tax-exempt status, making them highly attractive to a broad base of institutional investors[55].

Projects funded: These bonds decisively funded transit and commuter projects, marking a significant milestone in sustainable financing for public transport facilities. The proceeds were allocated to fund several key projects within MTA's extensive network, including[56]:

- *Renewal and expansion of rolling stock:* The acquisition of new, more energy-efficient subway cars and buses to replace older models that were less energy-efficient and incurred higher maintenance costs.
- *Infrastructure Upgrades:* Implement improvements to rail tracks, signaling, and stations to maximize energy efficiency and minimize the environmental impact of daily operations.
- *Energy Efficiency Projects:* Installation energy-efficient lighting and power systems in various facilities and upgradation heating and cooling systems to reduce energy consumption.

Impact: The Series 2016A green bond issuance had a number of immediate and long-term impacts, all of which were constructive and beneficial:

- *Environmental Impact:* The funded projects have significantly contributed to a reduction in the MTAs overall carbon footprint through the enhancement of energy efficiency and the phasing out of older, less environmentally friendly technologies.
- *Economic Impact:* Issuing green bonds allowed MTA to fund crucial infrastructure upgrades without depending solely on fare increases or traditional funding methods, which often result in higher costs or political challenges.
- *Market Perception:* The MTA has firmly established itself as a leader in the municipal green bond market, setting a precedent for other public transportation networks and municipalities to follow. The successful bond issuance demonstrates the market's confidence in the MTA's financial health and unwavering commitment to sustainable practices.

Challenges and Lessons Learned: The bond issuance was largely successful, despite the challenges faced by the MTA. These challenges included market timing, investor education on green bonds, and the rigorous process of obtaining and maintaining certification under environmental standards. These challenges underscore the importance of transparent reporting and communication with stakeholders regarding the use of proceeds and the environmental benefits of the projects funded[57].

Conclusion: The MTA's Series 2016A green bond represents a pioneering move in the intersection of public transportation financing and environmental sustainability. The bond issuance was a resounding success, attracting strong interest from environmentally conscious investors and setting a precedent for future issuances. This case study not only exemplifies the potential for green bonds to fund large-scale infrastructure projects but also reflect a growing trend in the capital markets towards investments that offer financial returns while also contributing to a sustainable future.

b. BlackRock's Strategy on ESG Integration Into Their Investment Portfolio

Overview and Background: BlackRock, the world's largest asset manager, has been a pioneer in integrating ESG criteria into its investment process. This case study examines how BlackRock has embraced ESG integration, the strategies they have implemented, and the impact of these strategies on their investment portfolio and broader market practices. With over USD 8 trillion in assets under management, BlackRock's investment decisions wield significant market influence. Recognizing the importance of sustainability, the firm has increasingly prioritized ESG factors as central components of its investment approach. This shift reflects a broader trend in the financial industry, where there is a growing recognition that ESG factors are crucial in identifying risks and opportunities that traditional financial analysis might overlook[58].

ESG Integration Strategy: BlackRock has proactively developed a comprehensive strategy for integrating ESG (Environmental, Social, and Governance) factors, covering a wide range of dimensions.:

a. *Corporate Strategy and Commitments:* In his annual letter, BlackRock's CEO Larry Fink emphasized the critical importance of sustainability and corporate purpose. Fink's letters consistently underscore the expectation for companies to transparently disclose their sustainability strategies and articulate how they plan to achieve them.
b. *Active Stewardship:* BlackRock actively utilizes its significant shareholder influence to collaborate with companies on ESG matters, promoting practices that drive long-term value creation. This proactive approach involves participating in shareholder meetings and exercising voting rights to shape company policies on critical issues such as climate action, diversity, and corporate governance.
c. *ESG-themed funds:* BlackRock offers a diverse array of ESG-themed investment products, such as mutual funds, ETFs, and custom portfolios specifically structured to drive sustainable outcomes. These products effectively attract capital to companies and projects that demonstrate high ESG performance.
d. *Risk Management:* BlackRock has fully integrated ESG considerations into its risk management framework. The company leverages advanced data analytics to assess the impact of ESG factors on financial performance and to uncover potential risks and opportunities within the investment portfolio.

Impact and Outcomes: BlackRock's dedication to ESG has some significant and diverse impacts[59].

a. *Market influence:* The strong stance on ESG integration by Blackrock, as a market leader, serves as a catalyst for other funds and companies to embrace similar practices. This positive influence creates a ripple effect, driving entire industries towards more sustainable practices.
b. *Financial Performance:* Based on BlackRock's report, it is evident that the performance of its sustainable investment funds has either equalled or surpassed that of traditional investment funds. This indicates that incorporating ESG criteria can potentially lead to competitive financial returns.
c. *Transparency and Disclosure:* BlackRock fervently advocates for heightened transparency and standardized reporting on ESG practices. This unwavering commitment has been instrumental in driving widespread enhancements in ESG disclosure across the market. Such advancements are indispensable for empowering investors to make well-informed decisions rooted in sustainability criteria.

Challenges: Despite the significant advancements, BlackRock continues to grapple with persistent challenges in effectively integrating ESG considerations[60].

a. *Data quality and availability:* In many sectors, there is still a need for high-quality, standardized data on ESG factors, which presents an opportunity to improve the accuracy of ESG performance assessments.
b. *Balancing Returns with ESG Goals:* Balancing high returns with ESG principles is often challenging, especially in sectors where sustainable practices are costlier or less developed.
c. *Criticism and Scrutiny:* As a leader in the industry, BlackRock is facing increased attention from investors advocating for higher ESG standards and critics questioning the firm's commitment to these principles[61].

Conclusion: BlackRock's successful integration of ESG criteria highlights the growing significance of sustainability in the financial sector. By incorporating ESG factors into its investment process, BlackRock not only strengthens its risk management but also contributes to a more sustainable global economy. The firm's approach and the obstacles it confronts offer valuable insights for other financial institutions aiming to elevate their commitment to sustainability. As the demand for ESG integration continues to escalate, BlackRock's experiences are poised to drive transformation across the investment landscape, shaping how companies and investors embrace and implement sustainable practices.

9. THE FUTURE OF GREEN FINANCIAL MARKETS

The green financial markets are poised for substantial expansion and evolution in the foreseeable future. This shift is propelled by a worldwide movement towards sustainability and the imperative to combat climate change. Factors such as technological progress, regulatory adjustments, shifting consumer inclinations, and a heightened awareness of the financial implications associated with environmental factors are all driving this upward trend. Governments and international entities are placing greater emphasis on addressing climate change, resulting in the development of more comprehensive regulatory structures to foster green finance. Anticipated to be embraced on a global scale, these regulatory endeavors are set to establish a more organized and encouraging landscape for green investments.

Cutting-edge technology will undoubtedly play a pivotal role in driving the evolution of the financial market. The anticipated advancements in data analytics, artificial intelligence, and blockchain will unquestionably enhance the assessment and management of environmental risks associated with investments (Rizzello, 2022).

ESG factors are now integral to investment decision-making processes. As awareness of climate risks grows, ESG factors are increasingly viewed as critical determinants of financial performance. Over time, ESG integration is expected to become a standard practice in investment portfolios, reshaping global investment strategies. This trend is strongly supported by an increasing amount of detailed and standardized ESG data, and the development of more sophisticated tools for ESG risk assessment and reporting.

In addition, institutional investors, such as pension funds, insurance companies, and sovereign wealth funds, are anticipated to assume a substantial role in the prospective expansion of green finance. Driven by regulatory imperatives and their fiduciary responsibility to assess long-term risks, these investors are allocating funds towards sustainable assets. The influence wielded by institutional investors has the potential to expedite the transition to a low-carbon economy by facilitating extensive funding for renewable energy ventures, sustainable infrastructure, and additional environmentally conscious initiatives. Their participation concurrently conveys a potent message to the financial markets regarding the paramount importance of sustainability and concerted endeavors to address climate change (Sangiorgi & Schopohl, 2021).

The demand for green financial products is expected to increase as consumer awareness of environmental issues grows. This shift is already visible in the popularity of sustainable investment funds and the growing interest in sustainable banking services. Financial services providers will likely respond by offering more products that allow consumers to align their financial activities with their environmental values.

The future of green financial markets is undeniably linked to broader economic, technological, and societal shifts toward sustainability. Despite persistent challenges, the trends unequivocally indicate a robust growth trajectory. As these markets mature, they will indisputably play a pivotal role in financing a sustainable future, highlighting the interconnectedness of environmental sustainability and economic viability. The ongoing evolution of green finance will not only help mitigate the impact of climate change but also undoubtedly offer new opportunities for investors and businesses to thrive in a green economy.

10. CONCLUSION

As the world battles climate change and environmental damage, mobilizing capital via green financing to support sustainable development and growth is vital. This chapter emphasizes the significance of green financing in meeting objectives such as reducing carbon footprint, promoting environmental sustainability, stimulating carbon-stranded assets, and supporting renewable energy projects. Globally, the financial markets have witnessed growth in green financing instruments such as green bonds, green funds, and other sustainable investments via different issuing players such as governments, corporations, financial institutions, and non-governmental institutions.

These markets are providing significant opportunities for investors, businesses, and societies. Investment in renewable energy sources, such as solar and wind, contributes to environmental sustainability. The development of green fintech and sustainable business models, such as green hydrogen, has brought innovation. Further, capital markets are being strengthened regionally and globally via green funds and bonds reserved explicitly for projects that deliver environmental benefits and a more sustainable economy. Green financial markets are essential for achieving all SDGs, as they provide significant funding for sustainable development initiatives that aim to combat global challenges.

Overcoming challenges to strengthen green sustainability efforts worldwide is paramount among these opportunities. It is crucial to build a comprehensive, standardized, and transparent reporting mechanism to avoid greenwashing risks. At present, the standards are globally fragmented. A higher quality and quantity of Environmental, Social, and Governance (ESG) information is required to allow investors proper analysis and comparison. Another challenge stems from the lack of data reliability and verification due to self-assessment and reporting. There should be transparent and clear stewardship plans for prominent asset managers and institutional investors in their engagement with executive boards on the reduction of climate intensity and commitment to emission targets, as well as remedial actions when issuers do not adhere to their stated plans.

REFERENCES

Aghion, P., Boneva, L., Breckenfelder, J., Laeven, L., Olovsson, C., Popov, A. A., & Rancoita, E. (2022). Financial markets and green innovation.

Allen, C., & Clouth, S. (2012). *A guidebook to the Green Economy*. UNDESA.

Allevi, E., Basso, A., Bonenti, F., Oggioni, G., & Riccardi, R. (2019). Measuring the environmental performance of green SRI funds: A DEA approach. *Energy Economics*, 79, 32–44. DOI: 10.1016/j.eneco.2018.07.023

Bhatnagar, S., & Sharma, D. (2022). Evolution of green finance and its enablers: A bibliometric analysis. *Renewable & Sustainable Energy Reviews*, 162, 112405. DOI: 10.1016/j.rser.2022.112405

Braga, S., Junior., Martínez, M. P., Correa, C. M., Moura-Leite, R. C., & Da Silva, D. (2019). Greenwashing effect, attitudes, and beliefs in green consumption. RAUSP Management Journal, 54(2), 226-241, https://www.emerald.com/insight/content/doi/10.1108/RAUSP-08-2018-0070/full/pdf

Chang, L., Taghizadeh-Hesary, F., Chen, H., & Mohsin, M. (2022). Do green bonds have environmental benefits? *Energy Economics*, 115, 106356. DOI: 10.1016/j.eneco.2022.106356

De Freitas Netto, S. V., Sobral, M. F. F., Ribeiro, A. R. B., & Soares, G. R. L. (2020). Concepts and forms of greenwashing: A systematic review. *Environmental Sciences Europe*, 32(1), 19. DOI: 10.1186/s12302-020-0300-3

El-Sayegh, S. M., Manjikian, S., Ibrahim, A., Abouelyousr, A., & Jabbour, R. (2018). Risk identification and assessment in sustainable construction projects in the UAE. *International Journal of Construction Management*, •••, 1–10.

Ferraz, D., & Pyka, A. (2023). Circular economy, bioeconomy, and sustainable development goals: A systematic literature review. *Environmental Science and Pollution Research International*. Advance online publication. DOI: 10.1007/s11356-023-29632-0 PMID: 37702868

Guang-Wen, Z., & Siddik, A. B. (2023). The effect of Fintech adoption on green finance and environmental performance of banking institutions during the COVID-19 pandemic: The role of green innovation. *Environmental Science and Pollution Research International*, 30(10), 25959–25971. DOI: 10.1007/s11356-022-23956-z PMID: 36350441

Gunduz, M., & Almuajebh, M. (2020). Critical Success Factors for Sustainable Construction Project Management. *Sustainability (Basel)*, 12(5), 1990. DOI: 10.3390/su12051990

Hwang, B.-G., Shan, M., Phua, H., & Chi, S. (2017). An Exploratory Analysis of Risks in Green Residential Building Construction Projects: The Case of Singapore. *Sustainability (Basel)*, 9(7), 1116. DOI: 10.3390/su9071116

IEA. (2024), Renewables 2023, IEA, Paris https://www.iea.org/reports/renewables-2023

Mansour, N. (2023). Green Technology Innovation and Financial Services System: Evidence from China. *Businesses*, 3(1), 98–113. DOI: 10.3390/businesses3010008

Matsumoto, T., & Yamada, Y. (2023). Multivariate Weather Derivatives for Wind Power Risk Management: Standardization Scheme and Trading Strategy. In: Caetano, N.S., Felgueiras, M.C. (eds) The 9th International Conference on Energy and Environment Research. ICEER 2022. Environmental Science and Engineering. Springer, Cham. DOI: 10.1007/978-3-031-43559-1_26

Naqvi, B., Rizvi, S. K. A., Hasnaoui, A., & Shao, X. (2022). Going beyond sustainability: The diversification benefits of green energy financial products. *Energy Economics*, 111, 106111. DOI: 10.1016/j.eneco.2022.106111

Purwanti, A. (2023). Green Investment Appraisal: A Comprehensive Framework for Evaluating Environmental and Financial Returns. Atestasi. *Jurnal Ilmiah Akuntansi*, 6(2), 797–810.

Rehman, A., Farooq, M., Lee, D. J., & Siddique, K. H. M. (2022). Sustainable agricultural practices for food security and ecosystem services. *Environmental Science and Pollution Research International*, 29(56), 84076–84095. DOI: 10.1007/s11356-022-23635-z PMID: 36258111

Rizzello, A. (2022). The Green Financing Framework Combining Innovation and Resilience: A Growing Toolbox of Green Finance Instruments. In: Green Investing. Palgrave Studies in Impact Finance, Palgrave Macmillan, Cham. DOI: 10.1007/978-3-031-08031-9_3

Roy, J. K. (2015). Green stocks: Green investment, green return. *The Financial Express*, 22, 255.

Sangiorgi, I., & Schopohl, L. (2021). Why do institutional investors buy green bonds: Evidence from a survey of European asset managers. *International Review of Financial Analysis*, 75, 101738. DOI: 10.1016/j.irfa.2021.101738

Xue, W. (2024). The influence of environmental and social criteria in green finance decision-making: Insights and trends. *Economic Change and Restructuring*, 57(3), 110. DOI: 10.1007/s10644-024-09686-x

KEY TERMS AND DEFINITIONS

Circular Economy: It revolves around a production and consumption framework that prioritizes reuse, recycling, waste reduction, and redesign.

Greenwashing: It refers to the misleading or exaggerated claims made by companies or financial institutions about the environmental benefits of their products, services, or investments.

Green Derivatives: The value of these financial products is derived from underlying indices, assets or rates that particularly focus on environment and sustainability.

Green futures: These are agreements to buy or sell a green asset at a predetermined price on a future date. These contracts help businesses mitigate price fluctuations in markets related to environmental goods or compliance obligations.

Green Bonds: Green Bonds are fixed-income instruments to raise funds for environmental and climate projects.

Catastrophe Bonds: These bonds are issued to provide insurance against natural disasters, such as hurricanes or earthquakes.

Renewable Energy Certificate (REC) Derivatives: Renewable energy certificates certify that the bearer owns one megawatt-hour of electricity. These tradable instruments can allow companies to hedge against price fluctuations in the energy markets.

ENDNOTES

1 Read: Regulating carbon markets - KPMG Global

2 Read: EU ETS Spot, Futures & Options (eex.com)

3 Read: ICE Carbon Futures Index Family

4 Read: Texas renewable energy credit markets advance; Green-e prices back off | S&P Global Market Intelligence (spglobal.com)

5 Read: Catastrophe Bonds: A Primer and Retrospective - Federal Reserve Bank of Chicago (chicagofed.org)

6 Read: EU taxonomy for sustainable activities - European Commission (europa.eu)

7 Read: Navigating the Sustainable Finance Disclosure Regulation (SFDR) for a Greener Future (rbinternational.com)

8 Read: Non-Financial Reporting Directive (NFRD) - Directive 2014/95/EU and the proposal for a Corporate Sustainability Reporting Directive (CSRD) | Green Finance Platform

9 Read: Federal Register:: Climate-Related Financial Risk

10 Read: SEC.gov | SEC Adopts Rules to Enhance and Standardize Climate-Related Disclosures for Investors

11 Read: People's Bank of China Green Bond Endorsed Project Catalogue (2020 Edition) | Green Finance Platform

12 Read: China launches world's largest carbon market: but is it ambitious enough? (nature.com)

13 Read: China Refines ESG Disclosure Rules For Listed Companies. - Conventus Law

14 Read: Green finance strategy - GOV.UK (www.gov.uk)

15 Read: The UK Green Taxonomy - KPMG Global

16 Read: Green-Bond-Principles-June-2021-100621.pdf (icmagroup.org)

17 Read: Green Growth Strategy Through Achieving Carbon Neutrality in 2050 / METI Ministry of Economy, Trade and Industry

18 Read: Climate-Related Information Disclosure Based on the TCFD Recommendations | JBIC Japan Bank for International Cooperation

19 Read: Our Approach to Sustainable Finance I DFSA I THE INDEPENDENT REGULATOR OF FINANCIAL SERVICES

20 Read: ADGM Sustainable Finance

21 Read: WHO WE ARE - Dubai Green Fund (DGF)

22 For more information, read: https://thesustainableagency.com/blog/greenwashing-examples/.

23 For more information, read: https://www.provenance.org/news-insights/5-food-and-drink-brands -called-out-for-greenwashing-and-the-lessons-we-can-learn.

24 Read: https://thesustainableagency.com/blog/greenwashing-examples/

25 Read: https://www.eco-business.com/news/22-brands-called-out-for-greenwashing-in-2023/

26 Read: https://www.eco-business.com/news/22-brands-called-out-for-greenwashing-in-2023/

27 Read: https://www.provenance.org/news-insights/5-food-and-drink-brands-called-out-for-greenwashing-and-the-lessons-we-can-learn.

28 Read: https://www.greenhive.io/blog/greenwashing-examples

29 Read: Two countries in Europe are powered by 100% renewable energy as wind capacity soars I Euronews.

30 Read: The 10 Largest Solar Power Parks In The World (2023) I SolarCtrl

31 Read: Deere's investment in InnerPlant is about changing Big Ag (agfundernews.com)

32 Read: The Future of Agriculture: Exploring AeroFarms Vertical Farming Technology — Sustainable Review

33 Read: Sustainable Development Bonds (worldbank.org)

34 Read: GreenCHOICE Mortgages® - Freddie Mac Single-Family

35 Read: MRSC - Green Building Incentives that Work: Follow the Roadmap

36 Read: DS Smith launches new recyclable container Greentote (nspackaging.com)

37 Read: GreenUp: Veolia launches its new strategic plan to accelerate ecological transformation to meet growing global demand I Veolia, ecological transformation: water management, energy and waste recycling

38 Read: How Grameen Bank in Bangladesh is Reducing Poverty - The Borgen Project

39 Read: Investors (ifad.org)

40 Read: The Water Crisis In India I Water.org

41 Read: US$2m solar project boosts rural education in Zimbabwe I UNICEF Zimbabwe.

42 Read: Access-to-Finance-of-Women-Owned-SMEs-in-Southeast-Asia-An-Assessment-of-Five-Countries1.pdf (womensworldbanking.org)

43 Read: The EIB Water Sector Fund

44 Read: World Bank Group Announces Major Initiative to Electrify Sub-Saharan Africa with Distributed Renewable Energy

45 Read: IFC and Bank BTPN Partner in Landmark Social and Green Bond Deal to Strengthen Climate Action and Boost Inclusive Development in Indonesia

46 Read: The Asian Development Bank's Green Finance Initiatives and Their Impacts (adb.org)

47 Read: https://www.hsbc.com/-/files/hsbc/investors/fixed-income-investors/green-and-sustainability-bonds/pdfs/191016-hsbc-sdg-bond-and-sukuk-report-2019.pdf?download=1&la=en-gb&hash=496F8247033D828FBD770A498A8D200601949856

48 Read: https://unfccc.int/climate-action/momentum-for-change/financing-for-climate-friendly-investment/c40

49 Read: How cities can fight food loss and waste | United Nations Development Programme (undp.org)

50 Read: GCF Spotlight: Small Island Developing States (SIDS) | Green Climate Fund

51 Read: Seychelles launches World's First Sovereign Blue Bond (worldbank.org)

52 Read: Norway announces R$250 million donation to the Amazon Fund at COP28 — Planalto (www.gov.br)

53 Read: How innovations in anti-corruption can build sustainable development | United Nations Development Programme (undp.org)

54 Read: Global Environment Facility | UNEP - UN Environment Programme

55 Read: New York MTA $500m Triple Treat of Firsts! First GB Issued by MTA! First US Muni bond Certified under CB Standard! Biggest US Certified Bond so Far! | Climate Bonds Initiative

56 Read: https://www.climatebonds.net/files/files/NY%20MTA%20Post-Issuance%20Verification%202019CD%202020AC.PDF

57 Read: https://nap.nationalacademies.org/read/26066/chapter/9#26

58 Read: blk-esg-investment-statement-web.pdf (blackrock.com)

59 Read: BlackRock: Emerging ESG Datasets and Insights Open Up New Avenues for Investment Outperformance - ESG Today

60 Read: BlackRock flags ESG scrutiny as risk factor in quarterly results for first time (responsible-investor.com)

61 Read: ESG Challenges: The Top 5 That Investors Face (visualcapitalist.com)

Chapter 10
Green Banking Innovations Navigating Environmental Solutions for Sustainable Finance in India

Piyush Kuman Jain
https://orcid.org/0009-0001-6953-0267
Christ University, India

ABSTRACT

This research paper investigates the landscape of green banking innovations in India, aiming to explore the role of financial institutions in fostering environmental sustainability. Through a comprehensive literature review and empirical analysis, this study evaluates the implementation of green banking practices, including innovative products, services, and technological advancements, within the Indian banking sector. Case studies of leading banks are examined to understand their strategies and impacts on environmental sustainability. Additionally, regulatory frameworks and policy implications are analysed to assess the effectiveness of existing measures and propose recommendations for enhancing green banking initiatives. Challenges and future directions are also discussed to provide insights for stakeholders and policymakers. The findings contribute to the growing body of literature on sustainable finance and offer practical implications for promoting environmental stewardship within the banking industry.

I. INTRODUCTION

A. Definition and Scope of Green Banking

Green banking is said to be a set of policies and processes by which a bank can prove its accountability to the environment by lessening carbon footprints and creating financial solutions that would benefit businesses dedicated to sustainability. The most critical programs consist of green loans, green bonds, and other products which are developed specifically to support environmentally friendly activities. They include sustainably cultivated agriculture and even renewable energy installations. Among them, the most important ones are these products. Green banking encompasses the energy processes being implemented

DOI: 10.4018/979-8-3693-5748-4.ch010

in the operations of banks, as well as the green policies and procedures that have to be implemented and the utilisation of technology in waste reduction and emissions (Climate Policy Initiative, 2024).

B. Importance of Green Banking in India

The Indian government is now staring in the face of quite numerous pressing environmental issues, and green banking stands out as a critically essential topic. According to Kumari et al. (2024), India is the third largest emitting country of greenhouse gases on Earth. It generates seven per cent of the overall emissions around the world, mainly due to economic activities such as agriculture, electricity generation, and transportation. It is of utmost importance to start adopting ecologically responsible practices as soon as possible to cut down emissions and bring the level of climate-related concerns under control. This may provide an opportunity to fund projects at various organisations, ranging from those concerning green finance to initiatives regarding renewable energy, sustainable agriculture, and energy efficiency. PwC India feels that these will be relevant initiatives and will help India in achieving its long-term goal of becoming net zero by the year 2070. Additionally, these efforts will prevent India from experiencing the effects of climate change.

The Indian government is now staring in the face of quite numerous pressing environmental issues, and green banking stands out as a critically essential topic. According to Kumari et al. (2024), India is the third largest emitting country of greenhouse gases on Earth. It generates seven per cent of the overall emissions around the world, mainly due to economic activities such as agriculture, electricity generation, and transportation. It is of utmost importance to start adopting ecologically responsible practices as soon as possible to cut down emissions and bring the level of climate-related concerns under control. This may provide an opportunity to fund projects at various organisations, ranging from those concerning green finance to initiatives regarding renewable energy, sustainable agriculture, and energy efficiency. PwC India feels that these will be relevant initiatives and will help India in achieving its long-term goal of becoming net zero by the year 2070. Additionally, these efforts will prevent India from experiencing the effects of climate change.

C. Objectives

1. Critically analyse and outline the concept of green banking and its relevance to ensuring environmental sustainability in India.
2. Evaluate the existing green banking practices implemented by Indian banks and their efficiency in promoting sustainable development.
3. Analyse the environmental, economic, and social impacts of green banking initiatives within the Indian context.
4. Identify the challenges banks face in conducting green banking operations.
5. Provide recommendations to enhance green banking activities and assess the potential of sustainable finance opportunities in India.

II. LITERATURE REVIEW

A. Evolution of Green Banking Practices Worldwide

Green finance has exponentially grown in the last few decades, driven by the growing concerns of environmental issues globally. The concept of green banking has been based on an agreement called the ESG factors involved in the banking process. The fundamental idea of green banking was to minimise the negative effects that conventional banking was giving to the environment. Green banking was designed to reduce the pollution which was being caused by the normal course of traditional banking. The Equator Principles is one of the landmark successes that fall under it, first published in 2003. It has helped create a framework for risk management up to the identification, evaluation, and managing environmental and social hazards related to projects according to Sharma and Choubey (2021). For instance, immediately after the Paris Agreement in 2015, it was an important turning point that became one of the greatest turning points back in time. After this pact on creating an international commitment to climate change, financial institutions have to be more aware of green financing and operations on sustainability investment.

Detailed Timeline of the Development towards Environmental Responsible Banking: step by step increase in commitment

Introduced for the very first time in 2003, the Equator Principles are.

This would result in a global mechanism of green funding as a consequence of the Paris Agreement in 2015. When 2020 comes, the frameworks of environment, society, and government will have gained more acceptance all over the continents of North America, Europe, and Asia.

Countries like Bangladesh, Brazil, and Indonesia have brought in environment-friendly legislation in the financial sectors, and the banking practices have been changed according to the needs of the local people. Now, it has become clear that the banks are not only worried about earning profits but also answerable to ensure that the economy continues to grow steadily while simultaneously reducing the risks and threats of climate change.

B. Green Banking Initiatives in India

This is the reason why the Reserve Bank of India undertakes such activity to lead Indian banking towards environmentally responsible practices. Strong support for reducing harm to the environment drives this activity, motivating the movement. The State Bank of India and ICICI Bank, among others, were some of the first financial institutes that used the technology. Besides that, it provided environmentally friendly financial services besides adopted environmentally responsible banking practices. Second, SBI sensitised customers to environmentally responsible banking practices. The ICICI Bank also was developing environmentally responsible banking practices. This is contributed to by more grassroots regional rural banks providing banking services and financing activities in rural neighbourhoods that are more sensitive to the ecological environment. Organic farming and alternative sources of energy are but a few examples of activities that fall under this category. Projects of this nature have the potential to assist in the achievement of rural sustainability.

Public sector banks, however, are the ones that are gradually bringing out the concept of sustainability into their core business environment as compared to public sector banks. This is quite different from the case with public sector banks. Public sector banks were also at the front in the banking industry while adopting green banking at first. For example, Gulzar et al. (2024) have found that such measures are critical in the pursuit of achieving sustainable development in India in the wake of the increased combat of environmental challenges and economic development.

C. Theoretical Frameworks in Green Banking

Several theoretical frameworks support the adoption of green banking practices:

1. This notion of CSR highlights the moral responsibility of banks to make sure that the greater good of society is accomplished through the conduct of and performance of environmentally friendly practices in the course of operating their businesses. Banks who engage in corporate social responsibility activities, such as green financing renewable energy projects, or simply reducing their carbon footprint through better operational management, would be expected to experience growth in goodwill to its stakeholders.
2. **Stakeholder Theory:** One of the practices that suggest an integrated approach addressing all stakeholders' interests, including the customers, communities, and workers, and considering the environment as well. The definition presented by N and Periasamy (2023) cited "Green Banking" to be the administration of benefits that accrue from a form of investment to both financial and social stakeholders. This leads to green banking having a positive impact on society.
3. **Triple Bottom Line (TBL):** Triple Bottom Line calls on financial institutions to measure not only their profit but also their impacts on the environment and society. It is one of the other definitions for the term "triple bottom line." Banks committed to the Triple Bottom Line approach are geared toward people, planet, and profit as they implement environmentally responsible policies and procedures.

The integration of ESG criteria into these frameworks, through the improvement of reputation and financial performance, gives more support to green banking. Using these frameworks enables financial institutions to realign their strategy with global sustainability goals, thus contributing to the development of a low-carbon economy (Gulzar et al.,2024).

III. METHODOLOGY

A. Research Design

This research study uses the descriptive method of research design based on secondary data to examine the most recent developments and practices in the field of green banking. Since the descriptive method of research does not alter any of the data that has been collected for this research thus far, it is very suitable for the kind of analysis that is currently undergoing. In doing so, it paints a more accurate and unbiased picture of the challenges green banking has faced throughout the length of its existence. To this effect, this design affords a combination of the outputs from reports and regulatory papers with the other previous works. This is to provide for a comprehensive review that shall ascertain the progress that has so far been achieved in India within the Indian context concerning green banking.

The relevance of the current availability of literature and sources from the industry for selecting secondary data is due to the nature of the research conducted. Yet, a few potential drawbacks associated with the use of secondary data do exist. These disadvantages include the poor quality of the data, the material being out of date, and, in some cases, an inherent bias that can be found in the source itself. In the case where secondary data is not as detailed or consistent as primary data, this may, in fact, have the effect of reducing the richness of the findings that were generated. Overcoming such restrictions requires making use of a wide range of trustworthy sources that are free to access and ensuring that data obtained from these sources is not only up to date-but also pertinent to the research conducted currently.

B. Data Collection Methods

In order to understand green banking, a comprehensive set of secondary data was obtained as follows:

1. **Academic Research Journals:** The journals chosen for the study have been peer-reviewed, and their contents have been appropriately validated through research in green banking. A bibliometric study on the trends of green banking by Riaz et al. from 2023 was acquired from academic sources to source material from a broad scope, including relevant ones and those based on the current advancement in the field of green banking.
2. **Industry Papers:** Reports from the banking industry. These papers have case studies and challenges arising while implementing green banking in real life. Consulting firms and industry groups compile industry reports. Almost all of these may be found on the internet. Selection of the literature above would be made on the basis that it focuses on practical applications and challenges that arise in sustainable banking practices. It may be obtained through websites of the industry and databases of financial information.
3. The reports published by the regulatory agencies in India, such as the Reserve Bank of India, are interpretations of regulations and processes and provide insights into the structural construction of green banking in India. The government considers these reports as official publications. This is why official publications from the government are further analysed to determine if they are relevant or not. This is to ensure that only the most current and relevant regulatory information is included (Bouteraa et al., 2023).
4. **Conference Proceedings:** Attending conferences is an excellent way of getting the most recent research and new approaches from professionals in the field of green banking. Because we could get access to conference materials from renowned financial and environmental organisations, we could contribute to a more comprehensive understanding of evolving worldwide trends and practices in green banking.

C. Data Analysis Techniques

Data collected were analysed with the use of the following analytical methods:

1. Content analysis is a technique for gathering and integrating data from different sources with the objective of finding themes, best practices, and issues in banking that are green. The process of content analysis helps trace the evolution of green banking practices and the ongoing trend of the industry (Bouteraa et al., 2023). This is achieved by putting together similar data points.
2. This qualitative approach examines trends within qualitative data, which will help in the detailed comprehension of relevant themes like consumer awareness, technological adoption, and regulatory difficulties. Thematic analysis is one technique used in qualitative research. With regard to the identification of crucial drivers and impediments toward green banking, thematic analysis is an extremely useful tool, especially when considered from the points of view of various players in the industry.
3. **Comparative Analysis:** This method focuses on the most effective practices and the learning experience by comparing green banking operations across several geographies and timelines. To gain further insight into the factors affecting green banking in diverse environments, a comparative study was conducted to identify the characteristics of green banking in India compared to global trends (Riaz et al., 2023).

IV. GREEN BANKING INNOVATIONS IN INDIA

A. Case Studies of Leading Banks

Green banking practices have been innovated by large banks in India, setting examples through their sustainable efforts, which focus on reducing their influence on the environment.

The "Green Policy" initiated by the State Bank of India encourages responsible lending through the consideration of the environment and reduction of carbon emissions to conserve energy in the branches. SBI invested financially in renewable energy sources such as solar and wind power to generate over 2,000 megawatts of electricity. By doing so, the bank is demonstrating its commitment to its deconcentrating with the use of traditional sources of energy and the development of environmentally friendly infrastructure. According to a paper, "A Study on Green Banking Initiatives in the Public and Private Sectors Banks of India," in 2023, the State Bank of India supported rural projects in energy-intensive areas with the help of its green funding. The rural communities thrive with sustainability in place because of this.

One new set of green financing products introduced by the ICICI Bank is placed under various guises, one being called the "Green Home Loan.". This loan product has special incentives for the customer to invest in a greenhouse with very favourable conditions and, by extension, responsible green consumption if they want to take advantage of it. These investments made by ICICI in firms of clean energy are towards building a sustainable future through sustainable development and enable India to meet the climate goals that it has for itself. (Khanna et al., 2022) The move that ICICI is undertaking in the implementation of these initiatives is a representation of how the company is doing the right thing by attempting to be proactive in the practice of incorporating sustainability into its financial business.

These are grants applied in support of initiatives toward renewable energy, such as solar farms and wind farms. HDFC Bank and Yes Bank This bank is the first private sector bank to begin green bonds, which raise capital for environmental activities. These are known as "green bonds." HDFC Bank is also one of the pioneering financial institutions that is offering green bonds. Similarly, HDFC has been encouraging green finance by offering green deposits and loans, which have opened up avenues of investment for retail and institutional customers in the name of sustainability.

B. Innovative Green Financial Products and Services

Indian banks are incorporating green financial solutions into their product portfolios to meet the needs of environmentally sensitive clients and national sustainability objectives.

1. "Green Loans" loans aimed to especially support initiatives which would have a good impact on the environment for the environment. Such initiatives may be fitting with solar energy systems, constructing energy-saving infrastructure, and launching sustainable agricultural management practices. Eco-friendly loans benefit customers, who could be either people or enterprises, because they can borrow at favourable rates of interest and on lenient terms, thereby making their customers behave in a friendly manner towards the environment.
2. In recent times, green bonds have increasingly been used as one of the most important financing tools aimed at environmentally responsible and sustainable activities. In the recent issue of green bonds by YES Bank and SBI, among other financial institutions, considerable money has been dedicated to initiatives such as waste management, renewable energy sources, etc. As noted by Khanna et al. (2022), these bonds are also contributing to the fulfilment of investors' need for ecologically responsible investment alternatives, which in turn is driving money into green sectors.
3. A set of financial institutions has been working on green credit cards that will encourage responsible spending for the environment. One way in which a consumer culture for sustainable development can be encouraged is by offering consumers advantages for making environmentally responsible purchases or making contributions to environmental organisations (Khanna et al., 2022). This growth can be achieved by offering customers benefits.
4. **Climate Change Loans:** Climate change loans are a new finance which would increase funding sources from green projects. For example, to encourage business clients to have more sustainable activities at the corporate level, the client's grant would either be switched on or the interest paid for the client's deposit, depending on or opposed to the customer's environmental targets.

C. Technology Adoption for Green Banking

Green finance has been completely transformed through these breakthrough technologies into blockchain, AI, and IoTs. All these boost efficiency and transparency with increased accountability.

1. **Blockchain:** The tech used in blockchain can further aid transparency since it is what allows the safe monitoring of funds channelled into environmentally positive projects. Traceability offers assurance to the stakeholders that monies invested into environmentally friendly activities will be applied in the direction intended; this reduces even the possibilities of green-washing. According to the report "A Study on Green Banking Initiatives in the Public and Private Sectors Banks of India, 2023, the decentralised nature of blockchain technology contributes to the development of confidence among investors that will ensure accountability in green financing.".

2. **Artificial Intelligence:** Artificial intelligence enables financial institutions to analyse customer data and make sustainable financial products based on that analysis, unique for every customer. Financial institutions can make predictions about the level of green financing demand, maximise loan take-up for green projects, and analyse environmental risks at every point of the credit assessment using the knowledge developed by artificial intelligence. Other applications of artificial intelligence are the speeding up of processes like a customer energy audit and carbon footprints and aiding customers in monitoring their energy use better.
3. **Internet of Things (IoT):** IoT technologies can assist in saving energy usage by the banking processes. For instance, smart sensors installed in the branches measure energy usage and automatically control lighting and HVAC controls, saving on energy usage. This includes having real-time data on output and consumption of energy, among other possible IoT applications, which can be vitally pivotal in remote monitoring of the environmentally friendly activities that are financed.

D. Collaborations and Partnerships

Collaboration among NGOs, Government agencies, and Financial Institutions has, in the last few years, significantly facilitated the growth and development of green banking.

1. Collaboration with the Government as the First Step To make green financing solutions integrate with the overall environmental objectives of the nation, the government of India has taken a partnership with financial institutions. In this kind of large-scale initiative, collaborations like this hold a specific place, especially when it is the expansion of renewable sources of energy, and the involvement of the government would most likely minimise the risk factors and attract private investors.
2. **The Indian Green Banking Council (IGBC):** The IGBC and other organisations of a similar kind serve as platforms for the exchange of knowledge and the formulation of guidelines for green banking. According to "A Study on Green Banking Initiatives in the Public and Private Sectors Banks of India," 2023, financial institutions may reap the benefits of such collaborations by gaining access to resources, best practices, and updated indicators pertaining to sustainability.
3. **Public-Private Partnership:** Environmental projects that banking and financial institutions cannot do themselves are increasingly being financed through public-private partnerships. Collaboration to finance environmentally friendly activities exists between banks and international financial banks and their counterpart businesses, where risks, as well as gains, are split proportionately between them.

V. IMPACT ASSESSMENT OF GREEN BANKING IN INDIA

A. Measurement and Evaluation Metrics

With present research validation, some measurements have been used to carry out an accurate analysis of the environmental impact of green banking activities in India. These are as follows:

1. **Carbon Footprint:** This measure is used to track the total greenhouse emissions coming from banking operations and financed projects, especially those that concentrate on high-emitting industries such as fossil fuel projects. Banks can set emission reduction targets and measure performance through the measurement of a bank's carbon footprint (Menon et al., 2017).
2. Another essential component of green banking is monitoring all the branches, automated teller machines, and data centres on energy consumption. This will help the credit unions measure the impact of energy-conserving updates. There are good indicators of sustainability solutions' effectiveness in terms of lower energy consumption.
3. **Water Use and Productivity:** Banks will be able to evaluate their water resource management by measuring the amount of water used in the branches of the financial institutions and in the projects that have been funded by them, especially in the agricultural industry. According to Tara et al. (2015), this is one of the indicators that is commonly used in the finance sector to assess sustainable practices.
4. The ratio of sustainably funded is a statistic that indicates what per cent of overall funding is on environmentally friendly efforts. According to Mir and Bhat (2022), the higher ratio is key to demonstrating a greater commitment to sustainability, and this, consequently, nudges the financial institution to have investments that cause the least amount of harm to the environment.
5. **Biodiversity:** An evaluation of the way in which sponsored projects affect the local ecosystems and biodiversity might provide essential knowledge about the environmental effects of business dealings with banks and other credit institutions.
6. **The Green Revenue Ratio:** This formula is an effort to get a percentage figure of the revenues generated through environmentally friendly financial products. This, of course, best represents the spirit of financially viable and sustainable banking as well as this shift away from traditional revenue sources of income.
7. **Impact Investment Number:** A number produced from an index that expresses the contribution of a bank's portfolio to the implementation of objectives of sustainable development. Its focus is on initiatives clearly and verifiably giving benefits to both environmental as well as social goals and objectives.

B. Quantitative Analysis of Green Banking Effectiveness

The application of quantitative methods analyses the effectiveness of green banking based on statistics, as follows:

- Cost saving is one of the economic benefits of sustainability efforts. The data on the cost savings of green technologies, like energy-efficient infrastructure, explain the economic benefits of sustainability projects. The association between investments in environmentally friendly infrastructure and cost savings may be understood based on the application of regression analysis.
- The banks can trace the volume and growth rate of the overall green loans, green bonds, and other sustainable financial products in which they have a growing stake. A better understanding of the rising pattern of green financing will, in turn, enable banks to have an increased share. According to Prabhu (2021), this means that one can undertake a time-series analysis of the statistical trends in green finance in an attempt to predict future growth and market prospects.

- **Impact of Investment Analysis:** Regression models are pretty useful in determining the effect of sustainable finance on environmental metrics such as carbon reduction or biodiversity scores. This goes a long way in quantifying the real impact that green investments create. For instance, the use of regression analysis to investigate the correlation between renewable energy investments and the decrease in emissions in portfolios.

C. Qualitative Assessment of Stakeholder Perceptions

Most importantly, any sustainability initiative requires an understanding of the viewpoints of different stakeholders about green banking.

1. **Interviews and Surveys of Stakeholders:** Qualitative research methods, through surveys and focus groups, are used to probe the views of bank customers, staff, and industry experts. A thematic analysis of comments reveals similar perspectives concerning customer satisfaction, awareness, and adoption of environmentally friendly banking products. Furthermore, this strategy offers a better understanding of the success of outreach and how well banks have communicated their green efforts.
2. **Thematic Analysis of Customer Awareness and Satisfaction:** Themes of customer awareness and satisfaction of green products can help find valuable insights into the acceptance rates and market demand for such products. A high degree of customer satisfaction is correlated with higher adoption of green financial products. In contrast, a low degree of knowledge may be indicative of areas where banks must increase outreach in many instances.
3. **Issues and Opportunities in Green Banking:** The views of stakeholders will identify all the issues that banks face, like low consumer awareness or resource constraints and suggest possible opportunities for advancement in sustainability practices. These insights also open a window of opportunity for green banking.

VI. REGULATORY FRAMEWORK AND POLICY IMPLICATIONS

A. Regulatory Measures in India

To this effect, RBI has enacted many laws that encourage environmentally responsible banking practices. From June 2023, green deposits will be the only source of funding for environment-friendly initiatives for all financial institutions. With this approach, financial institutions are forced to report periodically the distribution of their resources raised from green deposits. This is fundamentally aimed at bringing on accountancy and transparency in regard to the use of the funds (Finserv, 2023).

In addition, the banks' green deposit policy should receive authorisation from the board of directors, which demands this third-party evaluation to make sure it complies with the sustainability criteria. A major step taken by the Reserve Bank of India (RBI) is sectoral focus lending standards and enhancing incentives and support for sustainable agriculture as well as renewable energy within the banking sector. Due to such regulatory provisions in place, the RBI takes utmost care to integrate sustainability through the environment into banking India, thus helping Indian banks reach par with the climate aims of India.

B. Comparison with International Standards

The framework of green banking in India not only finds it very close to international norms but also retains some of its characteristics. For instance, the EU's Sustainable Finance Taxonomy has a detailed classification process that outlines sustainable economic activities in line with the EU goal of achieving net-zero emissions by the year 2050. In the European Union taxonomy, strict reporting standards are required to ensure there is fair distribution of money. Banks are required to publish the environmental impact of financed projects.

This implies that India is currently establishing its green banking system, aimed at meeting comparable transparency and efficiency standards. To date, there is no such fully-fledged green taxonomy that makes it challenging for India to compare and scale up its green finance industry to that of the European Union and other countries. This is despite the major step that the Reserve Bank of India's Green Deposit Framework provides. Bhattacharyya has opined that the country stands a chance to further cut this gap if it decides to adopt strict reporting criteria and increase the regulatory monitoring amount on the country.

C. Policy Recommendations

1. **Strengthen Compliance and Enforcement:** There should be increased tougher penalties for those not adhering to the set green banking standards. This should be coupled with regularly scheduled audits, which could guarantee that there is sustainability in banking practice.
2. The second step in creating a holistic Indian green taxonomy is establishing one that is different from other countries but compliant with the global standard and can clearly differentiate between green and non-green assets. A taxonomy of this nature can allow more investment capital because it will bring clarity and uniformity to the choices under green finance.
3. The participating banks and investors shall be rewarded with financial incentives, which might include tax exemption benefits as well as concessional interest rates. In return, this would enhance the tally of financial incentives accorded to renewable source energy investments. With these kinds of incentives in place, green banking is likely to become expansive, which subsequently may see more private investment being poured into green initiatives.
4. **Public-Private Partnership:** The government, banking, financial, and private sectors are supposed to collaborate towards creating green-friendly banking systems. In most cases, PPPs are going to pool resources as well as knowledge together; subsequently, the basic support shall be realised.
5. **Public Awareness Creation:** A national awareness campaign has to be launched to educate the general public, along with bank customers and employees, about the benefits of green banking and its role in sustainability through the adoption of digital content, seminars, and workshops, thus resulting in increased participation by the consumer and demand for green financial products.

VII. CHALLENGES AND FUTURE DIRECTIONS

A. Key Challenges

Challenges faced by green banking in India are numerous, very grave ones.

1. It is associated with a relatively low level of knowledge in place concerning the benefits of green banking both by clients and those in banking. Such clients are kept uninformed about sustainable alternatives like green loans and green bonds, and thus, a limited uptake of the various financial products green in nature is witnessed (Khanna et al., 2022). Due to a lack of awareness, limited adoption was recorded for green financial products.
2. **Lack of Finance:** Smaller and more regional financial institutions lack sufficient cash to implement environmentally friendly banking practices or instruments. Due to the probable high costs related to infrastructural upgrades, it may prove quite difficult for banks that are less resourceful to fully adopt green banking.
3. **Technological Limitations:** Advanced technology, whether it is blockchain or any artificial intelligence system that allows environmentally responsible banking, is readily available. Without such equipment, financial institutions cannot report on and track with precision what investments they are making in the environment, and due to this, green banking effectiveness decreases.
4. **Less Comprehensive Regulatory Exposure:** Over time, the regulatory framework has evolved but is yet to suffice in terms of comprehensiveness. Of the many rules that lack strict enforcement measures, some banking systems evade sustainable standards. There must be stricter penalties towards violation and uncovert laws if the same industry is to be prompted to comply entirely with green banking.
5. **Data Deficit:** There is a lack of information in terms of material scarcity concerning the environmental effects of the banking operation, which becomes a barrier to tracking the increase in these financial sectors concerning sustainability. According to Khanna et al. (2022), it becomes challenging to measure the environmental benefits of green banking or set achievable goals if proper data cannot be acquired.

B. Strategies for Overcoming Barriers

Among these, the following can be applied:

1. **Awareness Programmes:** Campaigns for focused education programs and campaigns may facilitate increasing the awareness of customers and employees at the level of banks about the advantages green banking can provide in the long run. The inclusion of in-branch information sessions, workshops, and online courses may have helped boost participation and adoption.
2. **Investment Incentives:** To encourage the investment in infrastructure and technology needed for sustainable practices by financial institutions, incentives may be offered in the form of grants, tax exemptions, or lower interest rates. Grants for financial institutions to use eco-friendly data centres or even renewable energy for branch operations would hasten the pace of green practices.
3. **Stringent Regulation:** Enhanced regulatory measures would add to better legislation with clear norms and procedures followed in enforcing such legislation that would result in enhanced consistency in the sector. A more streamlined green taxonomy, which will be distinct for India, would bring clarity on what actually constitutes a green asset so as to ensure consistent compliance and transparency.
4. **Enhanced Collection and Reporting of Data:** The development of better data collection and reporting will help financial institutions monitor their impact on the environment and just how far green activities have stretched. Real-time resource utilisation data collection is possible with the use of digital platforms and tools that enhance decision-making.

5. **International Benchmarking:** Research on developed models of green banking in other countries can help Indian banks get an idea of what practices would be followed. This can be achieved through international assessment. For example, an analysis of how the taxonomy of sustainable finance by the European Union is used may be helpful for India in building up its structure, thereby paving the way towards more effective green banking.

C. Future Trends and Opportunities

New trends and innovations in the banking sector offer new opportunities for responsible and environmentally friendly banks in India:

1. **Advanced Technologies Implementation:** Newer technologies like blockchain and artificial intelligence (AI) become more significant as the process of transforming green bank operations goes on. In fact, blockchain technology has a great prospect of ensuring transparency through the secure recording of the green money distribution as no money will be possible to embezzle with it. This capacity of AI enables it to grasp consumer data that would enable it to design bespoke, green financial products for customers. Such innovative finance options may help banks meet the demands of their customers more effectively.
2. **New Financial Instruments:** Green bonds and loans tied to sustainability are increasingly becoming popular alternatives for financing sustainable ventures. The use of these financial instruments presents attractive returns to investors while opening new avenues for initiatives that promote the environment, such as renewable energy sources and models of sustainable agriculture.
3. **Fintech Company Partnerships:** The partnerships of banks with fintech companies may further expand the gamut of green banking. Innovations from fintech firms would enhance customer experience and cater to green financial services. Some of these are data analytics, digital payment, and customer interaction.
4. **Global Partnership:** India may avail itself of international green funds if the country complies with international principles and partners with global financing organisations. India's green banking system will be enhanced if partnerships with international green finance platforms are launched in this country. These international partnerships can bring significant finance and know-how to India (Indriastuti & Chariri, 2021).
5. **Compliance with International Sustainability Principles:** The adoption of frameworks such as the United Nations' Sustainable Development Goals may make Indian banks an exclusive set of sustainability goals. This adherence to global principles has the potential to further investor trust, making India an even more attractive market for foreign sustainable finance enterprises.

REFERENCES

A Study on Green Banking Initiatives in the Public and Private Sectors Banks of India. (2023). In *International Conference on New Age Marketing - Catalysing Transformation Through Innovation, Technology, Analytics & Sustainability* [Conference-proceeding]. https://www.sdmimd.ac.in/mc2023/conferencepapers/IMC2269.pdf

Amuakwa-Mensah, F., & Näsström, E. (2021). Role of banking sector performance in renewable energy consumption. *Applied Energy*, 306, 118023. DOI: 10.1016/j.apenergy.2021.118023

Bhattacharyya, U. (2024, April 20). Understanding the Regulatory Framework for Sustainable Finance in India. *NLS BLR*. https://www.nlsblr.com/post/understanding-the-regulatory-framework-for-sustainable-finance-in-india

Bouteraa, M., Hisham, R. R. I. R., & Zainol, Z. (2022). Challenges affecting bank consumers' intention to adopt green banking technology in the UAE: A UTAUT-based mixed-methods approach. *Journal of Islamic Marketing*, 14(10), 2466–2501. DOI: 10.1108/JIMA-02-2022-0039

Chen, Z., Mirza, N., Huang, L., & Umar, M. (2022). Green Banking—Can Financial Institutions support green recovery? *Economic Analysis and Policy*, 75, 389–395. DOI: 10.1016/j.eap.2022.05.017 PMID: 35694019

Chhaochharia, M. (2021). Green Finance in India: Progress and Challenges. *Research Journal of Humanities and Social Sciences*, 223–226, 223–226. Advance online publication. DOI: 10.52711/2321-5828.2021.00039

Climate Policy Initiative. (2024, October 16). *A Roadmap for Green and Transition Finance in India - CPI*. CPI. https://www.climatepolicyinitiative.org/publication/a-roadmap-for-green-and-transition-finance-in-india/

Finserv, T. (2023, April 14). *RBI Framework for Green Deposits*. https://vinodkothari.com/2023/04/rbi-framework-for-green-deposits/

Ghosh, S., Nath, S., & Ranjan, A. (2021). *Green Finance in India: Progress and Challenges*. Reserve Bank of India. https://rbidocs.rbi.org.in/rdocs/Bulletin/PDFs/04AR_2101202185D9B6905ADD465CB7DD280B88266F77.PDF

Gulzar, R., Bhat, A. A., Mir, A. A., Athari, S. A., & Al-Adwan, A. S. (2024). Green banking practices and environmental performance: Navigating sustainability in banks. *Environmental Science and Pollution Research International*, 31(15), 23211–23226. Advance online publication. DOI: 10.1007/s11356-024-32418-7 PMID: 38413528

Indriastuti, M., & Chariri, A. (2021). The role of green investment and corporate social responsibility investment on sustainable performance. *Cogent Business & Management*, 8(1), 1960120. Advance online publication. DOI: 10.1080/23311975.2021.1960120

Jindal, A., Hishikar, S., & Shrimali, G. (2024). Examining the portfolio carbon footprint of Indian banking system. *Journal of Cleaner Production*, 142393, 142393. Advance online publication. DOI: 10.1016/j.jclepro.2024.142393

Kumar, K., & Prakash, A. (2018). Developing a framework for assessing the sustainable banking performance of the Indian banking sector. *Social Responsibility Journal*, 15(5), 689–709. DOI: 10.1108/SRJ-07-2018-0162

Kumari, K. G. S., Mukthar, K. P. J., & Rahhani, G. N. E. (2024). Green Finance in India: Driving Sustainable Development and Economic Growth. In *Studies in systems, decision and control* (pp. 283–293). https://doi.org/DOI: 10.1007/978-3-031-62106-2_23

Menon, D. G., Sreelakshmi, S. G., & Shivdas, A. (2017). Green banking initiatives: A review of Indian banking sector. *2017 International Conference on Technological Advancements in Power and Energy (TAP Energy)*. https://doi.org/DOI: 10.1109/TAPENERGY.2017.8397303

Mir, A. A., & Bhat, A. A. (2022). Green banking and sustainability – a review. *Arab Gulf Journal of Scientific Research*, 40(3), 247–263. DOI: 10.1108/AGJSR-04-2022-0017

Panse, P., & Sharma, D. K. (2023). Green Banking in India: A Sustainability Perspective. *MANTHAN Journal of Commerce and Management*, 10(1), 141–161. DOI: 10.17492/jpi.manthan.v10i1.1012308

Prabhu, G. N. (2021). GREEN BANKING PRACTICES – A Case Study on Indian Green Banking System. *International Journal of Research Publication and Reviews*, 2(3), 487–493. DOI: 10.5281/zenodo.4680016

PwC. (n.d.). Green bank deposits: A gateway to sustainable financing. In *PwC* (pp. 2–5). https://www.pwc.in/assets/pdfs/consulting/green-bank-deposits-a-gateway-to-sustainable-financing.pdf

Riaz, A., Riaz, N., Raza, H., & Mirza, F. (2023). Green Banking Practices: A Bibliometric Analysis and Systematic Literature Review. In *Emerald Publishing Limited eBooks* (pp. 299–317). https://doi.org/DOI: 10.1108/978-1-80455-678-820231016

Sahoo, B. P., & Singh, A. (2016). Adoption of Green Banking in India: Challenges and Prospects. In SGTB Khalsa College, *International Journal of Science and Research (IJSR)* (Vol. 5, Issue 8) [Journal-article]. https://www.ijsr.net/archive/v5i8/ART2016856.pdf

Sharma, M., & Choubey, A. (2021). Green banking initiatives: A qualitative study on Indian banking sector. *Environment, Development and Sustainability*, 24(1), 293–319. DOI: 10.1007/s10668-021-01426-9 PMID: 33967597

Tara, K., Singh, S., & Kumar, R. (2015). Green Banking for Environmental Management: A Paradigm Shift. *Current World Environment*, 10(3), 1029–1038. DOI: 10.12944/CWE.10.3.36

KEY TERMS AND DEFINITIONS

Green Banking: A set of rules and practices that bring the environment into the operations of the bank's financial services for a decrease in carbon footprint. These are, as such, used for a lot of eco-friendly projects through green loans, bonds, and other sustainable financing products.

Sustainable Finance: Financial activities that integrate ESG criteria into the process for long-term sustainable development by creating value and considering their impact on society and the wider environment.

Carbon Footprint: The amount of direct or indirect greenhouse-gas emissions that are generated in and financed by the operation of the bank, and expressed in CO2 equivalent units.

Green Loans: Loans that are specifically directed towards environment-friendly projects, such as renewable energy installations, energy-efficient buildings, and sustainable agriculture.

Blockchain in Green Banking: Implement blockchain in green banking to enhance the flow and operations of transactions and other financial transactions and operations in sustainable projects in terms of transparency, traceability, and accountability.

Public-Private Partnerships (PPPs): Colossus multilateral cooperation among government agencies, financial institutions, and private sector firms that promote and undertake green banking endeavours, taking and benefiting from the collective strengths and experience they can draw to realise sustainability objectives.

ESG Criteria: These are the environmental, social, and governance criteria used by investors to make an investment decision in relation to which companies to invest in, how they do their business, how they protect the environment, treat people, and govern themselves.

Chapter 11
Carbon Trading and Financial Sustainability:
Intersections, Implications, and Strategies

Nidhi Solanki
Banasthali Vidyapith, India

Priyanka Mathur
Prestige Institute of Management and Research, India

Harsh Purohit
Banasthali Vidyapith, India

ABSTRACT

In recent years the world has been focusing more on combating climate change and carbon trading has become a vital tool with an aim to lower the carbon emissions. The studies have shown that the financial market is essential for the success of carbon trading as they provide carbon related assets with liquidity, price discovery, and risk management. This chapter examines how carbon trading mechanisms affect the financial health of businesses and economies. It explores how emissions trading systems affect financial performance and its implications on company's profitability, liquidity and competitiveness. This study reflects how environmental regulations influence business operations, considering opportunities and challenges for achieving financial sustainability. By integrating theoretical frameworks with implication and strategies from different studies and reports, this chapter aims to inform policy makers, businesses, and investors on the role of carbon trading in promoting sustainable finance and building long-term economic resilience in the face of climate change challenges.

DOI: 10.4018/979-8-3693-5748-4.ch011

INTRODUCTION

Background of Chapter

Carbon Trading

Carbon Trading is a mechanism used for lowering the global emissions of greenhouse gases and particularly huge amount of carbon dioxide which contribute to the maximum extent among all the gases. Fossil fuels contribute to the largest source of carbon dioxide leading to climate change. Largest emissions from the burning of fossil fuels are mainly due to the requirements of electricity, heat and modes of transport. Burning of crude oil, natural gas, paraffin, manufacturing process of chemicals, oils are the major sources of carbon emissions responsible for directly releasing the harmful gases at the site of production. With the increase in global warming and climatic changes transition to a low carbon economy has now become a necessity worldwide. In this context carbon trading emerges as a unique financial instrument that can tackle the environmental challenges with the power of carbon markets. By introducing price tags on carbon emissions, it incentivizes businesses and industries to reduce their carbon footprint.

Financial Sustainability

Financial Sustainability is the ability of a company to design a strong business structure that is able to flourish the business, absorb uncertainties and can generate long term value creation with stakeholders. Motive of financial sustainability is to create a balance between profitability and responsible financial practices. A business project must be evaluated on the basis of its capability to generate funds to meet out its financial liabilities and maintain resource optimization (Govindaraju & Mabel, 2015). Carbon trading promotes financial sustainability by providing a platform to companies where they can purchase, sell and monetise their credits or allowances to reduce emissions. Companies can set their emissions up to a certain limit and can sell the excess credits available to those companies who are not able to meet their required limits of emissions. This process provides an additional revenue stream and also promotes the companies to remain environmentally responsible. This has created a shift towards the greener practices among the companies. By reducing carbon footprints and generating financial returns, carbon trading aligns economic development and environmental goals leading to a longer financial sustainability.

ROLE OF CARBON TRADING ROLE IN PROMOTING FINANCIAL SUSTAINABILITY

This study explores the concept of carbon trading and its potential to drive financial sustainability alongside environmental progress. Through a financial framework, entities that produce less than their allotted amount can sell the excess allowance they have (carbon credits) to others who are producing more than they can. This develops a market in which cutting emissions turns into a profitable tactic as well as an obligation to protect the environment (El Amri et. al., 2022). The money received from carbon credits can then be used to find the development of sustainable practices and clean technology, creating a win win situation for the environment and the state of the economy

However, strong rules, open market mechanisms, and maintaining environmental integrity are all necessary for carbon trading to be effective. While carbon trading holds immense promises yet several research gaps have not been addressed in the current researches. The current researches concentrate on reducing emissions overall. Understanding how components of market design, such as price caps, auction processes, and credit validation, affect market efficiency and long-term emission trends, is a critical knowledge gap. More research is needed to determine how carbon trading would affect investors' and companies' bottom lines.

Research can look at how financial models incorporate carbon pricing and how that affects choices about investments in sustainable infrastructure and clean technology. Investigation is required to determine if carbon trading will unfairly overload emerging economies or actually contribute to environmental injustice. Future research should examine approaches to promote fairness in carbon market participation and a fair transition for all actors towards low-carbon futures. This research paper intends to fill these gaps to fully reveal the dimensions of carbon trading and how it facilities financial sustainability, leads a way towards green future.

Origin of carbon trading began with numerous international agreements, policies and initiatives starting with United Nations Framework Convention on Climate Change (UNFCCC) in the year 1992 at Rio de Janeiro. This structured a framework for international cooperation related to climate change, highlighting the requirement for reduction of greenhouse gas emissions. Kyoto Protocol adopted in the year 1997 was the first binding international treaty on emission reduction meant for developed nations. Then the era came of Voluntary Carbon Markets in the early 2000s that provided an option to the companies to offset their carbon emissions by making investments in projects on reduction of greenhouse gases. Such projects include renewable energy generation, forest conservation, sustainable waste management and many more. Later in year 2005, The European Union Emissions Trading System was introduced which came up with setting a cap limit on total emissions to be followed by specific sector industries and it allowed the organizations to trade the emission allowances.

The Paris Agreement, adopted in 2015, further emphasized the role of market mechanisms in achieving global climate goals. While it doesn't mandate specific trading schemes, it encourages countries to implement carbon pricing policies, including emissions trading, as part of their national strategies (Tiwari, 2023).

The process or mechanism of purchase and sale of credits that allow a business or other organization to emit a specific quantity of carbon dioxide or other greenhouse gases is known as the carbon trade also known as carbon emission trading. Governments have approved the carbon trade and carbon credits with the intention of progressively lowering total carbon emissions and lessening their impact on climate change. Its aim is to prevent global warming by establishing a market with finite emission allowances. This might make fossil fuels less competitive and encourage investment in renewable energy sources like solar and wind power. The primary cause of climate change is fossil fuels which constitute around 89% of CO_2 and 68% of greenhouse gas emissions.

Carbon market is a platform for trading carbon credits which are essentially permits that allow emitting a specific amount of greenhouse gases typically one ton of carbon dioxide equivalent. These emissions can be categorised into two, Carbon credits and Carbon Offsets, both of which can be bought and sold on a carbon market. These two terms are used interchangeably however they are different.

Carbon credits are also called as carbon allowances. These act like permission slips for carbon emissions. A company needs to obtain permission from the Government and buys a carbon credit which permits to generate one ton of CO_2 emission. Carbon credits create a vertical stream from companies to regulators. Companies having surplus carbon credits can sell these credits to other companies who are falling short of carbon credits.

In contrast, Carbon Offsets generate horizontal revenue flow, trading carbon funds between businesses. When a company eliminates a unit of carbon from the atmosphere through its usual operations, it produces a carbon offset. Other companies can buy these offsets to lower their own carbon footprint (Poyser, 2024).

California in U.S. runs its own carbon market and provide credits to citizens based on gas and electricity consumption. Usually, annual credit allocations are determined by emissions targets. Credits are often distributed through a "cap-and-trade" scheme. Cap-and-trade is a mechanism intended to lower air pollution. A strict limit on pollution is the cap on greenhouse gas emissions, which cause global warming. With time, the cap becomes stricter. In the trade section, companies can purchase and sell emission permits, with supply and demand determining the price.

Types of Carbon Market

Carbon markets are of two types, one is Regulated market set by "cap-and-trade" regulations where the company operating under such scheme is issued a certain number of carbon credits each year. The companies who produce less emission may save their credits and sell the surplus carbon credits to other companies. The purchasing companies are the ones who have utilised all their credits but their emissions are more as compared to the credits allotted to them hence they purchase the credits from companies who have surplus available.

The other category is voluntary carbon market which are adopted by environmentally conscious companies which are not bound to do so but they want to demonstrate their responsibility towards environment and hence are choosing to offset their emissions (Hongtao and Nan, 2022)

A carbon tax directly sets a price on carbon by levying a tax on the carbon content of fossil fuels or on carbon emissions. The tax rate can be set per ton of CO_2 emitted. The cost of emitting carbon is predictable and fixed, providing businesses with certainty about the price they will pay for emissions. The government collects revenue from the tax reductions in other areas, or address social equity concerns. While the cost is predictable the exact amount of emission reduction is uncertain because it depends on how businesses and consumers respond to tax (Dias and Rocha, 2023). Carbon tax controls the price directly but leaves the quantity of emissions to market response. While Cap-and-trade controls the quantity of emissions directly but lets the market set the price. A carbon tax provides predictable costs for the business while cap and trade provide predictable environmental outcomes.

Intersection Between Carbon Trading and Financial Sustainability

Financial Sustainability can be defined as the capacity to use an ideal mix of investments and funding sources to create value for shareholders and maintain operations over the long term (the concept of continuity relates to the going concern principle of accounting). As a result, the scope of financial sustainability encompasses a number of financial factors that work together to provide the foundation for relationships between value and company continuance. Under specific business conditions, the best mix of these factors guarantees the maximum degree of financial sustainability. But the link between continuity and value deteriorates, with unclear consequences (Zabolotnyy & Wasilewski, 2019).

A company's capacity to be financially sustainable is greatly impacted by carbon trading due to a number of factors, including revenue creation, cost control, regulatory compliance, and competitive positioning. Businesses that purchase carbon credits for offsetting excess emissions sustain higher operating costs whereas, companies that sell excess credits make more income. While the process of following the cap and trade rules do incur additional costs but at the same time it also saves the company from fines and generate rewards. A company's market share and brand image also get improved attracting more clients and investors. Long-term financial gains provide a pathway for adopting more sustainable practices and energy-efficient technologies that also encourage innovation and cut down the expenses. Carbon trading also reduces environmental hazards and diversifies sources of income, which improves overall financial stability. All together, these elements strengthen a company's ability to adjust to changing regulatory frameworks and economic situations, preserving its long-term financial stability (Su and Chen, 2024).

Carbon Trading plays a very crucial role in assisting financial sustainability by adjusting environmental responsibility with economic viability. Utilisation of carbon credits and carbon trading incentivizes helps the companies to lower their greenhouse emissions, contributing to global climate goals. Companies that effectively lower their emissions can sell their excess allowances, creating a new revenue stream and encouraging innovation in clean technologies. The revenue generated from carbon credits can be reinvested into further sustainable projects, improving a company's long-term financial health and positively influencing its environmental impact. Besides, by participating in carbon trading companies can reduce their risks associated with regulatory changes and potential carbon taxes if any, ensuring compliance with the evolving market. Eventually, carbon trading creates a synergy between environmental governance and economic performance, demonstrating that responsible ecological practices can coexist with financial success (United Nations Conference on Trade and Development, 2023).

The introduction of carbon trading schemes has the efficiency to lower down the credit spreads for highly polluting corporations' bonds, thereby promoting and incentivizing green changes and bond financing within such companies (Mashari et. al., 2023). Further, there is a positive correlation between fossil CO_2 emissions and sustainability-linked financial markets, focussing the necessity for high-emission nations to issue sustainability-linked money for adherence to climate commitments (Liu et. al., 2023). In order to make the financial management driven towards sustainability it is necessary to mandate the product redesign and encouraging carbon-negative projects when the financial incentives are aligned with the disincentives for carbon emissions through methods such as the Carbon Equivalence Principle (Chidaushe & Njaya, 2022).

The study incorporates the influence of carbon trading on financial sustainability by studying the factors affecting the relationship between carbon trading and financial sustainability, these factors are market forces, environmental policies and regulations, macroeconomic factors, and corporate's environmental performance. Businesses who succeed in reducing their emissions can make money with this market-driven strategy, which can result in cost-effective carbon reductions (Bansal et. al., 2023).

Carbon trading can be one of the best stimulating investment options in green technologies and technologies as per the environmental rules and policies led by Government that are essential in forming these mechanisms and creating the legal environment in which carbon trading takes place. The effect of different level of micro as well as macro variable can adversely affect the general financial climate. These elements work together to produce a complicated interaction that establishes a company's ability to sustain its finances (Gopi et. al., 2023).

The effect of carbon trading programs on company's environmental performance differ at the global level, according to study. In general, carbon pricing regulations may result in output in areas where businesses have not substantially expanded their own innovation or embraced cutting-edge digital technology. But companies that are more productive at first tend to gain from more autonomous innovation, which is known as the "Porter Effect," which describes how increasingly stringent environmental laws stimulate innovation to gain a competitive edge. There are performance differences within industries, demonstrating the diverse effects of carbon trading. Inefficient resource allocation is the outcome of slow entry and leave times for inefficient businesses and potential new entrants. Finding a balance between economic growth and environmental conservation is made more difficult by this misallocation. The best way to increase the carbon trading initiatives is to value added prices to different international opportunities with better economic growth It is very important that environmental preservation with environment growth. With complete acceptance is given by corporates and big industries the new carbon trading schemes and technology will grow and outperform with the environmental growth (Sun et al., 2024). Based on the previous studies a Conceptual Framework has been developed to study the factors influencing the relationship between carbon trading and financial sustainability. These are outlined in figure 1 mentioned below.

Figure 1. Conceptual framework

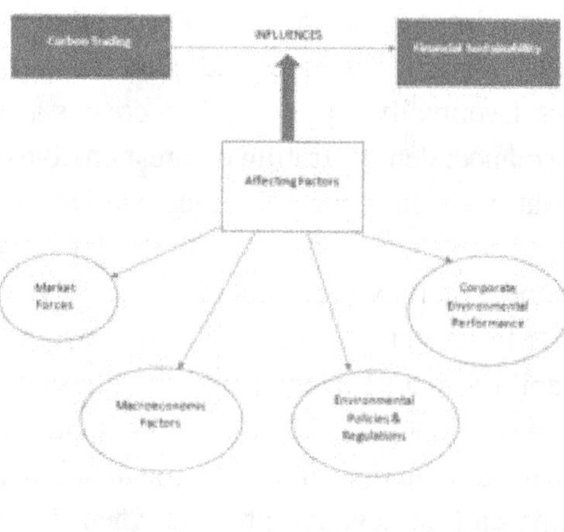

Source: Self-generated

THE CARBON MARKET IMPLICATIONS

Carbon markets are affected by both policies and financial domains. The trading method allow experts in the field to put expenses into the system and also affect the market of carbon trading with relevant policies of economic legislature. The goal of study is to investigate how carbon trading affects the market in large scales as well as in context with Indian economy. It explores the way in carbon trading affects asset price model, investment strategies, market liquidity and risk management for further economic ramification and investigation in this chapter. This section of study focused on carbon footprints on daily operations of business. It understands the influence of market dynamics of all sector and their function in advancing low carbon economy. The economic impacts of carbon trading are outlined in the figure 2 as shown below

Figure 2. Economic impacts of carbon trading

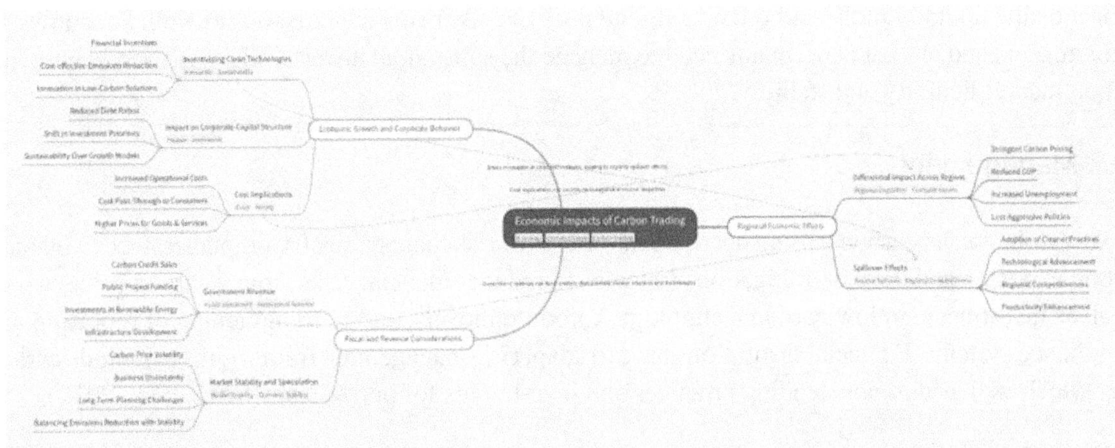

Source: Self-generated

Global Context

Valuation of Resources

Carbon trading has an impact, on the prices of assets. Businesses in industries like energy production and manufacturing that have high carbon emissions face expenses because of carbon pricing. This can affect their profits and the value of their stocks. On the hand companies working in energy and low carbon technologies might see an increase in their value as investors expect growth due to environmental rules and consumer interest, in eco-friendly practices. Carbon footprints are becoming increasingly important in the assessment of asset values. Are causing a revaluation of assets, in different industries (Engle et al., 2020).

Investing Techniques

Rather than using classic carbon trading approaches, financial investors have turned to a whole range of new funds and portfolios of significantly all those enterprises that are well placed to benefit most from the transition to a low-carbon economy – or are expected to be well-positioned to do so in the near-term. Green bonds, ESG funds and other climate-themed financial instruments have come onto the scene in large numbers – that is, as long as the problems with green washing can be kept at bay. Carbon exposure is also increasingly represented within investment portfolios, and this 'greening' approach to investment is quickly generating new financial products and indices based on sustainability indicators (Baker et al., 2018).

Volatility and Market Liquidity

There are several reasons why carbon trading could create market volatility and liquidity. Firstly, Substantial price swings can result from shifts in global energy markets, regulatory changes or new technical developments potentially changing the future demand for carbon credits, which might have an impact on the liquidity of the carbon markets and possibly other financial markets as well; Secondly, the policymakers and market participants need to manage the substantial uncertainties involved in ensuring carbon market liquidity and stability.

Risk Management

At a global scale, carbon trading opens up new fields of risk management. Companies need to manage carbon exposure and develop mitigation strategies to reduce financial risks from a carbon price, which involves investments in low-carbon technology, carbon removals, and a strengthening of reporting and compliance systems. Financial institutions have to adapt risk-management frameworks to include carbon risk, which will lead to new lending practices and investments for portfolios. (Busch et al., 2018).

Greater Economic Consequences

Other wider economic impacts of carbon trading include contributing to the low-carbon economy and thus providing support for sustainable development. Markets place a price on carbon and, through the reduction of greenhouse gas emissions, spur innovation and investments in clean energy. However, this transition does not come without challenges, particularly to those regions dependent on fossil fuel economies and companies that are highly carbon-producing. Support systems and policy formulation must, therefore, be painstakingly done since economic restructuring needed to lock in a just transition may cause disruptions or disturbances in the short run.

Indian Context

Cost of Assets

India's asset price is impacted by carbon trading in a way that is similar to the global market but has unique characteristics. The primary industries in India, such as power, steel, and cement, are major contributors to greenhouse gas emissions. The Perform, Achieve and Trade (PAT) system and the recently formed Indian carbon market have an effect on these sectors since they factor in the cost of carbon emissions in their evaluations. The companies in renewable energy or with lower emissions are highly benefitted with their share prices.

Investing Techniques

The techniques according to Baker et al. (2018), this tendency "encourage the development of new financial products and indexes based on ESG standards in accordance with global investment strategies." The climate of India is basically shaped by carbon trading. The growth of market expansion in green bonds and other financial instruments reflect the growing interest in sustainable investments. Big investors are looking forward to purchase stocks of low carbon emission companies which seems to be big shift.

Market and Voluntary Liquidity

The commencement of carbon trading may have high impact on the volatility and liquidity of Indian market exactly similar as International market. The technical advancement, shifts in national energy markets, and changes in regulations can all have a big impact on the cost of carbon credits and the PAT scheme in India. "These traits may affect the carbon market's liquidity and may have repercussions for other financial markets. Careful navigation of these uncertainties is necessary to preserve market stability and effective operation" (Benz & Hengelbrock, 2008).

Risk Mitigation

According (Busch et al., 2018) "The carbon risk must be taken into account for financial instruments of risk assessments in order to affect lending practices and investments decisions". Financial institutes and businesses in India have more options for risk management due to carbon trading. They need to control their carbon emission and need to implement plans to reduce carbon pricing. This include making investments in green technology, carbon offset schemes and improved compliance can work in wonders.

Wider Implication on Economy

Carbon Pricing reduces Green House Gas emission and stimulates investment in renewable energy. However, there are barriers to this transition, particularly for regions that rely heavily on fossil fuels and industries that emit large amounts of carbon. "The required economic restructuring may result in short-term disruptions, therefore support mechanisms and smart policy formulation are crucial to ensuring a fair transition." (Stavins, 2020).

STRATEGIES FOR CARBON TRADING FOR FINANCIAL SUSTAINABILITY

The major focus on strategies of carbon trading is to reduce the greenhouse gases emissions along with emphasizing on economic growth and development. Understanding these strategies involves re-viewing and analysing what has been accomplished, what is currently ongoing, and what future plans entail. Some of the recent and popular strategies are being discussed here.

Based on Market Forces Emission Reduction Mechanism

Through carbon trading industries can sell and purchase carbon credits, which are equivalent of the right to release a specific quantity of carbon dioxide. With this financial incentive the cap-and –trade system encourages businesses to cut their emissions. Those who can do so at lower cost can sell their excess allowances to those who find it more difficult to do so.

Financing Climate Financing

The carbon markets facilitate the raising of capital for initiatives that lower emissions. Most companies sponsor programs that might not be financially feasible without money by acquiring carbon credits.

Connecting with Green Finance

Green finance includes a range of financial products meant to encourage environmental sustainability related to carbon trading, financial measures to mitigate climate change are more effective overall because of the synergy between these two mechanisms.

International Cooperation and Standards

The goal of creating the world-wide frame work for carbon market in context of Paris Agreement and following COP conferences is to coordinate beyond boundaries. To foster trust among players and to increase credibility of market, the cooperation is crucial.

Facilitating Sustainable Development

In addition to lower emissions carbon trading can have a positive impact on the environment by protecting biodiversity, creating jobs and enhancing public health. Carbon trading very important tool for accomplishing both environment both nationally and internationally as it major role in sustainability.

Holistic Carbon Market System

The large industrial players in challenging industries anticipated to drive a sizable portion of the market for carbon credits in order to follow with compliance of carbon trading and fulfil net zero promises and satisfy investor demand. It is crucial for policy players to explicitly define laws, norms and market mechanism that let voluntary and compliance carbon market to coexist and complement one another as affirmed by primary respondent. The market will operate more effectively and can be seamlessly integrated into the NDCs' realization with clearer laws. The relationship between the domestic carbon market in India and the global carbon markets

THE CARBON MARKET (COMPLIANCE AND POLICY MATTER)

Carbon Border Adjustment Mechanism (CBAM)

One tool used by policy makers to internalize the environmental externalities linked to the production of imported commodities is the Carbon Border Adjustment Mechanism. CBAMs seek to reduce greenhouse gas emissions by attaching a monetary value to carbon emissions included in traded goods. This will aid in the worldwide endeavour to attain net-zero emissions. Net zero refers to a situation where total greenhouse gases would become equal to the emissions removed from the atmosphere by employing carbon offsets. To achieve this target governments of different nations have also introduced plans and regulations that can support to decarbonize different industries while adhering to set goals for reducing emissions. Emissions trading programs run the risk of just shifting the manufacturing of items with high carbon content to other places, or "pollution havens," where manufacturers might be exempt from stricter laws. Since the same amount of emissions are still created even if manufacturing may have shifted, this is frequently referred to as "carbon leakage." By imposing a fee on carbon emissions from the manufacture of imported carbon-intensive commodities like aluminium, cement, iron, and steel, a carbon-based alternate market mechanism (CBAM) seeks to guarantee fair treatment of both domestic and imported goods. Its goal is to stop carbon from leaking (Zhong & Pei, 2024).

CBAMs mandate the collection, reporting, and verification of information on emissions embedded in products so that customs officials can apply the relevant carbon tariff when the items are imported into the nation. The exporting nation may incur administrative expenses as a result of this operation. Any economic effects are probably going to be concentrated in the nations and the areas that are home to a large concentration of carbon intensive sectors. CBAMs must be created so that imported items are not treated less favourably than equivalent domestically produced commodities in order to comply with international trade regulations

The world's first CBAM is the EU CBAM, which is presently in a transitional phase and will be fully operational from January 1, 2026 (Perdana & Vielle, 2022). The UK government also declared the CBAM in December 2023 that will be implemented in the country. It will have consultations in 2024 over the UK CBAM's design. The UK CBAM will resemble the EU CBAM, however there will be variations in terms of scope and timeline, according to the government's declarations. The government has stated that implementing unilateral CBAMs is not the best way to address carbon leakage; instead, worldwide cooperation is the best approach, however it takes time. An effort to promote international collaboration on carbon leakage has been made with the recent formation of the "Climate Club," which was co-founded by Germany and Chile. However, there are certain concerns and uncertainties raised on CBAM. Exporters in developing nations that operate in the covered sectors—such as steel, aluminium, cement, and fertilizers—many of which are essential to the heavy industries and export revenue of these nations will see an increase in CBAMs have caused concern over their export markets. The EU CBAM has been referred to as a "green" trade barrier by China and Russia (Burnett et. al., 2024).

Carbon Nationalism: Article 6.2 & Article 6.4 of the Paris Agreement

The international carbon market is changing, and one prominent development is the rise of "carbon nationalism," in which nations strategically manage their carbon credits in accordance with Articles 6.4 and 6.2 of the Paris Agreement. Countries engage in this phenomenon by either holding or selling their carbon credits in order to optimize the advantages to the economy and environment. A number of nations are preparing to generate Article 6.4 credits, a system put in place by the Paris Agreement that allows nations to use internationally traded carbon credits to satisfy their climate commitments.

Some nations like Brazil, India and many African countries have variety of natural resources and hence they have a larger scope and potential for development and implementation of renewable energy projects. The major part of carbon credit is basically depending on whether to hold or buy carbon credits. It plays a significant role in the structure of carbon trading. Increase in the selling of carbon credits can be impacted by attracting more foreign investments. It can attract more financial and sustainability in the carbon market. Furthermore, holding credits would be better strategic move because it would allow countries to be benefitted from future price hike allowing them to use credits to meet their own climate goals.

The carbon markets will be significantly impacted by this policy. Selling credits allows countries to add to the global supply, which may lower or stabilize market prices. On the other hand, nations that retain their credits may cause a shortage in the market, raising costs and enticing investors to spend more on low-carbon technologies. According to the provisions covered under Article 6 of the Paris Agreement countries have the option to adopt voluntary cooperation in the implementation of Nationally Determined Contributions (NDC) where two or more companies can trade the allowances like one company can sell its excess emission allowances to other company requiring it to meet its emission targets.

Such collaboration in bilateral or multilateral agreements for trading carbon credits, create a flexible approach to fulfil sustainable goals at global level. To promote transparency and avoid double counting of emission reduction, policies established under law and tracking methods must be introduced at the national level in order to implement norms of Article 6.2. Leading nations which have formed the bilateral agreements include Peru, Japan, Switzerland, Ghana, South Korea, Sweden and Singapore. In the year 2022, Ghana Government set up a Ghana Carbon Registry for recording and approving carbon projects. Even before Paris Agreement, Japan has set up the Joint Credit Mechanism Concept in the year 2013 where it entered into bilateral agreements with 27 developing nations (so far) to finance their projects and in return for carbon credits acquired from supporting such projects. But all these efforts entail a strong monitoring, reporting, and verification (MRV) system for implementation to be successful. It is essential that national strategies should align with sustainable development goals to make sure that carbon trading brings broader social and economic benefits.

CORSIA: Carbon Offsetting and Reduction Scheme for International Aviation

Another global initiative known as was developed by the International Civil Aviation Organisation (ICAO). Purpose of introducing CORSIA was to address the increasing greenhouse gas emissions from international aviation. Aviation is one of the main sources of carbon dioxide. On the burning of fossil fuels in aircraft engines carbon dioxide, water vapor, nitrogen oxides, and other greenhouse gases are released leading to harmful emission in the climate. The white, splotchy lines that aircraft produce in the sky are known as contrails. They are composed of ice crystals that are created in high-altitude atmosphere by the condensation of water vapor from the aircraft's exhaust. Contrails trap the heat in the atmosphere of Earth leading to one of the causes for Global warming.

CORSIA provides a uniform framework for monitoring, reporting and verifying carbon emissions from international flights, ensuring consistency across countries and airlines. It requires airlines to purchase carbon offsets to compensate for any growth in CO_2 emissions above 2020 levels, promoting investments in renewable energy and any other carbon reduction projects (Almena et. al., 2024). As on 1 January 2024, total 126 countries have committed to participate in CORSIA.

In support of the temperature objective set forth in the UNFCCC Paris Agreement, the 41st ICAO Assembly established LTAG (Long-Term Aspirational Goals) for international aviation that targeted net-zero carbon emissions by 2050. The international organization has established a number of important aspirational targets to reduce aviation's carbon emissions and the effect they have on climate change. Among them are carbon neutral growth, net zero by 2050, and 2% yearly fuel efficiency improvements until 2050. The CORSIA Scheme is divided into three stages: The first phase (2024–2026) and the pilot (2021–2023) are optional, but the second phase (2027–2035) is required for all member states. India has made the decision to withdraw from the optional CORSIA phases because it believes that they are unfair to developing-nation airlines, which have the ability to grow significantly faster than those of industrialized nations. A significant number of international airlines, particularly Indian carriers, may also face financial hardship as a result of the CORSIA offsetting obligations.

Important steps taken by Ministry of Civil Aviation in India are adoption of Green Airports that have adopted sustainable methods to lessen their impact on the environment and support sustainable development. Green airports strive to decrease waste and emissions, minimize their carbon footprint, and preserve energy and water resources.

Another initiative is, the 2016 National Civil Aviation Policy (NCAP) creating a framework for sustainable aviation as one of its objectives; which encourages the use of energy-efficient aircraft, alternative fuels, and infrastructure, among other things. And lastly, Sustainable Aviation Fuel (SAF) to promote sustainable development and lower carbon emissions at airports, steps have been taken to increase the usage of SAF (Sen et. al., 2024).

The EU Emission Trading System

The Voluntary Carbon markets are regulated by market-based mechanism and by compliance to reduce the emission based carbon.EU trading system is very different stepping stone of European Union climate policy. The change is to use coal in place of gas in energy production is the important factor influencing EU ETS prices. For example, Germany's Energiewende program has increase the use of natural gas and renewable energy sources. The change from coal and fossil fuels to renewable sources reduces the emissions. Energy crises in 2021-2022 may lead the switch from coal to gas use. Due to variation in price of carbon it can be on investment planning and market stability.

The Market Stability Reserve (MSR) which automatically adjusts the supply of permits in response to market conditions. According to (Wei et al., 2024) "The addition of maritime emissions will increase demand for allowances, which will motivate more emission reductions." For example, the MSR lowers the quantity of allowances up for sale to stabilize prices during times of excess. When carbon prices jumped above €90 per ton in the beginning of 2022, this mechanism worked well to keep the market in balance and encourage long-term investments in low-carbon technologies.

It is anticipated that the EU ETS would grow to encompass new industries, like maritime transportation. The demand for allowances will rise in response to the inclusion of maritime emissions, which will encourage further emission reductions (Wei et. al., 2024). The aviation industry was one of the first to be included in 2012; as a result, airlines such as EasyJet and Lufthansa were required to buy carbon permits, which encouraged the industry to reduce emissions

Understanding the methodology behind carbon credit pricing is crucial considering factors such as project, location and co benefits. For example, credits from reforestation projects in the amazon may be priced differently from those from renewable energy projects in India due to low impacts. This means providing funds for various projects (such as sustainable energy and forestry) in various geographical locations (Laing et al., 2013).

Future Carbon Fund

The Future Carbon Fund (FCF) provides up front financial support for carbon trading pre-purchasing carbon credits from emission-reducing initiatives. By reducing project developer's risks and encouraging long-term investment in green technologies, this strategy provides financial sustainability and is in line with both sustainable development and climate goals. Figure 3 as shown below portrays the Future Carbon Fund Sustainable Plan for goals as derived from Asian Development Bank Reports

Figure 3. Future carbon fund sustainable plan for goals

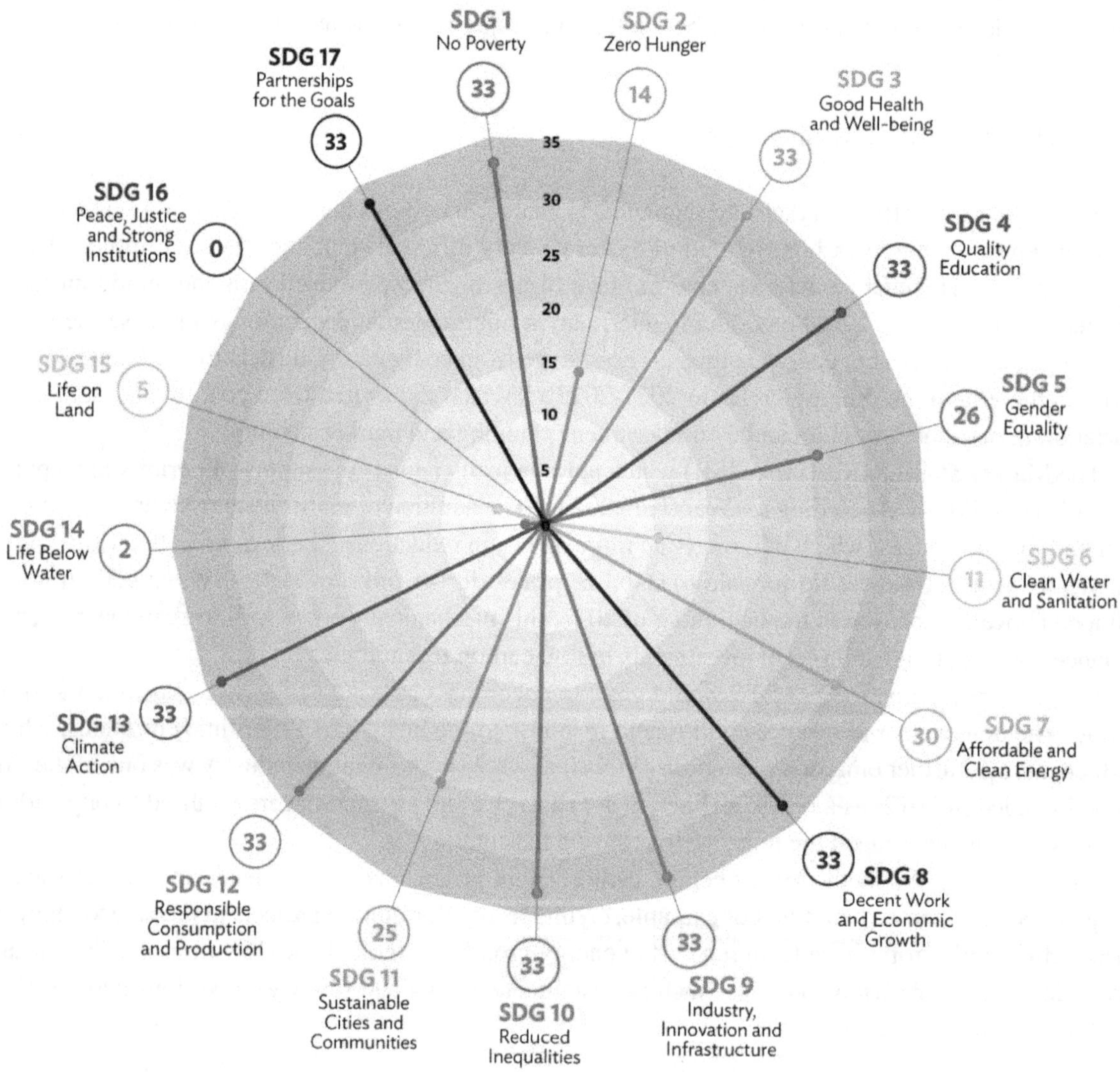

Source: Asian Development Bank Reports

The Asian Development Bank (ADB) is in charge of overseeing the Future Carbon Fund (FCF). It contributes significantly to the advancement of carbon financing by means of initiatives that lower greenhouse gas (GHG) emissions. Beyond monetary gains, these initiatives support sustainable development, especially in the Asia-Pacific area, by bringing about social, environmental, and economic co-benefits. It came into existence in year 2009

The FCF pre-purchases Certified emission reduction CERs or carbon credits a key example is the "Dagachhu hydropower project" a hydroelectric power plant in Bhutan which generates CERs from actual emission reductions and benefit from international carbon trading. The most important feature of FCF is to offer upfront payments combined with payments upon CER. Impact of the Future Carbon Fund (FCF) in promoting sustainability are shown below in figure 4. These have been sourced from the Report on Future Carbon Fund Delivering Co-Benefits for Sustainable Development.

Figure 4. Impact of the Future Carbon Fund (FCF) in promoting sustainability

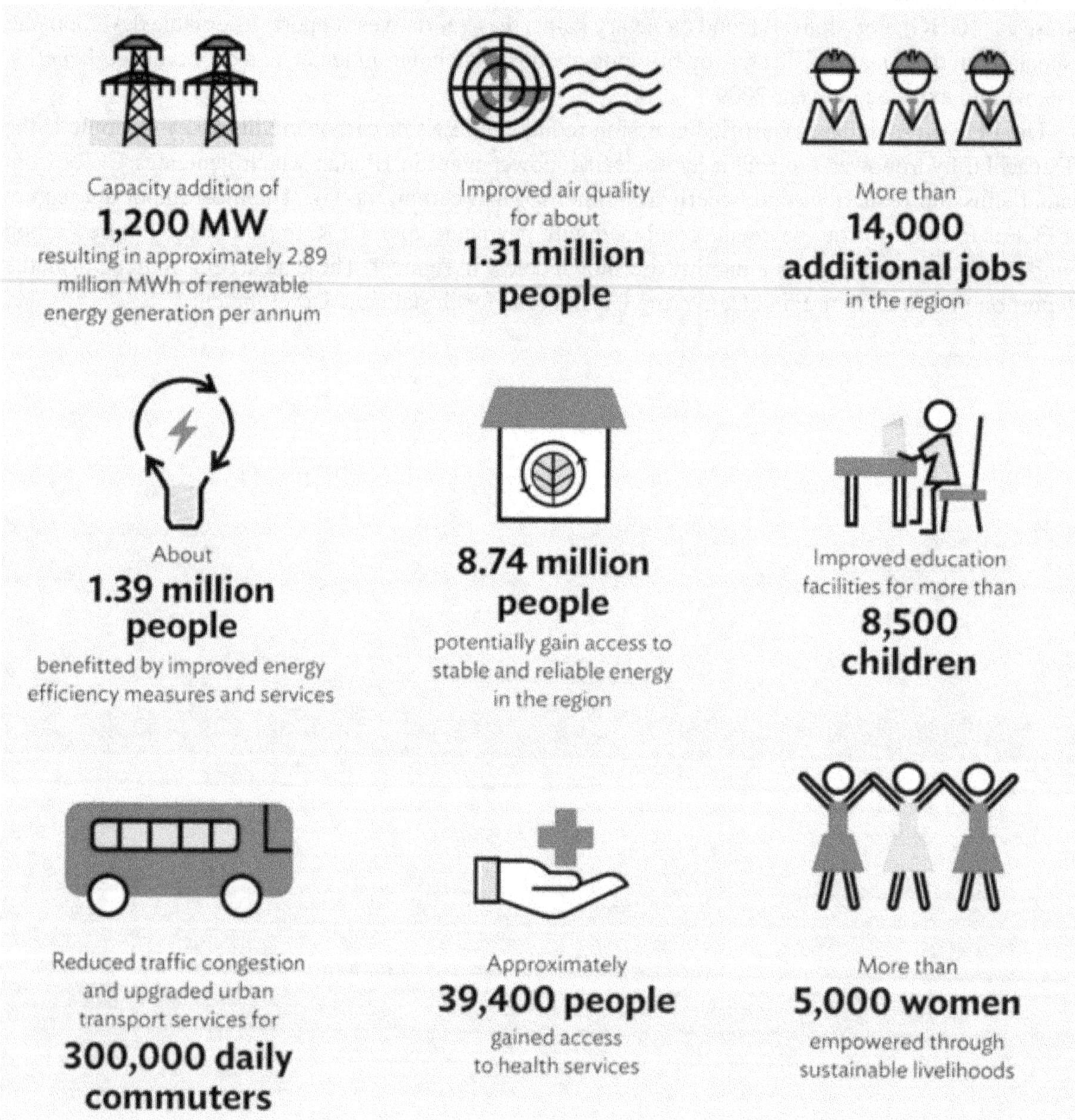

Source: Report on Future Carbon Fund Delivering Co-Benefits for Sustainable Development

Carbon as an Asset with Financial potential: A chance for companies to achieve financial sustainability

The concept that carbon credit is the best financial asset is growing pathways to open up for trade and investment opportunities for many nation and companies to reduce their emissions. The establishment of carbon trades hold immense potential to ignite innovation and technological advancement in transition sectors such as clean energy, transportation and agriculture. Both public and private sectors can unlock there potential to full extant for financial sustainability. The financial markets have been severely impacted by this shift in number of areas including price discover, supply and demand dynamics, risk and return assessments, and the inclusion of climate indicators in economic models.

THE IMPORTANCE OF PRICE DISCOVERY

Price discovery is most important establishment of carbon credits as a marketable asset. The EU ETS is one of the largest carbon markets, with trade determining prices. The strict emission limitations and increase in demand for carbon credits caused the cost of EU carbon permits to skyrocket up to more than 60 pounds by 2021.With the help of this pricing sign business may more effectively plan their investments in emission reduction systems.

Risk, Returns and Financial Impact

To evaluate the potential risk and benefits while investing in carbon markets is imperative. Investors are exposed to risk when it comes to the California Cap-and –Trade program, for instances, where prices have fluctuated as a result of market conditions and regularity changes. Nonetheless, there may be significant potential rewards to the same. An international Finance corporation (IFC) analysis is suggests that by 2030, investments in climate-smart initiatives in nation alone may return around $23 trillion. With this understanding of risk-return profiles helps investor's male well-informed decisions and can attract more money for carbon reduction programs.

Demand and Supply Dynamics

There is an important factor affecting the carbon market which is the dynamics of supply and demand. For example, the voluntary market has witnessed a substantial surge in demand for carbon credits due to the pledges made by companies such as Microsoft and Amazon to reach net zero emissions. Microsoft, for instance bought 1.3 million carbon credits to offset their emissions in 2020.On the supply side, countries like Brazil and India are increasing their output of carbon credits through reforestation and renewable energy efforts. Finding the balance between these variables is necessary to maintain a stable and productive market. With rising demand and tighter climate legislation encouraging more investment in environmentally friendly activities, it is anticipated that the value of carbon credits would rise.

Inclusion of Climate Indicators in Economic Models

In order to promote climate investments, climate indicators are increasingly being included in economic models. The Network for greening the financial system (NGFS), an association of central banks and financial supervisors, recommends, among other things, including climate-related risk into financial stability monitoring and economic forecasts. The model considers variables such as carbon prices, emissions reductions and climate change dangers, providing a more comprehensive picture of economic environment. By helping investors and politicians make decisions that support both the goals of economic growth and climate change mitigation. This inclusion encourages sustainable investments. Understanding price discovery methods, weighing the benefits and drawbacks of investing in carbon, analysing supply and demand dynamics and integrating climate indicators into economic models are necessary to recognize carbon as financial assets.

BENEFITS OF CARBON TRADING

In long term financial viability carbon trading is one of the best options. Majorly Businesses, Companies and industries are financially motivated to adopt cleaner technology and lessen the carbon footprint by assigning a market to carbon emissions. With the supportive investments in green technologies, renewable energy sources and energy efficient operations this support has been for long term financial sustainability and can eventually reduce the operating cost.

The projects in businesses that have successfully lower their emissions can also profit from the sales of excess carbon credits. The fiancé arrangement harmonizes environmental objectives with the economic expansion in addition to encouraging sustainable company practices. As demonstrated carbon trading provides an effective means of fusing financial sustainability with economic objective. Different Sectors such as airlines sector in EU ETS is a practical example where Airlines that operates inside the European Economic Area have their emissions or risk paying fines. When different airline companies participated in carbon trading like British airways, Lufthansa invested suffice in efficient aircraft fuel and sustainable aviation fuel policy to overcome the emission and to improve on expensive carbon credits.

The emission allowance issued to businesses can exchange excess allowances. The significant investments in cleaner technologies and renewable energy have resulted in different programs helping both pollution reduction and financial sustainability. The best illustration for this is California's Cap-and-trade Program which targets large polluters like power plants and oil refineries. Carbon trading facilitates the transition to low carbon technologies in both cases lowering operating expenses and allowing businesses to balance sound financial practices. The program encourages companies to invest in low carbon technologies. As recorded by (Mo et al., 2016) "this change not only contributes to a reduction in greenhouse gas (GHG) emissions but also has a favourable impact on employment and green investment."

As said by (Dong et al., 2022) "Better air quality results from two-fold advantage providing air quality, positive effect on public health. The approach lowers harmful pollutants such as sulphur dioxide, nitrogen, and particulate matter by reducing greenhouse gas emissions." In addition carbon trading stimulates the stock market and increases total factor productivity both of which contribute to economic growth (Wen et al ., 2021).It also put lot of focus on companies innovation and more energy sufficient approach which helps in profit increment. The compilation of benefits of carbon trading are shown below in figure 5.

Figure 5. Benefits of carbon trading

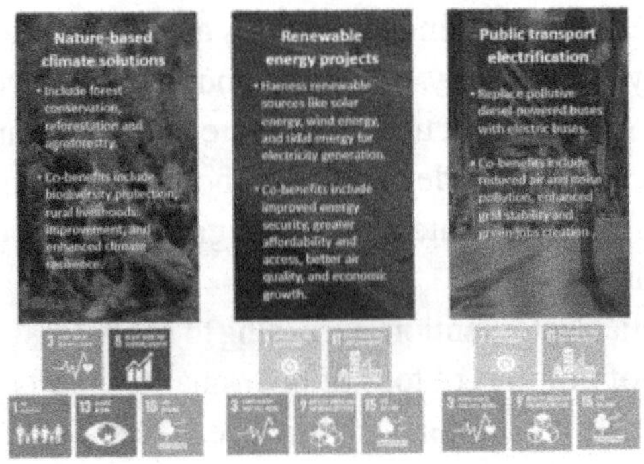

Source: Unsplash – Andreas Gluckhorn, Eyyoel Kahseeh, Jeyakumaran Mayooresan. Workers planting tree. Solar Panels. Singapore bus at night

CHALLENGES OF CARBON TRADING

With the lot of potential scope of carbon trading it also has some strong challenges Figure 6 reflects the major challenges associated with carbon trading. Initially it's critical to keep eye on reducing carbon emission which is the most important target. Though it is very clear but if targets to reduce it are set too high, pollution abatement will become too expensive and unpopular. And if reduction target is set too low then prices will be too low to encourage it. In order to guarantee participation reduction target must take into account broader economic and social effects.

Figure 6. Challenges faced by carbon markets

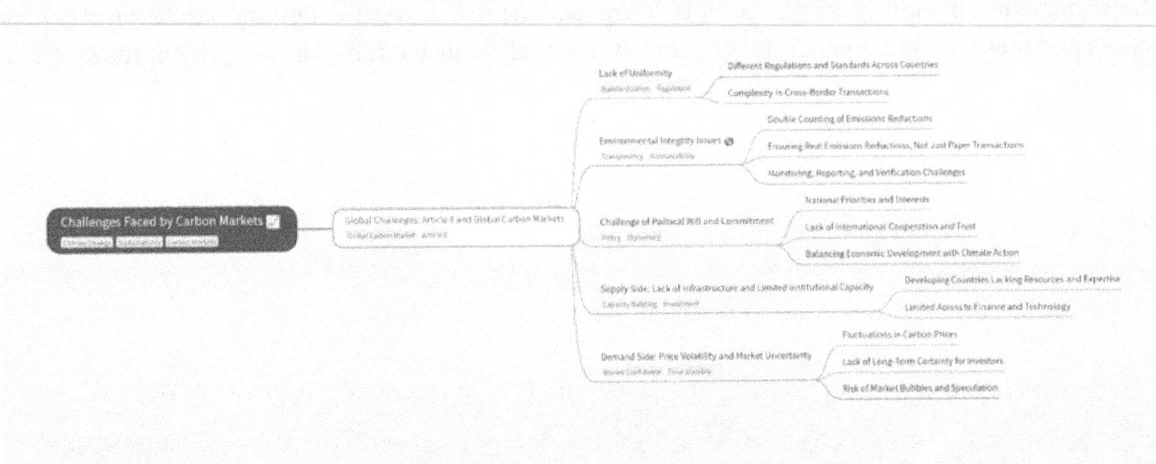

Source: Self-generated

Secondly there is another issue affecting carbon trading globally is no harmony among various market. The current compliance schemes are largely independent and there is currently no established global trading market. Third is the assessment of climates projects quality and authenticity involves figuring out how additionally, measurability, verifiability, durability and ability to avoid emission shifting are taken into consideration. There is also poor market transparency regarding the ambiguity about availability and accessibility of information on supply and demand of carbon credits, pricing transactions and impacts. Although the goal of Article 6 is to guarantee market alignment there is always a chance that carbon reductions will be double credited.

Fourthly the obstacle of political intervention, according to Mark Maslin of University college London "the market contains hundreds of millions of tons of poor-quality credits". Seeking to the complexity of carbon market and its structure we need robust governance is needed. Lack of accountability threatens the market creditability of market.

Fifth is limited institutional capacity which is another big obstacle impeding the emergence of carbon markets in poor nations. This makes more difficult for government agencies and regulatory to manage and control local markets. Meanwhile other elements that impede market development is lack of technical knowledge or insufficient enforcement procedures that erode investor confidence.

Overcoming Challenges in Carbon Trading

To overcome the challenges associated with carbon trading, it is necessary to address certain key issues as mentioned in figure 7, like establishing a standardised accounting system so as to ensure adoption of consistent emission measurement methods, uniform reporting standards and verification processes in auditing. Another aspect is a regulated governance which shall ensure establishment of independent oversight bodies, international agreements and regulations and implementation of enforcement mechanisms. Also, increasing transparency through public data on emissions and credits, easy traceability of carbon credits and real time market information helps to build trust in the system. Further, harmonizing different markets with interoperable trading platforms, standardised credit valuation methods and cross border trade facilitation supports in creating a unified global market. Finally, clear rules on carbon credits and offsets, including eligibility criteria for emission reduction projects, verification and validation of carbon credits and prevention of double counting are all essential for maintaining the integrity of the system.

Figure 7. Overcoming challenges of carbon trading

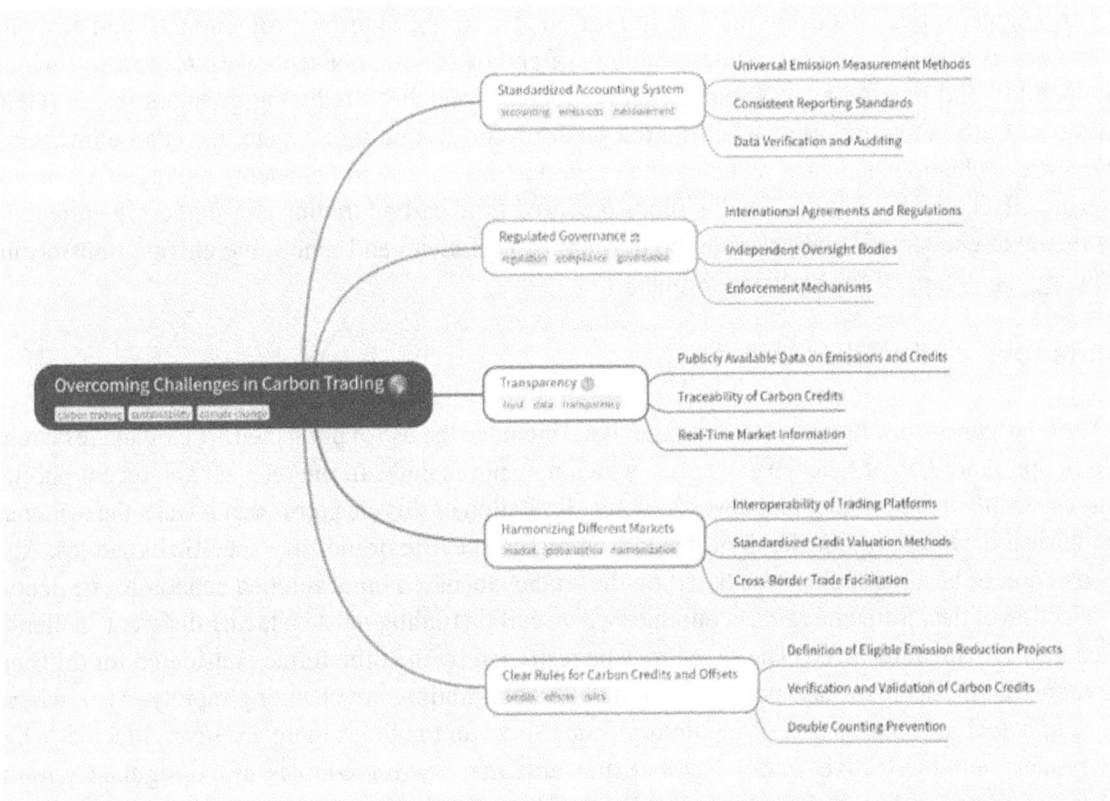

Source: Self-generated

CONCLUSION

The results of this chapter revealed the relationship between carbon trading and financial sustainability, analysing the effect of micro and macroeconomic variables, external market pressures, environmental laws, company environmental performance and carbon trading methods on business and economic viability. They showed how carbon pricing strategies, affect business profits, liquidity, and competitiveness in the market. They further explained potential opportunities and difficulties in reaching financial sustainability along with implications of regulatory norms. Moving ahead, productiveness carbon trading can be improved by fortifying legislative frameworks, accepting technology updates, and promoting international cooperation in order to maintain market integrity and prevention of fraud or manipulation. The significance of carbon trading can be maintained in cutting emissions and achieving financial sustainability on a global scale by adopting creative market designs and encouraging fair participation. The chapter also addressed the challenges and opportunities presented by the incorporation of carbon trading into existing financial structure including the development of new financial instruments and growth of specialized carbon focused investment funds. The discussions supported the broader implications of carbon trading for sustainable development goals. It considers the potential for carbon markets to mobilize capital for ecosystem supportive projects.

This chapter reflects valuable information for policymakers, industry professionals, and academic researchers as it shall broaden the understanding of carbon trading concepts with respect to financial sustainability and provide a solid foundation for future research on reducing greenhouse gas (GHG) emissions. Carbon trading offers a substantial platform for generating revenue that can be invested in low-carbon technologies, hence reducing the economic pressures and enhancing energy efficiency by lowering GHG emissions. Additionally, funds procured from carbon trading can support the protection of vulnerable ecosystems and promote the development of clean and renewable energy, transforming the energy landscape of the built environment.

Limitations of the Chapter

There are certain limitations of this chapter like it included the use of published articles and researches from Scopus and Web of Science databases, which may not include all the relevant and recent publications, especially those in other languages. Inherent limitation of this chapter is that it lacks the evidences of empirical data limiting the depth of insights on region specific or industry specific dynamics. Also, the selection of best researches was based on the studies focussed on developed economies neglecting the selection of data from emerging economies where carbon trading must be facing different challenges which may remain unresolved. The scope may be restricted by the time frame considered for this book chapter as the mechanisms and policies relating to carbon trading are changing rapidly. Also, when it comes to actual implementation of regulatory frameworks and policies there are several factors which may render them ineffective. Lack of global standards may leave loopholes in the regulatory norms creating regulatory arbitrage for companies. Weak institutional capacity in certain nations may lead to poor monitoring, verification and reporting of emissions. Additionally, political and economic pressures may attract penalties and compliance costs affecting the driving urge of companies to adopt sustainable practices. All these factors combined with significant market manipulations and speculative trading may weaken the implementation of carbon trading policies and thereby affecting the financial sustainability.

Future Recommendations

In the future coming years, carbon trading is expected to play a crucial role in advancing financial sustainability, particularly when there is a global pressure to mitigate climate change and support healthier environment. The future of carbon trading involves various aspects like expanding carbon markets, policies and financial systems, international cooperation to create a uniform and transparent system. As governments will introduce more stringent laws in the coming months carbon pricing will become an essential financial tool encouraging organizations to reduce harmful emissions and invest in sustainable technologies. To enhance financial sustainability, recommendations include promoting broader market access, majorly for emerging nations, and standardizing carbon credit systems to prevent frauds and manipulations. To gain a more complete knowledge of the impact of carbon trading on greenhouse gas emissions, it is recommended to consult with climate change stakeholders and professionals in the area of financial sustainability. Future research should also consider including articles from different languages or performing meta-analyses to broaden the insights. Future studies should also test and validate the constructs identified in this research.

REFERENCES

Akomea-Frimpong, I., Adeabah, D., Ofosu, D., & Tenakwah, E. J. (2021). A review of studies on green finance of banks, research gaps and future directions. *Journal of Sustainable Finance & Investment*, 12(4), 1241–1264. DOI: 10.1080/20430795.2020.1870202

Almena, A., Siu, R., Chong, K., Thornley, P., & Röder, M. (2024). Reducing the environmental impact of international aviation through sustainable aviation fuel with integrated carbon capture and storage. *Energy Conversion and Management*, 303, 118186. DOI: 10.1016/j.enconman.2024.118186

Asian Development Bank. (2017). Future carbon fund: Driving innovation and results through carbon finance. https://www.adb.org/sites/default/files/publication/389821/future-carbon-fund.pdf

Asian Development Bank. (2018). Fostering sustainable development through carbon markets. ADB Blogs. https://blogs.adb.org/blog/fostering-sustainable-development-through-carbon-markets

Baker, M., Bergstresser, D., Serafeim, G., & Wurgler, J. (2018). *Financing the response to climate change: The pricing and ownership of U.S. green bonds*. National Bureau of Economic Research., DOI: 10.3386/w25194

Bansal, S., Mukhopadhyay, M., & Maurya, S. (2023). Strategic drivers for sustainable implementation of carbon trading in India. *Environment, Development and Sustainability*, 25(5), 4411–4435. DOI: 10.1007/s10668-022-02205-w

Bansal, S., Mukhopadhyay, M., & Maurya, S. (2023). Strategic drivers for sustainable implementation of carbon trading in India. *Environment, Development and Sustainability*, 25(5), 4411–4435. DOI: 10.1007/s10668-022-02205-w

Benz, E., & Hengelbrock, J. (2008). Liquidity and price discovery in the European CO2 futures market: An intraday analysis. *Finance Research Letters*, 5(2), 114–124. DOI: 10.2139/ssrn.1283175

Burnett, N., Dillon, A., Jozepa, I., & Bolton, P. (2024, March 5). Carbon Border Adjustment Mechanism. House of Commons Library. https://commonslibrary.parliament.uk/research-briefings/cbp-9935/#:~:text=What%20is%20a%20Carbon%20Border,is%20to%20prevent%20carbon%20leakage

Busch, T., Johnson, M., Pioch, T., & Kopp, E. (2018). Corporate carbon performance data: Quo vadis? *Journal of Industrial Ecology*, 22(6), 1217–1233. DOI: 10.1111/jiec.13008

Chidaushe, W., & Njaya, T. (2022). Global Determinants of Sustainability Linked Financial Markets. *Journal of Economics and Sustainable Development*, 13(22), 202246. DOI: 10.7176/JESD/13-22-05

Climate Impact Partners. (2023, June 8). How carbon finance projects align to SDGs. Climate Impact Partners. https://www.climateimpact.com/news-insights/insights/how-carbon-finance-projects-align-to-sdgs/

Dias, P. J., & Rocha, A. (2023). Carbon Tax: Characterization of Research. In Taking on Climate Change Through Green Taxation (pp. 26-56). IGI Global. DOI: DOI: 10.4018/978-1-6684-8592-7.ch002

Dong, Z., Xia, C., Fang, K., & Zhang, W. (2022). Effect of the carbon emissions trading policy on the co-benefits of carbon emissions reduction and air pollution control. *Energy Policy*, 165, 112998. DOI: 10.1016/j.enpol.2022.112998

Economic Times. (2023, July 12). Regulated carbon markets are worth a trillion dollars a year and cover 25% of global emissions. The Economic Times. https://economictimes.indiatimes.com/news/et-evoke/regulated-carbon-markets-are-worth-a-trillion-dollars-a-year-and-cover-25-of-global-emissions-they-are-raising-climat

El Amri, A., Oulfarsi, S., Eddine, A. S., El Khamlichi, A., Hilmi, Y., Ibenrissoul, A., . . . Boutti, R. (2022). Carbon Financial Market: The Case of the EU Trading Scheme. In Handbook of Research on Energy and Environmental Finance 4.0 (pp. 424-445). IGI Global. DOI: DOI: 10.4018/978-1-7998-8210-7.ch017

Engle, R. F., Giglio, S., Kelly, B., Lee, H., & Stroebel, J. (2020). Hedging climate change news. *Review of Financial Studies*, 33(3), 1184–1216. http://hdl.handle.net/10.1093/rfs/hhz072. DOI: 10.1093/rfs/hhz072

Gopi, K., Sharma, A., Rani, M. J., Kamath, K. P., Manickam, T., Thangam, D., . . . Pol, N. (2024). Strategies to Achieve Carbon Neutrality and Foster Sustainability in Data Centers. In Computational Intelligence for Green Cloud Computing and Digital Waste Management (pp. 109-126). IGI Global. DOI: DOI: 10.4018/979-8-3693-1552-1.ch006

Govindaraju, P., & Mabel, M. M. (2015). Contingency Factors Impacting the Rural Information and Communication Technology Hubs. In Handbook of Research on Cultural and Economic Impacts of the Information Society (pp. 526-546). IGI Global. DOI: DOI: 10.4018/978-1-4666-8598-7.ch022

Kukah, A. S. K., Xiaohua, J., Osei-Kyei, R., & Perera, S. (2024). How carbon trading contributes to reduction in emission of greenhouse gases: A narrative literature review. *Journal of Facilities Management*. Advance online publication. DOI: 10.1108/JFM-08-2023-0094

Laing, T., Sato, M., Grubb, M., & Comberti, C. (2013). Assessing the effectiveness of the EU Emissions Trading System (Vol. 126). London: Grantham Research Institute on Climate Change and the Environment. http://www.jcfj.ie/wp-content/uploads/2016/02/WP106-effectiveness-eu-emissions-trading-system.pdf

Liu, X., Chen, Z., Yu, Q., Fang, L., Gong, J., & Wen, C. (2023). Carbon trading systems and Credit Spreads of Highly Polluting Corporate Bonds: An Empirical Study Based on a Triple Difference Model. DOI: 10.21203/rs.3.rs-2563261/v1

Mashari, D. P. S., Zagloel, T. Y. M., Soesilo, T. E. B., & Maftuchah, I. (2023). A bibliometric and literature review: Alignment of green finance and carbon trading. *Sustainability (Basel)*, 15(10), 7877. DOI: 10.3390/su15107877

Mashari, D. P. S., Zagloel, T. Y. M., Soesilo, T. E. B., & Maftuchah, I. (2023). A bibliometric and literature review: Alignment of green finance and carbon trading. *Sustainability (Basel)*, 15(10), 7877. DOI: 10.3390/su15107877

McKinsey & Company. (2022, April 26). The path to net zero: Investing in carbon markets. McKinsey & Company. https://www.mckinsey.com/featured-insights/future-of-asia/the-path-to-net-zero-investing-in-carbon-markets

Mo, J. L., Agnolucci, P., Jiang, M. R., & Fan, Y. (2016). The impact of Chinese carbon emission trading scheme (ETS) on low carbon energy (LCE) investment. *Energy Policy*, 89, 271–283. DOI: 10.1016/j.enpol.2015.12.002

Mohammed, S. D. (2020). Clean development mechanism and carbon emissions in Nigeria. Sustainability Accounting. *Management and Policy Journal*, 11(3), 523–551. DOI: 10.1108/SAMPJ-05-2017-0041

National Environment Agency Singapore. (n.d.). Benefits of carbon trading. Carbon Markets & Cooperation. https://www.carbonmarkets-cooperation.gov.sg/our-article-6-cooperation/benefits-of-carbon-trading/

Parhamfar, M., Sadeghkhani, I., & Adeli, A. M. (2024). Towards the net zero carbon future: A review of blockchain-enabled peer-to-peer carbon trading. *Energy Science & Engineering*, 12(3), 1242–1264. DOI: 10.1002/ese3.1697

Perdana, S., & Vielle, M. (2022). Making the EU Carbon Border Adjustment Mechanism acceptable and climate friendly for least developed countries. *Energy Policy*, 170, 113245. DOI: 10.1016/j.enpol.2022.113245

Poyser, A. (2024). Carbon Offsets and Green Finance: Understanding Corporate Demand for Carbon Offsets and Mitigation–A Global Literature Review. Sustainable Investments in Green Finance, 139-162. DOI: DOI: 10.4018/979-8-3693-1388-6.ch009

Sen, N., Ahuja, S., Dutta, S., & Chaudhary, K. (2024). The Application of an Emission Trade Scheme in the Indian Aviation Sector. *Air and Space Law*, 49(2), 165–190. Advance online publication. DOI: 10.54648/AILA2024020

Sharma, G. D., Verma, M., Shahbaz, M., Gupta, M., & Chopra, R. (2022). Transitioning green finance from theory to practice for renewable energy development. *Renewable Energy*, 195, 554–565. DOI: 10.1016/j.renene.2022.06.041

Shen, H., & Huang, N. (2022). Accounting for carbon emissions trading schemes. In *Handbook of Accounting and Sustainability* (pp. 365–380). Edward Elgar Publishing., DOI: 10.4337/9781800373518.00030

Shen, N., Zhao, Y., & Deng, R. (2020). A review of carbon trading based on an evolutionary perspective. *International Journal of Climate Change Strategies and Management*, 12(5), 739–756. DOI: 10.1108/IJCCSM-11-2019-0066

Stavins, R. N. (2020). The future of U.S. carbon-pricing policy. *Environmental and Energy Policy and the Economy*, 1(1), 7–32. DOI: 10.1086/706792

Su, Q., & Chen, L. (2024). Carbon financial trading risk based on multidimensional analysis of data flow from the perspective of low-carbon economy. *Environment, Development and Sustainability*, •••, 1–19. DOI: 10.1007/s10668-024-05078-3

Sun, C., Tie, Y., & Yu, L. (2024). How to achieve both environmental protection and firm performance improvement: Based on China's carbon emissions trading (CET) policy. *Energy Economics*, 130, 107282. DOI: 10.1016/j.eneco.2023.107282

Tiwari, D. S. (2022). Carbon Trading: A Tool to Control Global Warming. *International Journal of Health Sciences*, (I), 5713–5722. DOI: 10.53730/ijhs.v6nS1.6246

United Nations Conference on Trade and Development. (2023). Capital markets and sustainable finance. In World investment report (pp. 205–237). DOI: 10.18356/9789210027847c009

Wei, P., Zhou, J., Ren, X., & Huynh, L. D. T. (2024). Financialisation of the European Union Emissions Trading System European Union Emissions Trading System and its influencing factors in quantiles. *International Journal of Finance & Economics*. Advance online publication. DOI: 10.1002/ijfe.2950

Wen, F., Zhao, L., He, S., & Yang, G. (2020). Asymmetric relationship between carbon emission trading market and stock market: Evidences from China. *Energy Economics*, 91, 104850. DOI: 10.1016/j.eneco.2020.104850

Xiao, J., Li, G., Zhu, B., Xie, L., Hu, Y., & Huang, J. (2021). Evaluating the impact of carbon emissions trading scheme on Chinese firms' total factor productivity. *Journal of Cleaner Production*, 306, 127104. DOI: 10.1016/j.jclepro.2021.127104

Zabolotnyy, S., & Wasilewski, M. (2019). The concept of financial sustainability measurement: A case of food companies from Northern Europe. *Sustainability (Basel)*, 11(18), 5139. DOI: 10.3390/su11185139

Zhong, J., & Pei, J. (2024). Carbon border adjustment mechanism: A systematic literature review of the latest developments. *Climate Policy*, 24(2), 228–242. DOI: 10.1080/14693062.2023.2190074

ADDITIONAL READINGS

Asian Development Bank. (2017). Future carbon fund: Driving innovation and results through carbon finance. https://www.adb.org/sites/default/files/publication/389821/future-carbon-fund.pdf

Asian Development Bank. (2018). Fostering sustainable development through carbon markets. ADB Blogs. https://blogs.adb.org/blog/fostering-sustainable-development-through-carbon-markets

Economic Times. (2023, July 12). Regulated carbon markets are worth a trillion dollars a year and cover 25% of global emissions. The Economic Times. https://economictimes.indiatimes.com/news/et-evoke/regulated-carbon-markets-are-worth-a-trillion-dollars-a-year-and-cover-25-of-global-emissions-they-are-raising-climat

KEY TERMS AND DEFINITIONS

Carbon Trading: A market-based system where companies buy and sell carbon credits, representing the right to emit a specific amount of carbon dioxide, incentivizing emissions.

Carbon Credits: Permits allowing holders to emit a certain amount of carbon dioxide. Credits can be traded, providing financial rewards for reducing emissions.

Emissions Trading Scheme (ETS): A cap-and-trade system that limits total emissions while allowing companies to buy or sell emissions allowances within that limit.

Greenhouse Gases (GHG): Gases such as carbon dioxide (CO2), methane (CH4), and nitrous oxide (N20), that trap heat in the Earth's atmosphere, contributing to global warming.

Financial Sustainability: The capacity of a company to maintain long-term financial health, ensuring profitability while responsibly managing resources and risks.

Low Carbon Economy: An economy focused on minimizing carbon emissions through the adoption of cleaner energy sources and more efficient practices.

Carbon Footprint: The total amount of greenhouse gases emitted directly or indirectly by an individual, organization, or activity, typically measured in CO_2 equivalents.

Chapter 12
Building Climate Resilience Adaptation Strategies for Sustainable Development in the UAE

Rahul Jain
https://orcid.org/0000-0001-8572-7476
De Montfort University, Dubai, UAE

ABSTRACT

This research paper explores climate resilience and adaptation strategies in the United Arab Emirates (UAE) against the backdrop of increasing climate change impacts. As a nation highly vulnerable to climate risks such as rising temperatures, water scarcity, and extreme weather events, the UAE faces significant challenges in ensuring its sustainable development trajectory. Through a comprehensive review of existing literature, policy documents, and case studies, this paper identifies key climate resilience and adaptation measures undertaken by the UAE government and stakeholders. Furthermore, it examines the effectiveness of these strategies in enhancing the country's resilience to climate change impacts while fostering sustainable development. The analysis highlights the importance of integrating climate adaptation into national policies, urban planning, infrastructure development, and resource management practices. Moreover, it underscores the need for multi-sectoral collaboration, technology innovation, and community engagement to build a climate-resilient future in the UAE.

1. INTRODUCTION

Background Information on Climate Change and its Impacts on the UAE

Climate change can be described as long-term changes in temperature, precipitation patterns, and any other climatic variable primarily caused by human activities, mainly through greenhouse gas emissions (IPCC, 2014). Due to its location in the arid region of the Arabian Peninsula, the United Arab Emirates is

DOI: 10.4018/979-8-3693-5748-4.ch012

very vulnerable to climate change changes. These changes include higher temperatures, reduced rainfall, frequent extreme weather events, and an increased sea level (Al Hammadi et al., 2021).

The climate condition of the UAE has drastically changed in the past decades, with rising rates faster than that of the global scale (UAE State of Environment Report, 2017). Changes in precipitation patterns further resulted in shifts in water availability factors affecting agriculture and freshwater resources in the region (Abu Dhabi Global Environmental Data Initiative, 2020).

Importance of Climate Resilience and Adaptation Strategies

Climate resilience refers to the capacity of a system to absorb shocks and stresses associated with climate change, adapt to changing conditions, and remain functional (IPCC, 2012). In the UAE, climate resilience refers to the ability of the nation to maintain socio-economic stability, protect its infrastructure, preserve natural ecosystems, and, above all, ensure the safety and well-being of the population (Al-Maktoumi et al., 2020).

In the case of the UAE, which used agriculture, tourism, and energy, among others, this implies that rather effective adaptation strategies have to be employed to suppress the climate change impacts while enhancing resilience. These may include infrastructure development, water management tactics, agriculture innovations, as well as policies that will contribute to the reduction of vulnerability to climate-related risks (Al-Rashedi et al., 2019).

Purpose and Objectives of the Research Paper

The purpose of the research paper is to investigate climate resilience and adaptation strategies in the UAE and the importance of such directions in overcoming the challenges of climate change. The objectives of the research are as follows:

1. To describe the present situation of climate change and its impacts on the UAE with regard to changes in temperature, precipitation trends, and sea-level rise.
2. Integrating Appraisal: To assess the importance of climate resilience and adaptation practices to the socio-economic and environmental systems of the UAE.
3. To determine the actual climate resilience and adaptation programs started by governmental and non-governmental institutions in the UAE.
4. Whether those can effectively increase the responsiveness of the country toward climate change and reduce vulnerability.
5. Advise best practices on how to improve climate resilience and sustainably develop UAE from gaps and best-practice recommendations identified.

2. LITERATURE REVIEW

Overview of the Previous Work in the UAE on Climate Change Resilience and Adaptation

Climate change: In the past decade, literature regarding climate resilience and adaptation in the UAE has mushroomed significantly, showing growing recognition of climate change as an important challenge in this region. Some of the studies cover aspects of the resilience of climate conditions, strategies for adaptation, and impacts of change on various sectors in the UAE.

Al Hammadi et al. have introduced a comprehensive review of the impacts of climate change on the coastal zones within the UAE, which have focused on sea-level rise, coastal erosion, and saltwater intrusion. Their research highlighted the vulnerabilities of coastal communities and ecosystems to these changes, further emphasising the need for adaptation measures such as shoreline protection and management of coastal zones.

Key Concepts and Definitions Related to Climate Resilience and Adaptation

Climate resilience might be considered as the ability of a system, community, ecosystem, or infrastructure to withstand and recover from the influences of climate change with optimal retention of essential functions and services (Adger et al., 2005). Adaptation involves changing practices, behaviours, or structures to minimise vulnerability and maximise resilience to climate change (IPCC, 2014). Building climate resilience and implementing adaptation measures are crucial in coping with adverse effects on these sectors-in particular, water resources, agriculture, infrastructure, and human health in the UAE context.

Identification of Climate Change Impacts Specific to the UAE Context

Climate change is exerting integrated impacts on the UAE systems with multiple effects that impact both the natural and human systems. Increased temperature, decreased precipitation, and increased frequency of extreme climatic events demand urgent attention to water availability, food security, biodiversity, and human health in the region (UAE State of Environment Report, 2017). The sea level rise is a threat to the coastal regions and infrastructure, enhancing the risk of flooding and erosion along the coasts (Al Hammadi et al., 2021).

Review of Relevant Policies, Initiatives, and Strategies Implemented by the UAE Government and Other Stakeholders

The UAE Government has taken proactive steps in facing the challenge of climate change in its efforts to boost climate resilience through policies, initiatives, and strategies. Among the implemented policies, initiatives, and strategies in the UAE country are the UAE National Climate Change Adaptation Program; it aims to mainstream climate adaptation into national development plans and strategies of the country (UAE Ministry of Climate Change and Environment, 2020). On its part, UAE Vision 2021 and UAE Green Growth Strategy centre on sustainable development, renewable energy, and environmental conservation as the guiding principles to propel the future course of the country (UAE Ministry of Climate Change and Environment, 2018).

Other aspects include the NGOs, research entities, and private sector bodies in UAE that aid in building resilience and adaptation to climate change. For example, the Emirates Wildlife Society-WWF promotes biodiversity conservation and sustainable resource management with community and stakeholder engagement (EWS-WWF, n.d.).

3. METHODOLOGY

Research Approach

It applies a mixed-methods approach to the research: it carries a literature review and case study analysis to fully explore climate resilience and adaptation in the UAE. The literature review provides an essential base of existing knowledge and theories and informs case study analysis in terms of actual application or practice (Yin, 2014).

Data Collection Methods

1. Literature Review: A thorough systematic literature review shall be conducted to involve various academic articles and report submissions, governmental publications, and policy documents that would have relevance to climate resilience and adaptation in the UAE. This serves to provide the means of critically acquiring all existing knowledge in the domain by contextualising any research or policies initiated (Grant & Booth, 2009).
2. Document Analysis: Relevant government reports, policy briefs, and organisational publications will be useful in extracting related information on climate resilience initiatives, policies, and strategies implemented in the UAE (Bowen, 2009).
3. Case Study Analysis: A small set of case studies that represent the large diversity of climate resilience and adaptation initiatives in the UAE will be analysed. Case selection criteria include geographical representation, the sectoral focus of cases selected, engagement of stakeholders and innovation in adapting measures (Baxter & Jack, 2008).

Criteria for Selecting Case Studies

Case studies will be selected based on the criteria:

Geographical Representation: Representation from various regions of the UAE to capture regional disparities and challenges with adaptation.

Sectoral Diversity: Cases from diverse sectors such as water management, agriculture, infrastructure, and urban planning would be included for the overall comprehension of the efforts.

- **Stakeholder Engagement:** selection of cases that reflect cooperation among the government agencies, NGOs, community organisations, and the private sector in highlighting how climate resilience depends so much on the multi-stakeholder approach.

- **Innovation and Effectiveness:** selection of cases manifesting innovative adaptation methodologies with identified results on climatic-resilience building.

Limitations of the Research Methodology

Despite its merits, this research methodology does have limitations:

- **Selection bias in Literature review:** The literature review would tend to have selection bias as it relies on the availability of sources and databases.
- **Availability of data:** The report case studies could lack deepness in their analysis due to the unavailability of certain details.
- **Generalizability**: Case study results are limited in application to every corner and context within the UAE, primarily because of geographical variations, socioeconomic conditions, and governance structures.
- **Time and Resource Constraints:** Carrying out in-depth case study analysis entails extensive investments of time and resources, which could restrict the number of cases included in the research.

4. CLIMATE RESILIENCE AND ADAPTATION STRATEGIES IN THE UAE

Overview of the UAE's Climate Resilience Framework and Policy Landscape

The United Arab Emirates acknowledges and has done tremendous work in building climate resilience and adaptation in combating the various impacts of climate change. Toward this end, the country has developed an overall framework and policy landscape to enhance resilience across sectors. Among these, the key components of the overall framework include national strategies, action plans, and institutional mechanisms aimed at mainstreaming climate adaptation into development agendas (UAE Ministry of Climate Change and Environment, 2020).

The UAE's climate resilience framework emphasises incorporating climate concerns into sectoral policies and strategies- water resources management, agriculture, infrastructure development, and urban planning. More importantly, the UAE National Climate Change Adaptation Programme gives a strategy for priority adaptation areas to improve water security, protect coastal zones, and promote sustainable agriculture (UAE Ministry of Climate Change and Environment, 2020).

Analysis of Adaptation Strategies Across Different Sectors

Adaptation has been hampered in water management due to the scarcity and poor quality of water because the country is very dry with a relatively low freshwater source. Water management policies include investment in desalination technologies, wastewater treatment, and efforts to conserve water. The UAE initiated policies in the aspect of water reuse efficiency in agricultural and landscaping activities (Al-Hasani et al., 2021).

Agriculture: Climate change in the UAE creates a need for sustainable agriculture practices for food security and adapting to climate change. In agricultural adaptation, drought-resistant crops, precision irrigation technologies, and vertical farming technology have also been developed (Al-Rashedi et al.,

2020). Furthermore, the government of the UAE has taken initiatives to facilitate farmers in adopting climate-resilient agricultural practices, improvement of soil management and water management practices, etc. (UAE Ministry of Climate Change and Environment, 2018).

Infrastructure and Urban Planning: This is the adaptive infrastructure and the resilient urban planning. Adaptive infrastructure and resilient urban planning form core bases for reducing the impacts of climate change from built environments as well as ensuring continued functionality of critical services. Adaptation strategies in infrastructure and urban planning include Climate-responsive design, green infrastructure solutions, and resilient building codes (UAE Ministry of Infrastructure Development, 2019). The climate considerations have been incorporated into the urban planning of development within the UAE. Such considerations include enhancing the resilience of cities and communities to extreme weather events as well as sea-level rise (UAE Ministry of Climate Change and Environment, 2018).

Examination of Technological Innovations and Best Practices in Climate Adaptation

Technological advancement is amongst the factors that make climate adaptation in the UAE. Examples are advanced weather forecasting systems, remote sensing technology for monitoring environmental changes, and artificial intelligence applications for optimising resource utilisation (Al-Ansari et al., 2021). Another best practice related to climate adaptation is knowledge sharing, capacity building, and international cooperation in taping knowledge base and resources for efficient adaptation measures (Al-Maktoumi et al., 2020).

Case Studies Illustrating Successful Adaptation Initiatives in the UAE

1. Masdar City: Masdar City is an exemplary model for sustainable urban development and resilience in combatting climate change. This will be a low-carbon, zero-waste city infused with advanced technology and innovative design principles that minimise impacts on the environment while adapting to climate change.

Key Features:

- Integration of Renewable Energy: Masdar City generates solar power through photovoltaic panels and concentrated solar power systems, consuming significantly less fuel compared to fossil ones and reducing carbon emissions from the city (Masdar, n.d.).
- Water-Efficient Infrastructure: An efficient use of water for the community is also made by the implementation of greywater recycling, drip irrigation, and smart management of water to reduce water loss and increase efficiency (Masdar, n.d.).
- Green Building Design: The buildings in Masdar City are environmentally friendly, and they use sustainable materials and techniques. It has passive cooling, natural ventilation, and energy-efficient lighting to reduce energy consumption and increase thermal comfort.

Impact:

- Lower Carbon Footprint: Masdar City incorporates renewable energy and energy-efficient infrastructure that can send greenhouse gas emissions to an ultimate low level and contribute to the climate goals of the UAE (Masdar, n.d.).

- Hub for Innovation: The city has been a global hub for sustainable technology and innovation, thereby attracting business developers, researchers, and entrepreneurs to work together on solutions to mitigate and adapt to climate change (Masdar, n.d.).
- Sustainable Living: Masdar City allows residents to sustain good quality living in harmony with the environment, thus proving sustainable urban development is possible and economically feasible as well (Masdar, n.d.).

2. The Sheikh Zayed Desert Learning Centre: The Al Ain Sheikh Zayed Desert Learning Centre is the ultimate example of the best new approaches to desert conservation, education, and development. This flagship project of the Al Ain Zoo demonstrates the commitment of the UAE to biodiversity conservation as well as to climate resilience.

Key Features:

- Native Plant Landscaping: Using native desert plants during landscaping in the centre saves water, prevents reduced biodiversity, and enhances the potential for ecological resilience during arid environments (Al-Awar et al., 2021).
- Water Recycling: The facility uses water recycling and greywater treatment processes to reduce consumption rates and the dependence on freshwater supplies (Al-Awar et al., 2021).
- Renewable Energy Integration: This centre integrates solar panels and other sources of renewable energy into its design to offer clean energy, thereby reducing carbon emissions.
- The centre provides education programs and exhibitions to raise awareness among the audience about desert ecosystems, biodiversity conservation, and the exercise of sustainable living practices. It inspires its visitors to become part of climate resilience (Al-Awar et al., 2021).

Impact:

- Conservation Education: On its side, the Sheikh Zayed Desert Learning Centre plays an important role in educating the public about the benefits of desert conservation and bringing forth sustainable behaviour change among visitors, according to Al-Awar et al. (2021).
- Research and Innovation: The place is that of research excellence in desert ecology, renewable energy, and sustainable technologies. It shall focus on collaboration between scientists, educators, and conservationists in investigating their respective research interests (Al-Awar et al., 2021).
- Community Engagement: The centre engages the local communities in conservation activities, which allow the residents to be involved in environmental stewardship and climate resilience practices (Al-Awar et al., 2021).

3. Al Maha Desert Resort & Spa: Al Maha Desert Resort & Spa in the Dubai Desert Conservation Reserve is an eco-friendly resort that is differentiated from others as its core principle and ethos are embodied with conservation and sustainability. The resort seamlessly integrates itself within its natural surroundings through innovative practices to lessen its environmental footprint and contribute toward climate resilience.

Key Features:

- Sustainable Design: Al Maha Resort combines traditional Arabian architecture with sustainable building materials; its body is harmonious with the desert landscape and prevents environmental impacts (Al Maha Desert Resort & Spa, n.d.).
- Renewable Energy: The resort utilises sun energy in heating, cooling, and electricity. Thus, it reduces its use of fossil fuels greatly, thereby controlling greenhouse gas emissions as well (Al Maha Desert Resort & Spa, n.d.).
- Water Conservation: Implementation of rainwater harvesting, greywater recycling, and efficient irrigation practices to conserve as much water as possible and use as much of the scarce supplies of fresh water (Al Maha Desert Resort & Spa, n.d.).
- Wildlife Conservation: Al Maha Resort is very active in the process of conservational efforts towards endangered species and restoration of native habitats within the Dubai Desert Conservation Reserve, which is done positively in pursuit of biodiversity conservation and ecosystem resilience (Al Maha Desert Resort & Spa, n.d.).

Impact:

- Sustainable Tourism: Luxury tourism will definitely become eco-friendly, as demonstrated by Al Maha Resort, bringing guests through desert landscapes in an experience that aligns with the goals of conservation as well as eco-friendly living (Al Maha Desert Resort & Spa, n.d.).
- Community Engagement: The resort engages the local communities through cultural programs, educational activities, and employment opportunities. This fosters a sense of stewardship and pride in the conservation of the desert environment (Al Maha Desert Resort & Spa, n.d.).
- Economic Benefits: Al Maha Resort earns income and employment and stimulates the local economy and small businesses. Sustainable tourism is economically feasible and socially positive as well (Al Maha Desert Resort & Spa, n.d.).

4. Dubai Expo 2020 Sustainability Pavilion: The Dubai Expo 2020 Sustainability Pavilion expresses highly innovative solutions for sustainable development, climate resilience, and good environmental stewardship. Through these, the pavilion is perceived to be at the centre of the show at the Expo. It enlightens visitors on the idea of the urgency of the action needed to correct this trend in climate and inspires them toward embracing a sustainable life.

Key Features:

- Interactive Exhibits: The Sustainability Pavilion has interactive exhibitions, interactive displays, and multimedia presentations that expose the visitor to urgent environmental challenges and innovational solutions pertaining to sustainability (Expo 2020 Dubai, n.d.).
- Sustainable Design: The pavilion was designed under sustainable design considerations, such as passive cooling concepts, renewable energy sources, and recyclable materials, which mostly minimise energy consumption and face environmental footprints (Expo 2020 Dubai, n.d.).
- Biodiversity Conservation: The pavilion shows the ecosystems that are found in different parts of the world to reflect the importance of biodiversity conservation as well as the building of resilience through ecosystems in relation to climate change (Expo 2020 Dubai, n.d.).

- •Community Engagement: The Sustainability Pavilion interacts with visitors through workshops, events, and educational programs on how to make a difference in sustainability and climate resilience within their own lives and communities (Expo 2020 Dubai, n.d.).

Impact:

- Education and Awareness: The Sustainability Pavilion educates visitors on climate change and sustainability issues and empowers them to positively influence change and behavioural shifts, promoting sustainable action. Expo 2020 Dubai, n.d.,
- Collaboration and Innovation: The pavilion allows governments, businesses, civil society organisations, and individuals to collaborate and innovate in the way they receive and share knowledge to support innovation and partnerships toward sustainable development.
- Legacy and Inspiration: The Sustainability Pavilion will become a memory site for great sustainability and resiliency. Future generations should focus on protecting their earth and undertaking new sustainable and equitable development.

The case studies prove that the UAE is indeed committed to building resilience and adaptability toward climate change through innovative projects that integrate environmental, social responsibility, and economic prosperity.

5. ASSESSING EFFECTIVENESS AND CHALLENGES

Evaluation of the Effectiveness of Climate Resilience and Adaptation Strategies

In evaluating the efficiency of climate resilience and adaptation strategies in the UAE, one would have to compare the degree to which prescribed measures affected the reduction of vulnerability, enhancement of resilience, and mitigation of the impacts of climate change. These methods can then be pursued through various approaches such as performance indicators, case studies, stakeholder consultations, and assessments by experts.

- Performance Indicators: To assess the effectiveness of climate resilience and adaptation strategies, it is possible to use such performance indicators as lower levels of greenhouse gas emissions, efficient use of water, renewable energy adoption, and biodiversity conservation.
- Case Studies: Some selected adaptation projects or initiatives provide case studies, bringing insights into their outcomes, successes, challenges, and lessons learned. Relevant best practices, innovative approaches, and areas for improvement from case studies may be highlighted in climate resilience and adaptation efforts (Al Hammadi et al., 2021).
- Stakeholder Consultations: Government agencies, NGOs, community, and businesses would be involved in assessing climate resilience and adaptation strategies so diverse views and vital feedback are obtained based on how effective they are and their impact (Al-Maktoumi et al., 2020).

Identification of Barriers and Challenges to Implementation

Several barriers and challenges could prevent the successful implementation of climate resilience and adaptation strategies in the UAE:

- Available Data and Information: Unavailable information and a scarcity of data, combined with ignorance of what could happen in the future, diminishes the ability for informed and proactive decision-making and planning and enhances resilience to climate change by decreasing vulnerability (UAE State of Environment Report, 2017).
- Financial Constraint: There will be constraints on investments in climate resilience and adaptation measures due to a lack of resources, competing priorities and budget constraints, especially for infrastructure projects and long-term planning (Al-Rashedi et al., 2019).
- Policy and Institutional Challenges: It usually imposes several barriers in the coordination of actions and cooperation with stakeholders and sectors by government agencies due to inconsistent policies, regulation frameworks, and institutional capacities (UAE Ministry of Climate Change and Environment, 2018).

Analysis of Gaps in Current Approaches and Areas for Improvement

Though much progress has been made in climate-resilient and adaptation efforts, there are so many gaps in current approaches and areas of improvement.

- Lack of integration: Climate resilience and adaptation are often sector-specific, with little integration across sector scales or disciplines. Typically, integration among stakeholders and between sectors enhances the effectiveness and efficiency of adaptation efforts (Al-Maktoumi et al., 2020).
- Climate Change Vulnerability Hotspots: While some areas, communities, and ecosystems have a greater vulnerability to the impacts of climate change than others, they probably get less attention and adaptation resources. It is through vulnerability hotspots and their identification that prioritises adaptation measures and proper resource allocation; in this case, its emphasis is on climate conditions and impacts (Al Hammadi et al., 2021).
- Community Engagement and Participation: One of the significant drawbacks of adapting appropriately could be limited community engagement and public awareness about risks attributable to climate change and available adaptation alternatives. Public participation, education, and awareness-raising initiatives tend to enhance the ability of communities to take proactive action and build up resilience levels (Al-Hasani et al., 2021).

Lessons Learned from Past Experiences and Recommendations for Enhancing Resilience

Lessons of best practices as derived from past experiences may be drawn upon while making recommendations to enhance the climate resilience and adaptation of the UAE.

- Integrated Approaches: Adopt integrated, cross-sectoral approaches to climate resilience and adaptation planning. This can enhance synergies, maximise co-benefits, and serve multiple objectives simultaneously, as stated by the UAE Ministry of Climate Change and Environment in 2020.
- Building strength in the science-policy interface and linking knowledge exchange mechanisms can enhance uptake, informed decision-making capacity, and adaptive capacity at all levels (Al-Ansari et al., 2021).
- Adaptive governance-Extending adaptive capacity: This could be further developed adaptive governance structures and mechanisms that encourage flexibility, inclusiveness, and learning, thereby building the adaptive capacity of institutions and society to respond to transformations in climate change risk (Adger et al., 2005).

6. INTEGRATION AND COLLABORATION

Importance of Cross-Sectoral Collaboration and Stakeholder Engagement

In an effective climate adaptation and resilience-building process in the UAE, cross-sectoral collaboration and stakeholder involvement would be crucial. It is multifaceted and heterogeneous with respect to climate change impacts, touching different sectors such as agriculture, water resources, infrastructure, and health. Governments, public and private sector entities, civil society organisations, and local communities must, therefore, collaborate for holistic and coordinated responses to climate challenges (Adger et al., 2009).

Cross-sector collaboration is known to allow the aggregation of resources and expertise toward contributing knowledge from different stakeholders to pool their priorities together in the form of complete adaptation strategies. Furthermore, the engagement of stakeholders ensures that adaptation actions are inclusive, responsive to local needs, and culturally appropriate, which increases the effectiveness of adaptation actions and their acceptance within the concerned communities as well (Pelling et al., 2015).

Strategies to Integrate Climate Adaptation Into National Policies and Development Plans

Integrating climate adaptation into national policies and development plans is important in ensuring that resilience-building measures are systemically mainstreamed into decision-making processes and development agendas. A few strategies can help to make integration easier:

- Integration with national development strategy: Alignment of climate adaptation goals with existing national development strategies, environmental policies, and sustainable development agendas ensures coherence and consistency in adaptation planning and implementation (UAE Ministry of Climate Change and Environment, 2018).
- Institutional Coordination: Mechanisms for inter-ministerial coordination and cooperation will be in place for the integration of climate considerations into sectoral policies and programs. Coordination would result in synergistic and complementary adaptation efforts within any government agency or department.

- Capacity Building. Policymakers, planners, and practitioners need to be capacitated so that they may incorporate climate considerations into their decision-making. Capacity building, in this aspect, refers to some training programs and workshops or even knowledge-sharing initiatives to help them build awareness, understanding, and technical skills regarding adaptation to climate change.

International Cooperation and Partnerships in Enhancing Resilience to Climate Change

International cooperation and partnerships are, hence, key to building climate resilience. They promote knowledge exchange, resource mobilisation, and capacity building. International cooperative initiatives with organisations, donor agencies, and other countries enhance the effectiveness and impact of adaptation efforts in the UAE.

International cooperation unlocks the way for the distribution of best practices, lessons learned, and innovative solutions based on other international experiences and, therefore, enriches adaptive strategies and approaches by the UAE. International partners are also possible channels through which financial resources, technical capabilities, and technology transfer can be tapped in support of adaptation projects and programs in the UAE (Alam & Yamin, 2019).

Opportunities for Knowledge Sharing and Capacity Building

Knowledge dissemination and capacity building will enhance climate resilience and adaptation in the UAE. Opportunities for exchange, training, and collaboration will empower stakeholders to develop and implement effective adaptation strategies and practices.

These include workshops and seminars, training programs, education and awareness campaigns that are intended to improve awareness, skills, and competencies on climate resilience adaptation. Other enabling platforms can be knowledge-sharing platforms, networks, and communities of practice; some may create exchange opportunities for information, experience, and lessons learned among stakeholders in innovation and collaboration (Hallegatte et al., 2013).

These strategies, therefore, underscore the importance of integration and cooperation with others in building resilience and adaptation to the climate, further throwing emphasis on cooperative action and efforts locally, nationally, and internationally.

7. FUTURE DIRECTIONS AND RECOMMENDATIONS

Emerging Trends in Climate Resilience and Adaptation

Emerging trends in climate resilience and adaptation suggested that more holistic, proactive, and innovative approaches in the United Arab Emirates are being presented to deal with climate change challenges. Some of the key trends are as follows:

- Greater recognition of nature-based solutions is being gained about their role in enhancing the resilience of climate and ecosystem services through the application of green infrastructure, ecosystem restoration, and natural resource management.
- Increased use of modern technologies, including remote sensing, GIS, AI, and blockchain; these technologies can enhance monitoring, early warning systems, and decision support tools for climate adaptation (Al-Ansari et al., 2021).
- Community-Led Adaptation: Focus on adaptation approaches emphasising community-led adaptation, where the local community is empowered, and social cohesion and grassroots-level adaptive capacity are taken into consideration (Al-Hasani et al., 2021).

Long-Term Strategies for Enhancing Climate Resilience in the UAE

This means that long-term strategies for building climate resilience in the UAE need to focus on:

- Integrate planning and governance: Improving the frameworks of integrated planning and governance that mainstream climate considerations into national development strategies, policies, and sectoral plans (UAE National Climate Change Adaptation Program).
- Investment in Infrastructure: Investment in resilient infrastructure and climate-specific hazards, such as sea-level rise, extreme weather events, and water scarcity, can be environmentally friendly and ensure sustainable development (Al-Rashedi et al., 2019).
- Ecosystem-Based Adaptation: Promote the ecosystem-based adaptation approaches that maximise the protective and adaptive capacities of natural ecosystems for improving resilience as well as providing additional benefits to biodiversity, water security, and livelihoods (Al Hammadi et al., 2021).

Policy Recommendations for Policymakers, Practitioners, and Stakeholders

Some of the policy recommendations for improving climatic resilience in the UAE are:

- Policy Coherence: Policies on climate change, as well as development plans and sectoral strategies, must be coherent and aligned with each other in order not to have conflicting objectives and gain synergies. UAE Ministry of Climate Change and Environment, 2018.
- Inclusive Decision-making: This involves inclusive, participatory decision-making processes that involve stakeholders, including the previously marginalised communities, women, youth, and indigenous peoples, during the planning and implementation of climate adaptation (Al-Maktoumi et al., 2020).
- Flexibilities in Adaptation Pathways: Enabling adaptable governance frameworks that should evolve into iterative, flexible, and learning-oriented adaptation pathways in response to changing climate risks and uncertainties (Adger et al., 2005).

Areas for Future Research and Exploration

Prospects with scope for further exploration in the UAE include climate resilience and adaptation:

- There should be a comprehensive vulnerability assessment to identify priority areas, sectors, and communities most likely to be hit by climate change impacts and that, therefore require targeted adaptation interventions (Al-Ansari et al., 2021).
- Technological Innovation: Explore new technologies and solutions such as climate-smart agriculture, renewable energy, and resilient infrastructure. This will help in the enhancement of adaptive capacity and building resilience against climate change in the UAE (Al-Ansari et al., 2021).
- Policy evaluation: This is evaluating climate adaptation policies and measures regarding their effectiveness, equity, and co-benefits to allow for better evidence-based decision-making and policy refinement (Al-Mektoumi et al., 2020).

With these future directions and recommendations, there is a roadmap for the advancement of climate resilience and adaptation efforts in the UAE for sustainable development and well-being against the challenges of climate change.

8. CONCLUSION

In the final analysis, this paper has had an all-encompassing overview of climate resilience and adaptation efforts in the UAE. This has been through a number of case studies, existing literature, and policy analyses that have led to these findings and insights.

That is to say, there is a serious climate change problem affecting the UAE. Not just the boost of temperature but water scarcity, the rising sea, and extreme weather incidents threaten almost every sector of the UAE's economy, infrastructure, environment, and society.

By apparently praiseworthy efforts for developing and implementing climate resilience and adaptation strategies, the other crucial country is the UAE with a whole gamut of activities ranging from sustainable urban planning and water management to renewable energy deployment and biodiversity conservation.

Climate resilience and adaptation in the UAE thus must be assured through practising cooperation across sectors and also through stakeholder engagement. Climate considerations should further be ' mainstreamed' into national policy formulation and other development plans.

Such findings have huge implications for policy and practice. Policymakers and practitioners in the UAE need to put climate resilience and adaptation on top as an important aspect of sustainable development. It involves investing in resilient infrastructure, carrying out nature-based solutions, and encouraging community-based approaches towards adapting to the effects of climate.

Adaptive governance structures shall be needed to allow coordination, knowledge sharing, and building the capacity of relevant stakeholders. International cooperation and partnerships also form a vital role in supporting the efforts to build climate resilience in the UAE and to provide solutions to challenges that transcend national borders.

Conclusively, climate resilience is extremely important for the UAE, as it would be far more susceptible to the impacts of climate change. Through proactive adaptation measures and a culture of resilience, the UAE can benefit by averting the negative impacts of climate change and building towards a sustainable future for its people.

In light of the continuing challenges facing the climate, the UAE must remain the hub of enhancements in resilience, innovation, and collaboration at all levels. Through these efforts, the UAE will become an example of climate resilience and support global activities related to combating climate change.

REFERENCES

Abu Dhabi Global Environmental Data Initiative. (2020). *Water availability and use in the United Arab Emirates*. Retrieved from https://www.adgei.ae/en/theme/water-availability-and-use-in-the-united-arab-emirates/

Adger, W. N., Dessai, S., Goulden, M., Hulme, M., Lorenzoni, I., Nelson, D. R., & Wreford, A. (2009). Are there social limits to adaptation to climate change? *Climatic Change*, 93(3-4), 335–354. DOI: 10.1007/s10584-008-9520-z

Adger, W. N., Huq, S., Brown, K., Conway, D., & Hulme, M. (2005). Adaptation to climate change in the developing world. *Progress in Development Studies*, 3(3), 179–195. DOI: 10.1191/1464993403ps060oa

Al-Ansari, N., Al-Maktoumi, A., & Al-Nuaimi, H. (2021). Technological innovations for climate change adaptation in the United Arab Emirates. *International Journal of Environmental Research and Public Health*, 18(8), 4071. PMID: 33921547

Al-Awar, F., Al-Nuaimi, H., & Al-Maktoumi, A. (2021). Sustainable design and technology integration in the Sheikh Zayed Desert Learning Centre. *Journal of Arid Environments*, 186, 104434.

Al Hammadi, A., Al Darmaki, F., & Kotb, M. (2021). Climate change impacts on coastal zones of the United Arab Emirates: A review. *Water (Basel)*, 13(4), 542.

Al Maha Desert Resort & Spa. (n.d.). *Sustainability*. Retrieved from https://www.marriott.com/hotels/travel/dxbam-al-maha-a-luxury-collection-desert-resort-and-spa-dubai/

Al-Maktoumi, A., Al-Nuaimi, H., & Al-Awar, F. (2020). Climate change impact and adaptation in the UAE: A systematic review. *Sustainability*, 12(18), 7706.

Al-Rashedi, M., Al-Mazrouei, A., & Nisha, A. (2019). Climate change and its impact on the UAE economy. *International Journal of Scientific and Engineering Research*, 10(11), 1536–1542.

Alam, M., & Yamin, F. (2019). Enhancing climate resilience through regional cooperation in South Asia. *Regional Environmental Change*, 19(2), 353–366.

Baxter, P., & Jack, S. (2008). Qualitative case study methodology: Study design and implementation for novice researchers. *The Qualitative Report*, 13(4), 544–559.

Bowen, G. A. (2009). Document analysis as a qualitative research method. *Qualitative Research Journal*, 9(2), 27–40. DOI: 10.3316/QRJ0902027

Emirates Wildlife Society – World Wildlife Fund (EWS-WWF). (n.d.). *Our work*. Retrieved from https://uae.panda.org/our_work/

Expo 2020 Dubai. (n.d.). *Sustainability pavilion*. Retrieved from https://www.expo2020dubai.com/en/understanding-expo/expo-sustainability/sustainability-pavilion

Grant, M. J., & Booth, A. (2009). A typology of reviews: An analysis of 14 review types and associated methodologies. *Health Information and Libraries Journal*, 26(2), 91–108. DOI: 10.1111/j.1471-1842.2009.00848.x PMID: 19490148

Hallegatte, S., Green, C., Nicholls, R. J., & Corfee-Morlot, J. (2013). Future flood losses in major coastal cities. *Nature Climate Change*, 3(9), 802–806. DOI: 10.1038/nclimate1979

Intergovernmental Panel on Climate Change (IPCC). (2012). *Managing the risks of extreme events and disasters to advance climate change adaptation*. Retrieved from https://www.ipcc.ch/site/assets/uploads/2018/03/SREX-Chap3_FINAL-1.pdf

Intergovernmental Panel on Climate Change (IPCC). (2014). *Climate change 2014*: Impacts, adaptation, and vulnerability. Retrieved from https://www.ipcc.ch/report/ar5/wg2/

Masdar. (n.d.). *About Masdar City*. Retrieved from https://masdarcity.ae/en/about-masdar-city

Pelling, M., O'Brien, K., & Matyas, D. (2015). Adaptation and transformation. *Climatic Change*, 133(1), 113–127. DOI: 10.1007/s10584-014-1303-0

UAE Ministry of Climate Change and Environment. (2018). *UAE Green Growth Strategy*. Retrieved from https://www.moccae.gov.ae/assets/c11f96c6/UAE-Green-Growth-Strategy.pdf

UAE Ministry of Climate Change and Environment. (2020). *UAE National Climate Change Adaptation Program*. Retrieved from https://www.moccae.gov.ae/en/our-services/climate-change/national-climate-change-adaptation-program.aspx

UAE Ministry of Infrastructure Development. (2019). *Guidelines for climate-resilient infrastructure development*. Retrieved from https://www.infrastructure.ae/en/middleeast/infrastructure/news/2019-04-11-climate-resilient-infrastructure-guidelines-published-by-the-ministry-of-infrastructure-development-in-uae-5151725

UAE National Climate Change Adaptation Program. (n.d.). Retrieved from https://www.moccae.gov.ae/en/our-services/climate-change/national-climate-change-adaptation-program.aspx

UAE State of Environment Report. (2017). *Climate change*. Retrieved from https://www.ead.gov.ae/en/our-services/environmental-resources/climate-change.aspx

United Nations Environment Programme (UNEP). (2020). *Making peace with nature: A scientific blueprint to tackle the climate, biodiversity, and pollution emergencies*.

Yin, R. K. (2014). *Case study research: Design and methods*. Sage Publications.

KEY TERMS AND DEFINITIONS

Climate Change: Long-term changes in temperature, precipitation patterns, and other climatic variables mainly caused by human activities via greenhouse gas emissions

Climate Resilience: It refers to the ability of a system exposed to climate change-related shocks and stresses to adapt to new conditions while continuing to operate

Adaptation Strategies: These refer to measures and practices that reduce vulnerability and enhance resilience to climate change impacts.

Sustainable Development: A development that meets the needs of the present without compromising the ability of future generations to meet their own needs, taking into account economic, social, and environmental factors.

Greenhouse Gases: Gases that trap heat in the atmosphere and contribute to global warming and climate change, such as carbon dioxide, CO2 and methane, CH4.

Sea Level Rise: The increase in the average level of global sea primarily due to melting ice caps and glaciers as well as the thermal expansion of seawater, threatening to inundate lands near the coast.

Extreme Weather Events: Hurricanes, droughts, floods, and heat waves are extreme weather events that more frequently occur and are more intense than typical due to climate change.

Chapter 13
Environmental Sustainability Through Optimal Energy Consumption Using IoT-Based Edge-Computing and Image Processing

Shaheen Khatoon
University of East London, UK

Azhar Mahmood
https://orcid.org/0000-0002-1490-7389
University of East London, UK

Tariq Akhtar
https://orcid.org/0009-0004-8713-0007
University of East London, UK

Bintao Hu
https://orcid.org/0000-0003-4821-0448
Xi'an Jiaotong-Liverpool University, China

ABSTRACT

This chapter explored various techniques and modelling methodologies designed to estimate occupancy level in indoor environments. It also introduces an innovative image-based occupancy detection system that leverages edge computing and machine vision to accurately detect and classify occupants and other objects in an indoor environment which requires a certain thermal comfort level. By enabling real-time adjustments to HVAC operations based on actual occupancy, it can significantly reduce unnecessary energy consumption in unoccupied areas, thus improving overall energy management. The integration of edge computing allows for local data processing, which not only minimizes the computational load on centralized servers but also addresses privacy concerns by reducing the need for external data transmission. This is particularly important in environments where sensitive information about occupants

DOI: 10.4018/979-8-3693-5748-4.ch013

may be captured. A case study is presented in the end to demonstrate and examines the performance of several object detection models in the context of academic office occupancy detection.

CHAPTER OVERVIEW

Energy efficiency in buildings is a crucial component in tackling global sustainability challenges, particularly as Heating, Ventilation, and Air Conditioning (HVAC) systems are among the largest consumers of electricity. These systems often account for nearly half of a building's total energy consumption, making their optimization essential for reducing energy waste and minimizing the environmental impact of buildings. Occupancy-based control (OBC) systems offer innovative solutions to reduce energy consumption in buildings via demand-response control mechanism. By using sensors (e.g., temperature, humidity, light sensors, as well as specialized sensors like PIR sensors, smart meters, and thermal cameras) such systems can perceive environmental changes and estimate buildings' occupancy for adapting heating, ventilation and cooling in real time. Traditional systems have combined environmental sensors with intelligent techniques, such as machine learning and statistical modeling, to establish relationships between environmental parameters and occupancy levels.

This chapter explored various techniques and modelling methodologies designed to estimate occupancy level in indoor environments. It also introduces an innovative image-based occupancy detection system that leverages edge computing and machine vision to accurately detect and classify occupants and other objects in an indoor environment which requires a certain thermal comfort level. By enabling real-time adjustments to HVAC operations based on actual occupancy, it can significantly reduce unnecessary energy consumption in unoccupied areas, thus improving overall energy management. The integration of edge computing allows for local data processing, which not only minimizes the computational load on centralized servers but also addresses privacy concerns by reducing the need for external data transmission. This is particularly important in environments where sensitive information about occupants may be captured.

A case study is presented to demonstrate and examines the performance of several object detection models in the context of academic office occupancy detection. A custom dataset is created, consisting of 1,728 images from office settings, manually annotated to include eight object classes: person, cell phone, printer, mouse, computer, laptop, keyboard, and tablet. The models trained on this dataset include YOLOv8n, YOLOv9c, and YOLOv10n, as well as baseline models such as Faster R-CNN. To overcome challenges such as high computational loads, occlusion, and lighting inconsistencies, the dataset is thoroughly prepared, augmented, and annotated to enhance model training. By analyzing the strengths and weaknesses of these models, the study highlights their capabilities in accurately detecting occupants and while ensuring robust performance in real-world scenarios. Ultimately, this framework aims to provide a scalable and privacy-conscious solution for managing energy use in buildings, contributing to the broader objective of environmental sustainability.

1. INTRODUCTION

Energy efficiency is a significant challenge in today's world due to the growing global demand for energy and the environmental impacts associated with its production and consumption. As economies expand and populations increase, more energy is required to power industries, transportation, homes, and technology. Additionally, outdated infrastructure and resistance to adopting new, energy-efficient technologies further exacerbate the problem. Buildings are among the largest energy consumers in the U.S., with significant energy demand driven by their heating, ventilation, and air conditioning (HVAC) systems. In 2018, HVAC systems accounted for nearly 50% of the total electricity consumption in buildings across the country(Koebrich et al., 2019). This high energy usage results from the need to maintain comfortable indoor climates, especially in varying weather conditions. Also, a lot of energy is wasted in large indoor areas due to centralized air conditioning, irrespective of the presence of occupants.

With millions of residential, commercial, and industrial buildings relying on energy-intensive HVAC systems, this sector plays a major role in the overall energy footprint. Improving the efficiency of HVAC systems and adopting smarter building technologies could significantly reduce energy consumption, lower operational costs, and contribute to environmental sustainability by reducing greenhouse gas emissions. One way to reduce the energy consumption of HVAC is by efficiently controlling HVAC units via demand-response control mechanisms. Occupancy-based control (OBC) offer innovative solutions to reduce energy consumption in buildings, by optimizing HVAC system usage. By using sensors and cameras equipped with computer vision technology, buildings can detect real-time occupancy patterns, identifying when and where spaces are in use. This allows HVAC systems to adjust heating, cooling, and ventilation dynamically, supplying energy only to occupied areas and reducing waste in unoccupied ones. Occupancy detection can be integrated with advanced algorithms to predict occupancy trends and optimize energy use further, ensuring that systems operate only when necessary, lowering energy consumption and enhancing the overall energy efficiency of buildings.

There have been continuing advances in sensing technology for monitoring the interaction between the building system and occupants, e.g., motion sensors (Yun and Lee, 2014, Hashimoto et al., 1997, Agarwal et al., 2010), Infrared Proximity Sensors (PIR) (Rastogi and Lohani, 2020), CO2 concentration sensors (Wang et al., 1999), and thermal images (Beltran et al., 2013, Gomez et al., 2018, Griffiths et al., 2018). The motion and proximity sensor has limited range and fails to count people with increasing objects (Wahl et al., 2012). In addition, the detection likely fails for the static objects. CO2 concentration-based solution has a relatively poor real-time performance, and the measurement precision is likely to be significantly reduced when the doors or windows are opened. Thermal cameras usually cannot recognize the characteristics of the detected object (Anjomshoaa et al., 2018).

Video based occupancy detection methods are emerging as effective approaches for recognizing activity in buildings (Petersen et al., 2016). These methods utilize computer vision technology to analyze video, extracting information such as number of people, their location and even behaviors. Typically, video analysis involves detecting features like the head, face, body contour, or movement. However, these methods can face challenges, particularly in environments with many obstacles or when only parts of a person's body are visible. Detection methods that rely on a single body characteristic often struggle in real-world applications. Additionally, in terms of cost, users prefer to utilize existing surveillance footage rather than installing new cameras. A key challenge with video-based occupancy detection methods is the concern over privacy, as these systems capture sensitive personal data, raising issues around data protection, unauthorized access, and the need for anonymization and secure storage.

This chapter presents an image-based occupancy detection system utilizing edge computing, specifically designed to automatically monitor indoor spaces for energy-saving purposes. Using machine vision and RGB images for indoor occupancy detection offers several advantages, making it an effective solution for optimizing energy use and enhancing building management. RGB images provide rich visual data that can accurately identify occupants, differentiate between people and objects, and track movement within a space. This allows systems to make real-time adjustments to HVAC based on actual usage, reducing energy waste. Additionally, machine vision can work without requiring invasive or intrusive sensors on individuals, preserving privacy while maintaining functionality. By leveraging edge computing, the system processes data locally, ensuring faster response times and enhanced privacy protection by minimizing the need for data transmission to external servers. The system efficiently detects and tracks occupancy in real time, enabling dynamic adjustments to energy consumption of HVAC based on actual usage, thus optimizing energy efficiency in typical indoor environments such as offices, homes, and commercial buildings, enhancing overall building management and sustainability.

There are numerous machine vision models that are ready for deployment in edge computing environments, suitable for efficient occupancy detection once properly trained. In this chapter, a case study is conducted to examine the performance of various object detection models specifically for occupancy detection in office environment while also aiming to ensure the models' universal applicability across various environments. Previous machine vision approaches have faced several limitations, including high computational loads, occlusion challenges, difficulties in detecting small objects, lighting inconsistencies, camera placement issues, and difficulties in system generalization. To address these challenges, our main objectives is preparing a diverse set of data, accurately annotating this data to enhance model training, and training multiple machine vision models, including YOLOv8n, YOLOv9c, and YOLOv10n. We also compare their performance against baseline models like Faster R-CNN. This comprehensive approach aims to create robust, efficient, and adaptable occupancy detection systems in the next phase that can effectively manage energy consumption and improve building management in real-world scenarios.

2. OCCUPANCY DETECTION TECHNIQUES AND METHODS

Occupancy information in a building is critical in terms of indoor environmental quality, energy consumption and building energy simulation. In this section, we review the techniques and methods along modelling methodologies in occupancy estimation in buildings. Also discuss the advantages and limitations of these approaches proposed in recent past, providing insights for their potential application in Occupant-Based Control (OBC) systems.

2.1 Sensor-based Occupancy Detection Techniques and Methods

Most occupancy detection systems rely on the deployment of multiple environmental sensors, such as carbon dioxide (CO2), temperature, humidity, and light sensors, as well as specialized sensors like PIR sensors, smart meters, and thermal cameras (Rueda et al., 2020). Existing studies have combined these technologies with intelligent techniques, such as machine learning and statistical modeling, to establish relationships between environmental parameters and occupancy levels. For instance, Hobson et al. (2019) use Wi-Fi access point, CO2 sensors, PIR motion detector to collect data stream from an academic office environment to develop occupancy count estimation for HVAC control purpose. Various

ML based techniques such as linear regression and ANN are used to estimate occupancy information. In their case study linear regression models outperforms other methods. The study found Wi-Fi enabled device counts are useful for occupancy-count estimations achieving a mean R^2 of 80.1–83.0% when compared to ground truth counts during occupied hours.

Similarly, Dorokhova et al.(2020) used rule-based based system for occupancy forecasting using sensors such as Temperature, CO2, Humidity and Luminosity. It explores both supervised models with an accuracy up to 98.3 on cross validation and unsupervised algorithms with a cross-validation accuracy of 96.7 for predicting occupancy with high accuracy. In a case study of a mid-sized building in Portugal, a potential energy saving of 15.4% is demonstrated. The methodology includes extensive feature engineering to enhance data pre-processing and employs machine learning models like SVM, ANN, and LSTM for occupancy forecasting. The highest performing algorithms is LSTM for unsupervised and Feedforward ANN for supervised algorithms. Similarly, (Jiang et al., 2016) developed an indoor occupancy estimator to estimate real-time occupancy based on CO2 measurements, using a dynamic model of occupancy levels. The proposed Feature Scaled Extreme Learning Machine (FS-ELM), an improved version of the standard ELM, is introduced for this purpose. Since CO2 data often contains significant spikes, pre-smoothing the data is found to significantly enhance estimation accuracy. Given that real-time globally smoothed data isn't always available, the study proposed using locally smoothed data instead. The occupancy estimator, tested in an office setting with 24 cubicles and 11 open seats, achieved an accuracy of up to 94% with a tolerance of four occupants.

Dong and Lam (2014a) has presented a methodology reduce energy consumption based on prediction of occupant behaviour patterns and local weather conditions using CO2, acoustics, motion and lighting as input features. Adaptive Gaussian Process, Hidden Markov Model, Episode Discovery and Semi-Markov Model are modified and implemented into this study. Similarly, (Sayed et al.) developed an efficient and non-intrusive system for detecting occupancy in indoor spaces. The proposed method utilized an environmental sensing board to collect ambient data, which included temperature, humidity, pressure, light level, motion, sound, and CO2 levels. This data is then processed using a DL model, specifically a CNN, deployed on an edge device to ensure low-cost computing and enhanced data security. The accuracy achieved in this study is about 99.76%.

Abdel-Razek et al.(2022) used thermal comfort indices to estimate whether a room is occupied or empty. Data from fluctuations in light, CO2, and humidity levels are analyzed to assess the reliability of occupancy prediction. Additionally, kNN, ANNs, and DTs are applied as classification techniques. The results indicated that the kNN model outperformed both DT and ANN, achieving an accuracy of 99.50%.

Kou et al. (2021) investigated residential demand response (DR) implementations based on HVAC, focusing on effective control algorithms to coordinate the operating schedules of multiple HVAC devices. They developed both model-based and data-driven HVAC control strategies aimed at minimizing customers' electricity costs, discomfort, and utility-level load violations, using input data such as outdoor temperature (weather forecasts), indoor temperature, and non-HVAC electrical consumption. The model-based approach employed a thermal resistance-capacitance (RC) model and a distributed solution algorithm to establish optimal day-ahead HVAC operation schedules, achieving a 22% cost savings compared to the data-driven approach. In contrast, the data-driven approach, utilizing neural networks to interact with the environment during training for online decision-making, did not require outdoor temperature forecasts and is 46 times faster in computational speed, although it incurred higher electricity and discomfort costs.

2.2 Edge and Machine Learning Methods

There are studies using similar environmental sensors technologies as discussed in section 2.1, however, the deployment is conducted on the edge node. For instance, Rastogi et al. (2020) developed a methodology for estimating indoor occupancy using Linear and Quantile (QR) algorithms for indoor environment using CO2, temperature, relative humidity, and motion levels. Since the data generated by the setup is vast, sending it all to the cloud for processing may result in delayed prediction; hence, the suggested models are designed to be executed on an edge device.

In these studies, Zemouri et al. (Zemouri et al., 2019, Zemouri et al., 2018) addressed the effectiveness of machine learning algorithms used to predict human occupancy in closed office areas using temperature and humidity data as occupancy predictors. They employed a Raspberry Pi as an edge device to perform real-time occupancy detection. The findings revealed that kNN outperformed the other algorithms in all performance measures.

2.3 Vision-Based Techniques and Methods

Since mostly studies focused on sensing occupancy information through the count of occupants and their distribution in a certain environment, there is limited research and its material on sensing the actual objects which requires certain thermal comfort level. This is necessary to allow HAVC to dynamically adjust the heating and cooling requirements. Computer vision and AI- based occupancy detection solutions are emerging as effective approaches for recognizing activity in buildings using that can be implemented into building HVAC systems for higher accuracy monitoring and control. For Instance, Hu et al. (2022) presented a deep-learning-based approach for building occupancy detection using CCTV cameras. The method involves feature extraction through a deep convolutional neural network that constructs feature pyramids and a three-stage detection process using sequential detectors with increasing Intersection over Union (IoU) thresholds. Data is gathered by recording CCTV videos from a university building over five weeks, resulting in nearly 10,000 labeled images. Experimental results demonstrated that the proposed model achieved high detection accuracy with a Precision of 89.7%, a Recall of 89.2%, and an overall mean average precision (mAP) of 44.5%. The study highlights the effectiveness of this approach for real-time occupancy detection in complex indoor environments.

Tien et al.(2020) used thermal Cameras for detecting and predicting occupancy heat emissions to enable demand-driven control solutions in building energy management systems (BEMS). Developed a convolutional neural network (CNN) which detect and classify occupant activities from images such as sitting, standing, walking, and napping, generating real-time occupancy heat emission profiles that can optimize HVAC systems. Data is gathered by using images from online sources and captured images of various occupant activities in office spaces, with real-time live detection tests conducted in an office space in Nottingham, UK. The study achieved an average detection accuracy of 80.62%, with the highest detection accuracy being 89.39% for static images and about 93.7% for occupancy detection. This approach showed potential for better managing building energy loads and improving indoor environmental quality by accurately detecting occupant activities and adjusting HVAC systems accordingly.

Similarly, Wang et al.(2023) proposed indoor occupancy detection system using an advanced YOLOv5 model. The method involves improving the YOLOv5 algorithm with a decoupled prediction head to better manage classification and regression tasks separately, which enhances the detection accuracy. A self created dataset is prepared with various indoor scenes, and the training involved significant image preprocessing, resizing, and categorizing images into classes such as sitting, standing, walking, napping, and none. The machine learning algorithm used in this study is a modified version of YOLOv5, specifically the DFV-YOLOv5. The improved YOLOv5 model achieved high accuracy, with the best value being 86.3% for one of the key performance indicators. This result demonstrates significant improvements over other baseline models used for comparison in the study.

Lee et al., (2018) presented a method for counting people using a stereo camera mounted on the NVIDIA Jetson TX2 to improve accuracy in crowded environments by addressing occlusion issues. The process involves camera calibration, stereo matching using the Semi-Global Matching algorithm, background subtraction with a Gaussian Mixture Model (GMM), and Kalman filter-based tracking. Data is gathered using IMX 185 cameras installed at heights of 3 to 5 meters, recording 23 high-definition video sequences. The method achieved over 95% accuracy in most scenarios, with an overall accuracy of 98.95%, and demonstrated real-time processing capabilities with HD resolution at 12.5 frames per second, highlighting its effectiveness for surveillance and crowd monitoring.

Likewise, Sun et al. (2022) aims to enhance indoor occupancy measurement accuracy by combining motion detection and static estimation techniques. The method involves a four-stage algorithm: occupancy detection, static estimation using Fully Convolutional Head Detector (FCHD), motion detection at room entrances using GMM and Camshift tracker, and a fusion estimation method that utilizes the Kalman filter and Occupancy Frequency Histogram to integrate results from the two techniques. Data is gathered using two USB cameras installed at the room entrance and interior, capturing video data over two days. The motion detection events and static head detections are processed using Nvidia Jetson Nano, employing OpenCV for computer vision tasks and Pytorch for deep learning computations. The machine learning algorithms used are FCHD for head detection, GMM for background subtraction, Camshift tracker for motion tracking, and Kalman filter for fusing motion detection and static estimation results. The study achieved an occupancy detection accuracy of 97.8%, with an occupancy estimation score ranging from 78.52% to 79.18% and a mean square error (MSE) between 0.21 and 0.23, demonstrating high accuracy and efficiency in real-world applications.

Simliary, Paidi et al. (2020) used thermal cameras to acquire vehicle occupancy information in an open parking lot using deep learning techniques. Frames from these videos are extracted and manually labeled due to the lack of pre-labeled thermal images. Multiple deep learning networks, including Yolo, Yolo-conv, GoogleNet, ResNet18, and ResNet50, are evaluated for vehicle detection. Data is gathered using an Axis Q1942-E thermal camera installed on a two-storey building, capturing videos in different weather conditions such as snow, rain, darkness, and brightness. The videos are stored locally, and frames representing diverse conditions are manually labeled to prepare the dataset. The machine learning algorithms used include Yolo, Yolo-conv, GoogleNet, ResNet18, and ResNet50. These networks are tested for their efficiency in detecting vehicles in thermal images. Among these, ResNet18 performed the best, achieving an average precision of 96.16% and a log-average miss rate of 19.40. The study highlights the effectiveness of using thermal cameras and deep learning for real-time vehicle occupancy detection, suitable for varying illumination and environmental conditions.

Likewise, (Yu et al., 2023) presented YOLOv5s model for face mask recognition in heterogeneous IoT environments. The proposed method includes embedding a Coordinate Attention (CA) mechanism, integrating a Bidirectional Feature Pyramid Network (BiFPN) block, adding an Adaptive Spatial Feature Fusion (ASFF) layer, and merging Scaled Intersection over Union (SIoU) to improve detection accuracy and computational efficiency. Data is gathered from diverse sources, including the AIZOO face mask dataset, an internet-sourced dataset, and a self-built dataset incorporating photos taken on campus merged with the Real-World Masked Face Dataset (RMFD). The study employed the YOLOv5 model, enhanced with the aforementioned components to address specific challenges in face mask detection across different computing platforms. The improved YOLOv5s model achieved superior accuracy compared to baseline models, with a detection accuracy of up to 95.4% on heterogeneous IoT platforms. This indicates a significant improvement in face mask recognition, making it suitable for real-time applications in various environments.

3. NOVELTY AND GAPS IN KNOWLEDGE

Most of the existing studies on building occupancy estimation has used motion sensor (Yun and Lee, 2014, Hashimoto et al., 1997, Agarwal et al., 2010, Hobson et al., 2019), Infrared Proximity Sensors (PIR) (Rastogi and Lohani, 2020, Hobson et al., 2019), CO2 concentration sensors (Wang et al., 1999, Dorokhova et al., 2020), and thermal images (Beltran et al., 2013, Gomez et al., 2018, Griffiths et al., 2018). The motion and proximity sensor has limited range and fails to count people with increasing objects (Wahl et al., 2012). In addition, the detection likely fails for the static objects. CO2 concentration-based solution has a relatively poor real-time performance, and the measurement precision is likely to be significantly reduced when the doors or windows are opened. Thermal cameras usually cannot recognize the characteristics of the detected object (Anjomshoaa et al., 2018). Furthermore, all these methods are effective in determining whether a room is occupied but are unable to accurately estimate the exact number of occupants. Additionally, they are sensitive to environmental factors such as airflow and solar radiation. Environmental sensors like carbon dioxide and temperature sensors also have limitations in real-time detection, as their performance can be compromised by open windows and doors.

While vision-based methods (Yu et al., 2023, Wang et al., 2023, Tien et al., 2020, Paidi et al., 2020) for detecting the number and locations of occupants and recognizing their activities or behaviors show promise, they face significant challenges. Identifying individuals in complex academic office environments, where obstacles like furniture, equipment, and partitions are common, is difficult, especially, detecting occupant activities when parts of the body are obscured or occlusion issues in densely populated areas. Moreover, these methods raise serious concerns about data privacy. Many existing studies focus primarily on improving the performance and accuracy of deep learning models for human presence detection and activity classification. However, less attention has been given to addressing the issues of computational load and privacy protection. This leaves a gap in developing solutions that minimize unnecessary computational burdens and safeguard privacy.

Further exploration is necessary to refine these methods, addressing both privacy concerns and the computational demands of real-time processing. In response to these challenges, we propose a generalized edge-based framework that aims to reduce both privacy risks and computational load, making the system more efficient and secure. This work focous on an image-based deep learning framework for occupancy detection in typical indoor spaces, particularly for energy-saving purposes. The proposed end-to-end edge-based architecture utilizes RGB images from video streams along with edge-optimized deep learning models, to classify indoor objects into distinct categories. Since various objects, beyond people, also require specific thermal comfort conditions, the goal is to detect all objects that require such comfort such as people, computer, laptops, servers, etc. This will allow for the automated control of HVAC systems, optimizing energy use by ensuring that thermal comfort is maintained only where necessary.

By using edge-based processing, the system ensures that sensitive data is handled locally, reducing privacy risks associated with transmitting images to cloud servers. Additionally, this approach minimizes computational overhead by leveraging lightweight models and localized processing, making it ideal for real-time applications in energy-efficient building management.

4. DESIGN OF THE IOT-ENABLED OCCUPANCY-BASED HVAC CONTROL SYSTEM

The overall architecture of the IoT-based control system is shown in Figure. 1. It consists of a set of RPi Cameras that streams images to an edge node. We plan to set up four RPi Cam to capture the room from different angles. RPi Cam will be configured with the edge device to transmit image stream and run the DNNs module to predict a room occupancy in terms of people and other objects that require certain thermal comfort and their count. The analysis results will be sent to the IoT cloud. The HVAC control unit will deploy Model Predictive Control (MPC) to communicate with the IoT cloud server to receive occupancy prediction results and run the MPC algorithm. The HVAC modules modify the temperature of the HVAC unit according to the decisions taken by the MPC algorithm.

Figure 1. The overall architecture of the proposed IoT-based solution

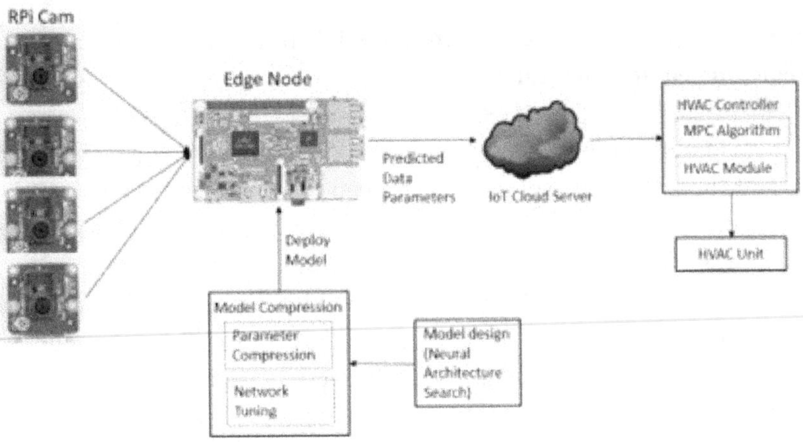

To implement embedded intelligence at the edge device following research questions will be addressed:

1. **What type of sensing technology is available for object identification in an indoor environment?**

Raspberry Pi Cameras (RPi Cam) for image streaming from an indoor space with high accuracy will be used. RPi Cam supports MIPI-CSI designed to use low power, smaller size, faster bandwidth, higher resolution and reduced latency. These features confirm to the resource constrained specification of the edge node.

2. **What type of edge technology can be used to deploy intelligent solutions?**

Previously researchers have used cloud-based systems for streaming raw data from an end device to the centralized servers or IoT cloud for analysis and send the results back to the end device (Ryu and Moon, 2016, Dong and Lam, 2014b, Carli et al., 2020). With the availability of low-cost and low-power IoT devices with edge computing features, it is possible to run advanced artificial intelligence algorithms at the edge where data originates. In the line of edge computing, previous studies have mainly designed edge nodes using few Micro Controller Units (MCUs) boards aiming to run AI algorithms at the local node (Metwaly et al., 2019). However, the configuration of the sensing device with the edge node is complex and needs manual calibration. Furthermore, cloud storage is required for intensive image computation. The adoption of Edge AI has led to developing specialized devices capable of performing AI inferencing efficiently, e.g., Nvidia Jetson Nano[1], Google Coral[2], and AWS DeepLens[3]. In the proposed edge computing architecture, the RPi Cam streams the data to an edge instead of uploading it directly to the cloud. The RPi Cam will be configured with the edge device using the NVIDIA Jetson Nano snapshot board. The Snapshot is the edge AI video capture device designed to run multiple neural networks in parallel for image classifications, object detection, and speech processing applications. Furthermore, the accompanying AGX Xavier Developer Kit provides tools and libraries to develop Edge AI applications. One device can support up to four RPi Cam connected via Wi-Fi or HDMI. Therefore, the snapshot board is more feasible and ready to use as an edge node to deploy deep learning-based occupancy estimation models.

3. **What are the most appropriate deep learning models that can be embedded at edge nodes for perceiving real-time environmental changes?**

We will develop and train multiple deep-learning models to detect occupancy and perceive environmental changes in real-time. We will explore various deep-learning-based models such as Feedforward Neural Networks (Moons et al., 2019), Convolutional Neural Networks (CNNs) (Gomez et al., 2018), Recurrent Neural Networks (RNNs) (Chung et al., 2014), and YOLO for image processing and object detections. At first, we will use GPU instances to train and test these models for realizing the proof of concept. Then we will use a tiny version of the most accurate model through model compression and network tuning so that the model can be deployed on the resource constraints microcontroller. To train any Neural Network, a large set of tagged training data is required. We plan to collect a dataset targeted at object recognition in the context of any office/classroom and tag data using Amazon Mechenical Turk[4], a crowdsourcing service that can be used to tag a huge amount of data.

5. CASE STUDY FOR OCCUPANCY DETECTION

In order to assess the performance of deep learning models, a case study is conducted, specifically for occupancy detection in an academic office environment. Previous machine vision approaches have faced several limitations, including high computational loads, occlusion challenges, difficulties in detecting small objects, lighting inconsistencies, camera placement issues, and difficulties in system generalization. To address these challenges, our main objectives is preparing a diverse set of data including people and other categories which requires certain thermal comfort, accurately annotating this data to enhance model training, and training multiple machine vision models, including YOLOv8n, YOLOv9c, and YOLOv10n and comparing their accuracy with baseline model, Faster R-CNN. This comprehensive approach aims to create robust, efficient, and adaptable occupancy detection systems in the next phase that can effectively manage energy consumption and improve building management in real-world scenarios. The dataset is prepared using Roboflow, which allowed us to efficiently create dataset by merging multiple office settings into a single cohesive dataset. For our purposes, the data needs to be annotated—a task easily accomplished with Roboflow. Additionally, Roboflow offers various functions for preprocessing the data.

5.1 The Occupancy Dataset Description

A custom dataset is used consisting of 1728 images from various office settings for the models training, with various possible classes for object recognition. The images are extracted at 1fps (frame per second) from 16 videos which are found on free websites like Pixabay. Afterwards the images are manually annotated into one of eight classes such as person, cell phone, printer, mouse, computer, laptop, keyboard and tablet. To prepare the dataset for training, the images are resized to 640x640, and a static crop is applied to retain the central 25%-75% for each image. Greyscale is applied and the images are tiled in 2 rows x 2 columns. Furthermore, the augmentation is applied to enhance the robustness, as shown in Table 1. After augmentation, the dataset is split into training, validation and testing. The annotated dataset is saved in both COCO and YOLO annotation, ensuring compatibility with multiple deep learning frameworks. The prepared dataset is then used to train the selected models, enabling them to recognize objects and occupants for effective HVAC system automation.

Table 1. Augmentation settings for the selected dataset

Augmentation Steps	Step Setting
Flip	Horizontal and Vertical
Noise	Of 0.1% of pixel
Bounding box rotation	-5% to +5%
Bounding box sheer	-10% to +10%

The image samples in the dataset are shown as an example in Figure.2. The dataset is created with the intention of overcoming limitations such as occlusion, camera angle issues, universal viability and ability to detect other heat emitting objects such as computers, cell phones and tablets. Looking through all the pictures they are all the same sorts of images but in different settings which should give the model a good understanding of how to perform in a variety of office layouts and thereby making the model universally viable. The matter of occlusion is tackled by carefully annotating the persons or object for example 1st image shows 2 full people and part of a 3rd person, image 2 shows 6 people 5 of them are clearly visible and the 6th is occluded. To tackle the issue of camera placement, when the dataset is created, we carefully chose different images where the camera is placed in different angles and positions. Lastly all other objects like computers, cell phones etc. where also annotated to pick these up whilst running object detection. Small devices such as cell phones or tablets are not likely to make a huge difference to the room temperature however, an office floor with a large number of computers could make a significant difference.

Figure 2. Examples of sample images in the dataset

5.2 Object Detection Models: Performance and Evaluation

First, its trained on different object detection such as YOLOv8n, YOLOv9c, and YOLOv10n against the Faster R-CNN baseline, emphasizing detection speed, computational efficiency, and small object detection. The trained models are evaluated based on two key metrics: Mean Average Precision (mAP) and computational efficiency. The following evaluations are conducted:

- **mAP:** The models are evaluated using mAP@50 and mAP@50-95 to measure their accuracy at different Intersection over Union (IoU) thresholds.
- **mAP@50:** Measures precision at a 50% IoU threshold, which is more lenient and captures larger objects.
- **mAP@50-95:** A stricter metric that measures precision across various IoU thresholds, providing insight into the model's performance with smaller or more occluded objects.
- **Detection Speed:** The time taken to process each image is measured to evaluate the model's suitability for real-time detection.
- **Resource Efficiency:** The models are compared in terms of the number of parameters and computational load, with a focus on reducing resource usage while maintaining high detection accuracy.

The performance of YOLOv8n, YOLOv9c, and YOLOv10n is compared against Faster R-CNN, which is chosen as baseline models due to its popularity in object detection research. Faster R-CNN, though accurate, is slower and more resource-intensive.

5.2.1 Faster R-CNN

The Faster R-CNN model achieved a mAP of 87.4% at IoU threshold 0.50 and 60% at IoU thresholds ranging from 0.50 to 0.95, indicating strong object detection performance, especially at lower IoU thresholds. Figure. 3 shows epochs along the x-axis, with mAP and mAP @50-95 on the y-axis, tracking the performance of the Faster R-CNN model as training progresses. Initially, at the lower epochs, both mAP and mAP@50-95 start relatively low, indicating that the model is still learning and hasn't achieved high precision yet. As the epochs increase, the lines representing mAP and mAP@50-95 gradually rise, demonstrating the model's improving ability to correctly detect and classify objects. Around the midpoint of the graph, you may notice a steady increase, showing that the model is refining its predictions and approaching its peak performance. Toward the later epochs, the graph begins to plateau, meaning the model has learned most of what it can from the data. At this point, the mAP and mAP@50-95 stabilize, indicating convergence. Any slight dips or oscillations in the curves might suggest overfitting The need for further tuning, but overall, the upward trend reflects the model's successful learning process over time.

Figure 3. mAP for Faster R-CNN

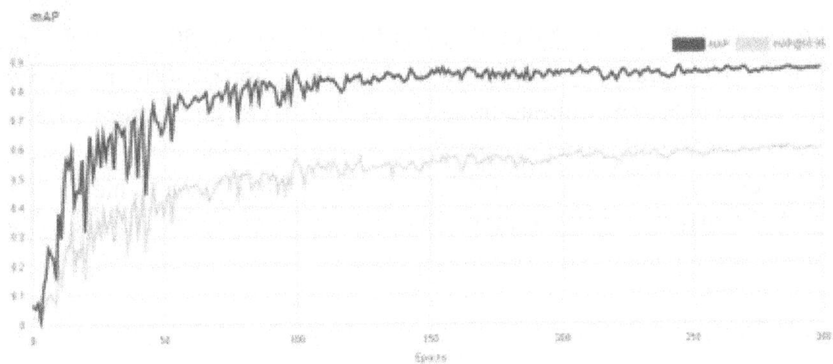

Figure. 4 provides a summary of the Faster R-CNN model's training and performance, showing various loss and metric trends across epochs. The train/box loss, train/cls loss, and train/dfl_loss all show a clear downward trend, indicating that the model is improving in its ability to predict object bounding boxes and classify objects accurately during training. However, on the validation side, both the val/box_loss and val/cls_loss show fluctuations, though with a slight overall decrease. This suggests that while the model is learning, its performance on unseen data remains inconsistent. Similarly, the val/dfl_loss fluctuates more significantly, indicating that the model struggles with object localization on validation data.

For the metrics, precision(B) and recall(B) show variable patterns but maintain relatively high values, reflecting good, though inconsistent, object detection and classification performance. The mAP50(B) and mAP50-95(B), which measure average precision, show an overall upward trend, suggesting improving detection accuracy, though with some variability.

Figure 4. Training visualization for Faster R-CNN

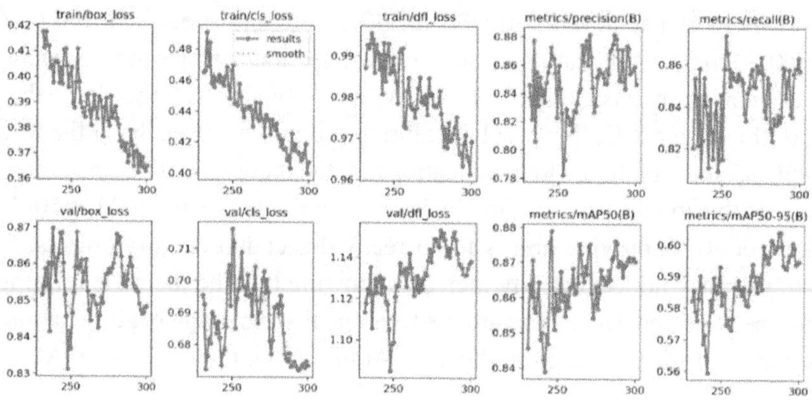

5.2.2 YOLO v8n

The YOLOv8n model achieved a mAP of 87.2% at IoU threshold 0.50 and 59.4% at IoU thresholds ranging from 0.50 to 0.95, demonstrating competitive object detection performance, particularly at lower IoU thresholds. Figure. 5 shows the training progress of YOLO V8 Nano across different metrics. The training box loss, classification loss, and distribution focal loss all exhibit sharp downward trends, particularly in the early epochs, indicating that the model is learning quickly and reducing errors in bounding box predictions and classifications. By around 100 epochs, the losses stabilize, suggesting the model has converged and learned the underlying patterns in the data. The precision and recall metrics improve steadily throughout training, with precision approaching 0.9 and recall nearing 0.85 by the final epoch, indicating that the model is becoming increasingly accurate and consistent in its predictions. Similarly, both mAP50 and mAP50-95 rise steadily, showing that the model's accuracy is improving not just at the 50% IoU threshold but across stricter IoU thresholds, reflecting robust performance. The validation losses follow a similar pattern as training, but with some fluctuations, suggesting occasional overfitting but generally strong generalization to unseen data.

Figure 5. Training Visualization for YOLO V8 Nano

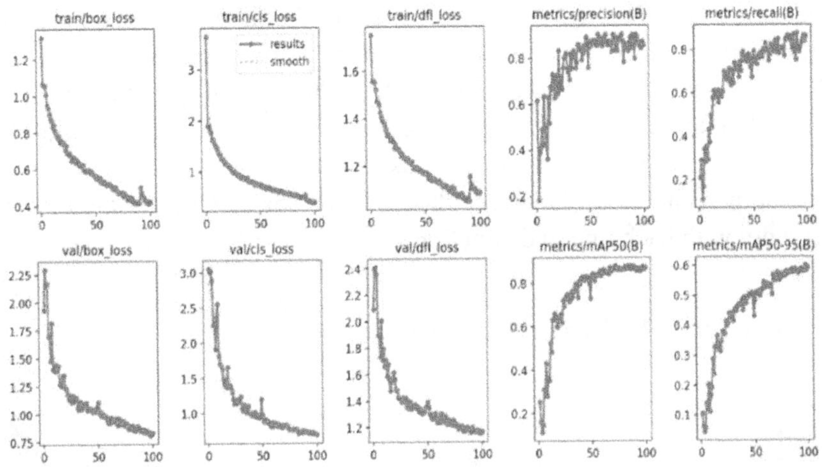

Figure.6. shows the model's precision across different recall levels for various object classes. Each line represents a different object class, with scores close to 1 indicating high precision and recall. The “Cell Phone” class performs the best, with almost perfect precision (0.988), showing that the model is highly accurate and consistent at detecting cell phones. “Laptop” and “Computer” also show strong performances with precision scores of 0.884 and 0.865, respectively. However, the “Tablet” class shows the lowest performance, with a precision of 0.761, indicating the model's difficulty in accurately detecting tablets, possibly due to confusion with similar objects. Overall, the average precision across all classes is 0.874 mAP@0.5, suggesting the model is performing well but has some difficulty with specific categories.

Figure 6. Precision and Recall Curve YOLO V8 Nano

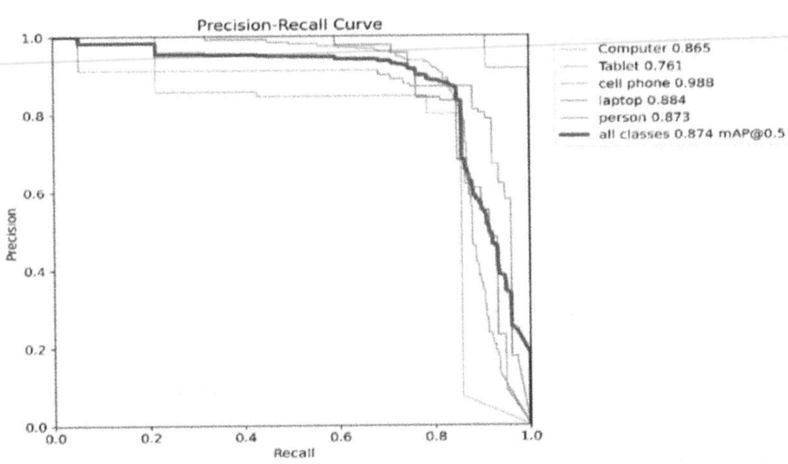

Figure.7. shows how well YOLO V8 Nano is able to classify various objects during testing. Along the diagonal, we observe high values, indicating strong performance in correctly classifying objects. For example, the model correctly classifies "Computer" 92% of the time, and "Tablet" 86% of the time. It perfectly classifies "Cell Phone" 100% of the time, which is a strong indication of the model's ability in that class. For "Person," there is some confusion, with an 86% classification accuracy, but also a 12% misclassification as "Background." This misclassification shows the model struggles in distinguishing between persons and backgrounds. Other small confusions are visible with categories like "Laptop" and "Mouse" where small percentages of incorrect classifications exist. Overall, the model performs well but has some difficulty with background distinction.

Figure 7. Normalized Confusion Matrix YOLO V8 Nano

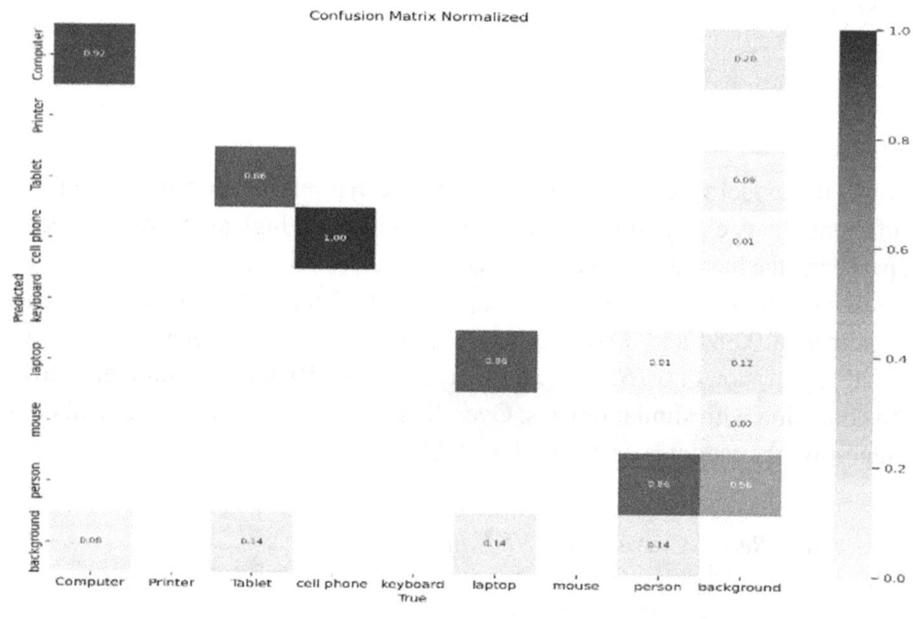

Figure. 8 depicts YOLO V8 Nano's predictions on a test set, where it successfully detects and labels various objects such as "person," "laptop," "tablet," and "cell phone." Each object is enclosed in a bounding box with a confidence score. The model correctly identifies multiple instances of people and laptops, with confidence scores ranging from 0.6 to 1.0. For instance, it detects a "person" with 0.9 confidence in the center and multiple "laptops" with confidence as high as 1.0. However, it seems to struggle with "tablet" and "cell phone" detections, with confidence scores as low as 0.3 to 0.5. This indicates the model is good at detecting larger, more distinctive objects (like people and laptops) but faces challenges with smaller or more similar-looking items like "tablets" and "cell phones."

Figure 8. YOLO V8 Nano prediction on Test Data

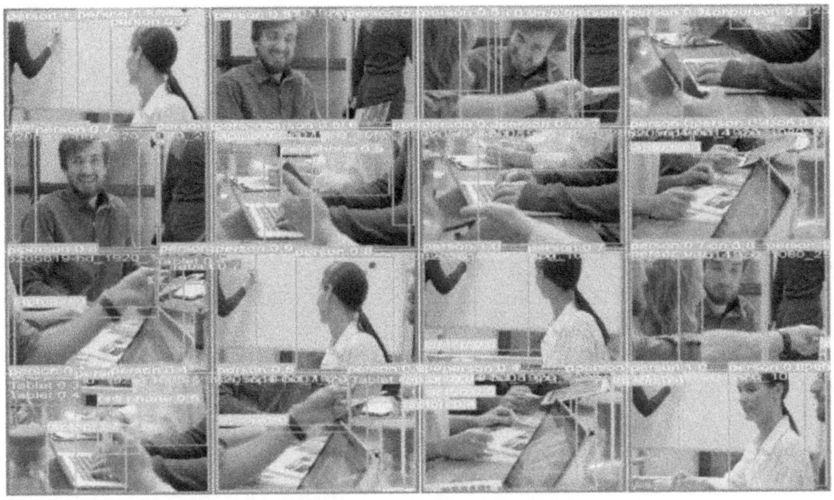

5.2.3 YOLO v9c

The YOLOv9c model achieved a mean Average Precision (mAP) of 88.0% at IoU threshold 0.50 and 59.8% at IoU thresholds ranging from 0.50 to 0.95, indicating strong object detection performance with a notable improvement at the lower IoU threshold compared to previous models. Figure. 9 shows various training metrics and loss curves over 100 epochs for YOLOv9c. The train/box_loss and train/cls_loss curves depict a sharp decrease, starting from approximately 1.2 and 2.5 respectively, and leveling out around 0.4 and 0.5. This indicates that the model's ability to predict bounding boxes and classify objects improves as training progresses. The train/dfl_loss curve, representing the distribution focal loss, follows a similar trend, starting around 1.6 and stabilizing near 1.1. The val/box_loss and val/cls_loss curves show more variability but decrease consistently, which suggests that the model is generalizing fairly well to unseen data, although some fluctuations may indicate slight overfitting. Metrics like precision, recall, and mAP (mean Average Precision) show steady improvement over epochs. Precision rises to around 0.88, and recall climbs above 0.75, demonstrating that the model's accuracy and ability to retrieve correct detections are improving over time. Both mAP_0.5 and mAP_0.5:0.95 curves also show significant growth, with mAP_0.5 peaking at around 0.8 and mAP_0.5:0.95, a stricter evaluation metric, peaking around 0.6, reflecting consistent improvements in detecting objects with more stringent overlap conditions.

Figure 9. Training visualization of YOLOv9c

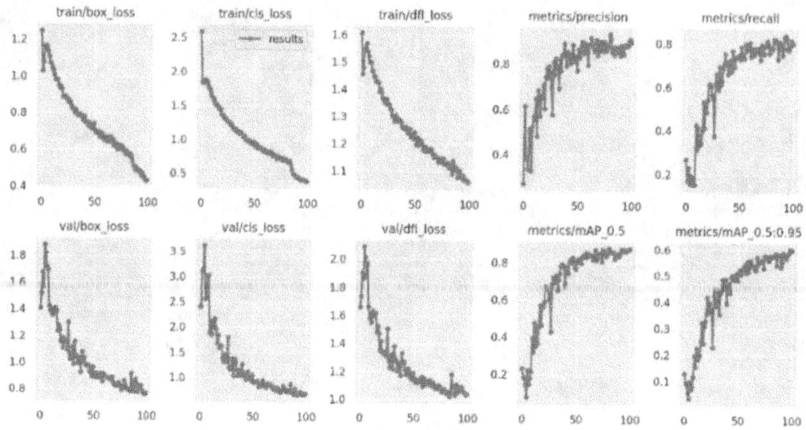

Figure.10 visualizes how YOLOv9c performs across different object classes. Each line represents a class, with performance evaluated based on precision versus recall. The cell phone class performs best, achieving near-perfect precision at 0.968, showing that the model is highly accurate in identifying cell phones. Other classes such as computer and laptop also exhibit strong performance, with precision values of 0.885 and 0.865, respectively. The tablet class performs the weakest, with a precision of 0.798, suggesting that the model has more difficulty detecting tablets compared to other objects. The overall mAP@0.5 for all classes is 0.874, which is a strong indication of the model's good performance across most categories, though some object classes still show room for improvement.

Figure 10. Precision and Recall curve of YOLOv9c

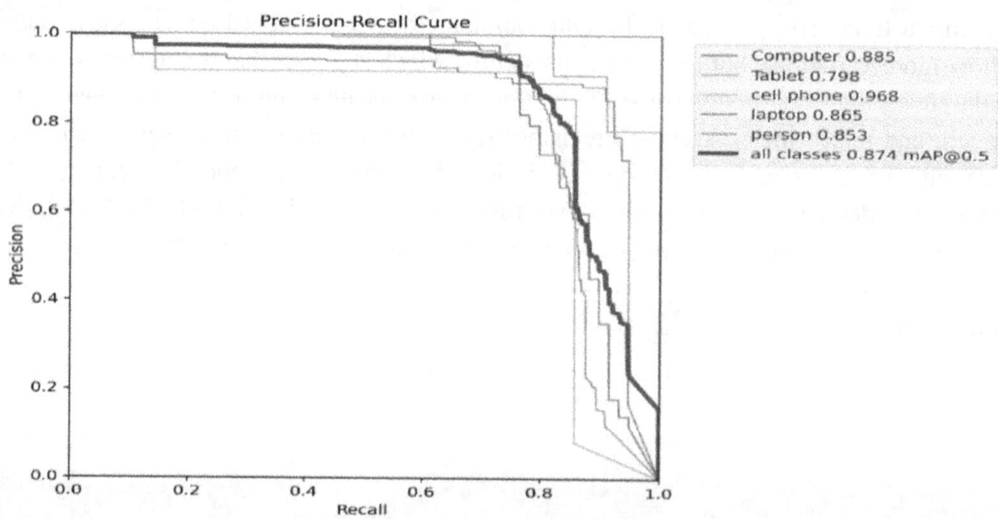

Figure. 11 shows how well YOLOv9c classifies different objects by comparing predicted and true labels. The diagonal values represent correct classifications, with the highest accuracies seen for "Computer" at 91%, "Cell Phone" at 91%, and "Person" at 82%. However, there is some confusion, particularly for the "Person" category, which is misclassified as "Background" 16% of the time. Additionally, the "Mouse" category has an accuracy of 80%, with some confusion possibly occurring with similarly shaped objects or backgrounds. For "Background," the model has a false positive rate of 18% when identifying it as "Person." These off-diagonal values indicate areas where the model could be improved to reduce misclassification, particularly for background and person distinctions.

Figure 11. Confusion Matrix of YOLOv9c

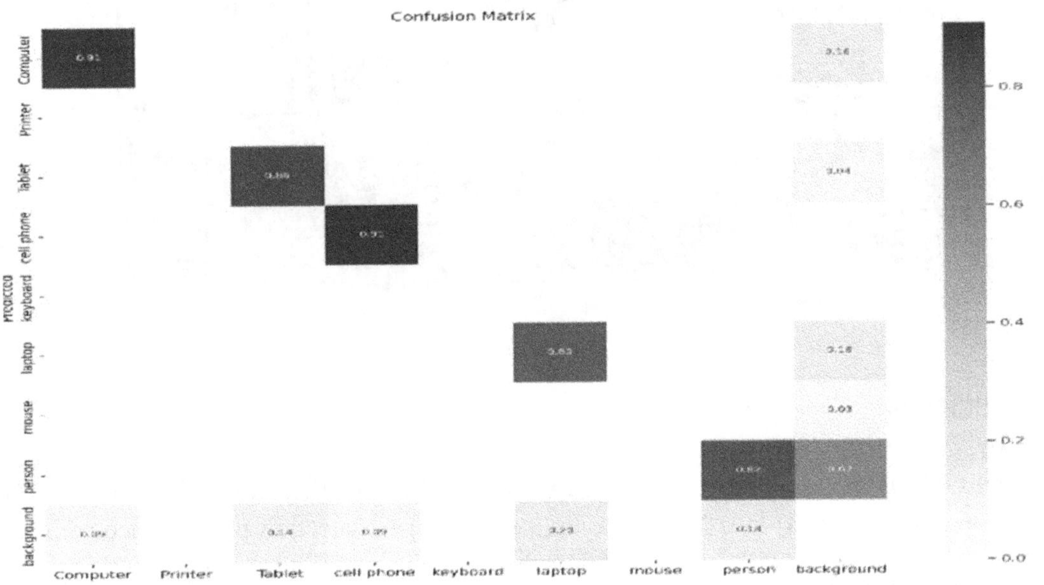

Figure. 12 shows the predictions made by YOLOv9c on a test dataset. The model correctly identifies objects such as "Computer," "Person," "Cell Phone," and "Laptop," with confidence scores generally ranging from 0.8 to 1.0. For example, multiple "Computer" instances are detected with high confidence (0.9), as well as "Person" and "Laptop" instances with confidence levels as high as 1.0 and 0.9. The "Cell Phone" class is identified with confidence scores around 0.8, indicating reasonably accurate detection. Overall, YOLOv9c demonstrates strong detection capabilities, with consistently high confidence scores across most objects, though occasional low-confidence detections (e.g., "Laptop" at 0.4) indicate the potential for further fine-tuning to improve recognition of specific items.

Figure 12. Prediction of YOLOv9c on Test Data

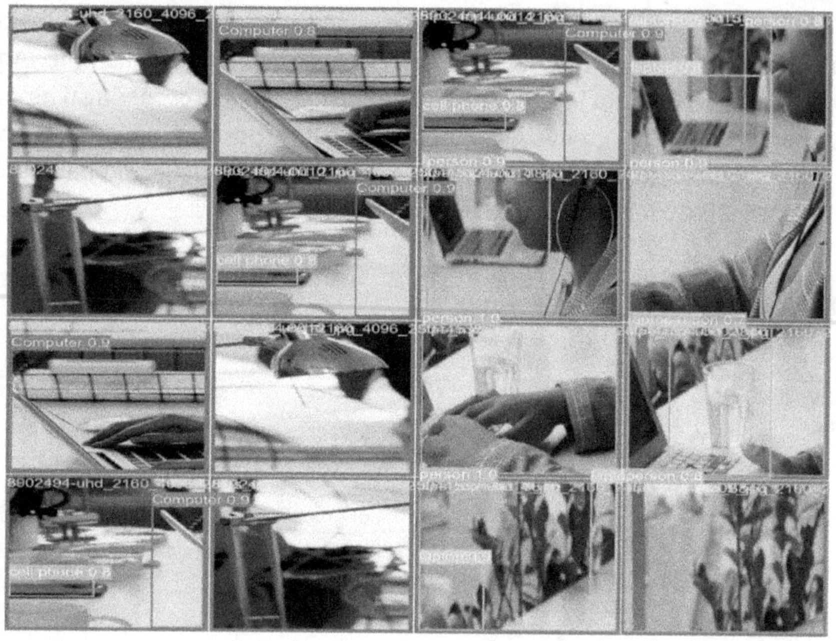

5.2.4 YOLO v10n

The YOLOv110n model achieved a mean mAP of 81.6% at IoU threshold 0.50 and 54.5% at IoU thresholds ranging from 0.50 to 0.95, indicating a lower object detection performance compared to earlier YOLO models, particularly at higher IoU thresholds. Figure.13 illustrates the YOLOv10n model's training progress over 100 epochs. The train/box_loss, train/cls_loss, and train/dfl_loss all exhibit significant decreases over time, reflecting that the model is effectively learning to localize objects and classify them correctly. Similarly, the metrics/mAP50(B) and metrics/mAP50-95(B) curves show steady improvements, with mAP50 approaching 0.8 and mAP50-95 (a stricter metric) rising to around 0.55. The metrics/precision(B) and metrics/recall(B) also improve consistently, with precision approaching 0.8 by the end of training and recall nearing 0.75. These results suggest that the model is learning effectively and improving in both precision and recall, although the performance can still be refined, particularly for challenging object classes.

Figure 13. Training Visualization for YOLOv10n

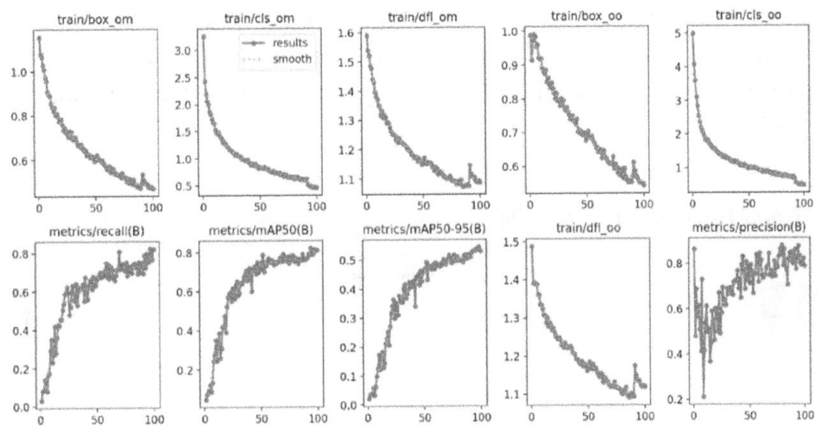

The precision-recall curve provides insight into the trade-off between precision and recall for YOLOv10n across different object categories. Person and Laptop classes perform well, with high precision and recall values of 0.862 and 0.855, respectively. Computer also performs well with a precision of 0.853. However, the Tablet class has the lowest precision at 0.679, indicating that the model struggles to consistently identify tablets. The overall mean Average Precision (mAP) across all classes at a 0.5 IoU threshold is 0.816, which shows that the model performs reasonably well but can be improved, especially for classes like "Tablet" and "Cell Phone."

Figure 14. Precision and Recall for YOLOv10n

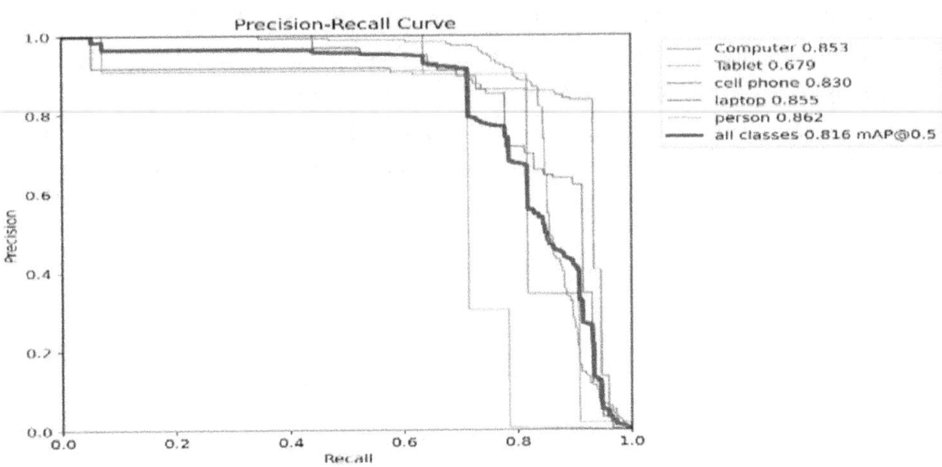

Figure. 15 shows a normalized confusion matrix which shows the model's performance in classifying objects during testing. The diagonal cells represent the percentage of correct predictions, and the non-diagonal cells represent misclassifications. YOLOv10n achieves 91% accuracy in detecting "Computer"

and 83% for "Person." However, it struggles with detecting "Tablet," with only 57% accuracy, and with "Background," where there is significant confusion, particularly for "Person," with a 17% misclassification rate. The model also confuses "Cell Phone" and "Tablet" quite often, with only 82% accuracy for "Cell Phone." This highlights areas where the model's performance could be improved, particularly in distinguishing between visually similar objects like "Tablet" and "Cell Phone," and in separating objects from the background.

Figure 15. Confusion Matrix for YOLOv10n

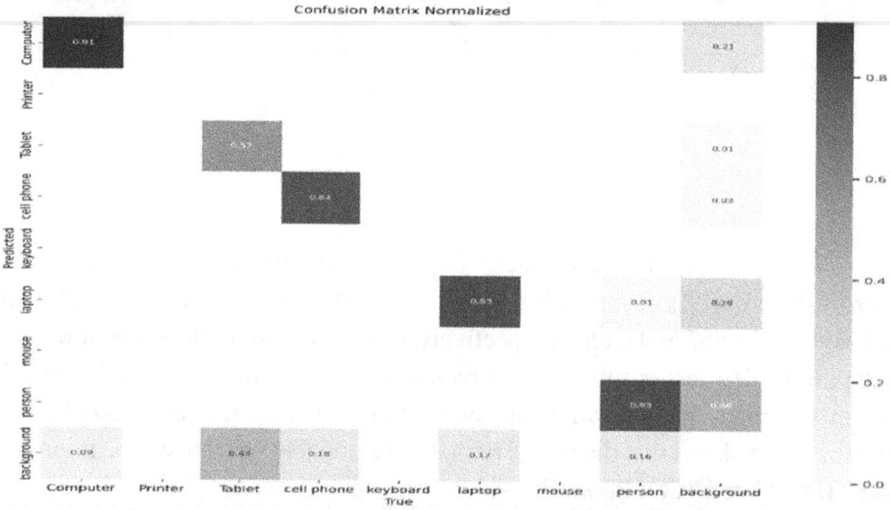

In Figure.16 YOLOv10n correctly detects and labels objects such as "Computer," "Cell Phone," and "Person." The confidence scores for these predictions range from 0.3 to 1.0. Notably, the model detects a "Computer" with high confidence (up to 1.0 in several instances) and consistently identifies "Person" with confidence ranging from 0.5 to 1.0. However, it struggles with identifying the "Cell Phone," with confidence scores as low as 0.4. This suggests that while YOLOv10n is effective at detecting larger objects like "Computer" and "Person," it faces challenges when detecting smaller or less distinct objects like "Cell Phone," leading to lower confidence scores in those instances.

Figure 16. Prediction of YOLOv10n on Test Data

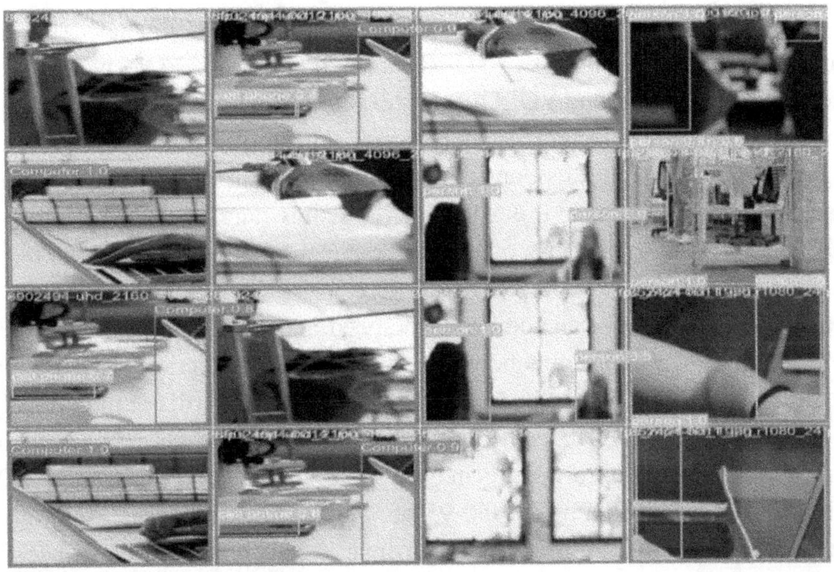

5.2.5 Comparison with Baseline Methods

In the result section, the performance of YOLOv8n, YOLOv9c, and YOLOv10n is compared against Faster R-CNN and YOLOv5, based on findings from recent studies. Faster R-CNN, as reported in studies like (Kou et al., 2021), typically achieves mAP@50 scores between 85-90% and an mAP@50 -95 around 40-60%. While these results demonstrate high precision, especially with small or complex objects, the model's significant drawbacks include its detection time of 150-200 milliseconds per image and a large model size of around 42 million parameters. This makes it unsuitable for real-time detection tasks, despite its accuracy.

On the other hand, YOLOv5 (Yu et al., 2023, Wang et al., 2023) achieves mAP@50 values between 87-90% and mAP@50-95 scores between 55-60%, with detection times ranging between 10-30 milliseconds per image. While this model is much faster and more efficient than Faster R-CNN, it still has a higher computational cost than newer models like YOLOv8n and YOLOv9c, requiring around 12 million parameters in its medium-sized configureurations.

In comparison, the models trained in this study shown remarkable improvements in both speed and resource efficiency. YOLOv8n achieves a mAP@50 of 87.2% and an mAP@50-95 of 59.4%, while requiring only 7 million parameters and having a detection time of 10-20 milliseconds per image, making it highly suitable for real-time applications. YOLOv9c, with a mAP@50 of 88.0% and a mAP@50-95 of 59.8%, offers the best balance between accuracy and efficiency, outperforming both Faster R-CNN and YOLOv5 in terms of precision and resource usage. YOLOv10n, though slightly less accurate with a mAP@50 of 81.6% and mAP@50-95 of 54.5%, still provides a reasonable trade-off between speed and precision, with 10 million parameters and detection times of 15-30 milliseconds.

6. DISCUSSION AND FUTURE WORK

The results of this study demonstrate significant advancements in occupancy detection models, particularly when comparing the performance of YOLOv8n, YOLOv9c, and YOLOv10n against established benchmarks such as Faster R-CNN and YOLOv5. Faster R-CNN, while recognized for its high precision in detecting small and complex objects, falls short in real-time applications due to its detection time of 150-200 milliseconds per image and its considerable model size of approximately 42 million parameters. Despite achieving mean Average Precision (mAP) scores ranging from 85-90% at IoU 0.50 and 40-60% at IoU 0.50-0.95, these drawbacks make it impractical for environments that require immediate occupancy detection.

YOLOv5 demonstrates a better balance between speed and accuracy, achieving mAP@50 scores between 87-90% and mAP@50-95 scores between 55-60%. With detection times ranging from 10-30 milliseconds, YOLOv5 is more efficient than Faster R-CNN, though it still computational intensive with approximately 12 million parameters in its medium-sized configureurations. YOLOv8n, YOLOv9c, and YOLOv10n—show remarkable improvements in both speed and resource efficiency. YOLOv8n stands out with a mAP@50 of 87.2% and a mAP@50-95 of 59.4%, requiring only 7 million parameters and achieving detection times of 10-20 milliseconds per image. This combination of accuracy and speed positions YOLOv8n as highly suitable for real-time occupancy detection applications. YOLOv9c further enhances this performance, boasting a mAP@50 of 88.0% and a mAP@50-95 of 59.8%, making it the model with the best balance between accuracy and efficiency in this study. This model outperforms both Faster R-CNN and YOLOv5 in terms of both precision and resource usage, solidifying its role as a preferred choice for real-time applications that require prompt decision-making. YOLOv10n shows slightly lower accuracy with a mAP@50 of 81.6% and a mAP@50-95 of 54.5%, it still provides a reasonable trade-off between speed and precision, making it a viable option in scenarios where computational resources are limited, or where the emphasis is placed on processing speed.

These findings show that further exploration in the field of occupancy detection should focus on several key areas to enhance the capabilities and applications of the models presented in this study. Integrating multi-modal data sources, such as PIR, infrared, CO2 sensors, alongside video analysis, can improve detection accuracy in complex environments. Additionally, prioritizing privacy protection through advanced techniques like data anonymization and federated learning will ensure occupant privacy while still harnessing valuable occupancy data. Long-term performance monitoring in diverse real-world settings will provide insights into occupancy patterns, informing better HVAC control strategies and maximizing energy savings. Implementing adaptive learning algorithms will enable models to continuously update and refine themselves based on new data, thereby enhancing their effectiveness over time. Finally, assessing the cost-effectiveness and scalability of these systems will be crucial for their widespread adoption, particularly in retrofitting existing buildings. By addressing these areas, future research can significantly contribute to the development of more robust, efficient, and sustainable occupancy detection systems, ultimately advancing the optimization of energy use in indoor environments and promoting environmental sustainability

FUNDING STATEMENT

This work is supported in part by XJTLU Research Development Fund under Grant No. RDF21-01-015.

REFERENCES

Abdel-Razek, S. A., Marie, H. S., Alshehri, A., & Elzeki, O. M. (2022). Energy efficiency through the implementation of an AI model to predict room occupancy based on thermal comfort parameters. *Sustainability (Basel)*, 14(13), 7734. DOI: 10.3390/su14137734

Agarwal, Y., Balaji, B., Gupta, R., Lyles, J., Wei, M., & Weng, T. (2010, November). Occupancy-driven energy management for smart building automation. In *Proceedings of the 2nd ACM workshop on embedded sensing systems for energy-efficiency in building* (pp. 1-6).

Anjomshoaa, A., Duarte, F., Rennings, D., Matarazzo, T. J., Desouza, P., & Ratti, C. (2018). City scanner: Building and scheduling a mobile sensing platform for smart city services. *IEEE Internet of Things Journal*, 5(6), 4567–4579. DOI: 10.1109/JIOT.2018.2839058

Beltran, A., Erickson, V. L., & Cerpa, A. E. (2013, November). Thermosense: Occupancy thermal based sensing for hvac control. In *Proceedings of the 5th ACM Workshop on Embedded Systems For Energy-Efficient Buildings* (pp. 1-8).

Carli, R., Cavone, G., Ben Othman, S., & Dotoli, M. (2020). Iot based architecture for model predictive control of hvac systems in smart buildings. *Sensors (Basel)*, 20(3), 781. DOI: 10.3390/s20030781 PMID: 32023965

Chung, J., Gulcehre, C., Cho, K., & Bengio, Y. (2014). Empirical evaluation of gated recurrent neural networks on sequence modeling. arXiv preprint arXiv:1412.3555.

Dong, B., & Lam, K. P. (2014, February). A real-time model predictive control for building heating and cooling systems based on the occupancy behavior pattern detection and local weather forecasting. []. Springer Berlin Heidelberg.]. *Building Simulation*, 7, 89–106.

Dong, B., & Lam, K. P. (2014, February). A real-time model predictive control for building heating and cooling systems based on the occupancy behavior pattern detection and local weather forecasting. []. Springer Berlin Heidelberg.]. *Building Simulation*, 7, 89–106.

Dorokhova, M., Ballif, C., & Wyrsch, N. (2020). Rule-based scheduling of air conditioning using occupancy forecasting. *Energy AI*, 2, 100022. DOI: 10.1016/j.egyai.2020.100022

Gomez, A., Conti, F., & Benini, L. Thermal image-based CNN's for ultra-low power people recognition. *Proceedings of the 15th ACM International Conference on Computing Frontiers*, 2018. 326-331. DOI: 10.1145/3203217.3204465

Griffiths, E., Assana, S., & Whitehouse, K. (2018). Privacy-preserving image processing with binocular thermal cameras. *Proceedings of the ACM on Interactive, Mobile, Wearable and Ubiquitous Technologies*, 1(4), 1–25. DOI: 10.1145/3161198

Hashimoto, K., Morinaka, K., Yoshiike, N., Kawaguchi, C., & Matsueda, S. (1997, June). People count system using multi-sensing application. In Proceedings of International Solid State Sensors and Actuators Conference (Transducers' 97) (Vol. 2, pp. 1291-1294). IEEE.

Hobson, B. W., Lowcay, D., Gunay, H. B., Ashouri, A., & Newsham, G. R. (2019). Opportunistic occupancy-count estimation using sensor fusion: A case study. *Building and Environment*, 159, 106154. DOI: 10.1016/j.buildenv.2019.05.032

Hu, S., Wang, P., Hoare, C., & O'Donnell, J. (2022). Building occupancy detection and localization using CCTV camera and deep learning. *IEEE Internet of Things Journal*, 10(1), 597–608. DOI: 10.1109/JIOT.2022.3201877

Jiang, C., Masood, M. K., Soh, Y. C., & Li, H. (2016). Indoor occupancy estimation from carbon dioxide concentration. *Energy and Building*, 131, 132–141. DOI: 10.1016/j.enbuild.2016.09.002

Koebrich, S., Chen, E. I., Bowen, T., Forrester, S., & Tian, T. (2019). *2017 Renewable Energy Data Book: Including Data and Trends for Energy Storage and Electric Vehicles. National Renewable Energy Lab*. NREL. DOI: 10.2172/1581221

Kou, X., Du, Y., Li, F., Pulgar-Painemal, H., Zandi, H., Dong, J., & Olama, M. M. (2021). Model-based and data-driven HVAC control strategies for residential demand response. *IEEE Open Access Journal of Power and Energy*, 8, 186–197. DOI: 10.1109/OAJPE.2021.3075426

Lee, G.-C., Lee, S.-H., & Yoo, J. (2018). 3-D people counting with a stereo camera on gpu embedded board. *Applied Sciences (Basel, Switzerland)*, 8(11), 2017. DOI: 10.3390/app8112017

Metwaly, A., Queralta, J. P., Sarker, V. K., Gia, T. N., Nasir, O., & Westerlund, T. (2019, October). Edge computing with embedded ai: Thermal image analysis for occupancy estimation in intelligent buildings. In *Proceedings of the INTelligent Embedded Systems Architectures and Applications Workshop 2019* (pp. 1-6).

Moons, B., Bankman, D., & Verhelst, M. (2019). *Embedded deep learning. Circuits for Always-on Neural Network Processing* (1st ed.). Springer.

Paidi, V., Fleyeh, H., & Nyberg, R. G. (2020). Deep learning-based vehicle occupancy detection in an open parking lot using thermal camera. *IET Intelligent Transport Systems*, 14(10), 1295–1302. DOI: 10.1049/iet-its.2019.0468

Petersen, S., Pedersen, T. H., Nielsen, K. U., & Knudsen, M. D. (2016). Establishing an image-based ground truth for validation of sensor data-based room occupancy detection. *Energy and Building*, 130, 787–793. DOI: 10.1016/j.enbuild.2016.09.009

Rastogi, K., & Lohani, D. (2020). IoT-based indoor occupancy estimation using Edge Computing. *Procedia Computer Science*, 171, 1943–1952. DOI: 10.1016/j.procs.2020.04.208

Rueda, L., Agbossou, K., Cardenas, A., Henao, N., & Kelouwani, S. (2020). A comprehensive review of approaches to building occupancy detection. *Building and Environment*, 180, 106966. DOI: 10.1016/j.buildenv.2020.106966

Ryu, S. H., & Moon, H. J. (2016). Development of an occupancy prediction model using indoor environmental data based on machine learning techniques. *Building and Environment*, 107, 1–9. DOI: 10.1016/j.buildenv.2016.06.039

Sayed, A. N., Bensaali, F., Himeur, Y., & Houchati, M. (2023). Edge-based real-time occupancy detection system through a non-intrusive sensing system. *Energies*, 16(5), 2388.

Sun, K., Zhao, Q., Zhang, Z., & Hu, X. (2022). Indoor occupancy measurement by the fusion of motion detection and static estimation. *Energy and Building*, 254, 111593. DOI: 10.1016/j.enbuild.2021.111593

Tien, P. W., Wei, S., Calautit, J. K., Darkwa, J., & Wood, C. (2020). A vision-based deep learning approach for the detection and prediction of occupancy heat emissions for demand-driven control solutions. *Energy and Building*, 226, 110386. DOI: 10.1016/j.enbuild.2020.110386

Wahl, F., Milenkovic, M., & Amft, O. A distributed PIR-based approach for estimating people count in office environments. 2012 IEEE 15th International Conference on Computational Science and Engineering, 2012. IEEE, 640-647.

Wang, C., Zhang, Y., Zhou, Y., Sun, S., Zhang, H., & Wang, Y. (2023). Automatic detection of indoor occupancy based on improved YOLOv5 model. *Neural Computing & Applications*, 35(3), 2575–2599. DOI: 10.1007/s00521-022-07730-3 PMID: 36068815

Wang, S., Burnett, J., & Chong, H. (1999). Experimental validation of CO2-based occupancy detection for demand-controlled ventilation. *Indoor and Built Environment*, 8(6), 377–391. DOI: 10.1177/1420326X9900800605

Yu, F., Zhang, G., Zhao, F., Wang, X., Liu, H., Lin, P., & Chen, Y. (2023). Improved YOLO-v5 model for boosting face mask recognition accuracy on heterogeneous IoT computing platforms. *Internet of Things : Engineering Cyber Physical Human Systems*, 23, 100881. DOI: 10.1016/j.iot.2023.100881

Yun, J., & Lee, S.-S. (2014). Human movement detection and identification using pyroelectric infrared sensors. *Sensors (Basel)*, 14(5), 8057–8081. DOI: 10.3390/s140508057 PMID: 24803195

Zemouri, S., Gkoufas, Y., & Murphy, J. (2019, August). A machine learning approach to indoor occupancy detection using non-intrusive environmental sensor data. In *Proceedings of the 3rd international conference on big data and internet of things* (pp. 70-74).

Zemouri, S., Magoni, D., Zemouri, A., Gkoufas, Y., Katrinis, K., & Murphy, J. (2018, September). An edge computing approach to explore indoor environmental sensor data for occupancy measurement in office spaces. In 2018 IEEE International Smart Cities Conference (ISC2) (pp. 1-8). IEEE.

KEY TERMS AND DEFINITIONS

OBC: Occupancy-based control.
HVAC: Heating, ventilation, and air conditioning.
PIR: Infrared pro proximity sensors.
RC: Resistance capacitance.

FCHD: Fully convolutional head detector.
mAP: Mean average precision.
BEMS: Building energy management systems.

ENDNOTES

1 https://www.gumstix.com/manufacturer/nvidia/nano-snapshot.html

2 https://coral.ai/

3 https://aws.amazon.com/deeplens/

4 https://www.mturk.com/

Chapter 14

Smart Integration of Renewable Energy for Sustainable Cities: Leveraging Technology for Urban Resilience

Sayed Sayeed Ahmad
https://orcid.org/0000-0001-6320-8720
De Montfort University, Dubai, UAE

Mohamed Ruknuddin Ghalib
https://orcid.org/0000-0002-2786-3370
De Montfort University, Dubai, UAE

Michael Gallimore
De Montfort University, Dubai, UAE

Atheer Al Mousa
De Montfort University, Dubai, UAE

Ihab Wattar
Cleveland State University, USA

ABSTRACT

This study analyzes the significance of integrating renewable energy in emerging urban areas that are highly supported for sustainability. Leveraging technology controls the impact of the environment and supports the advancement of urban life. An overview of the issues facing urban areas is given at the beginning of the chapter, including managing energy, reducing carbon emissions, and meeting public sector energy demands. Artificial intelligence (AI), the Internet of Things (IoT), and other intelligent integration techniques can significantly enhance the use of renewable energy. Moreover, this chapter highlights numerous case studies of smart integration of renewable energy in urban areas by implanting IoT-enabled smart grids, AI-based energy management, and abundant data-recorded management for urban development. It also provides an overview of the regulation and control of energy demand by technology- and data-driven methods to advance energy efficiency and environmental impact.

DOI: 10.4018/979-8-3693-5748-4.ch014

INTRODUCTION

Recently, many developed nations have looked to sustainable urban development to control and regulate energy management and create a green environment. Integrating renewable energy technologies will be the best alternative solution to monitor and normalise energy demand for the development of urban functions (Palanca et al., 2020; Shi et al., 2021). In the past century (Karim et al., 2022), there has been a consistent rise in the population of large cities and metropolitan areas. According to the United Nations (UN) population survey until 2021, over 60% of the global population resides in these urban hubs.

In the 21st century, the global population has become predominantly urban, with more than half residing in cities, marking the era of urbanisation. This trend has been driven by rapid advancements in information technology, which have accelerated urban growth, commonly referred to as the third wave of urbanisation. According to projections, urban areas are expected to accommodate more than 70% of the global population by 2050, solidifying their role as primary hubs for economic growth and inventive activity. The 600 largest cities worldwide generate approximately 80% of the global GDP.

Renewable energy, derived from sources such as sunlight, wind, water, geothermal heat, and organic matter, offers a path to mitigate the environmental impact of urbanisation while meeting the growing energy demands of expanding populations. The overview of renewable energy technologies reveals diverse solutions, each with unique advantages and applications.

i. Solar photovoltaics harnesses the abundant energy of sunlight to generate electricity.
ii. Wind turbines leverage the kinetic energy of moving air to produce power.
iii. Biomass, derived from organic materials such as wood, agricultural, and municipal waste, offers a versatile energy source for heating, electricity generation, and transportation fuels.
iv. Earth geothermal energy is abundant heat energy stored inside the earth, which is drawn by leveraging technology to provide a reliable and constant power source.
v. A hydroelectric power system is a massive water storage system in a catchment area that generates energy by flowing water from a reservoir to rivers, tides, or ocean currents.

In summary, presently, fossil fuels play a crucial role in meeting the current energy demand, which reveals massive deposition of water and air pollution, greenhouse gas emissions, and the production of fine particles of dust for prolonged periods, which affects health issues for human beings and the environment. In contrast, renewable energy is the superlative alternative energy for present technologies to control carbon emissions, grow the economy, and enhance environmental security.

However, the development and expansion of city boundaries are increasing gradually day by day, depending upon the mass of the population and employment opportunities. The consumption of energy is increasing based on the density of the population, the implementation of industrial applications, and the provision of transport facilities for the public and private sectors, among others. Due to this, the integration of renewable energy alone will not be a solution. Leveraging technology such as AI and IoT will greatly support predicting environmental demand in the current and future situation and highly support the challenges related to environmental degradation, climatic conditions, carbon emissions, etc.

Urban areas: Achieving sustainability in urban development requires adopting a new approach that emphasises the implementation of renewable energy technology in city construction, as suggested by Chibuike P.O. et al. (2024). The author suggests that by instigating these technologies, urban areas can be controlled through cleaner, more resilient energy systems that support consistent growth and augment the quality of resident lifestyle.

Advancement, which our study provides an overview of the potential of renewable energy and its advancement, which will highly support sustainability goals in urban areas by integrating AI and IoT Techniques. Through this, we can evaluate the challenges, opportunities, and suitable practices related to the integration of renewable energy in urban areas and researchers and. raisingMoreover, we focus on providing an informative collection to policymakers, researchers, and people about the benefits and potential outcomes of future sustainable urban.

CURRENT STATUS OF URBAN ENERGY SYSTEMS

On the development of urbanisation and infrastructure, the literature survey reveals that a negative impact has been recorded on the consumption of fossil fuels to meet energy demand and their impact on the environment. Numerous researchers investigated the possible ways and techniques to develop and support meeting energy demand in cities smartly by suggesting the integration of renewable energy. This chapter provides collective information on different strategies for promoting sustainability in urban energy. This chapter aims to provide overviews of substantial information in the following aspects:

- An effective strategies for integrating renewable energy
- Urban planning policies should be adapted to enhance the deployment of renewable energy
- Social, Economic, and Environmental benefits
- Resilience and Reliability of urban energy systems
- Community engagement and public participation should be enhanced to promote the acceptance and adoption of renewable energy

STRATEGIES FOR RENEWABLE INTEGRATION IN URBAN DEVELOPMENT

An innovative strategy, *"A smart city pilot programme,"* has been followed in China for urban growth and sustainability (Rui G. et al., 2024). It exposes that the government and people of China take the initiative to preserve resources and the environment to enhance the quality of the city lifestyle and create a more sustainable mode of life. They observed a high volume of people moving to urban areas for a better life, while cities faced high levels of pollution and health issues. To regulate and control this problem, the "Smart City pilot programme" will play a significant role in keeping the environment clean and friendly. Major consumption of renewable energy will support energy demand in cities by implementing "different in different models" and their nuances shaping the city in an effective and smart city pilot., especially in rich resource-urban centres. It's a role model to the world as an excellent sustainable urban planner.

Another innovative model for urban advancement has been introduced by Haiyan W. and Yanxia W. (2024) to provide the best solution to the challenges faced by urban. A comprehensive framework has been developed by the author to analyse the a) assessment of environmental impact, b) evaluation of cost efficiency, and c) aesthetic impact analyses to create a sustainable environment and liveable urban life in the name of "Digital Twin technology". The author conducted brief research on the location of a) carbon emission, b) optimising resource allocation, and c) fulfilling community satisfaction by using these dynamics and real-time dimensions simulation to predict the possible pollution control in selected locations for better urban planning. Besides, this data-driven methodology elevates the quality of the urban lifecycle and provides a correct solution for policymakers for designing landscapes urban landscapes and innovatively shaping cities.

I.A.Daniel I.A et al. (2023) are alert to the expanding urban borders owing to the density of the population and the energy required to meet demand by consuming renewable energy in the form of solar, wind, and ongoing progress in electrifying transportation systems. Ecuador,The author conducted a case study in Cuenca, Ecuador, to suggest a roadmap for the "Energy PLAN tool" to make a fully electrifying urban transport system. to implementA significant aim of this study is to implement leveraging technology to make a decentralised transport system by 2050 by integrating identified energy that was consumed automatically one by one season-based for energy production, and this process is monitored using IoT technology.

(Zhaiting Y et al., 2024) highlight the significance of multi-energy systems (MES) in accommodating renewable energy (RE) amid diminishing fossil fuel reserves. The authors suggest *"dynamic programming* and *machine learning techniques"* to predict power usage for future vehicles and reduce costs by integrating MES methods in a two-step scheme such as heat electricity and demand response (DR) to augment operational reliability. Moreover, power-to-gas (P2G) devices were linked with heat electric - DR to reduce the consumption of renewable energy and create a security system with economic benefits, especially in changing climate and socio-economic factors in urban energy consumption.

The research conducted by Ahmet S and Demet E. (2023) aims to reduce sustainability-related challenges in urban areas by integrating 2 or 3 renewable energy sources in a single plant as a hybrid renewable energy plant into spatial planning in Turkey to meet the energy demand. The author develops a unique spatial model used for site selection for Ericsson of a) wind, b) solar, and c) biomass energy sources in the Karatay district. They construct a new methodology framework in all closer locations to identify a correction location for installing the same setup by using the "fuzzy logic approach" and "*Geographic Information Systems*" (GIS) tools. Besides, the GAMMA operator is used to distinguish optimal sites and identify alternative scenarios for installation. Finally, they declared a combination of solar and biomass-hybrid power plants is the most suitable plant, and their power production efficiency is higher than other plants as per the location in Karatay district.

(Yueyan C. et al., 2024), The research focuses on providing a unique solution "Energy Hubs" (EHs) for resource scarcity and environmental issues in developing urban areas. This model highlights integrating high-power production plants in single units, such as renewable sources and thermal storage. These EHs aim to improve energy production and consumption depending on energy level to meet urban demand. However, they predict energy requirements in domestic and residential areas by using demand response programmes and data-driven strategies for the future too. This study provides a clear view of how integrating this power plant in a single location with possible techniques and how to collect information about energy supply in urban areas without any errors will advance sustainable urban energy.

- What are the most effective strategies for integrating renewable energy technologies into urban development plans?

The most effective methods are identified by numerous researchers by conducting analysis using various tools to enhance the production of renewable energy in a single location and predict the energy requirements for both domestic and resident purposes as a multifaceted approach to advance urban life with sustainability. Significant methods such as "Smart City Pilot Programs", exemplified by China's government and people's efforts focus on consuming limited / required renewable resources with high efficiency and preserving the environment. They develop a role model for urban life to limit the consumption of energy only required without polluting the environment. Moreover, "Digital twin technology" is a real-time decision-making leverage technology that helps to incorporate renewable energy strategies to support urban planning and the development of urban lifestyle. Besides, renewable energy supports fully electrifying mobility by using the "Energy Plan Tool" model for long-term roadmaps. It's a suitable solution for the present situation to control and reduce the consumption of fossil fuels and preserve environmental air pollution for a healthy lifestyle. Additionally, this study reports majorly support cold climatic cities and fulfill their energy demand all around. Finally, integrating multi-renewable energy systems such as spatial planning for hybrid renewable energy *"energy hubs"* is a unique solution to optimize production and consumption in urban energy management by using "ML and fussy logic approaches"

Renewable-Energy-Centric Urban Planning

(Mark M.A, et al., 2024) investigate stakeholders' awareness regarding urban form's impact on residential rooftop solar PV and its implications for solar urban planning in Ghana. Data was collected through surveys and numerous interviews with a0 households, b) real estate developers, c) licensed solar home system (SHS) installation companies, and d) government agencies. They highlight noteworthy factors influencing rooftop PV installation and augmentation of thermal performance, including roofs in different types, shapes, roofing materials, neighboring building height, tree positioning, etc. Despite these findings, households exhibit limited awareness of these urban form attributes' effects on rooftop solar PV. Notably, existing planning legislations do not mandate the inclusion of such systems in urban and building plans, suggesting a need for policy revisions to support effective solar urban planning

(Xianguo Z et al., 2021) aim to enhance the implementation of China's rural revitalization strategy by scientifically guiding beautiful rural construction and restoring rural vitality, using Shuanglongqiao Village in Sichuan Province. The authors conducted the on-site investigation and comparative analysis, which proposes strategies for beautiful rural planning aligned with the rural revitalization strategy's objectives. They are offering insights applicable to similar villages nationwide. This study employs a "multifaceted approach", such as combining a) a literature review, b) an on-site investigation, and c) a comparative analysis to explore the intersection of the rural revitalization strategy and attractive rural planning, culminating in tailored planning strategies for Shuanglongqiao Village.

Urban energy systems first need to analyze defining boundaries, often beyond municipal limits, considering renewable energy expansion's territorial implications stated by (Siamak H, et al., 2022). Utilizing the "PRISMI Plus toolkit" scenarios for Monachil Municipality's transition to 50% renewable energy involves neighboring areas, enhancing renewable penetration and diversification, while higher penetration incurs exponentially rising huge investment and high operation costs.

(Ismaila R.A and Maher S. A. 2023) presented a Sustainable urban planning is crucial for achieving low-carbon cities, particularly in the larger regions facing environmental challenges due to heavy reliance on fossil fuels. In this, various low-carbon schemes exist, and their effectiveness has been debated for controlling pollution. The UAE country leads in green buildings and renewables, but wastewater recycling rates vary significantly. Urban form and transportation infrastructure fall short of UN-Habitat recommendations, emphasizing the need for integrated energy-water demand reduction measures in land-use planning

- How can urban planning policies be adapted to incentivize the deployment of renewable energy solutions in cities?

Several key policies were suggested by various researchers, such as "leveraging pilot initiatives" as highlighted by (Rui G et al.2024), and can demonstrate the viability and benefits of integrating renewable energy into urban development, encouraging broader adoption through positive outcomes and economic inclusion. Secondly, incorporating "solar urban planning" considerations, as emphasized by (Mark M.A, et al.,2024) requires updating planning legislation to mandate the inclusion of rooftop solar PV systems in urban and building plans. This ensures that factors influencing PV installation, such as roof type and neighboring building height, are systematically considered in urban development projects. Then expanding the boundaries of energy system analysis, as demonstrated by (Siamak H et al.,2022) to encompass neighboring areas enables a more comprehensive approach to renewable energy expansion, enhancing penetration and diversification while managing investment and operation costs. Moreover, aligning rural revitalization strategies with renewable energy goals has been proposed by (Xianguo Z et al.,2021) which can guide beautiful rural construction and restore rural vitality by integrating renewable energy solutions into planning frameworks.

Alongside, in regions facing environmental challenges like the larger cities, as discussed by (Ismaila R.A and Maher S. A, 2023) integrated energy-water demand reduction measures should be prioritized in land-use planning to promote low-carbon cities, emphasizing the importance of sustainability in urban development policies. In general, these strategies highlight the significance of proactive policy adjustments to drive the widespread acceptance and implementation of renewable energy solutions in urban environments.

Social and Economic Benefits of Urban Renewable Energy Transition

A comprehensive framework has been presented by (Ana A, et al., 2023) for evaluating the impact of energy transition on the advance of economic growth, social welfare, and environmental sustainability from a triple perspective. It critically analyzes a) existing reports, b) policies, and c) initiatives, providing a roadmap for decision-making and assessing effectiveness. The study proposes indicators to evaluate the a) sustainable and circular potential, b) emphasizing environmental, and c) social, and economic dimensions, while suggesting policies to facilitate transition. Furthermore, it conducts a comparative analysis of European countries to identify areas for improving a) energy renewability, b) aiding stakeholders, c) citizens, and d) policymakers in enhancing efficiency, renewability, and technology development for a sustainable and circular energy transition.

(Mei Z et al., 2024) analyzed the integration of renewable energy in smart urban homes, emphasizing its role in balancing energy management and promoting sustainability for safe and secure urban life. Utilizing smart grid technologies and solar panels for renewable energy, it explores demand response strategies and introduces the "Polar Bear Optimization Algorithm" for efficient scheduling. In addition to "Leveraging neural network" (LNN) techniques for forecasting, the study offers insights into minimizing energy costs and enhancing environmental resilience.

Economic growth is closely related to carbon emissions stated by Florian M.N et al., 2024) particularly in industrial-centric structures. It is less energy efficient and creates more environmentally damaging than tertiary-focused ones. The author suggests "Feasible Generalized Least Squares" (FGLS) and "Panel-Corrected Standard Errors"(PCSEs) models, the study advocates for growth policies centered on sustainable urban communities and transitioning to a renewable energy-dominated energy mix to mitigate environmental degradation.

(Bjarnhedinn G, et al., 2023) evaluates the economic feasibility and societal benefits of transitioning to renewable energy-based energy systems through "cost-benefit analysis". The author conducted a case study using a low payback periods model such as (8.2 years for Barcelona, 2.8 years for St. Jean), which increased revenue by closer to (22% for St. Jean), decreased costs (4% for Barcelona), which enhanced energy system flexibility to (21% for St. Jean). Besides, it significantly emission reductions closer to (23.5% for St. Jean, and 4% for Barcelona). It highlighting the positive social and environmental impacts of energy transition.

- What are the social, economic, and environmental benefits of transitioning to renewable energy in urban areas?

Transitioning to renewable energy in urban areas offers a multitude of a) social, b) economic, and c) environmental benefits.

A) Socially, it fosters community engagement and empowerment by decentralizing energy production and promoting local ownership of renewable energy systems, as highlighted by (Ana A et al., 2023) Additionally, it improves public health by reducing air pollution and mitigating the adverse effects of fossil fuel combustion on respiratory illnesses and overall well-being in urban life.
B) Economically, the integration of renewable energy produces more job opportunities in sectors such as a) manufacturing, b) installation/, and c) maintenance of renewable energy infrastructure by remote and local site visits, contributing to local economic development and resilience. Furthermore, it controls the cost of energy for individual households and businesses, enhancing affordability and energy access while reducing reliance on volatile fossil fuel markets.
C) Environmentally, transitioning to renewable energy significantly reduces a) greenhouse gas emissions, b) mitigating climate change, and c) its associated impacts, such as extreme weather events and sea-level rise, as (Florian M.N et al., 2024. and Bjarnhedinn G et al., 2023) illustrate.

Furthermore, it conserves natural resources such as air and water pollution associated with fossil fuel extraction and combustion processes using various techniques, etc. It promotes ecological sustainability and biodiversity conservation for a natural lifestyle. Transforming to renewable energy in urban areas is highly necessary for achieving a) a sustainable, b) equitable, and c) resilient future for both present and future generations.

Resilience of Urban Renewable Energy Systems

(Yan Z and Yunqi L, 2024) Provide a solution to the unique problem: transmission congestion in urban areas and energy networks during climate challenges and cold climatic conditions. They are integrating photovoltaic (PV) solar panels in urban during these climatic conditions, and it is challenging to generate energy to meet energy demands. This author suggests implementing a digital twin framework such as "Gated Recurrent Unit" (GRU) - based forecasting, employing a "Lion Optimization Algorithm" (LOA) within this technique. Besides, it highlights the PV integrating sites with effective power production using GRU estimation and advancing network sustainability and reliability. The author demonstrates a simulation analysis report of the New England 39 bus system compared with traditional methods to prove that a positive outcome has been attained by managing congestion and boosting urban energy network resilience amidst climate change.

A "Graph theory" analysis has been conducted by (Anton R 2024) to identify substantial variables considered in the decentralized renewable energy market, which highlights a profit-driven approach and political policy agreement. He suggested a brief urban planning and customer engagement by implementing critical leverage techniques such as "Need for decentralized solutions," "Profit," and "Consumer interest". This study reveals a clear outcome of quick decision-making methods for interconnecting variable energy in a single market and revaluation procedure for the transition to decentralized renewable energy.

Regional Integrated Energy Systems (RIES) have been suggested by (Hongbo R et al., 2022), highlighting the significance of thermal energy storage in augmenting resilience and reliability. The study deliberates integrating a thermal energy system as a minimum energy required to maintain the urban demand as backup energy. It provides detailed economic scheduling for operating in critical conditions. Monte Carlo Simulation-based Markov process will support and enhance reliability and control energy supply interruption owing to

a) equipment failures,
b) climatic conditions,
c) energy fluctuation, etc.

Thermal energy storage will play a highly alternative backup energy to ensure the smooth operation of urban energy planning.

(Hui Z and Xinqun F., 2024) They proposed a digital twin technology, another novel two-stage methodology for simulating tidal currents. It highlights the significance of tidal energy in renewable energy-producing methods and supporting urban energy demand. The author implements "Leveraging fuzzy feature selection" and "Long short-term memory" (LSTM) methods for predicting the accuracy of energy demand in urban life. This methodology significantly supports cities closer to offshore and onshore, which develop tidal energy based on wind flow velocity over a seashore bed and are enhanced through evolutionary optimization. The report reveals the evaluation of tidal energy, and their data were validated with simulation results to check the accuracy level and compare it with traditional techniques. It indicates the potential accuracy rate of simulation results and highly supports predicting future data for energy production based on various monsoons.

The study measures the Energy Communities' performance by transitioning from individual buildings to a common centre using evaluation approach reports (Giuseppe A et al., 2023). Implementing simulation tools such as "DesignBuilder" and "EnergyPlus" significantly optimise energy consumption along with MATLAB software. It compared three retrofit models such as a) Primary Stage Consumption (PSC), b) CO_2 emissions, and c) running costs (RC). Besides, it provides an economic indicator for the investor to evaluate initial investment for energy production setup and the public about profitability. The report declared that three retrofit models offer a piece of information to policymakers and urban planning commissions in terms of economic benefit.

- What are the critical considerations for ensuring the resilience and reliability of urban energy systems powered by renewable energy?

Several significant vital points were identified to ensure the resilience and reliability of urban energy systems of renewable energy, such as the implementation of solar photovoltaic (PV) panels, tidal energy systems, etc., along with leverage techniques such as “Gated Recurrent Unit” (GRU) and “Long short-term memory” (LSTM), “Thermal Energy Storage” (TES) “Graph Theory Analysis”(GTA) etc were considered in the study to predict the energy consumption required to meet the urban life and regulate the energy demand production based on natural scenario.

A most potential concept is integrating a thermal energy storage system with an existing renewable energy plant model to play a backup energy supply to the urban and provide detailed economic scheduling for distributing electric load commonly to both commercial and residential. It will be most suitable for cold climatic challenges conditions and improve overall system resilences.

Moreover, it enables urban planners to implement and strategically deploy renewable energy resources to regulate and control the energy distribution in decentralized energy markets. This approach improves system flexibility and responsiveness, contributing to resilience in the face of altering energy demands and supply dynamics.

Finally, comprehensive evaluation frameworks are most noteworthy in assessing energy systems' performance and economic viability. The report reveals that considering three retrofit models such as a) primary energy consumption, b) CO_2 emissions, c) running costs, and d) economic indicators, the stakeholders can make individual decisions to optimize energy requirements, system design, and operation. It ensures the long-term reliability and sustainability of urban energy systems powered by renewable energy.

Enhancing Community Engagement for Urban Renewable Energy

The “Unified Theory of Acceptance and Use of Technology” (UTAUT) concept has been implemented in their research work by (Devine Q.A et al., 2023). It presented a concept of “Empowerment Theory” that intersects within a Partial Least Squares-Structural Equation Model (PLS-SEM) framework to interconnect familiar people with each other for consuming renewable energy to meet their energy demands. A survey was conducted in northern Ghana involving 613 volunteers about adapting renewable energy in urban life. The research highly supports the policymakers in Ghana's energy landscape rather than resonates with UN sustainable development goals. This report illustrates a clear picture of the high support from volunteers about acceptance of the importance of consuming renewable energy and the impact of reducing fossil fuel consumption on environmental development.

(Tabbi W, et al., 2024) discussed a comprehensive analysis of significant challenges and possible opportunities in renewable energy generation planning in urban areas and the evaluation and performance of simulation systems for predicting energy-generated accuracy levels. It deliberates the substantial processing costs associated with an existing model to achieve precise results for attaining effective power generation strategies. The authors focus on integrating renewable energy in buildings to enhance power production and maintain sustainability and resilience in urban energy systems. Moreover, they illustrate a) cluster methodology, b) influential factors, and c) modelling techniques that help to produce accurate results attained in renewable energy. Besides, it contributes a clear understanding of urban planning and economic benefits to achieve sustainable and efficient power generation.

Community engagement plays a significant role in the adaptation of urban renewable energy. Collecting opinions from stakeholders such as the public, industrialists, etc, are meaningful influences on energy policy development and market operation. However, practical implementation and a gap in understanding public perceptions of renewable energy support in developing nations are complex. This study was conducted in Kenya by (Sydney O et al., 2020), revealing widespread support for renewable energy technologies and highlighting the status of incorporating public perspectives into policymaking processes as the most effective energy transition.

(María-J.P, et al., 2022) Examine regional variations in community engagement with renewable energy projects in Portugal, Greece, and Israel. The author reveals that utilizing this region's multi-level and multi-actor perspectives plays a critical role in implementing the energy sector in urban areas. The research examines the significance of regional contexts and place attachment in promoting social acceptance and energy transition in various stages of large-scale wind and micro-grid solar projects. This research contributes to the PEARLS research on renewable energy landscapes and community engagement.

Public acceptance of sustainable energy initiatives is vital for achieving decarbonization goals, particularly in the EU. A significant outcome was attained in the greenhouse gas emitter. This paper by (Constantine S et al., 2022) presents a comprehensive framework "utilizing multilevel regression analysis" across E.U. nations to understand the factors motivating adoption and support for innovations like a) electric vehicles, b) solar panels, etc. The research findings indicate the importance of addressing a) personal attitudes, b) socio-political dimensions, and c) national policies to enhance adoption and support, advocating for a multifaceted approach to promote sustainable energy transition.

- How can community engagement and public participation be enhanced to promote the acceptance and adoption of renewable energy technologies in urban areas?

Through several strategies, it can be enhanced to promote the acceptance and adoption of renewable energy technologies in urban areas.

Initially, leveraging technologies and their theories, such as the "Unified Theory of Acceptance", "Use of Technology" (UTAUT), and the "Empowerment Theory" can help to understand, examine, and educate different factors influencing community attitudes and preparedness to adopt renewable energy. In particular, "Tailored policy" is highly recommended to create insights that align with local needs and provide significance and encouragement about the adoption of renewable energy in greater acceptance and engagement.

Moreover, conducting awareness workshops on gaming theory among the public is an initiative to educate and raise renewable energy consumption to advance urban life. Furthermore, it helps shape energy policy, urban planning, and legitimacy in energy transition efforts. In addition, multi-level and multi-actor perspectives support fostering community engagement through participatory approaches. It empowers stakeholders to enthusiastically contribute to implementing, accepting, and decision-making processes regarding renewable energy projects. Engaging communities at various stages, such as a) planning to implementation, b) enhancing social acceptance, and c) ownership, leading to converting the process into more successful and sustainable energy transitions.

Likewise, understanding regional variations in public participation and community "tailoring engagement strategies" will address a)personal attitudes, b) socio-political dimensions, and c) national policies through a multifaceted approach that can build agreement and organize support for renewable energy initiatives at both local and national levels.

Benefits of Integrating Renewable Energy Technologies

It was observed from the literature survey that integrating renewable energy in urban areas has a considerable possibility of controlling pollution to preserve the environment, monitoring and regulating energy production and supply to meet energy demand, and more. The details of the significant benefits are classified below for clear understanding.

Environmental Sustainability

It noticed that abundant solar energy, seasonally available wind energy, and the geographical location of marine energy are the most promising renewable energy sources highly suitable for integrating renewable energy in urban. Moreover, controlling and reducing a) greenhouse gas emissions, b) mitigating climate change, and c) preserving the environment are significant project benefits.

Economic Viability

Retrofit of renewable energy systems with traditional energy will be considered a cost-saving process by reducing the running time of traditional energy production. It indicates the advancement of economic support to increase investment in renewable energy plants and long-term financial benefits.

Energy Supply and Demand Management

The main objective is to minimize primary energy consumption and manage energy supply in urban. It is a critical process integrating renewable energy to monitor, regulate, and manage energy supply and demand. Detailed data-driven technology will support the storage and regulation of the energy production from the equipment to meet energy demand during peak times.

Community Empowerment

Collecting stakeholder (local ownership) opinions for adaptation and involvement in energy initiatives about integrating renewable energy and their policy decision-making opportunities will empower successful implementation in urban. It fosters community resilience and self-reliance in the face of energy challenges.

Efficiency Improvement

It can be achieved by integrating AI and IoT techniques into renewable energy production equipment. This retrofit process enhances energy efficiency more than traditional methods, which help regulate the energy supply in urban areas. This energy efficiency contributes to overall sustainability and optimizes energy resources for the future.

CHALLENGES AND BARRIERS

It observed that urban conflicts from a0 land owners, b0 technical issues in the power grid interconnection process, c) initial colossal investment, and d) central acceptance from the public community will play significant challenges in implementing renewable energy

Case Studies and Best Practices

The study illustrates a transparent methodology, possible implementation procedure, and significant outcome regarding the potential of renewable energy and its high contribution to the energy demand in urban life advancement. Here, we listed some successful case studies and their attainment.

Copenhagen, in (Denmark)

Copenhagen's government decided to become carbon neutral in 2025 by implementing various techniques. Among them, renewable energy contributions play a vital role in power production from wind energy, district-level heating systems from solar power, constructing only energy-efficient buildings, etc. Copenhagen has ambitious sustainability goals and a firm policy for promoting renewable energy from the public community and political wings, which will make this project successful.

Freiburg, in (Germany)

Freiburg is a role model sustainable city owing to its widespread application of renewable energy technologies. The city operates by consuming renewable energy sources such as a) solar, b) wind, and c) biomass integrated with traditional energy production methods. Freiburg has policies to elevate their adoption in combination with a) innovative planning, b) community involvement, and c) commitment to sustainability principles.

Reykjavik, in (Iceland)

Reykjavik has abundant geothermal resources, which can help manage the city's energy demand. The government and policymakers rely heavily on geothermal energy consumption for heating and electricity generation. This source is making the city one of the greenest in the world. Reykjavik's success strategy is consuming indigenous renewable energy sources and investing in geothermal infrastructure.to make the environment clean and pollution-free.

Vancouver, in (Canada)

Vancouver is one of the role model cities for making vital policy decisions regarding the implementation of renewable energy sources that will support the city's energy demand. Vancouver promotes green building standards and clear policy guidelines for urban development plans. They follow a comprehensive approach to sustainability, which includes a) transit-oriented development, b) green infrastructure, and c) community engagement by monitoring AI and IoT techniques.

Stockholm, in (Sweden)

Stockholm has executed many opportunities to promote renewable energy and sustainability in urban planning. The city has constructed completely integrating centralized a) district heating systems, b) renewable energy projects, and c) energy-efficient buildings. The major success of this project is a) long-term commitment to sustainability, b) strong partnerships between three critical stakeholders such as a) government, b) industry, and c) public-civil society. They focus on implementing this innovation and technology to advance the city.

Technological Solutions for Urban Renewable Energy Integration

The significant technological solutions to enhance energy performance in urban areas are listed below.

Smart Grids

It is a communication and control technology between the power production and consumption locations. Smart grids enable efficient energy transmission between long-range and managed energy storage systems and balanced supply and demand. It facilitates two-way communication by implementing Machine learning (ML) and IoT technologies.

Energy Storage

Possible energy storage systems such as a) Batteries (Wind), b) thermal energy storage chambers (Solar), and c) pumped hydro storage (Water) play a significant role in storing excess energy produced from renewable sources. It supports and ensures a reliable and stable energy supply backup for fossil fuels.

Distributed Generation

It is an energy hub that measures and records the data from a) rooftop solar panels, b) small wind turbines, and c) micro-hydro systems for predicting energy requirements for the future by using AI and IoT techniques. Using this leverage technology, it controls and regulates excess generated energy to demanded regions. It is a decentralized system that reduces transmission losses from individual plants and enhances grid resilience using a data-driven process.

Demand-Side Management

Researchers suggested this "*demand-side management*" method to policymakers about benefits and allowances provided to consumers based on their energy consumption. This process regulates energy fluctuations and indicates correct signals to the control room for altering energy production. It provides significant customer benefits, such as installing a) smart meters, b) smart home energy management systems, and c) time-of-use pricing schemes to provide financial benefits for energy efficiency and load shifting. Besides, it is an alternate method to control and reduce peak load demand and grid congestion.

Grid Integration Software

Advanced algorithms and modelling techniques were adopted to optimize energy efficiency by integrating renewable energy into grids. These tools support and predict energy demand, grid stability, load distribution based on regional requirements, etc. This tool highly supports minimizing costs and emissions.

Energy Management Systems (EMS)

EMS platforms monitor, regulate, and optimise energy systems within buildings, regions, or whole cities. These systems integrate data from various sources, including sensors, meters, and weather forecasts, to optimize energy use, reduce waste, and improve overall energy efficiency.

Policy and Regulatory Frameworks

Well-designed policies and regulations will stimulate innovation concepts to reduce a) investment costs, b) create a risk-free environment for investors, c) target security goals, and d) accelerate the deployment of energy, which contributes to climatic condition.

Examples of Supportive Policies, Incentives, and Regulations

It observed numerous supportive schemes offered to renewable energy-producing entrepreneurs from urban planning and government agencies, such as

Feed-in Tariff (FT): Providing a guaranteed income in a month to the energy producer,
Renewable Portfolio Standards (RPS): Creating market demand for producing minimum energy production,
Net Metering: Offering a chance to sell excess generated energy to the grid with demandable prices fixed by energy producers.
Tax Incentives and Rebates: Providing tax benefits to producer businesses to support the installation of renewable energy at subsidiary benefits.
Green Procurement Policies: Strongly suggested mode by policymakers maker to purchase a high volume of energy from renewable energy sources to meet city demand.
Building Codes and Standards: Integrating renewable energy in newly constructed buildings in the city limits enhances the quality of life by reducing emissions and creating a free polluted environment.

Evaluation of Effectiveness and Areas for Improvement

It observed that the "effectiveness" of employment opportunities is gradually increasing owing to the impact of implementing a power production policy from renewable energy and controlling CO_2 emission levels. Besides, it identified possible barriers such as a) lack of financial investment, b) inconsistency in energy policy, and c) irregular permission authorities were observed to be strengthened by implementing the a) suitable fund flow mechanism and b) policy regulation bodies to monitor the process to enhance the quality of implementation.

Challenges and Barriers to Widespread Adoption of Renewable Energy in Cities

Essential challenges in implementing renewable energy in urban we found three significant categories are

Socio-Economic Challenges

Conducting awareness programs through seminars, workshops, and local communication channels will support community engagement in accepting renewable energy in urban areas. This methodology supports clean energy initiatives, promoting sustainable urban development.

Technical Challenges

The foremost challenges in technical aspects are a) Intermittent and variability, b) grid constraints, c) technological limitations in storage, and d) grid management. To address this issue, we need to focus on implementing a) innovation in infrastructure, b) well-defined grid management, and c) energy storage technologies to confirm reliable and resilient renewable energy integration.

Institutional Challenges

We found significant concerns about communication gaps between stakeholders, regulatory frameworks, and industry resistance. The need for a structure/framework to create a clear correlation between these lines will accelerate the transition to renewable energy.

Future Directions and Emerging Trends

We observed that further technology significantly supports adaptation and augments the thermal performances of renewable energy in urban.

Blockchain Technology

It is an innovative energy market platform to keep peer-to-peer energy trading, transparent transactions between production units, grids, and end users, and secure data management in decentralized energy systems.

Decentralized Energy Systems

Implementing a centralized distributed system in power generation, microgrids, and smart grids augments a) resilience, b) energy production rate, and c) power station control. Moreover, this system has a) storage, b) demand response, and c) digital technologies, optimized energy management, and sustainability.

Community-Based Initiatives

Strong recommendations to enhance community ownership include active participation and collaboration to improve local economic development. We welcome contributions that promote grassroots approaches to a) energy planning, b) innovative design, and c) collaborative governance in order to establish a well-structured energy system that supports sustainable and inclusive transitions.

CONCLUSION

In conclusion, the chapter highlights the intermingling of technological advancement, policymaking support, and active community engagement in urban by leveraging IoT and AI to improve energy production, supply, and consumption. It noticed that implementing multifaceted technology improves energy efficiency, grid stability, and environmental sustainability. It articulates a proper framework for energy planning regulations and management through integrating AI and IoT-enabled systems. The involvement of the Stakeholders' alliance in deploying IoT sensors and AI algorithms for energy optimization is challenging and empowering for energy conservation.

REFERENCES

Ahmet, S., & Demet, E. (2023). Site selection modeling of hybrid renewable energy facilities using suitability index in spatial planning. *Renewable Energy*, 219, 119458. DOI: 10.1016/j.renene.2023.119458

Ana, A., Gumersindo, F., & Maria, T. M. (2023). Advancing the European energy transition based on environmental, economic and social justice. *Sustainable Production and Consumption*, 43, 77–93. DOI: 10.1016/j.spc.2023.10.013

Anton, R. (2024). Applying graph theory to find key leverage points in the transition toward urban renewable energy systems. *Applied Energy*, 361, 122854. DOI: 10.1016/j.apenergy.2024.122854

Chibuike, P. O., Salihu, A. R., & Ugbe, C. O. (2024). A comprehensive review of recent developments in smart grid through renewable energy resources integration. *Heliyon*, •••, 10. PMID: 38327483

Constantine, S., Miguel Angel, T. R., & Muireann, Á. L. (2022). Public acceptance of sustainable energy innovations in the European Union: A multidimensional comparative framework for national policy. *Journal of Cleaner Production*, 340, 130721. DOI: 10.1016/j.jclepro.2022.130721

Daniel, I. A., Francisco, J., & Marcos, T. V. (2023). Long-term planning for the integration of electric mobility with 100% renewable energy generation under various degrees of decentralization: Case study Cuenca, Ecuador. *Energy Reports*, 9, 4816–4829. DOI: 10.1016/j.egyr.2023.03.118

Divine, Q. A., Anthony, A. D., Bright, A. G., & Festus, V. B. (2023). Does psychological empowerment improve renewable energy technology acceptance and recommendation? Evidence from 17 rural communities. *Renewable Energy*, 219, 119449. DOI: 10.1016/j.renene.2023.119449

Florian, M. N., Ali, S., Olivier, J. A., Itbar, K., Muhammad, I., Alina, C. N., Alex, B. D., & Muhammad, A. (2024). The relationship among urbanization, economic growth, renewable energy consumption, and environmental degradation: A comparative view of European and Asian emerging economies. *Gondwana Research*, 128, 325–339. DOI: 10.1016/j.gr.2023.10.023

Giuseppe A, Fabrizio A, Nicola B, Gerardo M.M. (2023). Sustainability and energy communities: Assessing the potential of building energy retrofit and renewables to lead the local energy transition, *energy*, 282, 128377.

Haiyan, W., & Yanxia, W. (2024). Smart Cities Net Zero Planning considering renewable energy landscape design in Digital Twin. *Sustainable Energy Technologies and Assessments*, 63, 103629. DOI: 10.1016/j.seta.2024.103629

Hongbo R, Zipei J, Qiong W, Qifen L, Yongwen Y. (2022). Integrated optimization of a regional integrated energy system with thermal energy storage considering both resilience and reliability, *energy*, 261, 125333.

Hui, Z., & Xinqun, F. (2024). Reliability improvement and landscape planning for renewable energy integration in smart Cities: A case study by digital twin. *Sustainable Energy Technologies and Assessments*, 64, 103714. DOI: 10.1016/j.seta.2024.103714

Ismaila, R. A., & Maher, S. A. (2023). Urban planning schemes for developing low-carbon cities in the Gulf Cooperation Council region. *Habitat International*, 138, 102881. DOI: 10.1016/j.habitatint.2023.102881

Karim, M. R., Cochez, M., Zappa, A., Sahay, R., Rebholz-Schuhmann, D., Beyan, O., & Decker, S. (2022). Convolutional embedded networks for population scale clustering and bio-ancestry inferencing. *IEEE/ACM Transactions on Computational Biology and Bioinformatics*, 19(1), 369–382. DOI: 10.1109/TCBB.2020.2994649 PMID: 32750845

María-J, P., Ricardo, I. P., & Ángeles, B. (2022). Energy transition and community participation in Portugal, Greece and Israel: Regional differences from a multi-level perspective. *Energy Research & Social Science*, 87, 102467. DOI: 10.1016/j.erss.2021.102467

Mark, M. A., Mahesti, O., & Upalat, K. (2024). Stakeholders' awareness of urban form effects on rooftop solar photovoltaic in Ghana: Implications for integrated solar energy and urban planning. *Energy for Sustainable Development*, 78, 101377. DOI: 10.1016/j.esd.2024.101377

Mei, Z., Danting, Z., & Tingfeng, X. (2024). Balancing urban energy considering economic growth and environmental sustainability through integration of renewable energy. *Sustainable Cities and Society*, 101, 105178. DOI: 10.1016/j.scs.2024.105178

Palanca, J., Jordan, J., Bajo, J., & Botti, V. (2020). An energy-aware algorithm for electric vehicle infrastructures in smart cities. *Future Generation Computer Systems*, 108, 454–466. DOI: 10.1016/j.future.2020.03.001

Rui, G., Shan, X., Mirzat, U., & Peter, M. (2024). Assessing the impact of urban planning policies on renewable energy: A case a China using the DID estimation model. *Heliyon*. PMID: 38463897

Shi, H., Li, Z., Ye, H., Luo, H., Xiong, W., & Wang, X. (2021). First level 1 product results of the greenhouse gas monitoring instrument on the GaoFen-5 satellite. *IEEE Transactions on Geoscience and Remote Sensing*, 59(2), 899–914. DOI: 10.1109/TGRS.2020.2998729

Siamak, H., Benedetto, N., Daniele, G., & Davide, A. G. (2022). Exploring the penetration of renewable energy at increasing the boundaries of the urban energy system – The PRISMI plus toolkit application to Monachil, Spain. *Sustainable Energy Technologies and Assessments*, 54, 102908. DOI: 10.1016/j.seta.2022.102908

Sydney, O., Pankaj, L., Andres, S., & Neeraj, V. (2020). Assessment of public awareness, acceptance and attitudes towards renewable energy in Kenya. *Scientific African*, 9, e00512. DOI: 10.1016/j.sciaf.2020.e00512

Tabbi W, A.G. Olabi, Enas T.S, Montaser M, Abdul H. A, Mohammad A.A. (2024). The state of renewable energy source envelopes in urban areas. *The state of renewable energy source envelopes in urban areas*,. 21, 100581.

Xianguo, Z., Yanan, Z., & Zhiyong, C. (2021). Development and research of rural renewable energy management and ecological management information system under the background of beautiful rural revitalization strategy. *Sustainable Computing : Informatics and Systems*, 30, 100553. DOI: 10.1016/j.suscom.2021.100553

Yan, Z., & Yunqi, L. (2024). Statistical Evaluation of Sustainable Urban Planning: Integrating Renewable Energy Sources, Energy-Efficient Buildings, and Climate Resilience Measures. *Sustainable Cities and Society*, 101, 105160. DOI: 10.1016/j.scs.2023.105160

Yueyan, C., Maoli, T., Jianxin, X., Linda, Z., & Jinli, Y. (2024). Data-driven approaches to achieve energy transition: Statistical analysis of demand-side management and smart decision making in urban energy hub networks. *Sustainable Cities and Society*, 101, 105150. DOI: 10.1016/j.scs.2023.105150

Zemouri, S., Magoni, D., Zemouri, A., Gkoufas, Y., Katrinis, K., & Murphy, J. (2018, September). An edge computing approach to explore indoor environmental sensor data for occupancy measurement in office spaces. In 2018 IEEE International Smart Cities Conference (ISC2) (pp. 1-8). IEEE.

Zhaiting, Y., & Youwei, J. (2024). Quantifying resilient urban energy systems: Statistical analysis of climate adaptation, renewable integration, and socioeconomic dynamics. *Sustainable Cities and Society*, 101, 105153. DOI: 10.1016/j.scs.2023.105153

KEY TERMS AND DEFINITIONS

Renewable Energy: Energy derived from renewable natural resources such as solar energy, wind energy, tidal energy, and geothermal heat energy. Its incorporation into urban areas is critical for lowering environmental impact and fostering sustainable urban development.

Smart Grids: Integrating electrical networks that use technology, such as IoT and AI, to effectively distribute energy, particularly from renewable sources. Smart grids provide for real-time monitoring and energy optimisation in metropolitan contexts.

Artificial Intelligence (AI): Implementing computer algorithms to execute technical activities that normally need human intellect, such as optimising energy use. Artificial intelligence can greatly improve renewable energy management in urban settings by automating decision-making and enhancing system efficiency.

Internet of Things (IoT): A network of linked devices that interact and share data, allowing for improved monitoring and management of energy systems. The Internet of Things (IoT) plays an important role in the smart integration of renewable energy by increasing energy efficiency and dependability.

Energy Management: The strategic management and optimisation of energy production, storage, and consumption. In the context of urban sustainability, energy management entails combining AI and IoT technologies to increase the efficiency of renewable energy sources and minimise CO_2 emissions.

Policy Regulation: The government need to implement new rules and regulations with clear guidelines that govern how renewable energy and smart integration technologies are deployed in urban areas to ensure compliance with environmental and safety standards.

Carbon Emissions: Carbon dioxide emissions into the atmosphere are mostly caused by the combustion of fossil fuels. Control and reducing carbon emissions is a fundamental objective in the endeavour to include renewable energy in urban sustainability efforts.

Sustainable Development: Development that satisfies the demands of the present current scenario of urban population and energy demand without compromising the potential of future generations to meet their own needs, largely via responsible use of resources such as renewable energy.

Chapter 15
Michaung Cyclone 2023: Enhancing Preparedness and Response Strategies for Future Natural Disasters

K Poongodi
SR University, India

Geetha Manoharan
https://orcid.org/0000-0002-8644-8871
SR University, India

ABSTRACT

Natural catastrophes like Michaung Cyclone are well-studied. The disastrous cyclone's timing and socio-economic effects are examined from a humanistic perspective. The cyclone's route, damage, and development are analysed. The chapter discusses disaster response and recovery. For future climate change, the chapter reviews catastrophe management solutions and suggests stronger, robust, and adaptable systems. This highlights the intricate relationship between environment, civilisation, and extreme weather resilience. In a dynamic environment, disaster management and climate change mitigation and adaptation are essential. Chennai's disaster management and tragedy repairs are examined together. Planning and readiness measures, gaps, and pressing issues are covered in the chapter. This chapter uses the Chennai tragedy to discuss natural disasters, climate change, and urban resilience. Adaptable, adaptive, and sustainable catastrophe management systems help cities adapt to climate change.

INTRODUCTION TO NATURAL DISASTER RESEARCH ON MICHAUNG CYCLONE

Recent years, Tamil Nadu, one of the south state of India not able to get normal rain. As per Nammazvar, the love of nature, hereafter the rain for Tamil Nadu will be only due to cyclones and the main reason is deforestation. The hilly areas like Ooty and Kodaikanal are 3000 feet above from the sea level and before British period the trees are of minimum 300feets height. The clouds from the sea are restricted in these areas and able to get normal rainfall. But these trees were cut and instead of that the hilly areas are planted by coffee, tea, cauliflower and cabbages and potatoes which are shrubs. They are no way helps to protect the clouds and not able to give the normal rain fall. Because of this kind of shrub like

DOI: 10.4018/979-8-3693-5748-4.ch015

plants not able to get normal rainfall and erosion also not able to controlled and leads to landslides. The flow of perennial river Cauvery also affected and leads to the interstate issues. But Godavari and Krishna are not affected this much and still now able to maintain the perriniability because in those areas the deforestation does not takes place.

During summer Chennai capital of Tamil Nadu is seeking of drinking water and for water supply, the available schemes are Veeranam old and new projects, Telugu-Ganga river project, sea water desalination projects with the investment of multiple crores. But during rainy season 6 TMC of water is not stored properly and simply reached Bay of Bengal. The daily demand of water for Chennai is 1200 mld (Million Litres per Day) but able to receive only 985mld from various resources and they are desalination plants at Nemili, Minjur, Cauvery water from Veeranam lake, Krishna water from Andhra Pradesh. The reservoirs are also used to get water and they are Poondi, Redhills, Chembarambakkam and Sholavaram. But these reservoirs area and volume are reduced on daily basis by increase in population and made into plots with the approval of the Government bodies. If the 6TMC (Thousand Million Cubic feet) of rain water is properly handled then the flood and the drought are able to handle easily.

CHENNAI: TOPOGRAPHY AND GEOLOGICAL FEATURES

Chennai Location

Chennai, the old name was Madras is a metropolitan city at southern part of India, capital of Tamil Nadu State and the location is Latitude 13° 4' 2.7804" N and Longitude of 80° 14' 15.4212" E. It is the major centre of culture and education in India. The population of the Chennai in the year 2023 is around 1.2 crore. It is located in the Coromandel Coast and affected easily by cyclones. It has beautiful Marina Beach of 6 km Length. The mean sea level is 60 m. This is the Gateway to South India and weather is hot and humid throughout the year. It is one of the important tourist place located on the shores of Bay of Bengal. It is the biggest industrial centre in South India and also called as "Detroit of India" for its automobile industry.

Chennai, the old name was Madras, which is situated on Bay of Bengal of eastern India and is the capital of Indian state Tamil Nadu. The area of the city is 426 sq.km. The elevation is 6.7 m from the sea level. The normal weather condition is 25 degree celsius to 32 degree celsius with 79% humidity. The topography of the city includes coastal areas, rivers, canals, lakes with plains. It has the long coast line with stretch of around 19 Km. the popular beaches are Marina, Elliot and Covelong beach. It is one of the metropolitan cities in India and also the "Gateway to South India". The Chennai topography is almost flat in the central and northern part are influenced the urban development (Satpathy, et al., 2024) of the city. The city has the mixture landscape of residential, commercial and industrial areas. The ecological system of Chennai is due to the estuaries and backwaters where the rivers meet the Bay of Bengal. The annual rainfall is out 1250mm and most of the rain is due to North East Monsoon instead of South West Monsoon.

Upto 60% of the rainfall for Chennai and the east coast areas are due to North-East Monsoon between Octobers to December. The lowest recorded temperature is 13.9 degree celcious on December 11th of 1895 and 29th January of 1905.The low lying areas are prone to flooding during heavy rain. During 2015 heavy rainfall happened at Chennai due to El Nino effect. As per 1950 to 1955 data, there are 16 big canals, 500 lakes, 3 big rivers are at Chennai. Present scenario is the lakes are occupied with apartments, width of the rivers are reduced and used for other purposes. The Pallikarani Lake is the heart of Chennai and in the middle of the lake road construction is there. The canals are used as drainages. On December 4th and 5th due to cyclone Michaung heavy rainfall happens at Chennai and surrounding areas. Heavy rains cause city flooding, transport delays, and flooded suburbs. Heavy rain and strong winds are forecast from the intensifying cyclone. A storm caused heavy rains, uprooted trees, and cut power lines, flooding parts of India's southern coast. The heavy rains flooded streets and it knocked the doors of residential and people remained indoors.

Water Bodies in Around Chennai

Rivers of Chennai

Cooum River is 64 km in length and once upon a time it is the "Thames of South India" but now totally contaminated. It is 42 km in length and polluted. The Chennai River Restoration Trust (CRRT) takes revamp measures to reduce the pollution. The lesser polluted one is Kosasthalaiyar river and has the length of 136km.All the above rivers are mixed with the waste water and polluted and will be treated as dead.

Canals

The Buckingham canal is a man-made canal constructed during British period which runs parallel to the Coromandel Coast for the purpose of navigation and irrigation but nowadays it is used as drainage channel and storm water channel. Otteri Nullah runs through the heart of Chennai and used to prevent waterlogging in the central part of Chennai.

Lakes

Chennai has lot of lakes. Nearby Chennai, the Chengalpet district is called as land of thousand lakes. The important lakes at Chennai are Chembarambakkam at west, Puzhal at northwest, Sholavaram at north, Kanathur, Ambattur, Velachery, and Kumuthavalli. These lakes are the major sources of supplying drinking water to the city.

REASONS FOR FLOOD

Floods, the most common natural calamity, occur when water floods dry land. Heavy rainfall, quick snowmelt, or tropical cyclone or tsunami storm surges in coastal areas create floods. At an average speed of 9 feet per second, flash floods are the most common and possibly life-threatening type of flood. After storm waters recede, severe threats remain. After Cyclone Michaung, Chennai residents faced rain-related infections and unmet medical requirements (Sethuram et al., 2024). Many people got sick, and those with chronic illnesses couldn't get medicines because many areas were shut off from the city due to the floodwaters.

The basic health clinic in Thiyagaraya Nagar is seeing a consistent influx of patients with fever, nausea, and loose stools. Since the centre is fixed and known to people, many visit for medical help, according to the medical personnel. The floods can lead to various communicable diseases, damage the infrastructure and property, damage any plants or wildlife that come in its path and contaminate habitats. The types of flood are flash, coastal, urban, ricer and ponding flood.

The reasons for flood are as follows

- Heavy rainfall in minimum duration resulting from tropical weather disturbances.
- Deforestation.
- Improper agricultural practices.
- Inadequate design of drainage channels and structures.
- Inadequate maintenance of drainage facilities, blockage by debris brought by flood waters.
- Construction of settlements in flood plains.
- Rains, whenever there are more rains than the drainage system can take, they can cause floods.
- Rivers can overflow their banks to cause a flood.
- Strong winds in coastal areas.
- Dam breaking.
- Ice and snow melts.
- Emission of Greenhouse Gases.
- Thick clouds
- Climatic changes
- Others

Flood at Chennai

Chennai has seen floods before; in fact, the city was once impacted by a tsunami. Chennai is currently prepared for significant rainfall and possible flooding due to the early onset of the northeast monsoon season in 2024. The city has already made considerable preparations for the monsoon, which is predicted to arrive between mid-October and November. Authorities have erected community kitchens, relief centres, and rescue boats in locations that are prone to flooding, particularly in areas like Kosasthalaiyar and Kovalam. During the southwest monsoon, the city had record rainfall, which sparked worries of prolonged downpours.

Figure 1. NDRF team during disaster 2015, 2017 &2020 flood

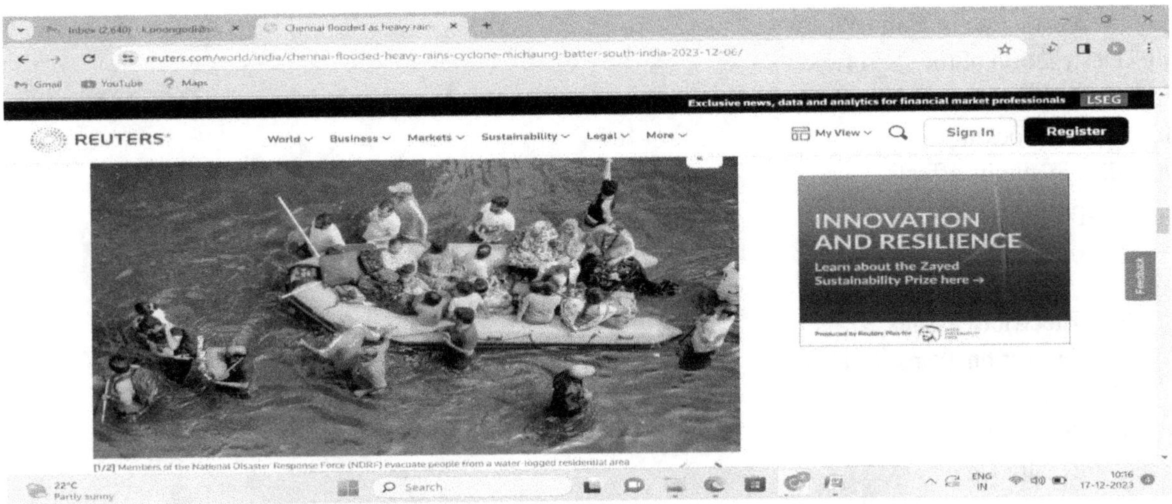

During December 2015 and 2017 the Chennai was flooded. 2015 was treated as heavy when compare with 2017.The reason for the flood is heavy rainfall in minimum duration. The city and the neighbouring areas also affected seriously. The Adyar and Cooum Rivers got flooded due to excessive rainfall, which submerged the low lying areas.

2023 Flood

The Chennai and the surrounding areas got nearly 400mm of average rainfall between the 4th December and 5th December of 24 hour period due to cyclone Michaung. The Adyar and Cooum river along with the lakes are over flowed due to the excessive rainfall and submerging the low lying areas. The roads and streets are flooded and the residential, apartments are seems like island with full of water. The National Disaster Response Force (NDRF) evacuates Chennai residents from a waterlogged residential area after storm Michaung rains in December 2023. Natural disasters (Manoharan, et al., 2024; Ganesan, et al., 2023) like floods can cause significant damage to lives and property. Rainwater collects, flooding populous areas. They can also kill many. It can destroy an entire village or city.

Types of issues:

a. Water
b. Damages for infrastructure
c. Announced disaster
d. Media and disaster
e. Electricity issues
f. Impact on transportation
g. Cargo and industrial area
h. Reptiles
i. HC

j. Infections and challenges
k. Armed Forces
l. Accumulation of Solid Waste
m. Destruction of the Agri field
n. Contamination
o. Water Blocking
p. Plastic and its effects
q. Property Damage

a. Electricity issues
c. Impact on Transportation
d. Cargo
e. Reptiles
f. Health Care
g. Infections and challenges
h. Armed forces

a. Water

During flood, the life threatening is water and water only in the form of flash flooding and sudden increase in the level of water level upto more than 10 feet. Canals, lakes, rivers are overflowing with the storm water and recalls the memories of Tsunami. The leading constructors constructed the multi storied apartments and sold to public by the name of catching word "Lake View" which are nearby lakes only. The fact is that lands are once upon a time lakes only. The example is Purvankara projects apartment nearby Pallikarani lake. It is a group of apartment with number of blocks. In one luxury apartment the over flowing flood destruct the compound wall and enters into the apartment like Tsunami and all the cars are floated irrespective of BMW and Audii cars. All the cars are damaged and few cars are not able to find out. As per estimation the loss is 100 crore due to damage of cars only in that particular area. Because of lake view lot of lakhs are lost. Flood water is mixed with pollutants, chemicals, plastics with debris and contaminating the water sources.

b. Damages for Infrastructure

This flood damages all the infrastructure of the Chennai like buildings, roads, railway routes, airports, harbor, water treatment plants, and sewage systems. It affects all the critical infrastructure and disrupting the supply of clean water and leading to sanitation issues. All the sumps are contaminated and people suffered a lot for getting good water even though they all fully surrounded by water. Most of them want to collect the rainwater at the top floor and survive. The household items like refrigerator, washing machine, Television, wooden cot and bed are kept outside in the heavily flooded area like Velachery. Each house the property damage is minimum of one lakh without considering damages for the cars.

c. Announced Disaster

So far: A super-cyclone named Michaung (pronounced mig-jaum) hit Nellore in Andhra Pradesh on December 5. In north Tamil Nadu, 150-200 mm of rain fell a day earlier. IMD observed a well-marked low pressure region in the southwest Bay of Bengal on November 29. By November 30, it would be a depression, then a deep depression on December 2, and a cyclonic storm on December 3. After that, the IMD predicted that the system would move north toward coastal Andhra Pradesh, bringing rain and severe gusts to north Tamil Nadu on December 3 and 4. The cyclonic storm was forecast to traverse land between Nellore and Machilipatnam with 80-90 km/hr winds gusting to 100 km/h.

d. Media and Disaster

The role of mass media on the latest cyclone in Chennai has been significant in providing updates and coverage of the disaster. News channels and social media platforms have played a critical role in disseminating information to the public, including live updates, rescue operations in progress, and tips for staying safe. The mass media has also been instrumental in mobilizing resources for the relief efforts and providing support to the affected communities. Various organizations and government agencies have provided information through press conferences, press releases, and social media updates. Overall, the media has played a vital role in creating awareness, galvanizing support, and providing critical information during the latest cyclone in Chennai. Chennai and flood sound strange together. To explain its oddity. From October, the North East monsoon brings rain to Chennai. Every year, low pressure bands above the Bay of Bengal turn around and impact Andhra Pradesh, Orissa, and West Bengal instead of Tamil Nadu. These regions have annual flooding, while Tamil Nadu suffers droughts. Just when the city was recuperating from the last onslaught, it rained nonstop for two days, overwhelming the drains and bringing the city to a halt on December 1. The entire city lost power for 2–5 days this time! Mobile networks became weak and landlines were cut, leaving the city dark and isolated. Never before had Chennai encountered such a devastating event, where people were utterly disconnected for so long. Without social media, the city would have been in the dark in many ways.

For the first few days, TC news and newspapers were unavailable due to low current and water. Due to poor radio transmissions, smartphone social media became the most popular way to keep up with municipal events. Even with restricted access, a charged phone was so important that power banks were supplied alongside food, water, and other relief materials. Skymet weather forecasts and Twitter updates on how to commute via water-receded roads. Social media helped people connect with loved ones When the cell signal improved, everyone tried to contact their loved ones by posting and commenting on Twitter, Facebook, etc., which happened instantly compared to message, allowing communication before the signal turned off. WhatsApp calls had lower drop rates than conventional calls, helping people stay in touch. Facebook's "Chennai Flooding safety check" lets you let friends know you're okay.

The Role of Social Media in Chennai Flood Relief Efforts

Social media enabled people to stay in touch with family and friends when normal communication channels were disrupted due to weak signals and power cuts. Platforms like Facebook facilitated the 'Chennai Flooding safety check' feature to let friends know about their safety.

Figure 2. Safety check during disaster through Facebook

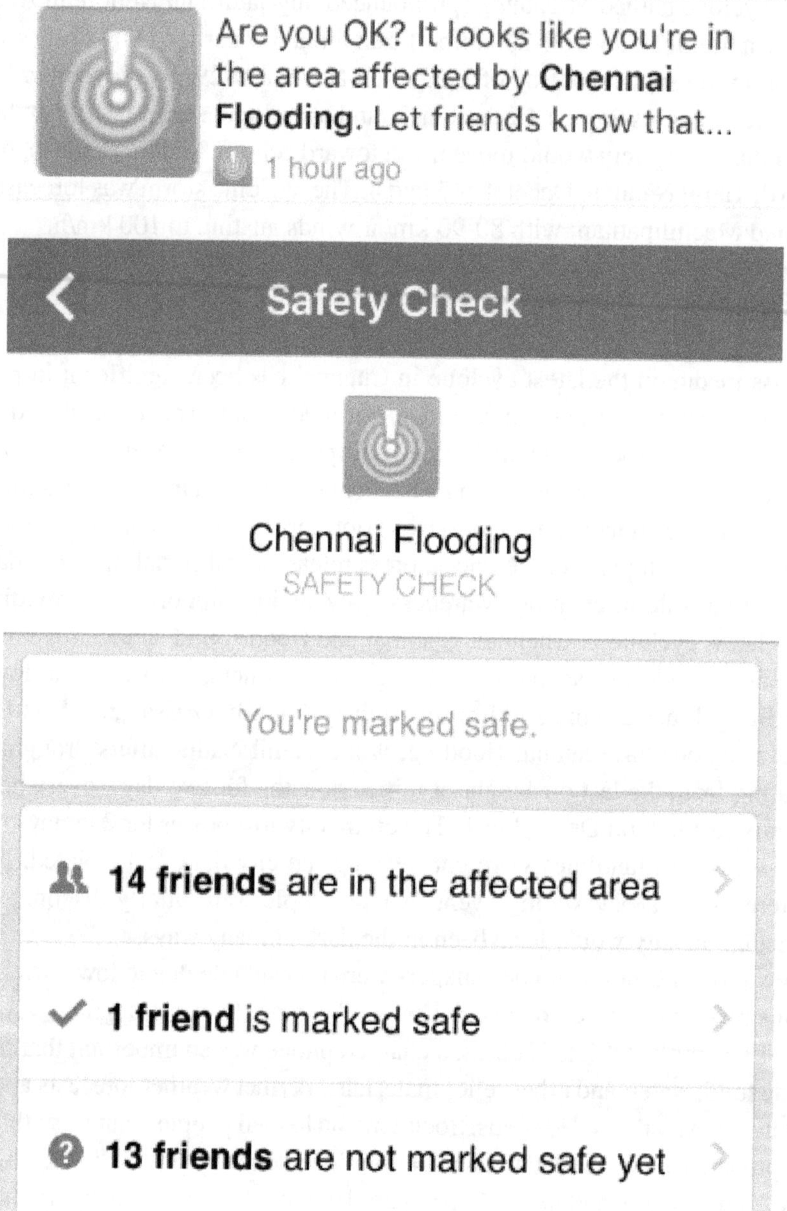

Figure 3. Graphical representation of relief efforts by Indian Navy

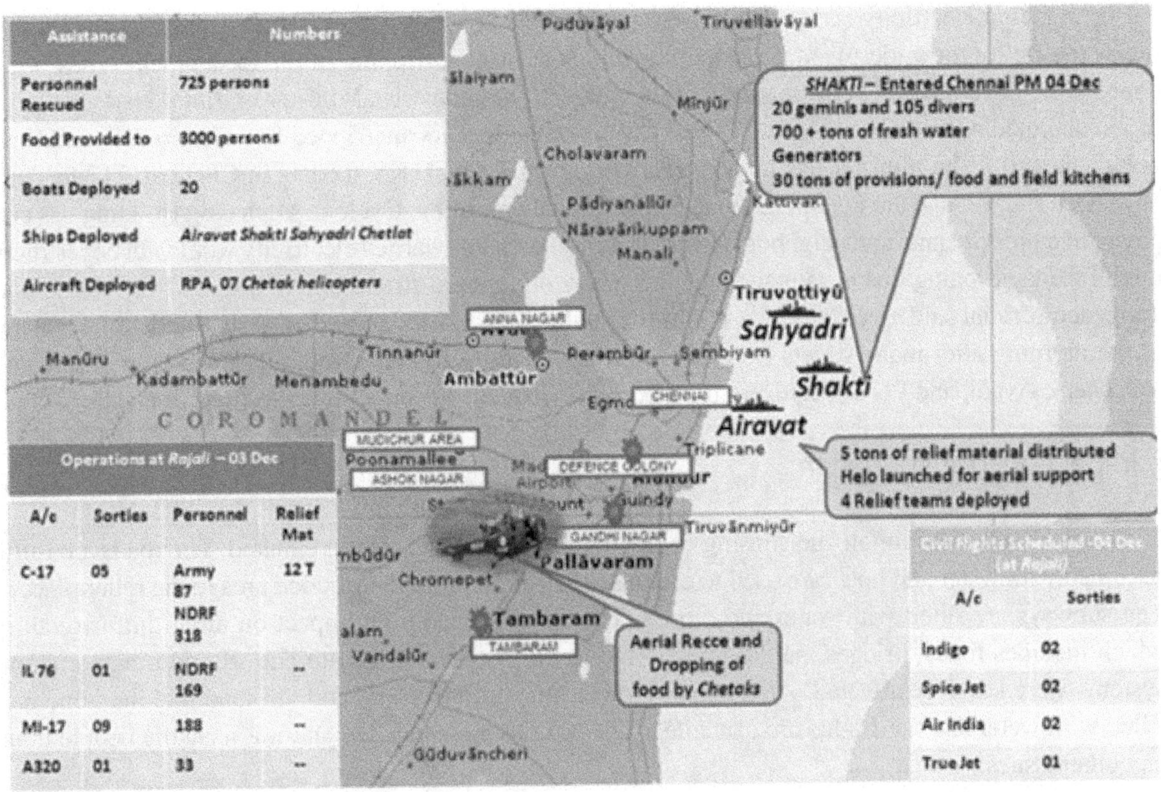

A GRAPHICAL REPRESENTATION OF RELIEF EFFORTS BY INDIAN NAVY DURING CHENNAI FLOODS

The people of Chennai faced lot of issues because of the Flood and, they are listed as follows

There were widespread property damages caused by the latest cyclone in Chennai, India. The cyclone hit the coastal region of Chennai on December 1, causing severe flooding, wind damage, and power outages. The damage was reported to be extensive, with many residential and commercial properties being destroyed or severely damaged. The cyclone impacted both low-income and high-income neighborhoods, with reports of property damage ranging from minor to major. It is recommended that property owners in the affected areas take immediate steps to assess and report the damages. In a statement, Association of Indian Entrepreneurs national President K. E. Raghunathan estimated that the massive rains that lashed Chennai and adjoining areas will cost over Rs.11, 000 crores. Most of the 2,100 units at Vichur, Ambattur (north phase), Perungudi, Thirumazhisai, Thirumudivakkam, and Villivakkam industrial estates were flooded days after the storm, damaging machines and finishes.

e. Electricity Issues

Recently, there have been electricity issues reported in Chennai due to a natural disaster. Power outages and blackouts have been reported in various parts of the city, causing in convenience to residents and businesses. The issue may be caused by damage to power lines or infrastructure, and Metroom officials are working to restore power as soon as possible. The affected areas are currently experiencing supply cuts, and residents are advised to be patient as electricity services are restored. As mentioned earlier, Chennai has been facing electricity issues and power cuts because of the 'severe cyclonic storm' Michaung. A major reason for the widespread power shutdown is the manual tripping of the feeders which has been done by the government to ensure safety of the people. The Industries Minister of Tamil Nadu has said in a social media (Kumari, et al., 2023) post, "Most of the feeders are tripped for safety reasons and are being charged." The minister has highlighted that out of 1,814 11KV feeders in Chennai, 712 feeders have been shut off and are not operational. Since Monday morning, Cyclone Michaung flooding has cut power in major Chennai and neighbouring areas. Power officials warned electricity would not be restored until Tuesday evening due to rain and stagnant water on internal streets and major roadways. A senior Tangedco official said electricity was cut in the south, west, north, and other parts of central Chennai in the morning after many streets flooded. Rainwater also infiltrated several substations in Taramani, Velachery, Avadi, and Porur, requiring the electricity officials to cease power to prevent electrocution.

f. Impact on Transportation

Because of heavy rainfall and causing of flood, the transport services are halted. The roads are fully occupied with water and boats are used to evacuate the people from the flooded area to the relief places. The subways are filled with water and closed. The flooding had high impact on urban infrastructure which includes roads, bridges, public transportation. Most of the areas are cut off and transportation systems were severely affected. At Chennai airport, the entire runway and airplane parking zone are filled with water and the flights are cancelled. Local trains are cancelled and it affects the people than any other issues.

g. Cargo and Industrial Area

(1) Effects on Industrial Area

The Chennai's industrial landscape is reeling under the bas taking effects of the cyclone which recently happened during the first week of December 2023. During the times of disaster it was the large volume of factors of production that got wasted. There are more than 110 companies in the Chennai that got its planned in the chemical and paint industry. During the end of the first week of December, the flood waters standing at an alarming height of 10 feet at many points. The submerged industrial units which employed thousands of workers and it has got the damage of rupees 160 crores. In few areas it was the hardest hit suffered with the loss of rupees 3 crores. Approximately 25 to 30 thunds of zinc Ash intended for recycling and it value that one and half crore rupees during this flood.

During this natural disaster the expensive electrical machinery acute with sensors has also been rendered useless by the flood waters. Anyways this is not Chennai's first flood but still the damage made is very huge. For many of the companies Revival is far from certain. Adding to this the absent of Electric Supply adds to the grim scenario casting a shadow of uncertainty over the companies Revival. After a. of the affected industries are also the IT industry (Deviprasad, et al., 2023) that has its impact which costs huge. Commercial establishments and industrial units operating in Chennai has also got its greater impact.

Among the bigger players it have taken the hardest hit with flooded facilities keeping employees out for the second day running where the IT companies have predominantly switched to work from home modes thinking the natural disaster scenario (Hanspal et al., 2024). Water logging is found in the larger companies and manufacturing units which has cost minimal damage when compared to the smaller industry and SMEs (Jaichandran, et al., 2023). Almost all the manufacturing industries in the Chennai region suffered extensive damage. The water logging has damaged crores worthy of machineries and materials in many manufacturing units. It has taken a minimum. of three weeks to Resume the operations in its own places. The chief minister has given the compensation of rupees 1000 crore for this cyclone.

With the Chennai airport entirely closed owing to heavy rains and unprecedented floods, 60 km away from Chennai, INS Rajali, Naval Air Station at Arakkonam, was immediately opened up for operation of flights carrying essential supplies and relief material. Under the special arrangements, national carrier Air India was the first to land an Airbus A320 carrying passengers from Hyderabad at INS Rajali at Arakkonam. Simultaneously INS Rajali, which has the longest runway in the country was transformed into hub of the relief operations in 'Operation Madad' and was increasingly used to operate commercial aircraft. INS Rajali was also used to fly in the National Disaster Response Force (NDRF) teams. Private airlines also joined the rescue operations in rain-battered Chennai, with some of the carriers transported hundreds of people from INS Rajali to Delhi, Hyderabad and Bengaluru. Over 600 people including more than 130 university students and an old French couple were taken to safety from Arakkonam Naval Base by a coordinated team of defense and NDRF personnel. Additional three warships carrying small boats, divers and other equipment were quickly deployed by the Eastern Naval Command, Visakhapatnam, to Chennai. The Indian Navy launched its relief efforts by providing fresh water and food packets to the people who were stranded for long due to devastating floods. Naval divers and rescue teams were deployed on fleet of Gemini boats to the inaccessible areas to transfer people to safe locations. Medical teams were also deputed to provide immediate succour to the stranded people. Amphibious ships were deployed for taking relief to the coastal areas for distribution the affected areas. The flood damaged almost 1000 crore of cargo at North Chennai itself.

h. Reptiles

During recent natural disasters in Chennai, such as Typhoon Hanna and Cyclone Ockhi, several reptiles were rescued and made safe by local authorities and animal welfare organizations. These included snakes, lizards, crocodile and turtles that were found in flooded areas or abandoned homes. Some of the reptiles were taken to wildlife sanctuaries or turned over to trained handlers who could transport them to safety. It is important to monitor the well-being of reptiles during and after natural disasters, as they can be vulnerable to stress and injuries.

i. Health Care

The role of healthcare on the latest cyclone in Chennai is to provide medical assistance to those affected by the disaster. This includes setting up medical camps, providing food and shelter, and administering vaccines. Doctors, nurses, and other healthcare providers are working tirelessly to save lives, treat the injured, and prevent the spread of disease. The government and non-governmental organizations are also playing a vital role by providing funding and resources to support the healthcare efforts.

It is important for people to stay informed about the latest updates on the situation and to support the healthcare workers in their efforts.

Figure 4. Medical camp and medication for the needy

Breathing issues and related symptoms are monitored in the medical camps and medication is prescribed after which monitoring of the illnesses too will be attended in the camps Pic Courtesy: Greater Chennai Corporation/Twitter

j. Infections and Challenges

In the last seven weeks, 16,516 flood-prone medical camps treated 7, 83,443 patients, according to the Tamil Nadu Department of Public Health. On December 16, medical camps in Chennai and nearby regions began giving rubella vaccinations to children aged nine months to 15 years. Greater Chennai Corporation (GCC) reported 3,270 fever cases and 1,972 cold and fever cases treated in the city. GCC is also giving each family in the four Cyclone Michaung-affected areas half a kg of bleaching powder to clean overhead and ground water tanks. Due to contaminated water and floods, water-borne diseases, vector-borne diseases like dengue, respiratory illnesses, colds, fevers, and allergies have increased. "A lot of cases of diarrhoea and vomiting were reported this week in all of the camps I worked. We do give preventive care and aftercare for every patient visiting the camp; in case of any symptoms that need more attention, we refer the patients to government hospitals," says HemaRosi, a staff nurse at a T Nagar medical camp (Kumar, et al., 2024). Stationed at the health centre of GCC's Department of Public Health during the day shifts, Hema awaits calls from camps. When staff are required, camp medical officers call for assistance.

Medical officer SivamJayraj details about health concerns of various age groups and the potential risk factors (Tyagi et al., 2024) in the coming weeks. "There are medical camps being arranged in schools as there are reported cases of various illnesses, especially stress-induced symptoms among school-going kids. At all camps, we provide necessary vaccinations, he says. Jayraj said that minor to large mishaps must be treated seriously and that metal diseases, sewage, or contaminated water will require medical attention at the camp. Nurses who provided vital healthcare support, emphasises, "Access to regular medications was hampered during the floods. Many individuals, especially the elderly faced challenges in managing their health. Our focus extended beyond immediate treatment to providing guidance on post-flood care and hygiene practices." Here are some post-cyclone cleanup health tips: Protect yourself from polluted water and debris by using gloves, boots, and masks while cleaning.

- Safely dispose of broken objects by sorting rubbish.
- Boil water before drinking to remove pollutants.
- Alternatives to boiling include water purification tablets and filters.
- Clean and disinfect water storage containers regularly.
- Discard flood-damaged perishable food.
- Clean and disinfect kitchen surfaces, utensils, and cookware.
- Eat only well-cooked meals to avoid foodborne infections.
- After touching floodwaters or hazardous surfaces, wash hands with soap and water.
- Cover cuts and wounds to prevent infections and visit the nearest camp/health centre.
- Use masks to avoid dust and mould inhalation during cleanup.
- Cleaning products must be ventilated to avoid respiratory concerns.
- If post-cyclone difficulties overwhelm you, seek help.
- Reduce stress and stay social.
- Provide clean water and hygienic living circumstances for children.
- Follow their health and seek medical attention for any symptoms.
- Maintain vaccines.

k. Armed Forces

North Chennai residents wade through floodwater at Sathyamoorthy Nagar, Vyasarpadi, after storm Michaung dumped heavy rain on December 4, 2023. (The Hindu Dec10, 2023)

Figure 5. Dumped Chennai

The Indian Air Force (IAF) dropped aid in cyclone-stricken Tamil Nadu in the first week of December. IAF stated rescue material was dropped near Adyar and Chennai Harbour. On Tuesday, authorities in Tamil Nadu reported eight deaths from the strong cyclonic storm Michaung, which flooded arterial roads and subways in Chennai and other major towns. In the evening, two IAF helicopters delivered food packets and rations to residents stranded in cyclone-hit Chennai regions cut off by water. IAF officials in New Delhi said more sorties would distribute relief supplies in other cyclone-affected districts. The entire operation is coordinated with Tamil Nadu State Government and Air Force Station, Tambaram. The officials claimed pregnant women, children, and the elderly were saved from dangerous regions, while families were evacuated. IAF Chetak helicopters are helping with flood relief, officials said. At least 162 aid centres operate in the state, 43 in Chennai. Flights at Chennai airport restarted after the cyclone halted them on Monday.

(2) Accumulation of Solid Waste

More than thousands of residents are affected by the flood. All the residents are affected seriously. The flood not only causes damages to properties but leads to the accumulation of solid waste also which includes the damaged cars, two wheelers, furniture, refrigerators, AC machines, washing machines, mixer grinders and plastics.

m. Destruction of Agriculture

PWD officials said 101 of the 369 irrigation tanks in the district have achieved full capacity, especially in farming areas like Nemili, Arakkonam, Sholinghur, and Walajah.

In Chennai, 369 tanks are maintained by the Public Works Department. Among these 101 tanks reached their full capacity and affected the agricultural land directly in the surrounding areas of Arakkonam. The flood water submerged the farmland, destroyed crops (Shaik, et al., 2023), and had a high impact on the reduction in productivity. The deposition of sediment also affects the top soils. More than 1600 acres of agricultural land and 600 farmers are affected by this cyclone. The affected areas Nemili, Thimiri, Arcot, Kaveripakkam, Wallajah, and Sholighur. The affected crops due to the inundation are paddy, banana, groundnut, and others. The horticulture was also affected seriously and the crops are brinjal of 50 acres, Bhendi of 65 acres, and green chilies 40 acres.

n. Contamination

The cyclone Michaung has brought a different issue for the Ennore people which is 25 km away from Chennai. The flood water was mixed with the oil in the refinery and spread over 20 Square kilometers. The fishery business is fully affected since the caught fish has the smell of crude oil (Mishra & Mohapatra, 2023). For other area people water is the only problem but for Ennore people, the water is mixed with crude oil. All electronic appliances and documents are damaged due to the contaminated water and removal of the sediment of crude oil will not be an easy job for the Ennore people.

o Plastic and its Effects

During the heavy floods caused by the cyclone, it has taken all the river banks to a cleaner look. The bulk generators were spotted dumping heat of waste and construction waste along with the river inside the city. The people living in the area under social activists (Shaikh, et al., 2023) pointed out the lack of boundary walls along the entire stretch of the river and it was seen that the poor and enforcement as the main reason for the dumping of waste and the shell of the river. The cyclone has damaged the stringent action against the river which is polluted. Because of this, the various areas were not able to see with their boundary walls. This seems to be like an open invitation given to the common people to dump their ways anywhere they see.

The water management expert says that people should continue to take it before they can dump the waste. It is said that that should be a behavioural change in this aspect. It is said that the riverside aspects of the dumping of waste should not be continued to see the clear and smart city (Manoharan, et al., 2023). If this continues then it is sacrile activity consideration taken by the municipal sets the expert. More than 100 tons of waste which includes plastic bottles slippers, come-out clothes, and many other items were collected by the Corporation during the clearance of the place. The waste was washed way by the river to the sea during the times of the lord and eventually was thrown back to the sea in the coastal areas said the expert. During the second week of December, the expert said that the corporation had finished removing the garbage which was near the beaches and the sea which also included many other items predominantly it was plastic items which is harmful to nature.

p. Insurance

Experts recommend a few things to consider while handling claims for flood-damaged automobiles, which are covered by comprehensive motor insurance. In the neighboring districts and in Chennai the latest cyclone has resulted in damage to multiple kinds of motor vehicles. It took more than two weeks to start working with the insurance companies regarding the clays that have to be taken care of for the motor vehicles. Comprehensive car insurance covers flood damage, with zero depreciation and engine protection suggested. The biggest insurance Broking Company manager said serious claims often result in entire losses. They claimed insurance claims might be filed if parked automobiles are washed away by a natural calamity. This kind of comprehensive coverage of the motor insurance policy both on damage to the insured vehicle and third-party liability is acid by the manager.

Repay the car owner for repairs or replacement under own damage coverage. This coverage usually applies whether the car was part of or running during floods. However, few other automobile insurance policy exclusions could effect claim payment. Instant some policies may not cover progressive deterioration or wear and tear, and Din Mein Ham Sab limits for tyre and electrical component damage. The manager added that if the flat damages the car, the park must follow the climbing method in the motor insurance policy. Insurance companies will always push filing a police complaint, documenting the FIR with images and videos. Notifying the insurance quickly was also crucial. The claim amount is supported by authorised repair shop estimates. All insurance surveyors inspect and gather data on damaged vehicles.

Admiral R.K. Dhowan, Chief of the Naval Staff, stated the Indian Navy has deployed many ships and tankers near Chennai to increase rescue efforts and open its base as an alternative airport as part of 'Operation Madad'. "We are using innovative methods like carrying out surveys using unmanned aerial vehicles to identify waterlogged areas and sharing the information with the civil administration to help them in rescue work," he said. To augment the relief efforts in the marooned areas aerial route was also adopted with several Chetak flights pressed into service from INS Adyar, Tambaram to airdrop food packets and relief material on the rooftops of the marooned citizens. INS Shakti and INS Sahyadri, two more ships from Visakhapatnam's Eastern Naval Command, brought 100 tonnes of relief items and seven lakh gallons of packaged water to Chennai. Ships carried planes, 25 inflatable boats, and 108 divers for special rescue.

DISASTER SUPPORT SYSTEM FROM PUBLIC

Weather Blogging Community and Crowd-Sourcing Efforts

A weather blogging community provided regular weather forecast updates and dispelled rumors during the floods. Crowd sourced efforts were put into mapping inundated roads, water-logged points, and flood relief camps in Chennai.

Global Help and Donations Through Social Media

People from other cities and around the world provided help through social media through online (Shameem, et al., 2023), coordinating efforts for Chennai relief with hashtags like #chennairains and #chennaifloods. Websites like Paytm and others became the go-to places for donating money towards flood relief efforts.

Awareness of the Power of Social Media

The Chennai floods made people aware of the vital role of social media in disaster management, prompting a reconsideration of the significance of maintaining and building networks beyond close friends.

CONCLUSION

In conclusion, the Michaung Cyclone of 2023 has underscored the importance of advancing our preparedness and response strategies to face future natural disasters. The lessons learned from this catastrophic event have been pivotal in shaping a more resilient approach towards disaster management. Recognizing the role of technology, community engagement, robust infrastructure, and proactive governance in mitigating disaster impacts, it is crucial that we continually refine, market digitally (Tripathi, et al., 2023) and evolve our strategies. The goal is to not only respond to these disasters effectively but also to mitigate their occurrence, where possible, and build capacities to bounce back swiftly and efficiently with evolving technology of artificial intelligence (Manoharan, et al., 2024; 2023). Through comprehensive planning, iterative learning, and collaborative efforts, we can ensure that we are better prepared for future natural disasters. The Michaung Cyclone 2023 has taught us that preparedness is not a choice, but a necessity in our rapidly changing world. We hope that the insights shared in this chapter will contribute to strengthening disaster management strategies globally and help build a safer, more resilient future for all.

REFERENCES

Deviprasad, S., Madhumithaa, N., Vikas, I. W., Yadav, A., & Manoharan, G. (2023). The Machine Learning-Based Task Automation Framework for Human Resource Management in MNC Companies. *Engineering Proceedings*, 59(1), 63.

Durai, S., Manoharan, G., Priya, T. S., Jayanthi, R., Razak, A., & Ashtikar, S. P. (2024). Quantifying the Impacts of Artificial Intelligence Implementations in Marketing. In Smart and Sustainable Interactive Marketing (pp. 120-144). IGI Global.

Ganesan, S., Manoharan, K. G., & Periyathambi, E. (2023). Utilising Deep Convolutional Neural Networks for Classifying Fire Disasters Through Surveillance: An Indoor and Outdoor Perspective to Predict Man-Made or Natural Disaster. *Revue d'Intelligence Artificielle*, 37(5).

Hanspal, M. S., & Behera, B. (2024). The Disaster Management Act, 2005: A Critical Review. *DME Journal of Law*, 5(01), 42–53. DOI: 10.53361/dmejl.v5i01.06

Jaichandran, R., Krishna, S. H., Madhavi, G. M., Mohammed, S., Raj, K. B., & Manoharan, G. (2023, January). Fuzzy Evaluation Method on the Financing Efficiency of Small and Medium-Sized Enterprises. In *2023 International Conference on Artificial Intelligence and Knowledge Discovery in Concurrent Engineering (ICECONF)* (pp. 1-7). IEEE.

Kumar, P., Maurya, N., Keerthiraj, S. H. K., Manoharan, G., & Bharti, A. (2024). Evaluating the Impact of Healthcare 4.0 on the Performance of Hospitals. *Human Cancer Diagnosis and Detection Using Exascale Computing*, 1, 1.

Kumari, S., Kumar, V., Sharmila, A., Murthy, C. R., Ahlawat, N., & Manoharan, G. (2023, August). Blockchain Based E-Analysis of Social Media Forums for Crypto Currency Phase Shifts. In 2023 5th International Conference on Inventive Research in Computing Applications (ICIRCA) (pp. 1222-1225). IEEE. doi:DOI: 10.1109/ICAISS58487.2023.10250754

Manoharan, G., Ashtikar, S. P., Smitha, V., Sundaramoorthi, S., & Krishna, I. M. (2023). Work-Life Balance Perceptions of Women in the IT and ITeS Sectors in Kerala: A Research Study. *Journal of Pharmaceutical Negative Results*, •••, 3363–3375.

Manoharan, G., Durai, S., Ashtikar, S. P., & Kumari, N. (2024). Artificial Intelligence in Marketing Applications. In Artificial Intelligence for Business (pp. 40-70). Productivity Press.

Manoharan, G., Durai, S., Rajesh, G. A., Razak, A., Rao, C. B., & Ashtikar, S. P. (2023). An investigation into the effectiveness of smart city projects by identifying the framework for measuring performance. In *Artificial Intelligence and Machine Learning in Smart City Planning* (pp. 71–84). Elsevier. DOI: 10.1016/B978-0-323-99503-0.00004-1

Manoharan, G., Razak, A., Rao, B. S., Singh, R., Ashtikar, S. P., & Nivedha, M. (2024). Navigating the Crescendo of Challenges in Harnessing Artificial Intelligence for Disaster Management. In Predicting Natural Disasters With AI and Machine Learning (pp. 64-94). IGI Global.

Mishra, S. P., & Mohapatra, S. (2023). Anthropocene Impacts on Marine Fishers: South Odisha Coast, India. *Current Journal of Applied Science and Technology, 42*(48), 176-196. https://en.wikipedia.org/wiki/Water_management_in_Chennai#:~:text=Chennai%20receives%20about%20985%20million,required%20amount%20of%201%2C200%20mld https://en.wikipedia.org/wiki/Water_management_in_Chennai https://www.thehindu.com/news/national/tamil-nadu/cyclone-michaung-over-1600-acres-of-crops-damaged-in-ranipet/article67607625.ece

Satpathy, A., Samal, A., Gupta, S., Kumar, S., Sharma, S., Manoharan, G., & Sharma, S. (2024). To Study the Sustainable Development Practices in Business and Food Industry. *Migration Letters : An International Journal of Migration Studies*, 21(S1), 743–747. DOI: 10.59670/ml.v21iS1.6400

Shaik, M. A., Manoharan, G., Prashanth, B., Akhil, N., Akash, A., & Reddy, T. R. S. (2022, May). Prediction of crop yield using machine learning. *In AIP Conference Proceedings (Vol. 2418, No. 1). AIP Publishing*. Sethuram, B. " Strengthening Local Governance for Cyclone Disaster Management: A Case Study of Puducherry District.

Shaikh, I. A. K., Kumar, C. N. S., Rohini, P., Jafersadhiq, A., Manoharan, G., & Suryanarayana, V. (2023, August). AST-Graph Convolution Network and LSTM Based Employees Behavioral and Emotional Reactions to Corporate Social Irresponsibility. In *2023 Second International Conference on Augmented Intelligence and Sustainable Systems (ICAISS)* (pp. 966-971). IEEE.

Shameem, A., Ramachandran, K. K., Sharma, A., Singh, R., Selvaraj, F. J., & Manoharan, G. (2023, May). The rising importance of AI in boosting the efficiency of online advertising in developing countries. *In 2023 3rd International Conference on Advance Computing and Innovative Technologies in Engineering (ICACITE) (pp. 1762-1766). IEEE.*

Tripathi, M. A., Tripathi, R., Effendy, F., Manoharan, G., Paul, M. J., & Aarif, M. (2023, January). An In-Depth Analysis of the Role That ML and Big Data Play in Driving Digital Marketing's Paradigm Shift. *In2023 International Conference on Computer Communication and Informatics (ICCCI)(pp.* 1-6*).* IEEE. DOI: 10.1109/ICCCI56745.2023.10128357

KEY TERMS AND DEFINITIONS

Disaster Management: Disaster management is the process of organizing and managing resources and responsibilities to deal with disasters.

Michaung Cyclone: Michaung is named after a suggestion by Myanmar. It is the sixth cyclone in the Indian Ocean and the fourth in the Bay of Bengal.

Resilience Management: Resilience Management refers to the practices and strategies implemented to help individuals, teams, and organizations better cope with adversity, uncertainty, and change.

Sustainability: It is a social goal that involves people co-existing on earth for over period of long years. It also has plans to follow for the global goodness.

Sustainable Disaster Management: It is a crucial approach to mitigating the impact of disaster on communities and the environment.

Solid Waste: solid waste is a discarded material that which is no longer will be used by people.

AI: It is the field of study that which helps to build the machines and computers and that can help to learn, reason and to act.

Chapter 16
Foundations of Sustainable Development:
Geospatial Data Perspective

Asmat Ali
https://orcid.org/0000-0002-8804-2285
Survey of Pakistan, Pakistan

Muhammad Hamza
PMAS-Arid Agriculture University, Pakistan

Hammad Hussain
https://orcid.org/0009-0009-0552-2549
Comsats University Islamabad, Pakistan

Rida Fatima
https://orcid.org/0009-0003-4867-6245
Comsats University Islamabad, Pakistan

ABSTRACT

The demand for geospatial information is growing rapidly across diverse sectors, including urban planning, agriculture, disaster management, and public health. Geospatial information has become a crucial requirement for sustainable development across all countries and sectors. This chapter aims to investigate and establish the rationale of geospatial data for sustainable development with the help of various use cases. The study also provides recommendations for enabling access to geospatial data.

INTRODUCTION

Sustainable development is a concept that has been used for more than 40 years(Shi et al., 2019). It has been defined in various ways(Scott & Rajabifard, 2017). Sustainable development, as defined by the World Commission on Environment and Development (or Brundtland Commission) is "Development which ensures the utilization of resources and the environment today does not damage prospects for use by future generations"(WCED, 1987). According to the United Nations, eradicating poverty is

DOI: 10.4018/979-8-3693-5748-4.ch016

a requirement for achieving sustainable development (UN, 2015). Mitoula and Papavasileiou (2023) find that mega physical infrastructures such as roads, rails, and metros are the cornerstone of sustainable development. Gunnarsdottir et al. (2020) argue that development of energy is an integral part of sustainable development. Guillaume (2024) asserts that sustainable development is all about human well-being, social inclusion and justice, environmental sustainability, global cooperation, and peace. A common way to explain the concept of sustainable development is by referring to the environmental, economic, and social as three distinct but interdependent pillars (Mitoula & Papavasileiou, 2023; World Bank, 2014, 2015). Sustainable development is an interdisciplinary challenge that demands balancing all three foundational pillars.

All developments require geospatial data (Rajabifard & Williamson, 2001). Many of the issues impacting sustainable development can be analyzed for evidence-based decision-making by making use of geospatial data (Benedek, 2023; Masser, 2019; Scott & Rajabifard, 2017). Geospatial data refers to any data that represents information about the location of objects on the Earth's surface(Rajabifard, A., Feeney, M. E. F., & Williamson, 2003). Spatial data is fundamental in geographic information systems (GIS) and is used in various fields such as urban planning, environmental science, agriculture, transportation, and many others to analyze, visualize, and understand spatial relationships and patterns. The Economist asserts that data, not oil, has become the world's most precious resource. Moreover, the United Nations emphasizes the importance of considering spatial data as a vital factor in decision-making procedures. Spatial data, which encompasses information with a geographic or locational component, plays a critical role in decision and policy formulation on a global scale. It is an indispensable component for carrying out planning, executing, monitoring, and evaluating socio-economic development initiatives. Over the past two decades, the advancement of GIS and associated technologies has revolutionized the management of spatial information produced by various public and private sector organizations worldwide. Spatial data provide foundation for analytical insights and informed decision-making processes crucial for addressing global challenges and driving sustainable development.

Geospatial information empowers decision makers to better understand the relationship between environmental, economic, and social settings(Manning, 1990). Geographic information systems (GIS) are used to integrate, and analyze geospatial data on these three pillars. However, GIS itself relies on underlaying geospatial data. This is the reason policymakers, and scientists demand unrestricted access to geospatial data. Sustainable development and sustainability are used as synonyms, even in the academic and scientific fields (Ruggerio, 2021).

This chapter aims to investigate and establish the rationale of geospatial data for sustainable development with the help of various use cases. The study also provides recommendations for enabling access to geospatial data. In the next section essential concepts of sustainable development are discussed for a better understanding of this study.

THEORETICAL BACKGROUND

This section presents a description of fundamental concepts relevant to this study such as economic, and social development and environmental protection being the foundations of sustainable development.

Sustainable Development

Sustainable development requires balancing the needs of present and future generations and has become a rapidly growing global concern(World Bank, 2014). For achieving sustainable development requires better geospatial information that empower decision makers to better understand the relationship between environmental, economic, and social settings(Manning, 1990). Geographic information systems are about integration and are therefore a natural tool to serve in dealing with sustainable development. This is the reason, geospatial data is needed by all governments and sectors for decision-making and development(UN-GGIM, 2020). The demand for geospatial information is growing rapidly across diverse sectors, including urban planning, agriculture, disaster management, and public health. Geospatial data development is on the rise thanks to rapid technological development. The data is used of geospatial information to address global challenges. The data plays a vital role in achieving the United Nations Sustainable Development Goals, such as poverty reduction, health, education, and environmental sustainability. Geospatial information offers insightful information that is useful for making decisions, formulating policies, and carrying out tasks pertaining to space and terrestrial issues, ultimately advancing sustainable development as a whole.

The critical review of the literature on sustainable development reveals that environmental, economic, and social development are the foundational pillars of sustainable development(World Bank, 2014). Figure 1 shows the three pillars of sustainable development.

Figure 1. Three pillars of sustainable development

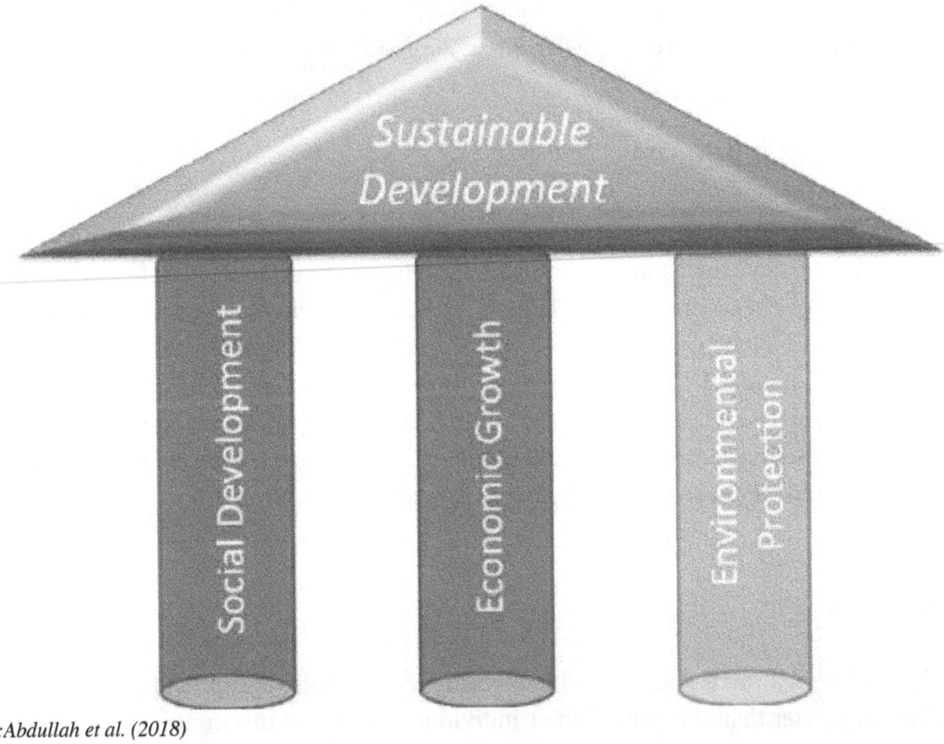

Source:Abdullah et al. (2018)

Economic Development

The term Economic Development refers to economic development aimed at balancing economic growth with environmental and social sustainability. However, (Barrier, 2017) argues that sustainable economic development primarily aims at ensuring access to secure livelihoods by the poor. Cornell et al. (2007) while discussing economic development in the context of India find that economic development means an increase in economic activity. For example, providing citizens with opportunities to earn so they can support themselves and live satisfying lives in their home communities. For achieving economic development concerted actions of policymakers and communities that improve the economic well-being and quality of life for a region's residents are required.

Economic development is a multifaceted concept that can include green economy, circular economy and digital modernization. Economists all over the world underscore the need to have a balance between economic development and sustainability. Despite this, it is important to mention that geospatial information has great value for the economy (Castelein et al., 2010; Krek, 2005).

Social Development

Social development refers to the process of improving the quality of life, well-being, and capabilities of individuals, families, communities, and societies as a whole. It encompasses various dimensions of human life beyond economic aspects, focusing on social justice, equity, inclusion, and empowerment. Social development adopts an approach that focuses on the need to "put people first" in development processes(World Bank, 2014). Social development includes education, employment, poverty reduction, gender equality, and social inclusion(Rahaman et al., 2024). Geospatial information has great value for the economy and society (Castelein et al., 2010; Krek, 2005) and is essential for aiding in socioeconomic growth.

Environmental Protection

Due to recent issues like climate change, population growth and urbanization, pollution, and an increase in natural disasters, environmental preservation has become increasingly important in recent years(Wang et al., 2024). The environmental challenges of today are many, complex and growing globally. Environmental impacts include greenhouse gas emissions, floods, degrading water and soil quality, air and noise pollution (Alipour & Dia, 2023). United Nations warns that environmental duration would be a great risk to human safety(United Nations, 1995). Environmental protection and development involve safeguarding natural resources, ecosystems, and biodiversity while promoting sustainable development practices that balance economic, social, and environmental considerations.

METHODOLOGY

This study employed a method that combines different elements or approaches in a way that their combined effect is greater than the sum of their individual effects. In this context, the study combined various methods or perspectives to analyze the subject matter more effectively. The study thoroughly reviewed existing research and literature related to sustainable development, economic development, environmental protection, and social development from both global and local sources. By doing so, we intend to gain a comprehensive understanding of the topic, considering insights and findings from various geographical contexts.

For making sustainable development, it is important to identify means and resources. To do so, this study focused on geospatial data as the means to explore its relevance for sustainable development. The study involved a detailed examination of relevant policy documents, regulations, and frameworks related to geospatial data specifically within the context of Pakistan. By analyzing these documents, we aim to understand the policy landscape and identify areas of alignment or divergence between policy objectives and on-the-ground realities.

The study also benefited from five use cases from Pakistan including the agriculture sector, environmental protection, health and energy sector, as well as natural resources management in the light of geospatial data. These use cases illustrate how the adopted synergistic approach in this study can be applied across diverse sectors in Pakistan to address pressing challenges in sustainable development.

Figure 2. Research Approach: integration of literature review and use cases

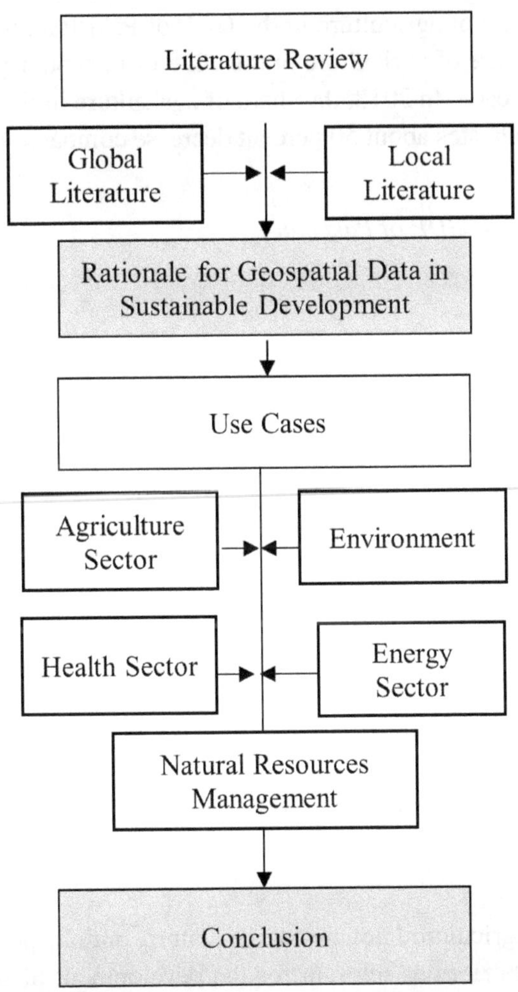

GEOSPATIAL DATA FOR SUSTAINABLE DEVELOPMENT: USE CASES FROM PAKISTAN

Agriculture Sector

Agriculture impacts us all as it is the source of our food, fiber, and jobs. The economy of several countries including Pakistan, India, China, Brazil, and Russia is dependent on agriculture. Robinson and Acemoglu (2012) find that the economy of some countries collapsed due to low agricultural productivity. However, the agriculture sector is facing multiple challenges including natural and human-induced such as climate change, floods, droughts, conversion of agricultural land into industrial and residential resulting in decreasing agricultural yield, increasing environmental degradation, the negative effect of extreme weather on soil and crop, and rapidly increasing population.

Agriculture sector had been a major contributor in the economy of Pakistan but dilemma is like many other sectors it is also on the decline for the last two decades. According to the World Bank, the share of agriculture in Pakistan's GDP was 43.7 percent in 1960, which dropped to 42.3 percent in 1961 and in 1962, it was 40.8 percent (Figure 3). In 1963, the GDP was 39.6 percent, which further dropped to 38.9 in 1964. It means that the share of agriculture in the GDP of Pakistan dropped 4.8 percent from 1960 to 1964. Unfortunately, the share of agriculture in the GDP of the country kept on declining (Rehman et al., 2019) in the coming decade. In 2018, the share of agriculture in the national GDP was observed equal to 24 percent, which indicates about 50 percent decrease compared to its share in 1960's.

Figure 3. Share of agriculture in GDP of Pakistan

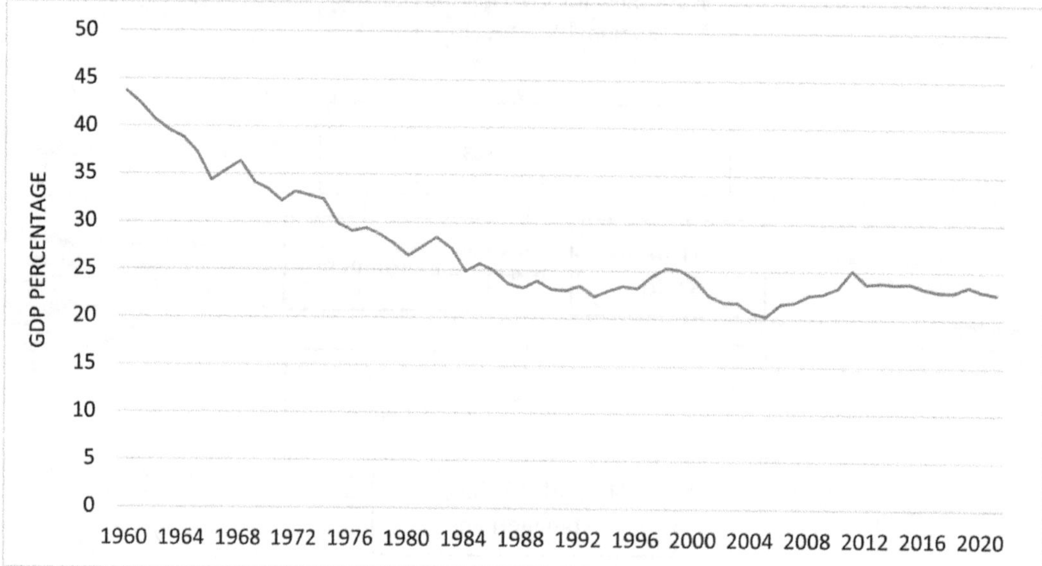

This scenario challenges agricultural policies of the country and the policy-making process as well. To understand issues and set priorities relevant for the development of the agriculture sector of the country, spatial data locked by various public sector organizations is a prerequisite. Spatial data is an important commodity that is used for numerous purposes such as policy making, development of agriculture, monitoring of natural resources and understanding causes of global climatic change, disaster management, homeland as well as food security. Geospatial data will facilitate monitoring of agricultural yield, tracking of soil and crop characteristics, reducing environmental degradation, natural and water resource management, and inform food security policies, agricultural yield improvement, and protecting the environment.

Environment Protection

For assessing and reducing environmental impacts on sustainable development, high quality and interoperable geospatial data are required as the data provide valuable insights for making informed decisions to address different environmental challenges (Afsari et al., 2022; Bačić et al., 2022; Casiano Flores & Crompvoets, 2023). The integration of geospatial data including remote sensing imagery, census data, the global navigation satellite system (GNSS), and volunteer geographic information (VGI), GIS makes it possible to analyze huge and complicated environmental data for making evidence-based policies. For reducing environmental impacts demand integration of planning, economics, and technological advancement (Banister, 2011).

Health Sector

The National Health Vision Pakistan 2016-2025 (Government of Pakistan, 2016) is a 10-year policy document that outlines the Government of Pakistan's priorities and strategies for enhancing the health and well-being of its citizens. This policy focuses on specific areas such as infectious disease control, maternal and child health, and non-communicable diseases. Additionally, the 12th Five Year Plan 2018-2023(Government of Pakistan, 2019) also outlines initiatives aimed at improving overall health outcomes and increasing access to healthcare, including strengthening the primary healthcare system, promoting preventive health measures, and addressing health disparities. It also includes measures to improve the quality and availability of medical facilities and equipment, as well as strategies to address specific health issues such as mental health, non-communicable diseases, and maternal and child health.

Pakistan is the only country in Asia that is still battling with Polio disease. The World Health Organization (WHO) recognizes polio as an international public health emergency that threatens to infect other countries with the disease. This is the reason, WHO has imposed travel restrictions on Pakistan in 2016. A Memorandum of Understanding (MoU) was signed between Survey of Pakistan (SoP) and National Emergency Operation Center (NEOC), Min of Heath to fight jointly against the disease. To support micro planning, the role of geographical maps will be instrumental and according to the signed MoU, the maps are being prepared by SoP.

Energy Sector

Energy sector is important (Aulia & W, 2023; Chen et al., 2020; Victor et al., 2015) as it plays vital role in socio-economic development of a country (GOP, 2022; Gunnarsdottir et al., 2020). The goal seven of the UN's Sustainable Development Goals (SDG), "Ensure access to affordable, reliable, sustainable and modern energy for all" underscores the important of the energy sector. The relationship between energy production and spatial information an important factor in planning (Lobosco et al., 2023). Spatial objects such as freshwater reservoirs, forests, crude oil, and natural gas fields have drawn the attention of governments all over the world due to their limited existence and availability on the one hand and their rapid consumption on the other hand. The increasing demand for energy can be attributed to three main factors i.e. surge in economic activities, population growth and rapid technological transformation (GOP, 2022). Therefore, there is need to employ efficient energy planning tools which require spatial data to integrate technical, and local information for supporting decision-making process (Ferrari et al., 2019; Urrutia-Azcona et al., 2021). Spatial data has the potential to identify, plan and distribute energy resources (Barik et al., 2019; Georgiadou & Reckien, 2018; Kotsev et al., 2021; Lobosco et al., 2023). For energy planning, data on location, quality, and quantity of energy resources, such as oil, gas, coal, and renewable energy sources, such as wind and solar, location of settlements, buildings, facilities such as academic institutions, public health and transport, environmental data, economic data, and data of administrative boundaries are required.

For many decades, Pakistan has been facing energy shortages, which have hindered its economic growth and development. The Ministry of Water and Power at the federal level issues energy policies on a regular basis, with the latest Energy Policy (2013-2018) outlined by Bhatti (2013). In 2017, the Ministry of Water and Power was dissolved into the Ministry of Water Resources and the Ministry of Energy (Power Division), with the Pakistan Water and Power Development Authority (WAPDA) working under the Ministry of Energy. The Ministry of Energy issued the National Electricity Policy in 2021. The energy sector contributes to the economic development of Pakistan by increasing power generation capacity, expanding the transmission and distribution network to rural and remote areas, creating job opportunities, attracting foreign investment, supporting exports, and enhancing energy security. Improving the energy sector is crucial to achieving the country's development goals, and stakeholders' needs and plans for the sector are presented in Table 1 by Millo et al. (2019) with respect to the NSDI Pakistan.

Natural Resources Management

Natural resources play a key role in the economy of a country such as Pakistan which depend more on land owned resources and less on industry for its economy. Pakistan's natural resources include but are not limited to natural gas, soil, oil, hydro power, coal and limestone etc. However, unplanned urbanization and industrialization, increasing frequency of natural hazards like floods and droughts as well as quickly growing population has caused serious damage to the natural resources of Pakistan. This scenario demands for management of the natural resources. To manage the resources, accessibility and availability to geospatial information of various kinds produced by many public sector organizations is a prerequisite. In Pakistan, no spatial information sharing mechanism exists, making it impossible to get required information for managing natural resources of the country. NSDI development will help to overcome this scary situation by enabling access to the information being produced by various government organizations.

ACCESS TO GEOSPATIAL DATA: THE CASE OF PAKISTAN

The availability and accessibility of such data on national level is challenging (Abramic et al., 2017) as the data are collected and maintained by various organizations that do not share and give access to their data (Ali & Imran, 2019) due to institutional issues (Ali et al., 2021). SDIs can help to enable sharing of standardized data (Ali & Imran, 2021; Flores & Crompvoets, 2020; Guidance, 2022; Masser, 2019, 2005; Tu et al., 2022) and therefore can inform and impact planning and development environmental policies and practices (Casiano Flores & Crompvoets, 2023; European Commission, 2022). SDI can be used to integrate social, economic and environmental data to support government decision making (Car et al., 2019).

Realizing the potential of spatial data for addressing environmental issues, European Union established an Infrastructure for Spatial Information in the European Community (INSPIRE). It aims to improve consistency, availability, and re-use of spatial information to support environmental policies (Bartha & Kocsis, 2011).

In Pakistan most of the geospatial data is produced by public sector organizations. The list of some government departments/ministries along with type of spatial data collected and maintained by them is presented as Table 1.

Table 1. Key custodians of geospatial data in Pakistan

Department/Ministry	Datasets
Survey of Pakistan (SoP)	Topographic maps
Geological Survey of Pakistan (GSP)	Geological Map
Ministry of Health	Health facilities
Space and Upper Atmosphere Research Commission (SUPARCO)	Satellite imagery
Soil Survey Departments	Land use and soil map
Water and Power Development Authority(WAPDA)	Power and Hydrology
Pakistan Metrological Department	Weather information
Pakistan Bureau of Statistics	Census information
Provincial Forest Departments	Protected areas & forests
National Highway Authority	Transportation network
Ministry of Tourism and Culture	Tourism & Cultural sites
Ministry of Climate Change	Environmental data
Zoological Survey Department	Animal species
National Wildlife Council	Wildlife
Department of Plant Protection (Pakistan)	Ecological zones
Pakistan Agricultural Research Council (PARC)	Land use/cover data

From Table 1, it is evident that practically it would be impossible for researchers, policymakers and practitioners working on various development projects to get all the needed datasets, ever.

Spatial data is a key strategic resource, commodity, and capital asset as finds (Masser, 2019). Geospatial information is a nation's 'digital currency' for evidence-based decision-making (UN, 2018). The importance and influence of spatial data have risen in all kinds of governmental and non-governmental processes (Sjoukema et al., 2019). Geospatial information has great value for the economy and society (Castelein et al., 2010; Krek, 2005).

CONCLUSION

Sustainable economic development is a multifaceted idea that includes social justice, environmental preservation, and economic prosperity. This has given rise to new concepts such as the green economy, digital modernization, environmental conservation, renewable resources, and the fight against climate change as the mandatory prerequisites for economic development.

There are many perspectives to consider when writing about sustainable economic development. Many authors have analyzed the relationship between environmental or social variables and various economic indicators such as GDP, income per capita, etc. Others have focused on the processes that enable a sustainable economic development, for example, sharing economy or intergovernmental programs, while others have explored the transformation of specific industries and the development of instruments such as green finance.

Geospatial data can be instrumental in analyzing agricultural landscapes, including factors such as soil quality, land use patterns, and water availability. By integrating this data with insights from global and local literature on sustainable agriculture practices, researchers can identify strategies to enhance productivity while minimizing environmental impact. Scrutinizing pertinent policy documents can further ensure that agricultural policies align with sustainable development goals, promoting practices such as organic farming, water conservation, and agroforestry.

Geospatial data can provide valuable information on environmental indicators such as air and water quality, deforestation rates, and habitat loss. By analyzing this data alongside relevant literature, researchers can identify areas of environmental concern and develop targeted conservation strategies. Scrutinizing policy documents can help assess the effectiveness of existing environmental regulations and advocate for stronger enforcement measures to protect ecosystems and biodiversity.

Geospatial data analysis can contribute to understanding health-related phenomena such as disease outbreaks, access to healthcare facilities, and environmental health risks. By integrating this data with literature on public health and epidemiology, researchers can identify interventions to improve healthcare access and prevent disease transmission. Scrutinizing policy documents can inform advocacy efforts for policies that promote public health initiatives, such as sanitation improvements and disease surveillance systems.

Geospatial data can inform decision-making in the energy sector by identifying suitable locations for renewable energy projects, assessing energy demand patterns, and analyzing infrastructure development needs. By integrating this data with insights from energy literature, researchers can develop strategies to promote renewable energy adoption and energy efficiency measures. Scrutinizing policy documents can help evaluate the effectiveness of energy policies and advocate for reforms to support sustainable energy transition.

Geospatial data analysis is essential for managing natural resources sustainably, including forests, water resources, and wildlife habitats. By analyzing spatial data alongside relevant literature, researchers can identify areas of conservation priority and develop management plans to mitigate resource depletion and environmental degradation. Scrutinizing policy documents can inform advocacy efforts for policies that promote sustainable resource management practices, such as watershed management and protected area designation.

REFERENCES

Abdullah, M. I., Sarfraz, M., Qun, W., & Javaid, N. (2018). Drivers of green supply chain management. *LogForum*, 14(4), 437–447. DOI: 10.17270/J.LOG.2018.297

Abramic, A., Kotsev, A., Cetl, V., Kephalopoulos, S., & Paviotti, M. (2017). A Spatial Data Infrastructure for Environmental Noise Data in Europe. *International Journal of Environmental Research and Public Health 2017, Vol. 14, Page 726, 14*(7), 726. DOI: 10.3390/ijerph14070726

Afsari, R., Shorabeh, S. N., Kouhnavard, M., Homaee, M., & Arsanjani, J. J. (2022). A Spatial Decision Support Approach for Flood Vulnerability Analysis in Urban Areas: A Case Study of Tehran. *ISPRS International Journal of Geo-Information 2022, Vol. 11, Page 380, 11*(7), 380. DOI: 10.3390/ijgi11070380

Ali, A., & Imran, M. (2019). The Evolution of National Spatial Data Infrastructure in Pakistan, Implementation Problems and the Way Forward. *International Journal of Spatial Data Infrastructures Research*, 14(0). Advance online publication. DOI: 10.2902/IJSDIR.V14I0.500

Ali, A., & Imran, M. (2021). National Spatial Data Infrastructure vs Cadastre System for Economic Development: Evidence from Pakistan. *Land (Basel)*, 10(2), 188. DOI: 10.3390/land10020188

Ali, A., Imran, M., Jabeen, M., Ali, Z., & Mahmood, S. A. (2021). Factors influencing integrated information management: Spatial data infrastructure in Pakistan. *Information Development*, 026666692110484. Advance online publication. DOI: 10.1177/02666669211048483

Alipour, D., & Dia, H. (2023). A Systematic Review of the Role of Land Use, Transport, and Energy-Environment Integration in Shaping Sustainable Cities. *Sustainability (Basel)*, 15(8), 6447. DOI: 10.3390/su15086447

Aulia, R. P., & Steelyana, E. (2023). A Study in Government Procurement System: Public-Private Partnership for Infrastructure in Energy Sector. In *E3S Web of Conferences* (Vol. 388, p. 01014). EDP Sciences.

Bačić, S., Tomić, H., Andlar, G., & Roić, M. (2022). Towards Integrated Land Management: The Role of Green Infrastructure. *ISPRS International Journal of Geo-Information 2022, Vol. 11, Page 513, 11*(10), 513. DOI: 10.3390/ijgi11100513

Banister, D. (2011). Cities, mobility and climate change. *Journal of Transport Geography*, 19(6), 1538–1546. DOI: 10.1016/j.jtrangeo.2011.03.009

Barik, R. K., Dubey, H., Mankodiya, K., Sasane, S. A., & Misra, C. (2019). GeoFog4Health: A fog-based SDI framework for geospatial health big data analysis. *Journal of Ambient Intelligence and Humanized Computing*, 10(2), 551–567. DOI: 10.1007/s12652-018-0702-x

Barrier, E. B. (2017). The Concept of Sustainable Economic Development. *The Economics of Sustainability*, 87–96. DOI: 10.4324/9781315240084-7

Bartha, G., & Kocsis, S. (2011). Standardization of geographic data: The European inspire directive. *European Journal of Geography*, 2(2), 79–89.

Benedek, J. (2023). Spatial Localization of the Sustainable Development Goals. *Public Finance Quarterly*, 69(1), 64–74. DOI: 10.35551/PFQ_2023_1_4

Car, N. J., Box, P. J., & Sommer, A. (2019). The location index: A semantic web spatial data infrastructure. *Lecture Notes in Computer Science (Including Subseries Lecture Notes in Artificial Intelligence and Lecture Notes in Bioinformatics), 11503 LNCS*, 543–557. https://doi.org/DOI: 10.1007/978-3-030-21348-0_35/COVER

Casiano Flores, C., & Crompvoets, J. (2023). Climate Change Adaptation: The Role of Geospatial Data in Sustainable Infrastructures. *ISPRS International Journal of Geo-Information 2023, Vol. 12, Page 68, 12*(2), 68. DOI: 10.3390/ijgi12020068

Castelein, W. T., Bregt, A. K., & Pluijmers, Y. (2010). The economic value of the Dutch geo-information sector *. *International Journal of Spatial Data Infrastructures Research*, 5, 58–76. DOI: 10.2902/1725-0463.2010.05.art2

Chen, K., Ren, Z., Mu, S., Sun, T. Q., & Mu, R. (2020). Integrating the Delphi survey into scenario planning for China's renewable energy development strategy towards 2030. *Technological Forecasting and Social Change*, 158, 120157. DOI: 10.1016/j.techfore.2020.120157

Cornell, S., & Policy, M. J.-N. C. of A. I., 2007, U., & Jorgensen, M. (2007). The nature and components of economic development in Indian country. *Citeseer, May*, 1–22. https://citeseerx.ist.psu.edu/viewdoc/download?doi=10.1.1.173.661&rep=rep1&type=pdf

European Commission. (2022). *Support to the evaluation of the implementation of the Directive 2007 / 2 / EC on Infrastructure for Spatial Information in the European Community*. INSPIRE., DOI: 10.2779/749002

Ferrari, S., Zagarella, F., Caputo, P., & Bonomolo, M. (2019). Assessment of tools for urban energy planning. *Energy*, 176, 544–551. DOI: 10.1016/j.energy.2019.04.054

Flores, C. C., & Crompvoets, J. (2020). Assessing the Governance Context Support for Creating a Pluvial Flood Risk Map with Climate Change Scenarios: The Flemish Subnational Case. *ISPRS International Journal of Geo-Information*, 9(7), 460. DOI: 10.3390/ijgi9070460

Georgiadou, Y., & Reckien, D. (2018). Geo-information tools, governance, and wicked policy problems. *ISPRS International Journal of Geo-Information*, 7(1), 21. DOI: 10.3390/ijgi7010021

GOP. (2022). *Pakistan Economic Survey 2021-22.*

Government of Pakistan. (2016). *National Health Vision Pakistan 2016-2025.*

Government of Pakistan. (2019). *12th FIVE YEAR PLAN (2018-2023)*. https://www.pc.gov.pk/uploads/annualplan2018/Annual.pdf

Guidance, B. A. (2022). *M EASURING THE ECONOMIC, SOCIAL AND ENVIRONMENTAL VALUE OF PUBLIC SECTOR LOCATION.*

Gunnarsdottir, I., Davidsdottir, B., Worrell, E., & Sigurgeirsdottir, S. (2020). Review of indicators for sustainable energy development. *Renewable & Sustainable Energy Reviews*, 133, 110294. DOI: 10.1016/j.rser.2020.110294

Kotsev, A., Minghini, M., Cetl, V., & En, E. U. R. (2021). *INSPIRE A Public Sector Contribution to the European Green Deal Data Space. A vision for the technological evolution of Europe's Spatial Data Infrastructures for 2030*. DOI: 10.2760/8563

Krek, A. (2005). Geographic information as an economic good. In *GIS for Sustainable Development* (pp. 85–103). CRC Press., DOI: 10.1201/9781420037845.ch6

Lafortune, G., Fuller, G., Kloke-Lesch, A., Koundouri, P., & Riccaboni, A. (2024). *European elections, Europe's future and the SDGs: Europe sustainable development report 2023/24*. SDG Transformation Center.

Lobosco, G., Tinti, L., Magagnoli, B., Mencarini, V., Mannucci, S., & Ferrero, M. (2023). Landscape as a Palimpsest for Energy Transition: Correlations between the Spatial Development of Energy-Production Infrastructure and Climate-Mitigation Goals. *Atmosphere 2023, Vol. 14, Page 921, 14*(6), 921. DOI: 10.3390/atmos14060921

Manning, E. W. (1990). Geographic information systems and sustainable development. *Government Information Quarterly*, 7(3), 329–342. DOI: 10.1016/0740-624X(90)90028-M

Masser, I. (2005). *GIS worlds : creating spatial data infrastructures* (Vol. 338). ESRI Press.

Masser, I. (2019). *From Geographic information systems to Spatial data infrastructures : a global perspective* (1st ed.).

Mitoula, R., & Papavasileiou, A. (2023). Mega infrastructure projects and their contribution to sustainable development: The case of the Athens Metro. *Economic Change and Restructuring*, 56(3), 1943–1969. DOI: 10.1007/s10644-023-09493-w

Rahaman, M. M., Islam, M. M., Polbitsyn, S. N., & Abbas, S. (2024). The electrification-social development nexus in Bangladesh: Do governmental policies matter? *Utilities Policy*, 86, 101690. DOI: 10.1016/j.jup.2023.101690

Rajabifard, A., Feeney, M. E. F., & Williamson, I. (2003). *Spatial data infrastructures: concept, nature and SDI hierarchy*. Taylor & Francis London.

Rajabifard, A., & Williamson, I. P. (2002). Spatial Data Infrastructures: an initiative to facilitate spatial data sharing. Global environmental databases-present situation; future directions, 2.

Robinson, J. A., & Acemoglu, D. (2012). *Why nations fail: The origins of power, prosperity and poverty*. Profile London.

Ruggerio, C. A. (2021). Sustainability and sustainable development: A review of principles and definitions. *The Science of the Total Environment*, 786, 147481. DOI: 10.1016/j.scitotenv.2021.147481 PMID: 33965820

Scott, G., & Rajabifard, A. (2017). Sustainable development and geospatial information: A strategic framework for integrating a global policy agenda into national geospatial capabilities. *Geo-Spatial Information Science*, 20(2), 59–76. DOI: 10.1080/10095020.2017.1325594

Shi, L., Han, L., Yang, F., & Gao, L. (2019). The Evolution of Sustainable Development Theory: Types, Goals, and Research Prospects. *Sustainability 2019, Vol. 11, Page 7158, 11*(24), 7158. DOI: 10.3390/su11247158

Sjoukema, J.-W., Bregt, A. K., & Crompvoets, J. (2019). Understanding Governance Dynamics: The Governing System of Spatial Data Infrastructures. *International Journal of Spatial Data Infrastructures Research*, 14(0). Advance online publication. DOI: 10.2902/IJSDIR.V14I0.495

Tu, C., Zang, C., Tan, Y., Zhou, Y., & Yu, C. (2022). Can information infrastructure development improve the health care environment? Evidence from China. *Frontiers in Public Health*, 10, 987391. Advance online publication. DOI: 10.3389/fpubh.2022.987391 PMID: 36091535

UN. (2015). *Transforming Our World: The 2030 Agenda for Sustainable Development* (p. 41). DOI: 10.1891/9780826190123.ap02

UN. (2018). National geospatial data and information systems. In *United Nations Committee of Experts on Global Geospatial Information Management,United Nations Economic and Social Council,E/C.20/2018/16/Add.1*. DOI: 10.1093/oxfordhb/9780199560103.003.0007

UN-GGIM. (2020). *Virtual Regional Seminar on Operationalizing the Integrated Geospatial Information Framework in Asia and the Pacific*.

United Nations. (1995). *Human Development Report 1995*. Human Development Reports.

Urrutia-Azcona, K., Usobiaga-Ferrer, E., De Agustín-Camacho, P., Molina-Costa, P., Benedito-Bordonau, M., & Flores-Abascal, I. (2021). ENER-BI: Integrating Energy and Spatial Data for Cities' Decarbonisation Planning. *Sustainability 2021, Vol. 13, Page 383, 13*(1), 383. DOI: 10.3390/su13010383

Victor, O. E., Aziz, N. A., & Jaffar, A. R. (2015). Privatization of Electricity Service Delivery in Developing Nations: Issues and Challenges. *International Journal of Built Environment and Sustainability*, 2(3). Advance online publication. DOI: 10.11113/ijbes.v2.n3.82

Wang, S., Huang, X., Liu, P., Zhang, M., Biljecki, F., Hu, T., Fu, X., Liu, L., Liu, X., Wang, R., Huang, Y., Yan, J., Jiang, J., Chukwu, M., Reza Naghedi, S., Hemmati, M., Shao, Y., Jia, N., Xiao, Z., & Bao, S. (2024). Mapping the landscape and roadmap of geospatial artificial intelligence (GeoAI) in quantitative human geography: An extensive systematic review. *International Journal of Applied Earth Observation and Geoinformation*, 128, 103734. DOI: 10.1016/j.jag.2024.103734

WCED. S. W. S. (1987). *World commission on environment and development. Our common future. 17*(1), 1–91. https://idl-bnc-idrc.dspacedirect.org/bitstream/10625/152/1/WCED_v17_doc149.pdf

World Bank. (2014). *Social development: putting people first*. https://documents.worldbank.org/curated/en/703901468147574810/Social-development-putting-people-first World Bank. (2015). *Bonds for Sustainable Development*. www.worldbank.org/debtsecurities

KEY TERMS AND DEFINITIONS

Sustainable Development: Sustainable development is a multifaced concept that refers to development that meets the needs of the present without compromising the ability of future generations to meet their own needs.

Economic Development: The term Economic Development refers to economic development aimed at balancing economic growth with environmental and social sustainability.

Social Development: Social development refers to the process of improving the quality of life, well-being, and capabilities of individuals, families, communities, and societies as a whole.

Environmental Protection: It refers to the practice of protecting the natural environment by individuals, groups and governments.

Geospatial Data: Geospatial data refers to the location of natural and manmade features on earth such as forests, roads, water bodies, health and education facilities, as well as human settlements. It plays important role in policymaking, implementation, and monitoring the policy outcomes. Access to geospatial data is crucial for achieving sustainable development policy outcomes.

Spatial Data Infrastructure: It is a legal and policy framework that is implemented to enable sharing of geospatial data among data user communities including researchers.

Natural Resources Management: It refers to protecting and saving the precious resources owned by the nature such as water, soil, and biodiversity from human exploitation and extinction. Geospatial data is used as evidence to monitor the impact of humans on natural resources.

Chapter 17
Towards Sustainable Health Systems:
Insights From Pakistan's Spatial Data Infrastructure

Munir Ahmad
https://orcid.org/0000-0003-4836-6151
Survey of Pakistan, Pakistan

Zubia Waqar
https://orcid.org/0000-0003-4354-3825
Indus University of Health Sciences, Pakistan

ABSTRACT

Numerous challenges are being faced by the healthcare system of Pakistan. It lacks universal access to quality healthcare, an inadequate supply of healthcare human resources, and healthcare sector inconsistencies. Incorporating the Spatial Data Infrastructure (SDI) in Pakistan can be a possible solution to tackle most of these problems. SDI is a platform that organizes and centralizes geographical information, such as the population, the placement of healthcare facilities, and disease incidence. By employing SDI, these challenges may be met efficiently by the planners. When implemented, spatial data infrastructure can support optimizing human resource development, investment in resources, and evidence-based decision-making. Furthermore, SDI can ensure the delivery of actual and timely information to stakeholders thus enhancing the aspect of accountability. This chapter discussed how SDI can help in coping with relative issues concerning the Pakistani health sector and thereby presented a way of creating a sustainable healthcare sector.

INTRODUCTION

Cities are expanding rapidly in Pakistan, and this is a tendency observed across the globe. However, the current situation of urban planning in Pakistan is utterly in vain. Being a part of urban centers the fact that out of 124 identified urban settlements in Pakistan, only 17% have a clear spatial plan and require urgent attention (ul Hussnain et al., 2020). Several issues resulting from the non-appearance of public

DOI: 10.4018/979-8-3693-5748-4.ch017

transport systems in various Pakistani cities underscore the need for governance efficiency, capacity enhancement, and physical planning in the context of urban development (Imran, 2009). Small and medium agri-food firm strategic development planning policies are reported to be lacking in Pakistan, so the five-year plans for agricultural industrialization in Pakistan's second-tier cities are hindered by financial reliance on the federal government and foreign assistance as noted by Kalwar et al. (2019).

Pakistanis are facing healthcare challenges that are transforming at a higher pace. According to research conducted by Memon et al. (2024) foreigners in Pakistan face barriers such as access to quality healthcare, navigating cultural differences in health beliefs, being a victim of healthcare disparity in the country, or developing complications that are likely to arise from previous health conditions. In the same manner,, Habib et al. (2021) highlighted culture and resource scarcity as barriers to accessing TB treatment for rural females in Pakistan. Women have limited control over their own money, they cannot move around independently, and they have to travel quite a distance to clinics. Moreover, healthcare itself is more inclined to male-centered care and there aren't enough women doctors to bridge this gap.

To overcome these challenges and tailor the healthcare delivery system to the beneficiary population's needs, techniques such as data analysis and statistical decision-making are paramount. Chakraborty et al. (2023) asserted that consumption of MCDM in the framework of the healthcare sector is increasing rapidly but such applications are mainly concentrating on waste disposal, supply chain management, and disease diagnosing. Furthermore, Pradhan et al. (2022) interestingly highlighted how such methods can be valuable for district health systems, especially in responding to challenges such as; a poor availability of female medical workforce, limited disaster management training, unequal reportage patterns across provinces, and absence of gender perspective in addressing during the floods. To sum it up, although Pakistan faces numerous challenges within the context of healthcare provision for its rapidly growing population, the utilization of data analytics and evidence-based practice can help the country adapt to new trends and deliver high-quality care to a population that is quickly expanding.

Geographic Information Systems (GIS) provide a variety of functionalities for spatial analysis, data management, and decision support, which is an important aspect of increasing planning demands. Decision makers may comprehend significant spatial relations and GIS in regard to land use, infrastructure, demography, education, healthcare, and health and conditions. Skelton (2014) argued that GIS can complement planning and decision-making in developed nations. Khashoggi & Murad (2020) noted that GIS can address two critical areas despite limitations in health data management: firstly, surveillance and early modeling of disease occurrences to control epidemic diseases. Secondly, assessing the allocation of health care facilities, to decide where to deliver a more sufficient supply of health care resources to satisfy the need.

In the context of health care planning, the role of SDI can be explained by the fact that it serves as the basic foundation of GIS. That includes the foundational guidelines and protocols for capturing, storing, and retrieving spatial health data in a wide variety of providers' networks and governmental/non-governmental organizations. This helps the integration of data and information exchange, thus leading to a better understanding of the general health trends as well as forwarding the best solutions to resource allocation. Poirier et al. (2024) pointed out that creating the data infrastructure for planning improves transparency and public participation, policy monitoring, and local coordination by enabling users to search terms across plans, track policy adoption, identify best practices, and compare plans for effectiveness. In many countries, SDI is a key integrative mechanism that enhances cooperation and exchange of data in the various sectors of the government in e-governance and g-governance, including ministries and departments such as the Ministry of Health. This further indicates that the applicability of evaluating SDI should be brought into consideration regarding Pakistani healthcare management. Within this context, this study outlines its primary objective to examine how the concept of SDI fits into the planning of health care in Pakistan.

In pursuit of the objective, the chapter is structured as follows: The following section focuses on defining the conceptual key ideas and approach within the context of SDI and healthcare planning. After this, Section 3 is used to explain the current status of SDI implementation in Pakistan. Section 4 unfolds a critical analysis of how SDI serves as the fundamental element contributing towards the formation of healthcare strategic planning in the Pakistani context. Finally, in the last part, a synthesis of the recommendations derived from the key findings is presented followed by the conclusions.

LITERATURE REVIEW

This section sets the foundation by introducing key concepts associated with spatial data infrastructure and healthcare planning.

Spatial Data Infrastructure

Spatial Data Infrastructure is a simplistic framework for facile access and usage of geographical information. SDI is a set of initiatives based on cooperation, business regulation, and IT platforms. SDI's primary purpose is to facilitate the acquisition of spatial data to ensure that it is readily available within regions as well as between provinces, states, or even countries (Cetl et al., 2009; Diaz et al., 2012). enables one to integrate social, environmental, and institutional databases. In other words, it is an interface between various types of information as it is described in the works of (Coleman & Nebert, 1998). McLaughlin & Nichols (1994) have underscored the importance of this factor for the development of a national SDI. A robust SDI with customized services leads to improved access and sharing of remote information resources for organizations, government agencies, and the general populace.

Earlier SDIs were more focused on services, but now more focuses on users (Granell et al., 2011). This evolution has emphasized enhancing the geospatial data exchange and access within its framework. SDI also enhance the efficiency of data exchange with stakeholders who are involved in spatial data production or usage (Rajabifard et al., 2002, 2006). Nedović-Budić et al. (2011), argued that SDIs are critical strategies in addressing data policy, technological standards, and interoperability issues. The area of SDI that is most connected to cost efficiency is interoperability because it increases the amount of spatial data available at lower costs (Crompvoets et al., 2018; Masser, 2019).

Benefits of SDI

SDI sustains the availability of timely and updated data which in turn enables preparedness of decision-makers at various levels in different fields like education, public health, disaster management, and agriculture among others to be prepared (Manfré et al., 2012). SDI datasets assist in defining the connections that may not be revealed or considered during a regular process of data analysis (Boerboom, 2010). For instance, using SDI datasets it is possible to find out which are the affected areas in the event of an earthquake and as a result, the disaster management authorities can be in a position to respond and initiate recovery activities much sooner (Manfré et al., 2012).

SDI is not just a way to collect data; in fact, it helps organizations reduce their expenses as well. These initiatives mainly reduce costs related to data collection, storage, and exchange since SDI helps to minimize the duplicate efforts in these processes (Bishop et al., 2000). Even greater efficiency is obtained when spatial data is available through web portals (Borzacchiello & Craglia, 2013). These cost savings free up resources for organizations to focus on what truly matters e.g. reviewing data and obtaining effective actionable knowledge.

SDI improves the efficiency in handling data by reducing the number of formalities that are likely to be encountered when working with data and the time taken in processing data (Borzacchiello & Craglia, 2013). SDI facilitates the automation of tasks such as identification, validation, and consolidation of data and is a time and cost-effective intervention as asserted by (Crompvoets et al., 2011). SDI also enhances organizational performance and helps organizations become more agile in the face of new conditions in a market. SDI datasets also help to collaborate public and private partnerships, governmental and non-governmental organizations, and even the ordinary public in the formulation and implementation of the policies and strategies (Bishop et al., 2000; Crompvoets et al., 2011). This common bus for information exchange increases transparency and accountability.

Sustainable Health Systems

Healthcare planning can be defined as the process of identifying aims and targets and selecting the most effective ways of using resources for creating and implementing a set of appropriate health services to serve a given community. Quantitative and qualitative information and the decisions derived from them can be very useful at every step in the planning of any health facility. Moreover, applying technological tools can enable the efficient acquisition, processing, and application of patient-related information for improved organizational decision-making (Batko & Ślęzak, 2022). Using data analysis tools, healthcare planners can identify those areas that require their attention, find out where they should invest, and determine whether the efforts are bearing fruit in the form of improving the health status of the targeted individuals. In this context, data relating to the users of primary care services is vital for planning healthcare resources availability based on the requirement of the population and in making policies regarding public healthcare (Yan et al., 2019).

In this regard, Horenstein & Heimberg (2020) pointed out that patients with anxiety disorders including health anxiety contribute to frequent utilization of health care services in different settings. Is also pointed out that anxiety can also manifest in times when a patient misses' appointments or comes irregularly or in cases where their visit to the healthcare center is inconsistent. To the extent that engaging patients in the process of healthcare planning and evaluation is considered as a potential approach to enhance the satisfaction of patients, a critical gap exists. To innovate and democratize patient-centered healthcare, therefore, healthcare planners must design patient engagement that is inclusive, and responsive to patient needs and capacity variations across diverse populations (Snow, 2022).

Wronski et al. (2021) described that in planning healthcare at the district level, the stakeholders emphasized a very large number of indicators classified into the following broad areas; nonhealth-related determinants, health status, usage of the health system, health system outcome, and provision of healthcare. However, Kornelsen et al. (2021) pointed out the need to incorporate essential local health care services such as emergency care, maternal care, senior care, primary care, specialty, and mental/substance use disorder treatment to plan for efficient health care.

AI can help realize massive data volumes, which contain information that can help in identifying diseases, selecting treatments, and performing lab tests. Combining all the above, it is possible to contribute to the modern techniques of the effective management of the health of entire populations by analyzing the general tendencies and the factors that can potentially carry certain risks (Alowais et al., 2023). Wang (2020) showed how the application of GIS is crucial in public health analysis for policy-relevant considerations.

The study mentioned that the prevailing health behaviours and healthcare resources are not the same across different geographic regions and hence, geographic location-specific policies should be recommended in matters of healthcare. These variations and the ability to subdivide can be studied with GIS and how various aspects of a person and the factors in his/her neighbourhood that affect health can be identified. GIS can also assist researchers and policymakers in identifying service delivery regions for fundamental healthcare services (such as primary care, hospitals, or cancer treatment) and envision where and how resources ought to be shared equally to increase the patients' hospitable access.

METHODOLOGY

This chapter offers Pakistan's NSDI analysis by employing a case study approach. This method allows the authors to obtain a holistic approach to the NSDI functioning in a particular context (Tasci et al., 2020). Thus, focusing on this particular case, the chapter provides meaningful insights into the common issues that can be encountered when setting up SDI for the whole of Pakistan. Also, it examines more detailed involvement of SDI in supporting the process of healthcare planning in the country.

Status of SDI in Pakistan

Pakistan has embarked on the task of creating an NSDI. This section moves directly into the activities already being undertaken to make this a possibility.

At the moment a feasibility study has been completed to draw up a strategy to build NSDI for the country. It consists of all facets needed to roll out the NSDI. Financial capital is a policy area that is currently being exercised carefully and cautiously, and legal structures alongside technologies; human capital is also being reinspected. It also seeks to understand how the NSDI will be sustained in the long term as an architecture, governance, management, and funding model. To ensure that the insights obtained for the feasibility study are as useful as possible, key stakeholders are engaged through workshops, surveys, and meetings. For instance, a workshop was conducted at COMSATS University Islamabad to investigate various facets of future Pakistani NSDI. Through partnership and with many voices to hear, they strive to design an NSDI that meets the requirements of key stakeholders.

Pakistan recognizes the importance of collaboration for the success of its National Spatial Data Infrastructure (NSDI). To foster this collaboration, the Survey of Pakistan has established a coordination committee comprising various stakeholders. This committee takes the lead in initiating and advancing the national NSDI, ensuring that all stakeholders contribute to this significant initiative.

CHALLENGES OF SUSTAINABLE HEALTHCARE IN PAKISTAN

This section addresses the challenges Pakistan encounters in healthcare system planning. Through a review of existing research, we will identify and outline the key hurdles. These challenges are delineated in Figure 1 for convenient reference.

To strengthen the healthcare scenario and to attain the Universal Health Coverage (UHC) in Pakistan, there is a desperate requirement of a makeover. This implies making health care accessible to all people, a broader provision of services and elimination of associated costs. For the achievement of UHC some fundamental strategies are imperative which include added healthcare financing, scaling up of the public sector, sustainable health protection, decreasing donor dependence, and improved accountability (Shaikh & Ali, 2023). Community health workers (CHWs are similar to the soldiers on the front line in the battle against dreadful health statuses prevalent in the less developed world today. They are very crucial in ensuring that minimal healthcare needs are met in the community. To support this view, the World Health Organization has advocated for the use of CHWs to support global health objectives. Several countries face difficulties in operating sustainable CHW programmes through emerging challenges like scarcity of human capital in the healthcare sector, fragile health systems, inadequate planning and coordination, and ineffective partnership models. However, there are potential countermeasures including international coordination, national policies and strategies, and strategic collaboration. Overcoming challenges as well as developing a proper partnership with CHWs, nations can capitalize on this workforce to work towards universal healthcare coverage and a better quality of life for all. Rizvi Jafree & Burhan (2020) also stated that mothers who are mostly taking care of their special needs children reported lots of mental and physical health issues. They consider issues within various aspects of mental and physical well-being. There is a need for partnership between social workers and healthcare providers to ensure that mothers receive the necessary healthcare support. Furthermore, cooperation with social workers, community members, and policymakers will enhance the availability of social and practical aid for such families.

Pakistan along with many other developing countries faces the problem of not being able to sign and then implement sustainable development policies. This is so true for government procurement, an aspect that has not been revolutionized by the Paris Agreement on climate changes. U. Ahmed et al. (2021) emphasized that the problem of sustainable procurement in the government requires the formation of a clear set of rules and requirements for sustainable purchasing in government agencies and institutions. Furthermore, implementing sustainable practices would require the engagement of senior leaders in supporting and./or enforcing those practices. This policy outline may be adopted as a guide to all the concerned policymakers and planners for establishing sustainable practices in all the sectors in Pakistan.

Imagine a disaster strikes! Will there be someone to assist the affected? Gillani et al. (2022) supported the importance of healthcare professionals across emergencies. These professionals must be knowledgeable, optimistic, and prepared for such events to happen. This study also revealed that a significant proportion of healthcare workers in Pakistan are not adequately knowledgeable, do not possess the right attitude, and are not well prepared in the event of a disaster. Floods are a familiar situation in Pakistan, as a consequence of the monsoon season. Of course, these floods can hinder healthcare services and endanger the lives of people. Pradhan et al. (2022) focused on five flood-affected districts to understand how the healthcare sector prepares and functions during such crises. The study found out that there were several challenges which include, a rise in the prevalence of diseases, rising incidents of violence against women, disruption in transportation, scarcity in medicine, shortage in the health workforce, lack of training opportunities, use of different reports, and lack of gender-sensitive response by Pakistan, if all these challenges are tackled, then Pakistan should be able to enhance its health care in responding to the needs of everyone during floods.

Despite commitments from countries like Pakistan to combat antibiotic resistance (AMR), implementing these plans remains challenging due to weak enforcement, significant opposition, and limited awareness. Global strategies fail to consider the local circumstances since they rely on potent rules that cannot be applied in lower-middle-income countries (LMICs) (Khan et al., 2020). Saleem et al. (2022) discussed challenges in Pakistan's National Action Plan (NAP) like insufficient equipment and funds in it. These hurdles can only be surmounted if NAP goals are continually assessed, and improved infection control measures are availed in healthcare facilities. Solving these problems will assist Pakistan in attaining the goals of NAP and controlling AMR. In their work, Fatima et al. (2020) pointed to the matter of safe surgeries, underlining the fact that in most countries, such an opportunity is unattainable.

The figures are concerned in Punjab, Pakistan where cancer incidence is gradually increasing. To cope with this increasing load, Khokhar et al. (2020) opined that there is a need to upgrade the healthcare system of the country, and to do so the following suggestions can be made: enhancement and development of education and training of the healthcare personnel, development of private-public sector partnership and collaborations, monitoring of cancer trends, early diagnosis program, awareness drive, and increase in research activities. Like Rattani et al. (2022) stress many cancer patients in Pakistan face very confronting situations influenced by financial problems, misconceptions about the disease, pain, cultural perception of health, irrelevant knowledge, and access to the healthcare centers. The study offered key recommendations for enhancing cancer care throughout the country: education of healthcare workers, fighting the corruption component, and attaining universal health insurance coverage.

However, vaccination remains a low priority in Pakistan even though it is among the world's most successful public health interventions as captured by Butt et al. (2020). This is in a way complicated since diseases that can be easily controlled in other regions continue to pose a threat to the people. The study presents barriers peculiar to Pakistan such as vaccine distribution, motivation of healthcare workers, health literacy, cultural and religious factors, and immunizing children in conflict areas which slows down Pakistan's averting to measles immunization goal. Polio negatively affects Pakistan and is still considered an Endemic disease affecting young children all over the world. According to the study conducted by Shabbir et al. (2022), Pakistan's ongoing polio challenge includes false beliefs, mistrust, low knowledge, insecurity, hard-to-reach communities, and inaccessible areas. The treatment of these problems is paramount in eradicating polio and protecting Pakistan's children from the ravages of this disease.

Kumar et al. (2022) highlighted a distressing issue in Baluchistan, Pakistan's largest province: poor primary health care, especially for newborns. Based on its size, it is rather a challenge for the province to adequately address the needs of the population because of the scarcity of services and equipment. Thus, national and local governments, healthcare workers, and communities should work together in the long-term process of sustained development of a strong healthcare system. Together, there are ways that Balochistan could improve its healthcare capacity as well as start the lives of its newborns on a healthy note. Das et al. (2020) analyzed the patterns of mothers' and children's healthcare access in conflict-affected Balochistan and Khyber Pakhtunkhwa and demonstrated that increased conflict intensity is associated with worse healthcare services. The research revealed a wide range of factors that limit the healthcare provision to the communities residing in these zones, such as poor planning and coordination, shortage of staff, weak infrastructure and information systems, finances, and supply instability. The essential sectors such as women's and children's health are respected as priorities, but due to limited resources and deadly conflicts, the provision of quality care is incomprehensible.

According to Zaidi et al. (2021), some of the trending strategic impediments in prompting sustainable healthcare objectives in Pakistan include; a lack of clarity on government policies, lack of unity from the leadership, and organizational factors like; resistance to change, involvement from outside, and inadequate access to green products. Childhood malnutrition especially in the poor and remote areas still poses a big challenge as far as health interventions are concerned. According to Ahmed et al. (2022), families struggle to attain treatment programmes owing to factors such as mothers' education levels; lack of adequate capital in healthcare; logistics, corruption, distance, and social disgrace. As for the shortcomings described by the study, the authors emphasized the need for government engagement to remove such barriers and provide equal children's access to all essential treatments.

In practical recommendations regarding the healthcare system strengthening in Pakistan, Khattak et al. (2023) stated the need for the Government of Pakistan to invest more in the health sector; they also pointed out the options for the implementation of UHC to make healthcare services accessible to all citizens. The measures proposed included improving public health facilities with an emphasis on preventive measures, developing partnerships between the public and private sectors, utilizing communication technologies for the dissemination of public health information and provision of telemedicine services, and bettering organizational performance. Other bureaucratic recommendations included transparency in reforms, leadership, and health policies that should also be learned from other developing countries. The employment of such measures could assist Pakistan in putting into place a stable health sector that supports the population.

Figure 1. Challenges of sustainable healthcare in Pakistan

ROLE OF SDI FOR HEALTHCARE PLANNING

Pakistan can utilize SDI to transform the methods of healthcare planning and provisioning. Since SDI holds data on population density and distribution, positions of health facilities, and rate of occurrences of diseases, among others, the planner is in a position to make informed decisions. This capability enables the right positioning of new centers, efficient resource utilization, and tackling health issues. SDI improves accountability because, through it, people get to see the data used in planning health system solutions. Considering SDI's application for important issues in planning health care needs, Table 1 provides a breakdown of how this infrastructure works in solving the outlined challenges.

Table 1. Potential role of SDI for sustainable healthcare in Pakistan

Challenges in Healthcare Planning	How SDI Address these Challenges
Limited Access to Quality Healthcare	Using SDI to overlay the current geographical placement of the healthcare facilities and alternatively use a geographical information system to determine areas with poor coverage of these services. It helps in identifying the locations that still lack access to health and assigning ways of how new centers or mobile clinics can be placed to improve the reach.
Limited Range of Services	Using SDI for demographic and healthcare geographical analysis to identify areas where certain services are missing, for instance, generalized or restricted care, or diagnostic capabilities. This means that funding can be directed properly as well as the services where there is demand for improvement.
Reduced Accountability	Adopting SDI in the field of healthcare to enhance clear revelations of results, cost, and resource allocation. This improves accountability as it is available to other stakeholders updated and accurate details concerning health facilities and services consumption.
Shortage of Healthcare Workers	Using SDI to enhance the understanding of the distribution of the population and density of health workers and patient-to-health officials' ratios. This procedure helps in utilizing the manpower most efficiently, finding out where there is a lack of HR to plan for more recruitment or redistribution of health care personnel.
Poor Planning	Applying form SDI as a tool for evidence-based planning by using demographic, health, and environmental data to determine where exactly the investment in health care and other infrastructure should be targeted. This also helps the policymakers to effectively fund and develop the plans to construct and organize the required healthcare structures.
Lack of Coordination and Collaboration	Applying the SDI concept to harmonize the healthcare data within and across the stakeholders' boundaries and ensure proper exchange of data among different agencies and partners. This fosters teamwork, communication, as well as interprofessional collaboration, hence results in improvement of the quality of health care provision and resources management.
Unsuccessful Partnerships	Applying SDI for spatial analysis to find areas that have redundant healthcare requirements and possible partnerships. Through matching healthcare resources, the demographic information of the suppliers and consumers of the services, and services' deficiencies and spaces, stakeholders can identify possible collaborative strategies to meet the existing challenges and provide efficient services depending on the geographic necessities.
Lack of National Planning	Creating the concept of a national SDI related to the use of integrated healthcare planning and policy purposes. This structure allows the assembly, normalization, and evaluation of healthcare data on the national level and assists in making governmental decisions and distributing resources most efficiently across the country.
Lack of Clear Policies and Regulations	Using SDI technologies in spatial analysis to determine geographical features in relation to health care delivery and policy making or monitoring and enforcing the laws. This enhances the chance of having data-driven policies and regulations on enhancing healthcare that respond to specific healthcare issues.

continued on following page

Table 1. Continued

Challenges in Healthcare Planning	How SDI Address these Challenges
Lack of Commitment from Top Leadership	Implementing SDI for displaying the spatial information to show results of health care projects and gaining management support. Combined with the findings thinking about the impacts of healthcare and overall perceptions of the state of healthcare, the visible representation of these impacts, service coverage, and resource distribution can assist stakeholders to successfully showcase the worth of healthcare investments to policymakers.
Lack of Knowledge and Awareness	Utilizing SDI for spatial data analysis in focused public health promotion crusades in parts of the geography with little or no knowledge or awareness. Through overlaying the health indicators, disease occurrence, and health institutions, intended information and education can be targeted most appropriately.
Lack of Preparedness	Applying SDI datasets for disaster preparedness context by evaluating the risk areas, the potentially existing health care facilities, and respective evacuation pathways. This makes it easy for the authorities to form accurate emergency preparedness measures, location of resources, and organization of health services during calamities to reduce avoidable loss of life and property.
Disruptions in Healthcare Services During Disasters	Successfully applying SDI in real-time response and cooperation of emergency health care in disaster situations will improve service delivery during disasters. By using the data acquired in the fields of transport, demography as well as health care, the authorities can identify the disruption of health care services, the needs, and the distribution of resources in the areas concerned.
Limited Enforcement	Applying SDI to manage compliance with the healthcare regulations as well as enforcement responsibilities. This involves physically tracking the healthcare facilities, licensing, quality assurance, and healthcare providers' conduct to ensure adherence to the legal requirements and promote responsibility in healthcare services provision.
Challenges in Implementing the National Action Plan	Integrating SDI for efficient execution and monitoring of national healthcare strategies. This includes spatially mapping goals, progress indicators, and healthcare resources of action plans to monitor implementation, identify obstacles, and adapt strategies for achieving healthcare objectives effectively.
Financial Hardship	Applying SDI for spatial analysis of distribution of the resource to identify the districts that require financial inputs for Health Infrastructure, Services & Programs. With the help of the location of factors such as healthcare expenditure, demography, and gaps in services, the government can direct resources more efficiently and prevent the concentration of healthcare services.
Conflict Zones	Assisting in the spatial mapping of areas of conflict where healthcare delivery and focused humanitarian activities are to be conducted safely. Identifying geographical locations of conflict, healthcare facilities, and approaching corridors allows for developing robust plans to protect caregivers and recipients, continue delivering primary care, and effectively bring in aid shipments.
Limited Facilities and Resources	Employing SDI datasets for efficient healthcare facility development with steering on population density, healthcare demand, and lack of infrastructural amenities. This can help the authorities identify the regions requiring new healthcare facilities, improve the infrastructure of existing structures, and direct resources to alleviate the lack of healthcare services.
Corruption	Implementation of SDI datasets for rectifying the problem of corruption in the allocation and tracking of resources in the health sector. When displayed geospatially, expenditures on healthcare and other physical infrastructural investments, as well as services delivered can alert stakeholders to instances of fraud, and enforce contractual responsibilities on anyone neglecting to use the health resources appropriately.
Lack of Logistics	Maximizing the role of SDI on healthcare supply chain management through mapping distribution channels, transport routes, and stockpiling. This capability allows the authorities to manage the transport and supply of pharmaceuticals and medical equipment more effectively and efficiently, especially to healthcare centers, hence improving healthcare resource availability.

This table shows that SDI is a great way to bring clarity and streamline cooperation in the area of healthcare planning. By utilizing SDI datasets, key decision-makers can respond effectively to urgent issues in the health segment. It shows that integrating SDI in a country's healthcare system such as Pakistan holds a massive potential to revolutionize the way several issues relating to healthcare planning and delivery are done. This is especially important to SDI as it is a methodology that, among other things, aims to reveal low-populated areas and to determine where new stations or mobile clinics can be built.

Concerning, it increases the delivery of additional services to areas that currently do not have such care, and secondly, it improves the accountability of the services by displaying the amount of funding and the services provided as well as the consumption by individual regions. Additionally, it backs up workforce planning as the distribution and proportion of population healthcare workers are comprehensively examined to help in recruitment and distribution of human resource in health. On a more strategic level, SDI can be used to provide an evidence-based approach to planning, which will help integrate the different stakeholders and their data in the effective coordination of development in healthcare. Using geographic information on health care requirements and organization SDI also helps in establishing partnerships between the private and public sectors, planning at the national level, and policy formulation as well. Notably, SDI improves the procedures for preparing to deal with disasters, implementing and obeying the legislation that governs healthcare, eradicating problems such as corruption, and inefficient supply chain management. Through these applications, SDI can also greatly contribute to enhancing Sustainable health care in Pakistan, equal distribution of services, better health care indicators, and optimal use of relevant resources across the country.

CONCLUSION

The integration of SDI presents a shift in health care planning in Pakistan where there is a lot of potential to improve. As a means for consolidating and processing geospatial data, SDI provides decision-makers with such information that enables addressing significant issues in healthcare. With the help of spatial mapping, analysis, and real-time monitoring SDI makes the decision-makers aware and accountable and also enhances the level of transparency in the planning of healthcare. It is important to emphasize that the analysis provided in this chapter demonstrates SDI as a versatile tool for addressing the broad spectrum of healthcare planning concerns that can either help improve the patient's access to high-quality doctors or act as a logistical advantage. This paper reveals how Pakistan can enhance the efficiency of its external resource mobilization and domestic service delivery to attain improved, person-centered health outcomes with SDI's support.

In the future, the application of SDI in establishing the healthcare plan has possible directions for improvements highlighted in the following section. Future work should investigate how to improve the SDI support structure and services at the national as well as provincial levels. This encompasses strengthening the data gathering, storage, and analysis processes and technologies for health care solutions. However, more importantly, government administrators, healthcare facilities, and other multifaceted technology partners do need to come forward for the real realization of societal and healthcare applications of SDI. To make spatial data credible and usable by decision-makers, data quality as well as data and software interoperability and availability need to be improved. Further, understanding novel use cases for the shifting and upcoming technologies of artificial intelligence, machine learning, and remote sensing in SDI is a way forward to harness new possibilities in handling multifaceted aspects related to healthcare. In conclusion, by adopting these future strategies, Pakistan has an opportunity to build a capable healthcare system that will effectively address the constantly shifting requirements of the population in the context of globalization.

REFERENCES

Ahmad, M. (2023). Leveraging Blockchain for Spatial Data Infrastructure: Challenges and Opportunities. In Ferreira, L., Cruz, M. R., Cruz, E. F., Quintela, H., & Cunha, M. C. (Eds.), *Supporting Technologies and the Impact of Blockchain on Organizations and Society* (pp. 177–194). IGI Global., DOI: 10.4018/978-1-6684-5747-4.ch011

Ahmad, M., & Ali, A. (2023a). A Review of Governance Frameworks for National Spatial Data Infrastructure. In Correia, A., & Agua, P. B. (Eds.), *Innovation, Strategy, and Transformation Frameworks for the Modern Enterprise* (pp. 186–207). IGI Global., DOI: 10.4018/979-8-3693-0458-7.ch008

Ahmad, M., & Ali, A. (2023b). Integrated Geospatial Information Management in the World of Metaverse: Challenges and Opportunities. In Lekhi, P., & Kaur, G. (Eds.), *Concepts, Technologies, Challenges, and the Future of Web 3* (pp. 248–270). IGI Global., DOI: 10.4018/978-1-6684-9919-1.ch013

Ahmad, M., Ali, A., & Khiyal, M. S. H. (2023). Fog Computing for Spatial Data Infrastructure: Challenges and Opportunities. In D. P. Acharjya & K. Ahmed P. (Eds.), *Multi-Disciplinary Applications of Fog Computing: Responsiveness in Real-Time* (pp. 152–178). IGI Global. DOI: 10.4018/978-1-6684-4466-5.ch008

Ahmad, M., Ali, A., Mobo, F. D., & Hussain, H. (2024). *Spatial Data Infrastructure for Information Governance: The Case Study From Pakistan.* https://doi.org/https://doi.org/10.4018/979-8-3693-0472-3.ch011

Ahmad, M., Ali, A., Nawaz, H., Arslan, M., & Kumar, N. (2024). Optimizing Integrated Spatial Data Management Through Fog Computing: A Comprehensive Overview. In Prakasha, G. S., Lapina, M., Balakrishnan, D., & Sajid, M. (Eds.), *Digital Technologies in Modeling and Management: Insights in Education and Industry* (pp. 263–283). IGI Global., DOI: 10.4018/978-1-6684-9576-6.ch013

Ahmad, M., Ali, A., Tin, T. T., & Ali, A. (2024). Harnessing the Power of Geospatial Data: The Convergence of SDI, Big Data, and Cloud Computing. In Darwish, D. (Ed.), *Emerging Trends in Cloud Computing Analytics, Scalability, and Service Models* (pp. 312–330). IGI Global., DOI: 10.4018/979-8-3693-0900-1.ch016

Ahmad, M., Khayal, M. S. H., & Tahir, A. (2022). Analysis of Factors Affecting Adoption of Volunteered Geographic Information in the Context of National Spatial Data Infrastructure. *ISPRS International Journal of Geo-Information*, 11(2), 120. Advance online publication. DOI: 10.3390/ijgi11020120

Ahmed, F., Malik, N. I., Malik, N., Qureshi, M. G., Shahzad, M., Shahid, M., Zia, S., & Tang, K. (2022). Key Challenges to Optimal Therapeutic Coverage and Maternal Utilization of CMAM Program in Rural Southern Pakistan: A Qualitative Exploratory Study. *Nutrients*, 14(13), 2612. Advance online publication. DOI: 10.3390/nu14132612 PMID: 35807795

Ahmed, S. U., Ali, A., Memon, A. H., & Hussain, A. (2021). Policy framework for implementation of sustainable procurement practices in healthcare industry. *Environment, Development and Sustainability*, 23(3), 3929–3945. Advance online publication. DOI: 10.1007/s10668-020-00750-w

Ali, A., Ahmad, M., Nawaz, M., & Sattar, F. (2024). Spatial data infrastructure as the means to assemble geographic information necessary for effective agricultural policies in Pakistan. *Information Development*, 0–0. DOI: 10.1177/02666669241244503

Ali, A., & Imran, M. (2021). National Spatial Data Infrastructure vs. Cadastre System for Economic Development: Evidence from Pakistan. *Land (Basel)*, 10(2), 188. DOI: 10.3390/land10020188

Alowais, S. A., Alghamdi, S. S., Alsuhebany, N., Alqahtani, T., Alshaya, A. I., Almohareb, S. N., Aldairem, A., Alrashed, M., Bin Saleh, K., Badreldin, H. A., Al Yami, M. S., Al Harbi, S., & Albekairy, A. M. (2023). Revolutionizing healthcare: the role of artificial intelligence in clinical practice. In *BMC Medical Education* (Vol. 23, Issue 1). DOI: 10.1186/s12909-023-04698-z

Batko, K., & Ślęzak, A. (2022). The use of Big Data Analytics in healthcare. *Journal of Big Data*, 9(1), 3. Advance online publication. DOI: 10.1186/s40537-021-00553-4 PMID: 35013701

Bishop, I. D., Escobar, F. J., Karuppannan, S., Williamson, I. P., Yates, P. M., Suwarnarat, K., & Yaqub, H. W. (2000). Spatial data infrastructures for cities in developing countries. Lessons from the Bangkok experience. *Cities (London, England)*, 17(2), 85–96. Advance online publication. DOI: 10.1016/S0264-2751(00)00004-4

Boerboom, I. (2010). Integrating spatial planning and decision support system infrastructure and spatial data infrastructure. *GSDI 12 World Conference Singapore*. http://www.gsdi.org/gsdiconf/gsdi12/papers/82.pdf

Borzacchiello, M. T., & Craglia, M. (2013). Estimating benefits of spatial data infrastructures: A case study on e-Cadastres. *Computers, Environment and Urban Systems*, 41, 276–288. Advance online publication. DOI: 10.1016/j.compenvurbsys.2012.05.004

Butt, M., Mohammed, R., Butt, E., Butt, S., & Xiang, J. (2020). Why have immunization efforts in Pakistan failed to achieve global standards of vaccination uptake and infectious disease control? In *Risk Management and Healthcare Policy* (Vol. 13). DOI: 10.2147/RMHP.S211170

Cetl, V., Ivić, S. M., & Tomić, H. (2009). Improvement of national spatial data infrastructure as a public project of permanent character. *Kartografija i Geoinformacije, 8*(11).

Chakraborty, S., Raut, R. D., Rofin, T. M., & Chakraborty, S. (2023). A comprehensive and systematic review of multi-criteria decision-making methods and applications in healthcare. In *Healthcare Analytics* (Vol. 4). DOI: 10.1016/j.health.2023.100232

Coleman, D. J., & Nebert, D. D. (1998). Building a North American Spatial Data Infrastructure. *Cartography and Geographic Information Systems*, 25(3), 151–160. Advance online publication. DOI: 10.1559/152304098782383098

Crompvoets, J., Vancauwenberghe, G., Bouckaert, G., & Vandenbroucke, D. (2011). Practices to develop spatial data infrastructures: Exploring the contribution to e-government. In *Practical Studies in E-Government.* Best Practices from Around the World., DOI: 10.1007/978-1-4419-7533-1_13

Crompvoets, J., Vancauwenberghe, G., Ho, S., Masser, I., & Timo De Vries, W. (2018). Governance of national spatial data infrastructures in Europe. *International Journal of Spatial Data Infrastructures Research*, 13, 253–285.

Das, J. K., Padhani, Z. A., Jabeen, S., Rizvi, A., Ansari, U., Fatima, M., Akbar, G., Ahmed, W., & Bhutta, Z. A. (2020). Impact of conflict on maternal and child health service delivery - How and how not: A country case study of conflict affected areas of Pakistan. *Conflict and Health*, 14(1), 32. Advance online publication. DOI: 10.1186/s13031-020-00271-3 PMID: 32514297

Diaz, L., Remke, A., Kauppinen, T., Degbelo, A., Foerster, T., Stasch, C., Rieke, M., Schaeffer, B., Baranski, B., Bröring, A., & Wytzisk, A. (2012). Future SDI - Impulses from Geoinformatics research and IT trends. *International Journal of Spatial Data Infrastructures Research*, 7. Advance online publication. DOI: 10.2902/1725-0463.2012.07.art18

Fatima, I., Shoman, H., Peters, A. W., Samad, L., & Nishtar, S. (2020). Pakistan's National Surgical, Obstetric, and Anesthesia Plan: An adapted model for a devolved federal-provincial health system. *Canadian Journal of Anaesthesia*, 67(9), 1212–1216. Advance online publication. DOI: 10.1007/s12630-020-01708-2 PMID: 32409943

Gillani, A. H., Li, S., Akbar, J., Omer, S., Fatima, B., Ibrahim, M. I. M., & Fang, Y. (2022). How Prepared Are the Health Care Professionals for Disaster Medicine Management? An Insight from Pakistan. *International Journal of Environmental Research and Public Health*, 19(1), 200. Advance online publication. DOI: 10.3390/ijerph19010200 PMID: 35010475

Granell, C., Schade, S., & Hobona, G. (2011). Linked Data: Connecting Spatial Data Infrastructures and Volunteered Geographic Information. In *Geospatial Web Services: Advances in Information Interoperability*.

Habib, S. S., Jamal, W. Z., Zaidi, S. M. A., Siddiqui, J. U. R., Khan, H. M., Creswell, J., Batra, S., & Versfeld, A. (2021). Barriers to access of healthcare services for rural women— Applying gender lens on tb in a rural district of sindh, pakistan. *International Journal of Environmental Research and Public Health*, 18(19), 10102. Advance online publication. DOI: 10.3390/ijerph181910102 PMID: 34639403

Horenstein, A., & Heimberg, R. G. (2020). Anxiety disorders and healthcare utilization: A systematic review. In *Clinical Psychology Review* (Vol. 81). DOI: 10.1016/j.cpr.2020.101894

Hussnain, M. Q. U., Waheed, A., Wakil, K., Pettit, C. J., Hussain, E., Naeem, M. A., & Anjum, G. A. (2020). Shaping up the future spatial plans for urban areas in Pakistan. *Sustainability*, 12(10), 4216.

Imran, M. (2009). Public transport in Pakistan: A critical overview. *Journal of Public Transportation*, 12(2), 53–83. DOI: 10.5038/2375-0901.12.2.4

Kalwar, S., Sahito, N., Memon, I. A., Hwang, J., Mangi, M. Y., & Lashari, Z. A. (2019). National Planning Strategies for Agro-based Industrial Development in Secondary Cities of Sindh Province, Pakistan. *Sustainability (Basel)*, 11(24), 7066. Advance online publication. DOI: 10.3390/su11247066

Khan, M. S., Durrance-Bagale, A., Mateus, A., Sultana, Z., Hasan, R., & Hanefeld, J. (2020). What are the barriers to implementing national antimicrobial resistance action plans? A novel mixed-methods policy analysis in Pakistan. *Health Policy and Planning*, 35(8), 973–982. Advance online publication. DOI: 10.1093/heapol/czaa065 PMID: 32743655

Khashoggi, B. F., & Murad, A. (2020). Issues of healthcare planning and GIS: A review. In *ISPRS International Journal of Geo-Information* (Vol. 9, Issue 6). DOI: 10.3390/ijgi9060352

Khattak, A. F., Rahman, A. U., Khattak, M., Qazi, M., Gilani, H., & Khan, A. (2023). Toward Sustainable Healthcare Systems: A Low and Middle-Income Country's Case for Investing in Healthcare Reforms. *Cureus*. Advance online publication. DOI: 10.7759/cureus.39345 PMID: 37351239

Khokhar, M. A., Ali, M. M., Liaqat, S., Moin, A., Sarwar, H. A., & Sarwar, M. Z. (2020). A review of access to cancer facilities in Punjab, Pakistan. In *Cancer Reports* (Vol. 3, Issue 3). DOI: 10.1002/cnr2.1245

Kornelsen, J., Carthew, C., Míguez, K., Taylor, M., Bodroghy, C., Petrunia, K., & Roberts, D. (2021). Rural citizen-patient priorities for healthcare in British Columbia, Canada: Findings from a mixed methods study. *BMC Health Services Research*, 21(1), 987. Advance online publication. DOI: 10.1186/s12913-021-06933-z PMID: 34537027

Kumar, V., Ali, B. S., Choudry, E., Khan, S., Baig, K., Durrani, N. U. R., & Ali, S. R. (2022). Quality of Neonatal Care: A Health Facility Assessment in Balochistan Province, Pakistan. *Cureus*. Advance online publication. DOI: 10.7759/cureus.22744 PMID: 35386481

Manfré, L., Hirata, E., Silva, J. B., Shinohara, E. J., Giannotti, M., Larocca, A. P. C., & Quintanilha, J. (2012). An Analysis of Geospatial Technologies for Risk and Natural Disaster Management. *ISPRS International Journal of Geo-Information*, 1(2), 166–185. DOI: 10.3390/ijgi1020166

Masser, I. (2019). Geographic Information Systems to Spatial Data Infrastructure. In *Geographic Information Systems to Spatial Data Infrastructure*. DOI: 10.1201/9780429505904

McLaughlin, J., & Nichols, S. (1994). Developing a national spatial data infrastructure. *Journal of Surveying Engineering*, 120(2), 62–76. DOI: 10.1061/(ASCE)0733-9453(1994)120:2(62)

Memon, A. A. Q., Osama, M., Wei, C. R., Rasool, G., Bhurgri, R. S., Siyal, D. R., & Siyal, F. J. (2024). Common Health Challenges For Foreigners In Pakistan. *Migration Letters : An International Journal of Migration Studies*, 21(S8), 131–142.

Nedović-Budić, Z., Crompvoets, J., & Georgiadou, Y. (2011). Spatial data infrastructures in context: North and South. In *Spatial Data Infrastructures in Context*. North and South. DOI: 10.1201/b10857

Poirier, L., Antonio, D., Dettmann, M., Eng, T., Ganata, J., Ghosh, S., Lopez, M., Karma, R., Natekal, A., & Brinkley, C. (2024). Making plans findable, accessible, interoperable, and reusable with data infrastructure: A search engine for constructing, analyzing, and visualizing planning documents. *Environment and Planning. B, Urban Analytics and City Science*, 51(8), 1913–1930. Advance online publication. DOI: 10.1177/23998083241227471

Pradhan, N. A., Najmi, R., & Fatmi, Z. (2022). District health systems capacity to maintain healthcare service delivery in Pakistan during floods: A qualitative study. *International Journal of Disaster Risk Reduction*, 78, 103092. Advance online publication. DOI: 10.1016/j.ijdrr.2022.103092

Rajabifard, A., Binns, A., Masser, I., & Williamson, I. (2006). The role of sub-national government and the private sector in future spatial data infrastructures. *International Journal of Geographical Information Science*, 20(7), 727–741. Advance online publication. DOI: 10.1080/13658810500432224

Rajabifard, A., Feeney, M. E. F., & Williamson, I. P. (2002). Future directions for SDI development. *International Journal of Applied Earth Observation and Geoinformation*, 4(1), 11–22. DOI: 10.1016/S0303-2434(02)00002-8

Rattani, S. A., Dahlke, S., & Cameron, B. (2022). Cancer Care in Pakistan: A Descriptive Case Study. *Global Qualitative Nursing Research*, 9, 23333936221080988. Advance online publication. DOI: 10.1177/23333936221080988 PMID: 35237708

Rizvi Jafree, S., & Burhan, S. K. (2020). Health challenges of mothers with special needs children in Pakistan and the importance of integrating health social workers. *Social Work in Health Care*, 59(6), 408–429. Advance online publication. DOI: 10.1080/00981389.2020.1781738 PMID: 32614737

Saleem, Z., Godman, B., Azhar, F., Kalungia, A. C., Fadare, J., Opanga, S., Markovic-Pekovic, V., Hoxha, I., Saeed, A., Al-Gethamy, M., Haseeb, A., Salman, M., Khan, A. A., Nadeem, M. U., Rehman, I. U., Qamar, M. U., Amir, A., Ikram, A., & Hassali, M. A. (2022). Progress on the national action plan of Pakistan on antimicrobial resistance (AMR): a narrative review and the implications. In *Expert Review of Anti-Infective Therapy* (Vol. 20, Issue 1). DOI: 10.1080/14787210.2021.1935238

Shabbir, H., Saeed, S., Farhan, M., Abbas, K., Rehman, M. E., Gul, F., & Basit, J. (2022). Poliomyelitis in Pakistan: Challenges to polio eradication and future prospects. *Annals of Medicine and Surgery (London)*, 80. Advance online publication. DOI: 10.1016/j.amsu.2022.104274 PMID: 36045811

Shaikh, B. T., & Ali, N. (2023). Universal health coverage in Pakistan: Is the health system geared up to take on the challenge? *Globalization and Health*, 19(1), 4. Advance online publication. DOI: 10.1186/s12992-023-00904-1 PMID: 36635698

Skelton, C. (2014). *The implementation and education of geographic information systems in a local government for municipal planning: A case study of Dangriga, Belize.*

Snow, M. E. (2022). Patient engagement in healthcare planning and evaluation: A call for social justice. *The International Journal of Health Planning and Management*, 37(S1), 20–31. Advance online publication. DOI: 10.1002/hpm.3509 PMID: 35643909

Tasci, A. D. A., Wei, W., & Milman, A. (2020). Uses and misuses of the case study method. *Annals of Tourism Research*, 82, 102815. Advance online publication. DOI: 10.1016/j.annals.2019.102815

Wang, F. (2020). Why public health needs GIS: a methodological overview. In *Annals of GIS* (Vol. 26, Issue 1). DOI: 10.1080/19475683.2019.1702099

Wronski, P., Koetsenruijter, J., Ose, D., Paulus, J., Szecsenyi, J., & Wensing, M. (2021). Healthcare planning across healthcare sectors in Baden-Wuerttemberg, Germany: A stakeholder online survey to identify indicators. *BMC Health Services Research*, 21(1), 510. Advance online publication. DOI: 10.1186/s12913-021-06514-0 PMID: 34039315

Yan, S., Seng, B. J. J., Kwan, Y. H., Tan, C. S., Quah, J. H. M., Thumboo, J., & Low, L. L. (2019). Identifying heterogeneous health profiles of primary care utilizers and their differential healthcare utilization and mortality - A retrospective cohort study. *BMC Family Practice*, 20(1), 54. Advance online publication. DOI: 10.1186/s12875-019-0939-2 PMID: 31014231

Zaidi, S. A. H., Shahbaz, M., Hou, F., & Abbas, Q. (2021). Sustainability challenges in public health sector procurement: An application of interpretative structural modelling. *Socio-Economic Planning Sciences*, 77, 101028. Advance online publication. DOI: 10.1016/j.seps.2021.101028

KEY TERMS AND DEFINITIONS

Spatial Data: Refers to any data that is related to or describes a specific location on Earth. This data can include geographical coordinates (like latitude and longitude), addresses, or any information that can be mapped, such as the locations of hospitals, population density, or environmental features. Spatial data is essential in fields like urban planning, transportation, and environmental management.

Spatial Data Infrastructure (SDI): Refers to the framework of technologies, policies, standards, and human resources necessary to acquire, process, store, distribute, and improve the utilization of spatial data. It includes the tools and platforms that facilitate the sharing and use of spatial data across different organizations and sectors. SDI is crucial for enabling effective decision-making in areas like disaster management, urban planning, and environmental conservation.

Healthcare Planning: Involves the strategic organization and management of health services and resources to meet the health needs of a population. This includes planning for hospitals, clinics, emergency services, and public health initiatives. Spatial data and Geographic Information Systems (GIS) play a vital role in healthcare planning by helping to analyze health trends, determine optimal locations for facilities, and ensure equitable access to healthcare.

Geographic Information System (GIS): A technology that allows users to capture, store, analyze, and visualize spatial or geographic data. It combines location data with other types of data to create maps and models that can be used to understand patterns, relationships, and trends. In healthcare, GIS is used for tasks such as mapping disease outbreaks, planning healthcare services, and analyzing environmental health risks.

Sustainability: Refers to the practice of meeting the needs of the present without compromising the ability of future generations to meet their own needs. It involves balancing environmental, social, and economic considerations in decision-making. In the context of spatial data and healthcare, sustainability might involve planning healthcare infrastructure that minimizes environmental impact, ensuring equitable access to health services, and using resources efficiently.

Compilation of References

A Study on Green Banking Initiatives in the Public and Private Sectors Banks of India. (2023). In *International Conference on New Age Marketing - Catalysing Transformation Through Innovation, Technology, Analytics & Sustainability* [Conference-proceeding]. https://www.sdmimd.ac.in/mc2023/conferencepapers/IMC2269.pdf

Abd-Elaal, E.-S., Gamage, S. H., & Mills, J. E. (2022). Assisting academics to identify computer generated writing. *European Journal of Engineering Education*, 47(5), 725–745. DOI: 10.1080/03043797.2022.2046709

Abdel-Razek, S. A., Marie, H. S., Alshehri, A., & Elzeki, O. M. (2022). Energy efficiency through the implementation of an AI model to predict room occupancy based on thermal comfort parameters. *Sustainability (Basel)*, 14(13), 7734. DOI: 10.3390/su14137734

Abdou, A. H., Hassan, T. H., & Moustafa, M. E. D. (2020). A description of green hotel practices and their role in achieving sustainable development. *Sustainability (Basel)*, 12(22), 9624. DOI: 10.3390/su12229624

Abdullah, M. I., Sarfraz, M., Qun, W., & Javaid, N. (2018). Drivers of green supply chain management. *LogForum*, 14(4), 437–447. DOI: 10.17270/J.LOG.2018.297

Abeydeera, S. (2021). Teaching and Learning Sustainability during and beyond COVID-19: A Review of Business and Management Studies. *University of Colombo Review*, 2(1), 138–154. DOI: 10.4038/ucr.v2i1.43

Abid, E. (2019). The optimum tilt angle and orientation for Solar panels. Journal Port Science Research, 259-263.

Abowardah, E. S., Labib, W., Aboelnagah, H., & Nurunnabi, M. (2024). Students' Perception of Sustainable Development in Higher Education in Saudi Arabia. *Sustainability, 16*(4, 1483), 1–23. DOI: 10.3390/su16041483

Abramic, A., Kotsev, A., Cetl, V., Kephalopoulos, S., & Paviotti, M. (2017). A Spatial Data Infrastructure for Environmental Noise Data in Europe. *International Journal of Environmental Research and Public Health 2017, Vol. 14, Page 726, 14*(7), 726. DOI: 10.3390/ijerph14070726

Abu Dhabi Global Environmental Data Initiative. (2020). *Water availability and use in the United Arab Emirates*. Retrieved from https://www.adgei.ae/en/theme/water-availability-and-use-in-the-united-arab-emirates/

Adam, A. (2019). Optimum Thermal Insulation Thickness for Building Under Different Climate Regions - A Review. *Thermal Science*, 59, 254–268.

Adamopoulos, E., Rinaudo, F., & Ardissono, L. (2021). Geo-Information A Critical Comparison of 3D Digitization Techniques for Heritage Objects †. *ISPRS International Journal of Geo-Information*, 10(1), 10. DOI: 10.3390/ijgi10010010

Adger, W. N., Dessai, S., Goulden, M., Hulme, M., Lorenzoni, I., Nelson, D. R., & Wreford, A. (2009). Are there social limits to adaptation to climate change? *Climatic Change*, 93(3-4), 335–354. DOI: 10.1007/s10584-008-9520-z

Adger, W. N., Huq, S., Brown, K., Conway, D., & Hulme, M. (2005). Adaptation to climate change in the developing world. *Progress in Development Studies*, 3(3), 179–195. DOI: 10.1191/1464993403ps060oa

Adilov, N., Alexander, P. J., & Cunningham, B. M. (2018). Corrigendum to "An economic 'Kessler Syndrome': A dynamic model of earth orbit debris" [Econom. Lett. 166 (2018) 79–82]. *Economics Letters*, 170, 185. DOI: 10.1016/j.econlet.2018.04.012

Afsari, R., Shorabeh, S. N., Kouhnavard, M., Homaee, M., & Arsanjani, J. J. (2022). A Spatial Decision Support Approach for Flood Vulnerability Analysis in Urban Areas: A Case Study of Tehran. *ISPRS International Journal of Geo-Information 2022, Vol. 11, Page 380, 11*(7), 380. DOI: 10.3390/ijgi11070380

Agarwal, Y., Balaji, B., Gupta, R., Lyles, J., Wei, M., & Weng, T. (2010, November). Occupancy-driven energy management for smart building automation. In *Proceedings of the 2nd ACM workshop on embedded sensing systems for energy-efficiency in building* (pp. 1-6).

Aghion, P., Boneva, L., Breckenfelder, J., Laeven, L., Olovsson, C., Popov, A. A., & Rancoita, E. (2022). Financial markets and green innovation.

Agovino, M., Casaccia, M., Ciommi, M., Ferrara, M., & Marchesano, K. (2019). Agriculture, climate change and sustainability: The case of EU-28. *Ecological Indicators*, 105, 525–543. DOI: 10.1016/j.ecolind.2018.04.064

Aguilar-Santana, J., Jarimi, H., Velasco-Carrasco, M., & Riffat, S. (2020). Review on window-glazing technologies and future prospects. International Journal of Low-Carbon Technologies. *The International Journal of Low Carbon Technologies*, 15(1), 112–120. DOI: 10.1093/ijlct/ctz032

Ahmad, M., Ali, A., & Khiyal, M. S. H. (2023). Fog Computing for Spatial Data Infrastructure: Challenges and Opportunities. In D. P. Acharjya & K. Ahmed P. (Eds.), *Multi-Disciplinary Applications of Fog Computing: Responsiveness in Real-Time* (pp. 152–178). IGI Global. DOI: 10.4018/978-1-6684-4466-5.ch008

Ahmad, M., Ali, A., Mobo, F. D., & Hussain, H. (2024). *Spatial Data Infrastructure for Information Governance: The Case Study From Pakistan*. https://doi.org/https://doi.org/10.4018/979-8-3693-0472-3.ch011

Ahmad, A. F., & Karadas, G. (2021). Managers' perceptions regarding the effect of leadership on organizational performance: Mediating role of green supply chain management practices. *SAGE Open*, 11(2), 21582440211018686. DOI: 10.1177/21582440211018686

Ahmadkhani, B. (2011). Windcatcher: passive and low energy cooling system in Iranian vernacular architecture. Tech Phys Problems Eng., 3.

Ahmad, M. (2023). Leveraging Blockchain for Spatial Data Infrastructure: Challenges and Opportunities. In Ferreira, L., Cruz, M. R., Cruz, E. F., Quintela, H., & Cunha, M. C. (Eds.), *Supporting Technologies and the Impact of Blockchain on Organizations and Society* (pp. 177–194). IGI Global., DOI: 10.4018/978-1-6684-5747-4.ch011

Ahmad, M., & Ali, A. (2023a). A Review of Governance Frameworks for National Spatial Data Infrastructure. In Correia, A., & Agua, P. B. (Eds.), *Innovation, Strategy, and Transformation Frameworks for the Modern Enterprise* (pp. 186–207). IGI Global., DOI: 10.4018/979-8-3693-0458-7.ch008

Ahmad, M., & Ali, A. (2023b). Integrated Geospatial Information Management in the World of Metaverse: Challenges and Opportunities. In Lekhi, P., & Kaur, G. (Eds.), *Concepts, Technologies, Challenges, and the Future of Web 3* (pp. 248–270). IGI Global., DOI: 10.4018/978-1-6684-9919-1.ch013

Ahmad, M., Ali, A., Nawaz, H., Arslan, M., & Kumar, N. (2024). Optimizing Integrated Spatial Data Management Through Fog Computing: A Comprehensive Overview. In Prakasha, G. S., Lapina, M., Balakrishnan, D., & Sajid, M. (Eds.), *Digital Technologies in Modeling and Management: Insights in Education and Industry* (pp. 263–283). IGI Global., DOI: 10.4018/978-1-6684-9576-6.ch013

Ahmad, M., Ali, A., Tin, T. T., & Ali, A. (2024). Harnessing the Power of Geospatial Data: The Convergence of SDI, Big Data, and Cloud Computing. In Darwish, D. (Ed.), *Emerging Trends in Cloud Computing Analytics, Scalability, and Service Models* (pp. 312–330). IGI Global., DOI: 10.4018/979-8-3693-0900-1.ch016

Ahmad, M., Khayal, M. S. H., & Tahir, A. (2022). Analysis of Factors Affecting Adoption of Volunteered Geographic Information in the Context of National Spatial Data Infrastructure. *ISPRS International Journal of Geo-Information*, 11(2), 120. Advance online publication. DOI: 10.3390/ijgi11020120

Ahmed, F., Malik, N. I., Malik, N., Qureshi, M. G., Shahzad, M., Shahid, M., Zia, S., & Tang, K. (2022). Key Challenges to Optimal Therapeutic Coverage and Maternal Utilization of CMAM Program in Rural Southern Pakistan: A Qualitative Exploratory Study. *Nutrients*, 14(13), 2612. Advance online publication. DOI: 10.3390/nu14132612 PMID: 35807795

Ahmed, M., Guo, O., Qureshi, M. A., Raza, S. A., Khan, K. A., & Salam, J. (2021). Do green HR practices enhance green motivation and proactive environmental management maturity in hotel industry? *International Journal of Hospitality Management*, 94, 102852. DOI: 10.1016/j.ijhm.2020.102852

Ahmed, S. U., Ali, A., Memon, A. H., & Hussain, A. (2021). Policy framework for implementation of sustainable procurement practices in healthcare industry. *Environment, Development and Sustainability*, 23(3), 3929–3945. Advance online publication. DOI: 10.1007/s10668-020-00750-w

Ahmet, S., & Demet, E. (2023). Site selection modeling of hybrid renewable energy facilities using suitability index in spatial planning. *Renewable Energy*, 219, 119458. DOI: 10.1016/j.renene.2023.119458

Ahsan, M. J. (2024). Unlocking sustainable success: Exploring the impact of transformational leadership, organizational culture, and CSR performance on financial performance in the Italian manufacturing sector. *Social Responsibility Journal*, 20(4), 783–803. DOI: 10.1108/SRJ-06-2023-0332

Aitichya Chandra, A. P. (2020). Designing Zoning Systems for Freight Transportation Planning: A GIS-based approach for Automated Zone Design using Public Data Sources. *Transportation Research Procedia*, 48, 605–619. DOI: 10.1016/j.trpro.2020.08.063

Akçay, K., Altinay, F., Altınay, Z., Daglı, G., Shadiev, R., Altinay, M., Adedoyin, O. B., & Okur, Z. G. (2024). Global Citizenship for the Students of Higher Education in the Realization of Sustainable Development Goals. *Sustainability (Basel)*, 16(4), 1–13. DOI: 10.3390/su16041604

Akers, A. (2012). To Infinity and Beyond: Orbital Space Debris and how to clean it up. *University of La Verne Law Review, 33*(2).

Akomea-Frimpong, I., Adeabah, D., Ofosu, D., & Tenakwah, E. J. (2021). A review of studies on green finance of banks, research gaps and future directions. *Journal of Sustainable Finance & Investment*, 12(4), 1241–1264. DOI: 10.1080/20430795.2020.1870202

Al Hammadi, A., Al Darmaki, F., & Kotb, M. (2021). Climate change impacts on coastal zones of the United Arab Emirates: A review. *Water (Basel)*, 13(4), 542.

Al Maha Desert Resort & Spa. (n.d.). *Sustainability*. Retrieved from https://www.marriott.com/hotels/travel/dxbam-al-maha-a-luxury-collection-desert-resort-and-spa-dubai/

Alain Starke, M. W. (2021). Promoting Energy-Efficient Behavior by Depicting Social Norms in a Recommender Interface. *ACM Transactions on Interactive Intelligent Systems*, 11(3-4), 1–32. DOI: 10.1145/3460005

Alam, M., & Yamin, F. (2019). Enhancing climate resilience through regional cooperation in South Asia. *Regional Environmental Change*, 19(2), 353–366.

Al-Ansari, N., Al-Maktoumi, A., & Al-Nuaimi, H. (2021). Technological innovations for climate change adaptation in the United Arab Emirates. *International Journal of Environmental Research and Public Health*, 18(8), 4071. PMID: 33921547

Alarie, B., & Cockfield, A. (2021). Will machines replace us?: Machine-authored texts and the future of scholarship. *Law. Technology and Humans*, 3(1), 5–11. DOI: 10.5204/lthj.2089

Al-Awar, F., Al-Nuaimi, H., & Al-Maktoumi, A. (2021). Sustainable design and technology integration in the Sheikh Zayed Desert Learning Centre. *Journal of Arid Environments*, 186, 104434.

Alberton, A., Kieling, A. P., Lyra, F. R., Hoffmann, E. M., Lopez, M. P. V., & Stefano, S. R. (2022). Competencies for sustainability in hotels: Insights from Brazil. *Employee Relations*, 44(3), 555–575. DOI: 10.1108/ER-01-2019-0093

Alemão, C. (2019). Environmental sustainability practices in the supply chain of hotels and the consumers' perception, Master Dissertation in Management. UCP, Lisbon.

Alfathimy, D. H. A. (2019). Application of Sustainability Concept to Near-Earth Space as an Integral Part of Earth-system in the Context of Sustainable Development Goals (SDGs). *A I.*

Algarve, C. C. D. R. (2020). Observatório das dinâmicas regionais. https://www.ccdr-alg.pt/site/info/observatorio-das-dinamicas-regionais

Alhawari, A. &. (2018). Thermal bridges in building envelopes – An overview of impacts and solutions. International Review of Applied Sciences and Engineering, 31-40.

Ali, A., Ahmad, M., Nawaz, M., & Sattar, F. (2024). Spatial data infrastructure as the means to assemble geographic information necessary for effective agricultural policies in Pakistan. *Information Development*, 0–0. DOI: 10.1177/02666669241244503

Ali, A., & Imran, M. (2019). The Evolution of National Spatial Data Infrastructure in Pakistan, Implementation Problems and the Way Forward. *International Journal of Spatial Data Infrastructures Research*, 14(0). Advance online publication. DOI: 10.2902/IJSDIR.V14I0.500

Ali, A., & Imran, M. (2021). National Spatial Data Infrastructure vs Cadastre System for Economic Development: Evidence from Pakistan. *Land (Basel)*, 10(2), 188. DOI: 10.3390/land10020188

Ali, A., Imran, M., Jabeen, M., Ali, Z., & Mahmood, S. A. (2021). Factors influencing integrated information management: Spatial data infrastructure in Pakistan. *Information Development*, 026666692110484. Advance online publication. DOI: 10.1177/02666669211048483

Alipour, D., & Dia, H. (2023). A Systematic Review of the Role of Land Use, Transport, and Energy-Environment Integration in Shaping Sustainable Cities. *Sustainability (Basel)*, 15(8), 6447. DOI: 10.3390/su15086447

Allen, C., & Clouth, S. (2012). *A guidebook to the Green Economy*. UNDESA.

Allevi, E., Basso, A., Bonenti, F., Oggioni, G., & Riccardi, R. (2019). Measuring the environmental performance of green SRI funds: A DEA approach. *Energy Economics*, 79, 32–44. DOI: 10.1016/j.eneco.2018.07.023

Al-Maktoumi, A., Al-Nuaimi, H., & Al-Awar, F. (2020). Climate change impact and adaptation in the UAE: A systematic review. *Sustainability*, 12(18), 7706.

Almeida, J. (2016). Sustentabilidade na hotelaria. Uma análise de infusão/difusão em hotéis de Lisboa. Master Dissertation, Universidade Europeia. Lisbon, Portugal.

Almena, A., Siu, R., Chong, K., Thornley, P., & Röder, M. (2024). Reducing the environmental impact of international aviation through sustainable aviation fuel with integrated carbon capture and storage. *Energy Conversion and Management*, 303, 118186. DOI: 10.1016/j.enconman.2024.118186

Alowais, S. A., Alghamdi, S. S., Alsuhebany, N., Alqahtani, T., Alshaya, A. I., Almohareb, S. N., Aldairem, A., Alrashed, M., Bin Saleh, K., Badreldin, H. A., Al Yami, M. S., Al Harbi, S., & Albekairy, A. M. (2023). Revolutionizing healthcare: the role of artificial intelligence in clinical practice. In *BMC Medical Education* (Vol. 23, Issue 1). DOI: 10.1186/s12909-023-04698-z

Al-Rashedi, M., Al-Mazrouei, A., & Nisha, A. (2019). Climate change and its impact on the UAE economy. *International Journal of Scientific and Engineering Research*, 10(11), 1536–1542.

Alvarez-Risco, A., Estrada-Merino, A., & Perez-Luyo, R. (2020). Sustainable development goals in hospitality management. In Ruël, H., & Lombarts, A. (Eds.), *Sustainable Hospitality Management. Advanced Series in Management, 24* (pp. 159–178). Emerald Publishing., DOI: 10.1108/S1877-636120200000024012

Amoako, G.K., Obuobisa-Darko, T., Dartey-Baah, K. and Sedalo, G. (2023), "A conceptual framework on finding the nexus between sustainability and desired outcomes for smart cities – the moderating role of green leadership", *International Journal of Tourism Cities*, Vol. ahead-of-print No. ahead-of-print. .DOI: 10.1108/IJTC-12-2021-0238

Amuakwa-Mensah, F., & Násström, E. (2021). Role of banking sector performance in renewable energy consumption. *Applied Energy*, 306, 118023. DOI: 10.1016/j.apenergy.2021.118023

Ana, A., Gumersindo, F., & Maria, T. M. (2023). Advancing the European energy transition based on environmental, economic and social justice. *Sustainable Production and Consumption*, 43, 77–93. DOI: 10.1016/j.spc.2023.10.013

Angelaki, M. E., Bersimis, F., Karvounidis, T., & Douligeris, C. (2024). Towards more sustainable higher education institutions: Implementing the sustainable development goals and embedding sustainability into the information and computer technology curricula. *Education and Information Technologies*, 29(4), 5079–5113. DOI: 10.1007/s10639-023-12025-8

Anil Aswani, N. M. (2012). Reducing Transient and Steady State Electricity Consumption in HVAC Using Learning-Based Model-Predictive Control. *Proceedings of the IEEE*. DOI: 10.1109/JPROC.2011.2161242

Anjomshoaa, A., Duarte, F., Rennings, D., Matarazzo, T. J., Desouza, P., & Ratti, C. (2018). City scanner: Building and scheduling a mobile sensing platform for smart city services. *IEEE Internet of Things Journal*, 5(6), 4567–4579. DOI: 10.1109/JIOT.2018.2839058

Ansari, Z. N., & Kant, R. (2017). A state-of-art literature review reflecting 15 years of focus on sustainable supply chain management. *Journal of Cleaner Production*, 142, 2524–2543. DOI: 10.1016/j.jclepro.2016.11.023

Anton, R. (2024). Applying graph theory to find key leverage points in the transition toward urban renewable energy systems. *Applied Energy*, 361, 122854. DOI: 10.1016/j.apenergy.2024.122854

Antony, J., Swarnakar, V., Gupta, N., Kaur, J., Jayaraman, R., Tortorella, G. L., & Cudney, E. (2023). Critical success factors for operational excellence initiatives in manufacturing: A meta-analysis. *Total Quality Management & Business Excellence*, 34(9-10), 1152–1172. DOI: 10.1080/14783363.2022.2157714

Arbiv, O. E. (2024). *ChatGPT and Sustainability in Universities: Exploring Environmental and Educational Impacts*. University of Twente.

Archibald, M. M., & Clark, A. M. (2023). *ChatGTP: What is it and how can nursing and health science education use it?* Wiley Online Library.

Arora, N. K. (2018). Agricultural sustainability and food security. *Environmental Sustainability*, 1(3), 217–219. DOI: 10.1007/s42398-018-00032-2

Arun Kumar Shukla, A. K. (2023). Active and passive methods for cooling load reduction in a tropical building: A case study. *Energy Conversion and Management*, 293(117490).

Asadi, S., Pourhashemi, S. O., Nilashi, M., Abdullah, R., Samad, S., Yadegaridehkordi, E., Aljojo, N., & Razali, N. S. (2020). Investigating influence of green innovation on sustainability performance: A case on Malaysian hotel industry. *Journal of Cleaner Production*, 258, 120860. DOI: 10.1016/j.jclepro.2020.120860

Asian Development Bank. (2017). Future carbon fund: Driving innovation and results through carbon finance. https://www.adb.org/sites/default/files/publication/389821/future-carbon-fund.pdf

Asian Development Bank. (2018). Fostering sustainable development through carbon markets. ADB Blogs. https://blogs.adb.org/blog/fostering-sustainable-development-through-carbon-markets

Atney-Yurdin, I. (1991). Space Debris Legal Research Guide. *Pace International Law Review*, 3(1), 167. DOI: 10.58948/2331-3536.1019

Aulia, R. P., & Steelyana, E. (2023). A Study in Government Procurement System: Public-Private Partnership for Infrastructure in Energy Sector. In *E3S Web of Conferences* (Vol. 388, p. 01014). EDP Sciences.

Aurelija Stonkuvienė, R. B. (2021). The impact of connector's thermal and geometrical characteristics on the energy performance of facade systems. *Journal of Building Engineering*.

Awan, U., & Khan, S. A. R. (2021). Mediating role of sustainable leadership in buyer-supplier relationships: an supply chain performance: an empirical study. *LogForum, 17(1).*

Azadegan, K., & Dooley, K. (2021). Dooley, A typology of supply network resilience strategies: Complex collaborations in a complex world. *The Journal of Supply Chain Management*, 57(1), 17–26. DOI: 10.1111/jscm.12256

Bačić, S., Tomić, H., Andlar, G., & Roić, M. (2022). Towards Integrated Land Management: The Role of Green Infrastructure. *ISPRS International Journal of Geo-Information 2022, Vol. 11, Page 513, 11*(10), 513. DOI: 10.3390/ijgi11100513

Bag, S., Wood, L., Xu, L., Dhamija, P., & Kayikci, Y. (2019). Big data analytics as an operational excellence approach to enhance sustainable supply chain performance. *Resources, Conservation and Recycling*. Advance online publication. *10.1016/*j.resconrec.2019.104559

Baker, M., Bergstresser, D., Serafeim, G., & Wurgler, J. (2018). *Financing the response to climate change: The pricing and ownership of U.S. green bonds*. National Bureau of Economic Research., DOI: 10.3386/w25194

Banister, D. (2011). Cities, mobility and climate change. *Journal of Transport Geography*, 19(6), 1538–1546. DOI: 10.1016/j.jtrangeo.2011.03.009

Bansal, S., Mukhopadhyay, M., & Maurya, S. (2023). Strategic drivers for sustainable implementation of carbon trading in India. *Environment, Development and Sustainability*, 25(5), 4411–4435. DOI: 10.1007/s10668-022-02205-w

Barik, R. K., Dubey, H., Mankodiya, K., Sasane, S. A., & Misra, C. (2019). GeoFog4Health: A fog-based SDI framework for geospatial health big data analysis. *Journal of Ambient Intelligence and Humanized Computing*, 10(2), 551–567. DOI: 10.1007/s12652-018-0702-x

Barrier, E. B. (2017). The Concept of Sustainable Economic Development. *The Economics of Sustainability*, 87–96. DOI: 10.4324/9781315240084-7

Barry, K. (2022). Space Debris Mitigation and Remediation: Historical Best Practices and Lessons Learned for Economically Preserving and Utilizing Common Areas. *New Space*, 10(3), 246–258. DOI: 10.1089/space.2021.0022

Bartha, G., & Kocsis, S. (2011). Standardization of geographic data: The European inspire directive. *European Journal of Geography*, 2(2), 79–89.

Bas Knoll, W. B. (2018). Improving the usability and performance of heat recovery ventilation systems in practice. aivc.org.

Bastas, A., & Liyanage, K. (2018). Sustainable supply chain quality management: A systematic review. *Journal of Cleaner Production*, 181, 726–744. DOI: 10.1016/j.jclepro.2018.01.110

Batko, K., & Ślęzak, A. (2022). The use of Big Data Analytics in healthcare. *Journal of Big Data*, 9(1), 3. Advance online publication. DOI: 10.1186/s40537-021-00553-4 PMID: 35013701

Batle, J., Sintes, F. O., & Moon, C. J. (2018). Environmental management best practices: Towards social innovation. *International Journal of Hospitality Management*, 69, 14–20. DOI: 10.1016/j.ijhm.2017.10.013

Baxter, P., & Jack, S. (2008). Qualitative case study methodology: Study design and implementation for novice researchers. *The Qualitative Report*, 13(4), 544–559.

Bayoumi, M. (2017). Energy saving method for improving thermal comfort and air quality in warm humid climates using isothermal high-velocity ventilation. *Renewable Energy*, 114, 502–512. DOI: 10.1016/j.renene.2017.07.056

Béal, S., Deschamps, M., & Moulin, H. (2020). Taxing congestion of the space commons. *Acta Astronautica*, 177, 313–319. DOI: 10.1016/j.actaastro.2020.07.036

Bec, A., Moyle, B., Timms, K., Schaffer, V., Skavronskaya, L., & Little, C. (2019). Management of immersive heritage tourism experiencs: A conceptual model. *Tourism Management*, 72, 117–120. DOI: 10.1016/j.tourman.2018.10.033

Bednarova, P. V. (2011). *An energy-efficient building for the Arctic climate Is a passive house a sensible solution for Greenland?*

Bekele, M. K., Pierdicca, R., Frontoni, E., Malinverni, E. S., & Gain, J. (2018). A survey of augmented, virtual, and mixed reality for cultural heritage. [JOCCH]. *Journal on Computing and Cultural Heritage*, 11(2), 1–36.

Beltran, A., Erickson, V. L., & Cerpa, A. E. (2013, November). Thermosense: Occupancy thermal based sensing for hvac control. In *Proceedings of the 5th ACM Workshop on Embedded Systems For Energy-Efficient Buildings* (pp. 1-8).

Benedek, J. (2023). Spatial Localization of the Sustainable Development Goals. *Public Finance Quarterly*, 69(1), 64–74. DOI: 10.35551/PFQ_2023_1_4

Benz, E., & Hengelbrock, J. (2008). Liquidity and price discovery in the European CO2 futures market: An intraday analysis. *Finance Research Letters*, 5(2), 114–124. DOI: 10.2139/ssrn.1283175

Bernhard, P., Deschamps, M., & Zaccour, G. (2023). Large satellite constellations and space debris: Exploratory analysis of strategic management of the space commons. *European Journal of Operational Research*, 304(3), 1140–1157. DOI: 10.1016/j.ejor.2022.04.030

Bevilacqua, M. G., Russo, M., Giordano, A., & Spallone, R. (2022). 3D Reconstruction, Digital Twinning, and Virtual Reality: Architectural Heritage Applications. *Proceedings - 2022 IEEE Conference on Virtual Reality and 3D User Interfaces Abstracts and Workshops, VRW 2022*, 92–96. DOI: 10.1109/VRW55335.2022.00031

Bhatnagar, S., & Sharma, D. (2022). Evolution of green finance and its enablers: A bibliometric analysis. *Renewable & Sustainable Energy Reviews*, 162, 112405. DOI: 10.1016/j.rser.2022.112405

Bhattacharyya, U. (2024, April 20). Understanding the Regulatory Framework for Sustainable Finance in India. *NLS BLR*. https://www.nlsblr.com/post/understanding-the-regulatory-framework-for-sustainable-finance-in-india

Bhuiyan, B. A. (2018). An Overview of Game Theory and Some Applications. *Philosophy and Progress*, 111–128. DOI: 10.3329/pp.v59i1-2.36683

Bishop, I. D., Escobar, F. J., Karuppannan, S., Williamson, I. P., Yates, P. M., Suwarnarat, K., & Yaqub, H. W. (2000). Spatial data infrastructures for cities in developing countries. Lessons from the Bangkok experience. *Cities (London, England)*, 17(2), 85–96. Advance online publication. DOI: 10.1016/S0264-2751(00)00004-4

Bitter, C. & Loney, E. (2015). Deeper Learning: Improving Student Outcomes for College, Career, and Civic Life. Policy Brief. *Education Policy Center at American Institutes for Research.*

Blake, J. A. (2022). Looking out for a sustainable space. *Astronomy & Geophysics, 63*(2), 2.14-2.20. DOI: 10.1093/astrogeo/atac022

Boerboom, I. (2010). Integrating spatial planning and decision support system infrastructure and spatial data infrastructure. *GSDI 12 World Conference Singapore*. http://www.gsdi.org/gsdiconf/gsdi12/papers/82.pdf

Boggaram, C., Omotosho, O. & Tenghjiao, Z. (2023). Plagiarism Checker Based on Machine Learning and OpenAI.

Bohlmann, U. M., & Petrovici, G. (2019). Developing planetary sustainability: Legal challenges of Space 4.0. *Global Sustainability*, 2, e10. DOI: 10.1017/sus.2019.10

Bongers, A., & Torres, J. L. (2023). Star Wars: Anti-Satellite Weapons and Orbital Debris. *Defence and Peace Economics*, •••, 1–20. DOI: 10.1080/10242694.2023.2208020

Borowy, I. (2013). *Defining sustainable development for our common future: A history of the World Commission on Environment and Development (Brundtland Commission).* Routledge. DOI: 10.4324/9780203383797

Borzacchiello, M. T., & Craglia, M. (2013). Estimating benefits of spatial data infrastructures: A case study on e-Cadastres. *Computers, Environment and Urban Systems*, 41, 276–288. Advance online publication. DOI: 10.1016/j.compenvurbsys.2012.05.004

Bouteraa, M., Hisham, R. R. I. R., & Zainol, Z. (2022). Challenges affecting bank consumers' intention to adopt green banking technology in the UAE: A UTAUT-based mixed-methods approach. *Journal of Islamic Marketing*, 14(10), 2466–2501. DOI: 10.1108/JIMA-02-2022-0039

Bowen, G. A. (2009). Document analysis as a qualitative research method. *Qualitative Research Journal*, 9(2), 27–40. DOI: 10.3316/QRJ0902027

Braccini, A. M., & Margherita, E. G. (2019). Exploring organizational sustainability of industry 4.0 under the Triple Bottom Line: The case of a manufacturing company. *Sustainability (Basel)*, 11(1), 36. DOI: 10.3390/su11010036

Braga, S., Junior., Martínez, M. P., Correa, C. M., Moura-Leite, R. C., & Da Silva, D. (2019). Greenwashing effect, attitudes, and beliefs in green consumption. RAUSP Management Journal, 54(2), 226-241, https://www.emerald.com/insight/content/doi/10.1108/RAUSP-08-2018-0070/full/pdf

Brandenburg, M., Gruchmann, T., & Oelze, N. (2019). Sustainable Supply Chain Management—A Conceptual Framework and Future Research Perspectives. *Sustainability (Basel)*, 11(24), 7239. DOI: 10.3390/su11247239

Brown, T., Mann, B., Ryder, N., Subbiah, M., Kaplan, J., Dhariwal, P., Neelakantan, A., Shyam, P., Sastry, G., & Askell, A. (2020). Language Models are Few-Shot Learners. *Advances in Neural Information Processing Systems*, 33.

Buerkle, A., O'Dell, A., Matharu, H., Buerkle, L., & Ferreira, P. (2023). Recommendations to align higher education teaching with the UN sustainability goals – A scoping survey. *International Journal of Educational Research Open, 5*(April, 100280), 1–19. DOI: 10.1016/j.ijedro.2023.100280

Burawat, P. (2019). The relationships among transformational leadership, sustainable leadership, lean manufacturing and sustainability performance in Thai SMEs manufacturing industry. *International Journal of Quality & Reliability Management*, 36(6), 1014–1036. DOI: 10.1108/IJQRM-09-2017-0178

Burnett, N., Dillon, A., Jozepa, I., & Bolton, P. (2024, March 5). Carbon Border Adjustment Mechanism. House of Commons Library. https://commonslibrary.parliament.uk/research-briefings/cbp-9935/#:~:text=What%20is%20a%20Carbon%20Border,is%20to%20prevent%20carbon%20leakage

Burton, E., & Mitchell, L. (2006). *Inclusive urban design: Streets for life*. routledge.

Busch, T., Johnson, M., Pioch, T., & Kopp, E. (2018). Corporate carbon performance data: Quo vadis? *Journal of Industrial Ecology*, 22(6), 1217–1233. DOI: 10.1111/jiec.13008

Butt, M., Mohammed, R., Butt, E., Butt, S., & Xiang, J. (2020). Why have immunization efforts in Pakistan failed to achieve global standards of vaccination uptake and infectious disease control? In *Risk Management and Healthcare Policy* (Vol. 13). DOI: 10.2147/RMHP.S211170

Calheiros, L. (2021). A importância das práticas sustentáveis e a sua utilização como estratégia de marketing nos hotéis de 4 estrelas da região Norte de Portugal. Master Dissertation, Politécnico do Porto, Portugal.

Car, N. J., Box, P. J., & Sommer, A. (2019). The location index: A semantic web spatial data infrastructure. *Lecture Notes in Computer Science (Including Subseries Lecture Notes in Artificial Intelligence and Lecture Notes in Bioinformatics), 11503 LNCS*, 543–557. https://doi.org/DOI: 10.1007/978-3-030-21348-0_35/COVER

Carli, R., Cavone, G., Ben Othman, S., & Dotoli, M. (2020). Iot based architecture for model predictive control of hvac systems in smart buildings. *Sensors (Basel)*, 20(3), 781. DOI: 10.3390/s20030781 PMID: 32023965

Carolina de los Ríos, J. E. (2018). Persistence of methodological, taxonomical, and geographical bias in assessments of species vulnerability to climate change: A review. *Global Ecology and Conservation*, 15, e00412. Advance online publication. DOI: 10.1016/j.gecco.2018.e00412

Carter, C. R., & Washispack, S. (2018). Mapping the Path Forward for Sustainable Supply Chain Management: A Review of Reviews. *Journal of Business Logistics*, 2018(39), 242–247. DOI: 10.1111/jbl.12196

Casiano Flores, C., & Crompvoets, J. (2023). Climate Change Adaptation: The Role of Geospatial Data in Sustainable Infrastructures. *ISPRS International Journal of Geo-Information 2023, Vol. 12, Page 68, 12*(2), 68. DOI: 10.3390/ijgi12020068

Castelein, W. T., Bregt, A. K., & Pluijmers, Y. (2010). The economic value of the Dutch geo-information sector *. *International Journal of Spatial Data Infrastructures Research*, 5, 58–76. DOI: 10.2902/1725-0463.2010.05.art2

Catan, M. (2018). Environmental practices in hospitality industry for a more sustainable tourism. Master Dissertation, Escola Superior de Hotelaria e Turismo do Estoril. Lisbon, Portugal.

Cavazotte, F., Mansur, J., & Moreno, V. (2021). Authentic leadership and sustainable operations: How leader morality and selflessness can foster frontline safety performance. *Journal of Cleaner Production*, 313, 127819. DOI: 10.1016/j.jclepro.2021.127819

Cetl, V., Ivić, S. M., & Tomić, H. (2009). Improvement of national spatial data infrastructure as a public project of permanent character. *Kartografija i Geoinformacije, 8*(11).

Chakraborty, S., Raut, R. D., Rofin, T. M., & Chakraborty, S. (2023). A comprehensive and systematic review of multi-criteria decision-making methods and applications in healthcare. In *Healthcare Analytics* (Vol. 4). DOI: 10.1016/j.health.2023.100232

Chams, N., & García-Blandón, J. On the importance of sustainable human resource management for the adoption of sustainable development goals, *Resources, Conservation and Recycling*, Volume 141, 2019, Pages 109-122, ISSN 0921-3449, DOI: 10.1016/j.resconrec.2018.10.006

Chang, L., Taghizadeh-Hesary, F., Chen, H., & Mohsin, M. (2022). Do green bonds have environmental benefits? *Energy Economics*, 115, 106356. DOI: 10.1016/j.eneco.2022.106356

Chang, Y., & Ke, J. (2024). Socially responsible artificial intelligence empowered people analytics: A novel framework towards sustainability. *Human Resource Development Review*, 23(1), 88–120. DOI: 10.1177/15344843231200930

Chaokun Hong, H. L. (2020). Review of connections for engineered bamboo structures. *Journal of Building Engineering*.

Charles Ryan, J. (2014). Analogue Angels and Digital Diamonds: Tracing the Origins of New Media Art. In *Philosophy Study* (Vol. 4, Issue 6).

Chen, K., Ren, Z., Mu, S., Sun, T. Q., & Mu, R. (2020). Integrating the Delphi survey into scenario planning for China's renewable energy development strategy towards 2030. *Technological Forecasting and Social Change*, 158, 120157. DOI: 10.1016/j.techfore.2020.120157

Chen, S. L., Su, Y. S., Tufail, B., Lam, V. T., Phan, T. T. H., & Ngo, T. Q. (2023). The moderating role of leadership on the relationship between green supply chain management, technological advancement, and knowledge management in sustainable performance. *Environmental Science and Pollution Research International*, 30(19), 56654–56669. DOI: 10.1007/s11356-023-26304-x PMID: 36920606

Chen, Z., Mirza, N., Huang, L., & Umar, M. (2022). Green Banking—Can Financial Institutions support green recovery? *Economic Analysis and Policy*, 75, 389–395. DOI: 10.1016/j.eap.2022.05.017 PMID: 35694019

Chen, Z., Wu, M., Chan, A., Li, X., & Ong, Y. (2023). Survey on AI sustainability: Emerging trends on learning algorithms and research challenges. *IEEE Computational Intelligence Magazine*, 18(2), 60–77. DOI: 10.1109/MCI.2023.3245733

Chhaochharia, M. (2021). Green Finance in India: Progress and Challenges. *Research Journal of Humanities and Social Sciences*, 223–226, 223–226. Advance online publication. DOI: 10.52711/2321-5828.2021.00039

Chibuike, P. O., Salihu, A. R., & Ugbe, C. O. (2024). A comprehensive review of recent developments in smart grid through renewable energy resources integration. *Heliyon*, •••, 10. PMID: 38327483

Chidaushe, W., & Njaya, T. (2022). Global Determinants of Sustainability Linked Financial Markets. *Journal of Economics and Sustainable Development*, 13(22), 202246. DOI: 10.7176/JESD/13-22-05

Chin-Chi Cheng, D. L. (2019). Artificial Intelligence-Assisted Heating Ventilation and Air Conditioning Control and the Unmet Demand for Sensors: Part 1. Problem Formulation and the Hypothesis. *Sensors (Basel)*, 19(5), 1131. Advance online publication. DOI: 10.3390/s19051131 PMID: 30845664

Choi, L. (2014). The strengths and weaknesses of research methodology: Comparison and complimentary between qualitative and quantitative approaches. *Journal of the Humanities and Social Sciences*, 19(4), 99–104. DOI: 10.9790/0837-194399104

Chu, H. (2015). Research methods in library and information science: A content analysis. *Library & Information Science Research*, 37(1), 36–41. DOI: 10.1016/j.lisr.2014.09.003

Chung, J., Gulcehre, C., Cho, K., & Bengio, Y. (2014). Empirical evaluation of gated recurrent neural networks on sequence modeling. arXiv preprint arXiv:1412.3555.

Chung, K. C. (2020). Green marketing orientation: Achieving sustainable development in green hotel management. *Journal of Hospitality Marketing & Management*, 29(6), 722–738. DOI: 10.1080/19368623.2020.1693471

CIS – Community Innovation Survey (2022). Principais Resultados (2018-2020). https://www.dgeec.mec.pt/np4/207/

Climate Impact Partners. (2023, June 8). How carbon finance projects align to SDGs. Climate Impact Partners. https://www.climateimpact.com/news-insights/insights/how-carbon-finance-projects-align-to-sdgs/

Climate Policy Initiative. (2024, October 16). *A Roadmap for Green and Transition Finance in India - CPI*. CPI. https://www.climatepolicyinitiative.org/publication/a-roadmap-for-green-and-transition-finance-in-india/

Clormann, M., & Klimburg-Witjes, N. (2022). Troubled Orbits and Earthly Concerns: Space Debris as a Boundary Infrastructure. *Science, Technology & Human Values*, 47(5), 960–985. DOI: 10.1177/01622439211023554

Coalition for Green Growth. (2015). Compromisso para o Crescimento Verde. https://www.crescimentoverde.gov.pt/compromisso/

Coffey, L. (2024). Professors Cautious of Tools to Detect AI-Generated Writing. *Inside Higher Ed (IHE)*.

Coleman, D. J., & Nebert, D. D. (1998). Building a North American Spatial Data Infrastructure. *Cartography and Geographic Information Systems*, 25(3), 151–160. Advance online publication. DOI: 10.1559/152304098782383098

Constantine, S., Miguel Angel, T. R., & Muireann, Á. L. (2022). Public acceptance of sustainable energy innovations in the European Union: A multidimensional comparative framework for national policy. *Journal of Cleaner Production*, 340, 130721. DOI: 10.1016/j.jclepro.2022.130721

Cop, S., Alola, U. V., & Alola, A. A. (2020). Perceived behavioral control as a mediator of hotels' green training, environmental commitment and organizational citizenship behavior: A sustainable environmental practice. *Business Strategy and the Environment*, 29(3), 3495–3508. Advance online publication. DOI: 10.1002/bse.2592

Çop, S., Olorunsola, V. O., & Alola, U. V. (2021). Achieving environmental sustainability through green transformational leadership policy: Can green team resilience help? *Business Strategy and the Environment*, 30(1), 671–682. DOI: 10.1002/bse.2646

Cornell, S., & Policy, M. J.-N. C. of A. I., 2007, U., & Jorgensen, M. (2007). The nature and components of economic development in Indian country. *Citeseer, May*, 1–22. https://citeseerx.ist.psu.edu/viewdoc/download?doi=10.1.1.173.661&rep=rep1&type=pdf

Cornet, S., Barpanda, S., Guidi, M. A. D., & Viswanathan, P. K. (2023). Sustainability education and community development in higher education using participatory and case based approaches in India. *International Journal of Sustainability in Higher Education*, 25(9), 78–93. DOI: 10.1108/IJSHE-07-2022-0242

Costa, C. M., Barbosa, J. C., Gonçalves, R., Castro, H., Del Campo, F. J., & Lanceros-Méndez, S. (2021). Recycling and environmental issues of lithium-ion batteries: Advances, challenges and opportunities. *Energy Storage Materials*, 37, 433–465. DOI: 10.1016/j.ensm.2021.02.032

Counsell, G., Edney, G., & Dick, S. (2024). Improving awareness on sustainable palm oil: Measuring the effectiveness of a repeat-engagement zoo outreach programme. *Environmental Education Research*, 30(2), 153–169. DOI: 10.1080/13504622.2023.2225805

Crompvoets, J., Vancauwenberghe, G., Bouckaert, G., & Vandenbroucke, D. (2011). Practices to develop spatial data infrastructures: Exploring the contribution to e-government. In *Practical Studies in E-Government.* Best Practices from Around the World., DOI: 10.1007/978-1-4419-7533-1_13

Crompvoets, J., Vancauwenberghe, G., Ho, S., Masser, I., & Timo De Vries, W. (2018). Governance of national spatial data infrastructures in Europe. *International Journal of Spatial Data Infrastructures Research*, 13, 253–285.

Dai, D. (2021). *Future Courtyard 2050.* UNITEC Institute of Technology.

Dale, R. (2021). GPT-3: What's it good for? *Natural Language Engineering*, 27(1), 113–118. DOI: 10.1017/S1351324920000601

Damjanov, K. (2017). Of Defunct Satellites and Other Space Debris: Media Waste in the Orbital Commons. *Science, Technology & Human Values*, 42(1), 166–185. DOI: 10.1177/0162243916671005

Dan Zhu, M. K. (2015). Analysis of Sustainable Energy Systems in Ecovillages: A Review of Progress in BedZED and Masdar City. *Low Carbon Economy*, 6(1).

Daniel, I. A., Francisco, J., & Marcos, T. V. (2023). Long-term planning for the integration of electric mobility with 100% renewable energy generation under various degrees of decentralization: Case study Cuenca, Ecuador. *Energy Reports*, 9, 4816–4829. DOI: 10.1016/j.egyr.2023.03.118

Das, J. K., Padhani, Z. A., Jabeen, S., Rizvi, A., Ansari, U., Fatima, M., Akbar, G., Ahmed, W., & Bhutta, Z. A. (2020). Impact of conflict on maternal and child health service delivery - How and how not: A country case study of conflict affected areas of Pakistan. *Conflict and Health*, 14(1), 32. Advance online publication. DOI: 10.1186/s13031-020-00271-3 PMID: 32514297

De Freitas Netto, S. V., Sobral, M. F. F., Ribeiro, A. R. B., & Soares, G. R. L. (2020). Concepts and forms of greenwashing: A systematic review. *Environmental Sciences Europe*, 32(1), 19. DOI: 10.1186/s12302-020-0300-3

De Sampaio, C. P. (2022). THE CHALLENGE OF TEACHING DESIGN FOR SUSTAINABILITY IN A PANDEMIC CONTEXT: A DIDACTICAL-PEDAGOGICAL EXPERIENCE. *MIX Sustentável*, 8(2), 121–132. DOI: 10.29183/2447-3073.MIX2022.v8.n2.121-132

De Vries, A. (2023). The growing energy footprint of artificial intelligence. *Joule*, 7(10), 2191–2194. DOI: 10.1016/j.joule.2023.09.004

Dean, D. (2002). Museum Exhibition. *Theory into Practice.*

Dell'Era, C., & Landoni, P. (2014). Living Lab: A Methodology between User-Centred Design and Participatory Design. *Creativity and Innovation Management*, 23(2), 137–154. DOI: 10.1111/caim.12061

DeQuante Rashon Mckoy, R. T. (2023). Review of HVAC Systems History and Future Applications. *Energies*, 16(17), 6109. Advance online publication. DOI: 10.3390/en16176109

Despeisse, M. (2018). Teaching Sustainability Leadership in Manufacturing: A Reflection on the Educational Benefits of the Board Game Factory Heroes. *Procedia CIRP*, 69, 621–626. DOI: 10.1016/j.procir.2017.11.130

Deviprasad, S., Madhumithaa, N., Vikas, I. W., Yadav, A., & Manoharan, G. (2023). The Machine Learning-Based Task Automation Framework for Human Resource Management in MNC Companies. *Engineering Proceedings*, 59(1), 63.

Dewsbury, M. (2015). The Environmental Measurement of Residential Buildings A Technical Guide. Dewsbury, Mark: The Environmental Measurement of Residential Buildings. DOI: 10.13140/RG.2.1.1520.7840

Di Pippo, S. (2019). The contribution of space for a more sustainable earth: Leveraging space to achieve the sustainable development goals. *Global Sustainability*, 2, e3. DOI: 10.1017/sus.2018.17

Dias, P. J., & Rocha, A. (2023). Carbon Tax: Characterization of Research. In Taking on Climate Change Through Green Taxation (pp. 26-56). IGI Global. DOI: DOI: 10.4018/978-1-6684-8592-7.ch002

Díaz, C. G., Zambrana-Vasquez, D., & Bartolomé, C. (2024). Building resilient cities: A comprehensive review of climate change adaptation indicators for urban design. *Energies*, 17(8), 1959. DOI: 10.3390/en17081959

Diaz, L., Remke, A., Kauppinen, T., Degbelo, A., Foerster, T., Stasch, C., Rieke, M., Schaeffer, B., Baranski, B., Bröring, A., & Wytzisk, A. (2012). Future SDI - Impulses from Geoinformatics research and IT trends. *International Journal of Spatial Data Infrastructures Research*, 7. Advance online publication. DOI: 10.2902/1725-0463.2012.07.art18

Divine, Q. A., Anthony, A. D., Bright, A. G., & Festus, V. B. (2023). Does psychological empowerment improve renewable energy technology acceptance and recommendation? Evidence from 17 rural communities. *Renewable Energy*, 219, 119449. DOI: 10.1016/j.renene.2023.119449

Dong, B., & Lam, K. P. (2014, February). A real-time model predictive control for building heating and cooling systems based on the occupancy behavior pattern detection and local weather forecasting. [). Springer Berlin Heidelberg.]. *Building Simulation*, 7, 89–106.

Dong, Z., Xia, C., Fang, K., & Zhang, W. (2022). Effect of the carbon emissions trading policy on the co-benefits of carbon emissions reduction and air pollution control. *Energy Policy*, 165, 112998. DOI: 10.1016/j.enpol.2022.112998

Dorbayani, M. (2019). INVESTIGATING VANCOUVER'S HOUSING CRISIS. Manchester: University of Salford. DOI: 10.17866/rd.Salford.17124347.v1

Dore, C., a. M. (2012). Integration of Historic Building Information Modeling (HBIM) and 3D GIS for recording and. *18th International Conference on Virtual Systems and Multimedia*. DOI: 10.1109/VSMM.2012.6365947

Dorokhova, M., Ballif, C., & Wyrsch, N. (2020). Rule-based scheduling of air conditioning using occupancy forecasting. *Energy AI*, 2, 100022. DOI: 10.1016/j.egyai.2020.100022

Dubey, R., Gunasekaran, A., & Ali, S. S. (2015). Exploring the relationship between leadership, operational practices, institutional pressures and environmental performance: A framework for green supply chain. *International Journal of Production Economics*, 160, 120–132. DOI: 10.1016/j.ijpe.2014.10.001

Durai, S., Manoharan, G., Priya, T. S., Jayanthi, R., Razak, A., & Ashtikar, S. P. (2024). Quantifying the Impacts of Artificial Intelligence Implementations in Marketing. In Smart and Sustainable Interactive Marketing (pp. 120-144). IGI Global.

E. Rossi, E. S. (2016). Costruire il Tempio. Alla ricerca del progetto di Baldassarre Peruzzi per il Duomo di Carpi. *APM edizioni, Carpi,* .

Eagan, P., Cook, T., & Joeres, E. (2002). Teaching the importance of culture and interdisciplinary education for sustainable development. *International Journal of Sustainability in Higher Education*, 3(1), 48–66. DOI: 10.1108/14676370210414173

Economic Times. (2023, July 12). Regulated carbon markets are worth a trillion dollars a year and cover 25% of global emissions. The Economic Times. https://economictimes.indiatimes.com/news/et-evoke/regulated-carbon-markets-are-worth-a-trillion-dollars-a-year-and-cover-25-of-global-emissions-they-are-raising-climat

El Amri, A., Oulfarsi, S., Eddine, A. S., El Khamlichi, A., Hilmi, Y., Ibenrissoul, A., . . . Boutti, R. (2022). Carbon Financial Market: The Case of the EU Trading Scheme. In Handbook of Research on Energy and Environmental Finance 4.0 (pp. 424-445). IGI Global. DOI: DOI: 10.4018/978-1-7998-8210-7.ch017

El-Sayegh, S. M., Manjikian, S., Ibrahim, A., Abouelyousr, A., & Jabbour, R. (2018). Risk identification and assessment in sustainable construction projects in the UAE. *International Journal of Construction Management*, •••, 1–10.

Elshafei, G., Vilcekova, S., Zelenakova, M., & Negm, A. M. (2021). Towards an Adaptation of Efficient Passive Design for Thermal Comfort Buildings. *Sustainability (Basel)*, 13(17), 9570. DOI: 10.3390/su13179570

El-Shawa, S., Cattani, B. M., Innocenti, L., & Delaval, J. (2021). *From Cradle to Grave: ESA Clean Space's Approach to Space Sustainability*.

Emad Mushtaha, T. S. (2021). The impact of passive design strategies on cooling loads of buildings in a temperate climate. *Case Studies in Thermal Engineering*, 28(101588), 101588. DOI: 10.1016/j.csite.2021.101588

Emanuelli, M., Federico, G., Loughman, J., Prasad, D., Chow, T., & Rathnasabapathy, M. (2014). Conceptualizing an economically, legally, and politically viable active debris removal option. *Acta Astronautica*, 104(1), 197–205. DOI: 10.1016/j.actaastro.2014.07.035

Emirates Wildlife Society – World Wildlife Fund (EWS-WWF). (n.d.). *Our work*. Retrieved from https://uae.panda.org/our_work/

Engle, R. F., Giglio, S., Kelly, B., Lee, H., & Stroebel, J. (2020). Hedging climate change news. *Review of Financial Studies*, 33(3), 1184–1216. http://hdl.handle.net/10.1093/rfs/hhz072. DOI: 10.1093/rfs/hhz072

Esangbedo, C. O., Zhang, J., Pérez, P. B., & Skitmore, M. (2024). Sustainable performance and supply chain leadership in logistic firms: The role of corporate sustainability strategies and digital supply chain. *Supply Chain Management*, 29(6), 963–977. DOI: 10.1108/SCM-02-2024-0131

Euchner, J. (2023). Generative AI. *Research Technology Management*, 66(3), 71–74. DOI: 10.1080/08956308.2023.2188861

European Commission. (2022). *Support to the evaluation of the implementation of the Directive 2007/2/EC on Infrastructure for Spatial Information in the European Community*. INSPIRE., DOI: 10.2779/749002

Evans, J., Jones, R., Karvonen, A., Millard, L., & Wendler, J. (2015). Living labs and co-production: University campuses as platforms for sustainability science. *Current Opinion in Environmental Sustainability*, 16, 1–6. DOI: 10.1016/j.cosust.2015.06.005

Expo 2020 Dubai. (n.d.). *Sustainability pavilion*. Retrieved from https://www.expo2020dubai.com/en/understanding-expo/expo-sustainability/sustainability-pavilion

Farinha, F., Huertas, D. H., Pinheiro, M. D., Silva, E. M. J., Lança, R., Oliveira, M. J., & Batista, R. (2021). Sustainable competitiveness of tourism in the Algarve region. Critical stakeholders' perception of the supply sector. *Sustainability (Basel)*, 13(11), 6072. DOI: 10.3390/su13116072

Farinha, F., Oliveira, M. J., Silva, E. M. J., Lança, R., Pinheiro, M. D., & Miguel, C. (2019). Selection process of sustainable indicators for the Algarve region - OBSERVE project. *Sustainability (Basel)*, 11(2), 444. DOI: 10.3390/su11020444

Fatima, I., Shoman, H., Peters, A. W., Samad, L., & Nishtar, S. (2020). Pakistan's National Surgical, Obstetric, and Anesthesia Plan: An adapted model for a devolved federal-provincial health system. *Canadian Journal of Anaesthesia*, 67(9), 1212–1216. Advance online publication. DOI: 10.1007/s12630-020-01708-2 PMID: 32409943

Faye, V., & McNeill, R. C. (2022). Room-level ventilation in schools and universities. *Atmospheric Environment: X*, •••, 13. PMID: 35098105

Feng, F., Kunwar, N., Cetin, K., & O'Neill, Z. (2021). A critical review of fenestration/window system design methods for high-performance buildings. *Energy and Building*, 248, 248. DOI: 10.1016/j.enbuild.2021.111184

Ferrari, S., Zagarella, F., Caputo, P., & Bonomolo, M. (2019). Assessment of tools for urban energy planning. *Energy*, 176, 544–551. DOI: 10.1016/j.energy.2019.04.054

Ferraz, D., & Pyka, A. (2023). Circular economy, bioeconomy, and sustainable development goals: A systematic literature review. *Environmental Science and Pollution Research International*. Advance online publication. DOI: 10.1007/s11356-023-29632-0 PMID: 37702868

Ferreira, S. (2021). Sustentabilidade ambiental e os antecedentes da intenção comportamental dos consumidores: Uma abordagem aos hotéis ecológicos em Portugal. Master Dissertation, Universidade do Minho. Braga, Portugal.

Ferreira, S., Pereira, O., & Simões, C. (2021). Sustentabilidade ambiental na indústria hoteleira: Uma perspetiva dos gestores de hotéis ecológicos em Portugal. *International Journal of Marketing. Communication and New Media*, 9, 150–170. DOI: 10.54663/2182-9306

Feuerriegel, S., Hartmann, J., Janiesch, C., & Zschech, P. (2024). Generative ai. *Business & Information Systems Engineering*, 66(1), 111–126. DOI: 10.1007/s12599-023-00834-7

Fiksel, J. (2006). Sustainability and resilience: toward a systems approach. *Sustainability: Science. Practice and Policy*, 2(2), 14–21.

Finserv, T. (2023, April 14). *RBI Framework for Green Deposits*. https://vinodkothari.com/2023/04/rbi-framework-for-green-deposits/

Flores, C. C., & Crompvoets, J. (2020). Assessing the Governance Context Support for Creating a Pluvial Flood Risk Map with Climate Change Scenarios: The Flemish Subnational Case. *ISPRS International Journal of Geo-Information*, 9(7), 460. DOI: 10.3390/ijgi9070460

Flores, G. T. (2023). Fostering Awareness of Integrating Education for Sustainable Development in Teacher Education Institutions across the Caraga Region, Philippines. *International Journal of Membrane Science and Technology*, 10(3), 1277–1284. DOI: 10.15379/ijmst.v10i3.1700

Florian, M. N., Ali, S., Olivier, J. A., Itbar, K., Muhammad, I., Alina, C. N., Alex, B. D., & Muhammad, A. (2024). The relationship among urbanization, economic growth, renewable energy consumption, and environmental degradation: A comparative view of European and Asian emerging economies. *Gondwana Research*, 128, 325–339. DOI: 10.1016/j.gr.2023.10.023

Følstad, A. & Brandtzæg, P. B. (2017). Chatbots and the new world of HCI. *interactions,* 24, 38-42.

Foris, D., Crihalmean, N., & Foris, T. (2020). Exploring the environmental practices in hospitality through booking websites and online tourist reviews. *Sustainability (Basel)*, 12(24), 10282. DOI: 10.3390/su122410282

Foris, D., Crihălmean, N., & Panoiu, T. M. (2020). The new technologies and sustainable practices in hospitality. *Bulletin of the Transilvania University of Brasov. Series V, Economic Sciences*, 13(62), 2. DOI: 10.31926/but.es.2020.13.62.2.7

Furze, L. (2024). *AI Detection in Education is a Dead End* [Online]. Available: https://leonfurze.com/2024/04/09/ai-detection-in-education-is-a-dead-end/ [Accessed].

Gabler, C. B., Landers, V. M., Agnihotri, R., & Morgan, T. R. (2023). Environmental orientation on the frontline: A boundary-spanning perspective for supply chain management. *Journal of Business Logistics*, 44(3), 369–386.

Galloway, J. F. (2004). Game theory and the law and policy of outer space. *Space Policy*, 20(2), 87–90. DOI: 10.1016/j.spacepol.2004.02.006

Gamage, K. A. A., Ekanayake, S. Y., & Dehideniya, S. C. P. (2022a). Embedding Sustainability in Learning and Teaching: Lessons Learned and Moving Forward-Approaches in STEM Higher Education Programmes. *Education Sciences*, 12(3), 225. Advance online publication. DOI: 10.3390/educsci12030225

Ganesan, S., Manoharan, K. G., & Periyathambi, E. (2023). Utilising Deep Convolutional Neural Networks for Classifying Fire Disasters Through Surveillance: An Indoor and Outdoor Perspective to Predict Man-Made or Natural Disaster. *Revue d'Intelligence Artificielle*, 37(5).

García-González, E., Jiménez-Fontana, R., & Goded, P. A. (2020). Approaches to teaching and learning for sustainability: Characterizing students' perceptions. *Journal of Cleaner Production*, 274, 1–9. DOI: 10.1016/j.jclepro.2020.122928

Garzón-García, R., Florido-Trujillo, G., & Ramírez-López, M. L. (2023). The integration of sustainability and the Sustainability Development Goals (SDGs) in the geographical teaching of tourism: A didactic and methodological proposal for University. *Investigaciones Turísticas*, 26(26), 86–113. DOI: 10.14198/INTURI.22411

Georgallis, P., & Bruijn, K. (2022). Sustainability teaching using case-based debates. *Journal of International Education in Business*, 15(1), 147–163. DOI: 10.1108/JIEB-03-2021-0039

Georgiadou, Y., & Reckien, D. (2018). Geo-information tools, governance, and wicked policy problems. *ISPRS International Journal of Geo-Information*, 7(1), 21. DOI: 10.3390/ijgi7010021

GERE. C. (2006). Digital Culture (second edition). *London: Reaction books.*

Gevme (2023). The 3 pillars of sustainability. https://www.gevme.com/en/blog/the-three-pillars-of-sustainability/

Ghosh, S., Nath, S., & Ranjan, A. (2021). *Green Finance in India: Progress and Challenges*. Reserve Bank of India. https://rbidocs.rbi.org.in/rdocs/Bulletin/PDFs/04AR_2101202185D9B6905ADD465CB7DD280B88266F77.PDF

Gillani, A. H., Li, S., Akbar, J., Omer, S., Fatima, B., Ibrahim, M. I. M., & Fang, Y. (2022). How Prepared Are the Health Care Professionals for Disaster Medicine Management? An Insight from Pakistan. *International Journal of Environmental Research and Public Health*, 19(1), 200. Advance online publication. DOI: 10.3390/ijerph19010200 PMID: 35010475

Giuseppe A, Fabrizio A, Nicola B, Gerardo M.M. (2023). Sustainability and energy communities: Assessing the potential of building energy retrofit and renewables to lead the local energy transition, *energy*, 282, 128377.

Gligor, D. M., & Holcomb, M. C. (2012). Understanding the role of logistics capabilities in achieving supply chain agility: A systematic literature review. *Supply Chain Management*, 17(4), 438–453. DOI: 10.1108/13598541211246594

Goh, C. S., Chong, H. Y., Jack, L., & Faris, A. F. M. (2020). Revisiting triple bottom line within the context of sustainable construction: A systematic review. *Journal of Cleaner Production*, 252, 119884. DOI: 10.1016/j.jclepro.2019.119884

Goh, E., Muskat, B., & Tan, A. H. (2017). The nexus between sustainable practices in hotels and future Gen Y hospitality students' career path decisions. *Journal of Teaching in Travel & Tourism*, 17(4), 237–253. DOI: 10.1080/15313220.2017.1362971

Gomez, A., Conti, F., & Benini, L. Thermal image-based CNN's for ultra-low power people recognition. *Proceedings of the 15th ACM International Conference on Computing Frontiers*, 2018. 326-331. DOI: 10.1145/3203217.3204465

Gong, R., Xue, J., Zhao, L., Zolotova, O., Ji, X., & Xu, Y. (2019). A Bibliometric Analysis of Green Supply Chain Management Based on the Web of Science (WOS) Platform. *Sustainability (Basel)*, 11(12), 3459. DOI: 10.3390/su11123459

GOP. (2022). *Pakistan Economic Survey 2021-22.*

Gopi, K., Sharma, A., Rani, M. J., Kamath, K. P., Manickam, T., Thangam, D., . . . Pol, N. (2024). Strategies to Achieve Carbon Neutrality and Foster Sustainability in Data Centers. In Computational Intelligence for Green Cloud Computing and Digital Waste Management (pp. 109-126). IGI Global. DOI: DOI: 10.4018/979-8-3693-1552-1.ch006

Government of Pakistan. (2016). *National Health Vision Pakistan 2016-2025.*

Government of Pakistan. (2019). *12th FIVE YEAR PLAN (2018-2023).* https://www.pc.gov.pk/uploads/annualplan2018/Annual.pdf

Govindan, K. (2018). Sustainable consumption and production in the food supply chain: A conceptual framework. *International Journal of Production Economics*, 195, 419–431. DOI: 10.1016/j.ijpe.2017.03.003

Govindaraju, P., & Mabel, M. M. (2015). Contingency Factors Impacting the Rural Information and Communication Technology Hubs. In Handbook of Research on Cultural and Economic Impacts of the Information Society (pp. 526-546). IGI Global. DOI: DOI: 10.4018/978-1-4666-8598-7.ch022

Granell, C., Schade, S., & Hobona, G. (2011). Linked Data: Connecting Spatial Data Infrastructures and Volunteered Geographic Information. In *Geospatial Web Services: Advances in Information Interoperability.*

Grant, M. J., & Booth, A. (2009). A typology of reviews: An analysis of 14 review types and associated methodologies. *Health Information and Libraries Journal*, 26(2), 91–108. DOI: 10.1111/j.1471-1842.2009.00848.x PMID: 19490148

Greengard, S. (2019). *Virtual reality.* MIT Press. DOI: 10.7551/mitpress/11836.001.0001

Griffiths, E., Assana, S., & Whitehouse, K. (2018). Privacy-preserving image processing with binocular thermal cameras. *Proceedings of the ACM on Interactive, Mobile, Wearable and Ubiquitous Technologies*, 1(4), 1–25. DOI: 10.1145/3161198

Grzelka, Z., & Wagner, J. (2019). Managing Satellite Debris in Low-Earth Orbit: Incentivizing Ex Ante Satellite Quality and Ex Post Take-Back Programs. *Environmental and Resource Economics*, 74(1), 319–336. DOI: 10.1007/s10640-019-00320-3

Guang-Wen, Z., & Siddik, A. B. (2023). The effect of Fintech adoption on green finance and environmental performance of banking institutions during the COVID-19 pandemic: The role of green innovation. *Environmental Science and Pollution Research International*, 30(10), 25959–25971. DOI: 10.1007/s11356-022-23956-z PMID: 36350441

Guidance, B. A. (2022). *M EASURING THE ECONOMIC, SOCIAL AND ENVIRONMENTAL VALUE OF PUBLIC SECTOR LOCATION.*

Gulzar, R., Bhat, A. A., Mir, A. A., Athari, S. A., & Al-Adwan, A. S. (2024). Green banking practices and environmental performance: Navigating sustainability in banks. *Environmental Science and Pollution Research International*, 31(15), 23211–23226. Advance online publication. DOI: 10.1007/s11356-024-32418-7 PMID: 38413528

Gunduz, M., & Almuajebh, M. (2020). Critical Success Factors for Sustainable Construction Project Management. *Sustainability (Basel)*, 12(5), 1990. DOI: 10.3390/su12051990

Gunnarsdottir, I., Davidsdottir, B., Worrell, E., & Sigurgeirsdottir, S. (2020). Review of indicators for sustainable energy development. *Renewable & Sustainable Energy Reviews*, 133, 110294. DOI: 10.1016/j.rser.2020.110294

Habib, S. S., Jamal, W. Z., Zaidi, S. M. A., Siddiqui, J. U. R., Khan, H. M., Creswell, J., Batra, S., & Versfeld, A. (2021). Barriers to access of healthcare services for rural women— Applying gender lens on tb in a rural district of sindh, pakistan. *International Journal of Environmental Research and Public Health*, 18(19), 10102. Advance online publication. DOI: 10.3390/ijerph181910102 PMID: 34639403

Hafez, F. S., Sa'di, B., Safa-Gamal, M., Taufiq-Yap, Y. H., Alrifaey, M., Seyedmahmoudian, M., & Mekhilef, S. (2023). Energy efficiency in sustainable buildings: A systematic review with taxonomy, challenges, motivations, methodological aspects, recommendations, and pathways for future research. *Energy Strategy Reviews*, 45, 101013.

Hahn, T., Figge, F., Pinkse, J., & Preuss, L. (2018). A Paradox Perspective on Corporate *Sustainability: Descriptive, Instrumental, and Normative Aspects.Journal of Business Ethics*, 148(2), 235–248. DOI: 10.1007/s10551-017-3587-2

Hahn, T., Pinkse, J., Preuss, L., & Figge, F. (2015). Tensions in Corporate Sustainability: Towards an Integrative Framework. *Journal of Business Ethics*, 127(2), 297–316. DOI: 10.1007/s10551-014-2047-5

Haiyan, W., & Yanxia, W. (2024). Smart Cities Net Zero Planning considering renewable energy landscape design in Digital Twin. *Sustainable Energy Technologies and Assessments*, 63, 103629. DOI: 10.1016/j.seta.2024.103629

Hall, L. (2014). *The History of Space Debris*. Space Traffic Management Conference. 19. https://commons.erau.edu/stm/2014/thursday/19

Hallegatte, S., Green, C., Nicholls, R. J., & Corfee-Morlot, J. (2013). Future flood losses in major coastal cities. *Nature Climate Change*, 3(9), 802–806. DOI: 10.1038/nclimate1979

Hamza, A. (2023). Renewable Energy Integration.

Hanspal, M. S., & Behera, B. (2024). The Disaster Management Act, 2005: A Critical Review. *DME Journal of Law*, 5(01), 42–53. DOI: 10.53361/dmejl.v5i01.06

Haroun, F., Ajibade, S., Oladimeji, P., & Igbozurike, J. K. (2021). Toward the Sustainability of Outer Space: Addressing the Issue of Space Debris. *New Space*, 9(1), 63–71. DOI: 10.1089/space.2020.0047

Hashimoto, K., Morinaka, K., Yoshiike, N., Kawaguchi, C., & Matsueda, S. (1997, June). People count system using multi-sensing application. In Proceedings of International Solid State Sensors and Actuators Conference (Transducers' 97) (Vol. 2, pp. 1291-1294). IEEE.

Hasibuan, F. U., & Judijanto, L. (2023). The Influence of Education, Consumer Awareness, and Government Regulation on Sustainable Consumption Behavior: A Case Study on Ciawitali Market in Garut Regency. *West Science Journal Economic and Entrepreneurship*, 1(03), 109–119. DOI: 10.58812/wsjee.v1i03.406

Hasim Altan, M. H. (2016). Passive Design. ZEMCH: Toward the Delivery of Zero Energy Mass Custom Homes. DOI: 10.1007/978-3-319-31967-4_8

Hassani, A. R., Domenighini, P., Belloni, E., Ihara, T., & Buratti, C. (2022). Evaluation of the solar heat gain coefficient of innovative aerogel glazing systems: Experimental campaigns and numerical results. *Journal of Building Engineering*, 62, 62. DOI: 10.1016/j.jobe.2022.105354

Hassan, T. H., Shehata, H. S., El-Dief, M., & Salem, A. (2020). The social responsibility of tourism and hotel establishments and their role in sustainable tourism development in Al-Ahsa, Saudi Arabia. *Geo Journal of Tourism and Geosites*, 33(4), 1564–1570. DOI: 10.30892/gtg.334spl18-609

Haydn, T. (2013). *Using New Technologies to Enhance teaching and learning in History*. Routledge. DOI: 10.4324/9780203075593

Heidari, A., Taghipour, M., & Yarmahmoodi, Z. (2021). The Effect of Fixed External Shading Devices on Daylighting and Thermal Comfort in Residential Buildings. *Journal of Daylighting*, 8(2), 165–180. DOI: 10.15627/jd.2021.15

Helgesson, G. & Eriksson, S. (2014). Plagiarism in research. Medicine Health Care and Philosophy. doi.

Hobson, B. W., Lowcay, D., Gunay, H. B., Ashouri, A., & Newsham, G. R. (2019). Opportunistic occupancy-count estimation using sensor fusion: A case study. *Building and Environment*, 159, 106154. DOI: 10.1016/j.buildenv.2019.05.032

Hongbo R, Zipei J, Qiong W, Qifen L, Yongwen Y. (2022). Integrated optimization of a regional integrated energy system with thermal energy storage considering both resilience and reliability, *energy*, 261, 125333.

Hong, L., Xinyan, K., Sheng, L., & Yangmengliu, Z. (2023). Studying Tian Hock Keng's architectural heritage through the lens of interaction through digital design. *Usability and User Experience*, 110. Advance online publication. DOI: 10.54941/ahfe1003222

Hoque, F., Yasin, R. M., & Sopian, K. (2022). Revisiting Education for Sustainable Development: Methods to Inspire Secondary School Students toward Renewable Energy. *Sustainability (Basel)*, 14(14), 1–18. DOI: 10.3390/su14148296

Horenstein, A., & Heimberg, R. G. (2020). Anxiety disorders and healthcare utilization: A systematic review. In *Clinical Psychology Review* (Vol. 81). DOI: 10.1016/j.cpr.2020.101894

Hosseini, S. M., Fathpour, F., & Chanda, S. (2021). Space Debris Mitigation: Some Lessons from International Environmental Law. *Environmental Policy and Law*, 51(6), 391–401. DOI: 10.3233/EPL-210015

Houmam, L., & Aomar, I. (2023). Effects of School-based Environmental Clubs: Fostering Environmental Awareness and Strengthening Eco-citizen Behaviors Among Students. *Journal of Law and Sustainable Development*, 11(12), e1775. Advance online publication. DOI: 10.55908/sdgs.v11i12.1775

Huang, Z., Peng, A., Yang, T., Deng, S., & He, Y. (2020). A design-based learning approach for fostering sustainability competency in engineering education. *Sustainability (Basel)*, 12(7), 1–17. DOI: 10.3390/su12072958

Hui, Z., & Xinqun, F. (2024). Reliability improvement and landscape planning for renewable energy integration in smart Cities: A case study by digital twin. *Sustainable Energy Technologies and Assessments*, 64, 103714. DOI: 10.1016/j.seta.2024.103714

Huo, B., Wang, K., & Zhang, Y. (2021). The impact of leadership on supply chain green strategy alignment and operational performance. *Operations Management Research : Advancing Practice Through Research*, 14(1-2), 152–165. DOI: 10.1007/s12063-020-00175-8

Hu, S., Wang, P., Hoare, C., & O'Donnell, J. (2022). Building occupancy detection and localization using CCTV camera and deep learning. *IEEE Internet of Things Journal*, 10(1), 597–608. DOI: 10.1109/JIOT.2022.3201877

Hussnain, M. Q. U., Waheed, A., Wakil, K., Pettit, C. J., Hussain, E., Naeem, M. A., & Anjum, G. A. (2020). Shaping up the future spatial plans for urban areas in Pakistan. *Sustainability*, 12(10), 4216.

Hutagalung, J. M., Tobing, C. I., Debastri, J., & Amanda, R. T. (2020). Space debris as environmental threat and the requirement of Indonesia's prevention regulation. *IOP Conference Series. Earth and Environmental Science*, 456(1), 012081. DOI: 10.1088/1755-1315/456/1/012081

Hwang, B.-G., Shan, M., Phua, H., & Chi, S. (2017). An Exploratory Analysis of Risks in Green Residential Building Construction Projects: The Case of Singapore. *Sustainability (Basel)*, 9(7), 1116. DOI: 10.3390/su9071116

Ibrahim, M., Labaki, M., Giraudon, J., & Lamonier, J. (2020). Hydroxyapatite, a multifunctional material for air, water and soil pollution control: A review. *Journal of Hazardous Materials*, 383, 121139. DOI: 10.1016/j.jhazmat.2019.121139 PMID: 31520935

Ibrahim, U. (2020). *Design Strategies and Smart Technologies that Contribute to Well-Being in the Workplace; The Case Studies of the Atlas and the Edge in the Netherlands*. Research in Urbanism and Architecture.

ICOMOS. (2003). *ICOMOS PRINCIPLES FOR THE PRESERVATION AND CONSERVATION-RESTORATION OF WALL PAINTINGS (2003) Introduction and Definition.*

IEA. (2024), Renewables 2023, IEA, Paris https://www.iea.org/reports/renewables-2023

Ijjada & Nayaka, R. (2022). Review on properties of some thermal insulating materials providing more comfort in the building. *Materials Today: Proceedings.*

Imran, M. (2009). Public transport in Pakistan: A critical overview. *Journal of Public Transportation*, 12(2), 53–83. DOI: 10.5038/2375-0901.12.2.4

Indriastuti, M., & Chariri, A. (2021). The role of green investment and corporate social responsibility investment on sustainable performance. *Cogent Business & Management*, 8(1), 1960120. Advance online publication. DOI: 10.1080/23311975.2021.1960120

Informação, S. (2021). Águas do Algarve lança nova campanha de sensibilização. https://www.sulinformacao.pt/2021/08/aguas-do-algarve-lanca-nova-campanha-de-sensibilizacao

INSTITUTE. I. L. (2022). KHOO TECK PUAT HOSPITAL. Retrieved from https://living-future.org/: https://living-future.org/case-studies/award-winner-khoo-teck-puat-hospital/

Intergovernmental Panel on Climate Change (IPCC). (2012). *Managing the risks of extreme events and disasters to advance climate change adaptation*. Retrieved from https://www.ipcc.ch/site/assets/uploads/2018/03/SREX-Chap3_FINAL-1.pdf

Intergovernmental Panel on Climate Change (IPCC). (2014). *Climate change 2014*: Impacts, adaptation, and vulnerability. Retrieved from https://www.ipcc.ch/report/ar5/wg2/

Ismaiel, M. &.-N. (2021). Thermal resistance of masonry walls: A literature review on influence factors, evaluation, and improvement. *Journal of Building Physics*, •••, 528–567.

Ismaila, R. A., & Maher, S. A. (2023). Urban planning schemes for developing low-carbon cities in the Gulf Cooperation Council region. *Habitat International*, 138, 102881. DOI: 10.1016/j.habitatint.2023.102881

Jaichandran, R., Krishna, S. H., Madhavi, G. M., Mohammed, S., Raj, K. B., & Manoharan, G. (2023, January). Fuzzy Evaluation Method on the Financing Efficiency of Small and Medium-Sized Enterprises. In *2023 International Conference on Artificial Intelligence and Knowledge Discovery in Concurrent Engineering (ICECONF)* (pp. 1-7). IEEE.

Janker, J., & Mann, S. (2020). Understanding the social dimension of sustainability in agriculture: A critical review of sustainability assessment tools. *Environment, Development and Sustainability*, 22(3), 1671–1691. DOI: 10.1007/s10668-018-0282-0

Jaykumar, P. (2020). The triumphs & challenges in the path of green hotel housekeeping. *Journal of Services Research*, 20, 21–37.

Jeknic, O. (2014). New media or new interfaces: The concept of interface in Lev Manovich's' theory of new media. *Kultura*, 145(145), 188–197. DOI: 10.5937/kultura1445188j

Jia, F., Gong, Y., & Brown, S. (2019). Multi-tier sustainable supply chain management: The role of supply chain leadership. *International Journal of Production Economics*, 217, 44–63. DOI: 10.1016/j.ijpe.2018.07.022

Jiang, C., Masood, M. K., Soh, Y. C., & Li, H. (2016). Indoor occupancy estimation from carbon dioxide concentration. *Energy and Building*, 131, 132–141. DOI: 10.1016/j.enbuild.2016.09.002

Jindal, A., Hishikar, S., & Shrimali, G. (2024). Examining the portfolio carbon footprint of Indian banking system. *Journal of Cleaner Production*, 142393, 142393. Advance online publication. DOI: 10.1016/j.jclepro.2024.142393

Jiyoung, L. E. E. K. S. (n.d.). Passive Design Techniques applied To green Buildings as an aesthetic and spatial Design concept. Journal of Green Building, 10(2), 79-109. Retrieved from https://www.researchgate.net/publication/281169946_Passive_design_techniques_applied_to_green_buildings_as_an_aesthetic_and_spatial_design_concept

Jones, P., Hillier, D., & Comfort, D. (2013). Sustainability in the global hotel industry. Emerald Insight. *International Journal of Contemporary Hospitality Management*, 26(1), 5–17. DOI: 10.1108/IJCHM-10-2012-0180

Jornal de Negócios. (2022). O negócio do mercado de carbono. https://www.jornaldenegocios.pt/sustentabilidade/ambiente---descarbonizacao/detalhe/o-negocio-do-mercado-de-carbono

Kafetzopoulos, D., & Gotzamani, K. (2022). The effect of talent management and leadership styles on firms' sustainable performance. *European Business Review*, 34(6), 837–857. DOI: 10.1108/EBR-07-2021-0148

Kalwar, S., Sahito, N., Memon, I. A., Hwang, J., Mangi, M. Y., & Lashari, Z. A. (2019). National Planning Strategies for Agro-based Industrial Development in Secondary Cities of Sindh Province, Pakistan. *Sustainability (Basel)*, 11(24), 7066. Advance online publication. DOI: 10.3390/su11247066

Karim, M. R., Cochez, M., Zappa, A., Sahay, R., Rebholz-Schuhmann, D., Beyan, O., & Decker, S. (2022). Convolutional embedded networks for population scale clustering and bio-ancestry inferencing. *IEEE/ACM Transactions on Computational Biology and Bioinformatics*, 19(1), 369–382. DOI: 10.1109/TCBB.2020.2994649 PMID: 32750845

Kasim, A., Gursoy, D., Okumus, F., & Wong, A. (2014). The importance of water management in hotels: A framework for sustainability through innovation. *Journal of Sustainable Tourism*, 22(7), 1090–1107. DOI: 10.1080/09669582.2013.873444

Kasim, H., & Antwi, S. K. (2015). Qualitative and Quantitative Research Paradigms in Business Research: A Philosophical Reflection. *European Journal of Business and Management*, 220, 217–226.

Kasim, H., & Antwi, S. K. (2015). Qualitative and quantitative research paradigms in business research: A philosophical reflection. *European Journal of Business and Management*, 7(3), 217–226.

Kasneci, E., Sessler, K., Küchemann, S., Bannert, M., Dementieva, D., Fischer, F., Gasser, U., Groh, G., Günnemann, S., Hüllermeier, E., Krusche, S., Kutyniok, G., Michaeli, T., Nerdel, C., Pfeffer, J., Poquet, O., Sailer, M., Schmidt, A., Seidel, T., & Kasneci, G. (2023). ChatGPT for good? On opportunities and challenges of large language models for education. *Learning and Individual Differences*, 103, 102274. DOI: 10.1016/j.lindif.2023.102274

Kaunda, R. B. (2020). Potential environmental impacts of lithium mining. *Journal of Energy & Natural Resources Law*, 38(3), 237–244. DOI: 10.1080/02646811.2020.1754596

Kaushal, A. (2022). *The Tower at PNC Plaza by Gensler*. Rethinking the Future.

Kessler, D. J. (1993). A Partial History of Orbital Debris: A Personal View. *Orbital Debris Monitor*, 6(3), 16–20.

Khairy Sayed, H. A. (2018). Building Energy Management Systems (BEMS). In H. A. Khairy Sayed, Energy Conservation in Residential, Commercial, and Industrial Facilities (pp. 15-81). DOI: 10.1002/9781119422099.ch2

Khalid, A. (2021). Conservation Challenges and Emerging Trends of Digital Preservation for UNESCO Architectural Heritage, Pakistan. *Conservation*, 2(1), 26–37. DOI: 10.3390/conservation2010003

Khalid, M. R., Alam, M. S., Sarwar, A., & Asghar, M. J. (2019). A Comprehensive review on electric vehicles charging infrastructures and their impacts on power-quality of the utility grid. *ETransportation*, 1, 100006. DOI: 10.1016/j.etran.2019.100006

Khalil, M., & Er, E. (2023). Will ChatGPT G et You Caught? Rethinking of Plagiarism Detection. *International Conference on Human-Computer Interaction*, 475-487.

Khana, M. P., Talibb, N. A., & Kowangc, T. O. (2020). The development of a sustainability framework via lean green six sigma practices in SMEs based upon RBV theory. *Development*, 12(5), 135–156.

Khan, M. S., Durrance-Bagale, A., Mateus, A., Sultana, Z., Hasan, R., & Hanefeld, J. (2020). What are the barriers to implementing national antimicrobial resistance action plans? A novel mixed-methods policy analysis in Pakistan. *Health Policy and Planning*, 35(8), 973–982. Advance online publication. DOI: 10.1093/heapol/czaa065 PMID: 32743655

Khashoggi, B. F., & Murad, A. (2020). Issues of healthcare planning and GIS: A review. In *ISPRS International Journal of Geo-Information* (Vol. 9, Issue 6). DOI: 10.3390/ijgi9060352

Khattak, A. F., Rahman, A. U., Khattak, M., Qazi, M., Gilani, H., & Khan, A. (2023). Toward Sustainable Healthcare Systems: A Low and Middle-Income Country's Case for Investing in Healthcare Reforms. *Cureus*. Advance online publication. DOI: 10.7759/cureus.39345 PMID: 37351239

Khokhar, M. A., Ali, M. M., Liaqat, S., Moin, A., Sarwar, H. A., & Sarwar, M. Z. (2020). A review of access to cancer facilities in Punjab, Pakistan. In *Cancer Reports* (Vol. 3, Issue 3). DOI: 10.1002/cnr2.1245

Kılıç, T. &. (2017). An Investigation on Internet of Things Technology (IoT) In Smart Houses. Uluslararası Muhendislik Arastirma ve Gelistirme Dergisi, 196-207.

Kim, M. S., & Thapa, B. (2018). Relationship of ethical leadership, corporate social responsibility and organizational performance. *Sustainability (Basel)*, 10(2), 447. DOI: 10.3390/su10020447

Kim, Y. H., Barber, N., & Kim, D. K. (2019). Sustainability research in the hotel industry: Past, present, and future. *Journal of Hospitality Marketing & Management*, 28(5), 576–620. DOI: 10.1080/19368623.2019.1533907

Kirimtat, A., Koyunbaba, B. K., Chatzikonstantinou, I., & Sariyildiz, S. (2016). Review of simulation modeling for shading devices in buildings. *Renewable & Sustainable Energy Reviews*, 53, 23–49. DOI: 10.1016/j.rser.2015.08.020

Klima, R., Bloembergen, D., Savani, R., Tuyls, K., Hennes, D., & Izzo, D. (2016). Space Debris Removal: A Game Theoretic Analysis. *Games*, 7(3), 20. DOI: 10.3390/g7030020

Klima, R., Bloembergen, D., Savani, R., Tuyls, K., Wittig, A., Sapera, A., & Izzo, D. (2018). Space Debris Removal: Learning to Cooperate and the Price of Anarchy. *Frontiers in Robotics and AI*, 5, 54. DOI: 10.3389/frobt.2018.00054 PMID: 33500936

Koebrich, S., Chen, E. I., Bowen, T., Forrester, S., & Tian, T. (2019). *2017 Renewable Energy Data Book: Including Data and Trends for Energy Storage and Electric Vehicles. National Renewable Energy Lab*. NREL. DOI: 10.2172/1581221

Kollarova, D. (2024). Forest Pedagogy and Its Didactic Approach in Pre-Primary and Primary Education. *Slavonic Pedagogical Studies Journal*, 13(1), 61–74. DOI: 10.18355/PG.2024.13.1.6

Kooi, Y. K. (2022). Adaptability and Flexibility in Architecture Concepts & Theories Applied in Residential Architecture to Achieve Adaptability.

Kopnina, H. (2020). Education for the future? Critical evaluation of education for sustainable development goals. *The Journal of Environmental Education*, 51(4), 280–291. DOI: 10.1080/00958964.2019.1710444

Kornelsen, J., Carthew, C., Míguez, K., Taylor, M., Bodroghy, C., Petrunia, K., & Roberts, D. (2021). Rural citizen-patient priorities for healthcare in British Columbia, Canada: Findings from a mixed methods study. *BMC Health Services Research*, 21(1), 987. Advance online publication. DOI: 10.1186/s12913-021-06933-z PMID: 34537027

Kotsev, A., Minghini, M., Cetl, V., & En, E. U. R. (2021). *INSPIRE A Public Sector Contribution to the European Green Deal Data Space. A vision for the technological evolution of Europe's Spatial Data Infrastructures for 2030*. DOI: 10.2760/8563

Kou, X., Du, Y., Li, F., Pulgar-Painemal, H., Zandi, H., Dong, J., & Olama, M. M. (2021). Model-based and data-driven HVAC control strategies for residential demand response. *IEEE Open Access Journal of Power and Energy*, 8, 186–197. DOI: 10.1109/OAJPE.2021.3075426

Kozma, R. B. (1991). Computer-based writing tools and the cognitive needs of novice writers.

Krek, A. (2005). Geographic information as an economic good. In *GIS for Sustainable Development* (pp. 85–103). CRC Press., DOI: 10.1201/9781420037845.ch6

Kritzinger, W., Karner, M., Traar, G., Henjes, J., & Sihn, W. (2018). Digital Twin in manufacturing: A categorical literature review and classification. *IFAC-PapersOnLine*, 51(11), 1016–1022.

Kryvuts, S., Gonchar, O., Skorokhodova, A., & Radomskyi, M. (2021). The phenomenon of digital art as a means of preservation of cultural heritage works. *Muzeologia a Kulturne Dedicstvo*, 9(1), 145–156. DOI: 10.46284/mkd.2021.9.1.9

Kuczynski, T., & Staszczuk, A. (2020). Experimental study of the influence of thermal mass on thermal comfort and cooling energy demand in residential buildings. *Energy*, 195(1), 116984. DOI: 10.1016/j.energy.2020.116984

Kuhlman, T., & Farrington, J. (2010). What is sustainability? *Sustainability (Basel)*, 2(11), 3436–3448. DOI: 10.3390/su2113436

Kukah, A. S. K., Xiaohua, J., Osei-Kyei, R., & Perera, S. (2024). How carbon trading contributes to reduction in emission of greenhouse gases: A narrative literature review. *Journal of Facilities Management*. Advance online publication. DOI: 10.1108/JFM-08-2023-0094

Kumari, K. G. S., Mukthar, K. P. J., & Rahhani, G. N. E. (2024). Green Finance in India: Driving Sustainable Development and Economic Growth. In *Studies in systems, decision and control* (pp. 283–293). https://doi.org/DOI: 10.1007/978-3-031-62106-2_23

Kumari, S., Kumar, V., Sharmila, A., Murthy, C. R., Ahlawat, N., & Manoharan, G. (2023, August). Blockchain Based E-Analysis of Social Media Forums for Crypto Currency Phase Shifts. In 2023 5th International Conference on Inventive Research in Computing Applications (ICIRCA) (pp. 1222-1225). IEEE. doi:DOI: 10.1109/ICAISS58487.2023.10250754

Kumar, K., & Prakash, A. (2018). Developing a framework for assessing the sustainable banking performance of the Indian banking sector. *Social Responsibility Journal*, 15(5), 689–709. DOI: 10.1108/SRJ-07-2018-0162

Kumar, P., Maurya, N., Keerthiraj, S. H. K., Manoharan, G., & Bharti, A. (2024). Evaluating the Impact of Healthcare 4.0 on the Performance of Hospitals. *Human Cancer Diagnosis and Detection Using Exascale Computing*, 1, 1.

Kumar, V., Ali, B. S., Choudry, E., Khan, S., Baig, K., Durrani, N. U. R., & Ali, S. R. (2022). Quality of Neonatal Care: A Health Facility Assessment in Balochistan Province, Pakistan. *Cureus*. Advance online publication. DOI: 10.7759/cureus.22744 PMID: 35386481

Lafortune, G., Fuller, G., Kloke-Lesch, A., Koundouri, P., & Riccaboni, A. (2024). *European elections, Europe's future and the SDGs: Europe sustainable development report 2023/24*. SDG Transformation Center.

Laing, T., Sato, M., Grubb, M., & Comberti, C. (2013). Assessing the effectiveness of the EU Emissions Trading System (Vol. 126). London: Grantham Research Institute on Climate Change and the Environment. http://www.jcfj.ie/wp-content/uploads/2016/02/WP106-effectiveness-eu-emissions-trading-system.pdf

Lambach, D., & Wesel, L. (2021). TACKLING THE SPACE DEBRIS PROBLEM: A GLOBAL COMMONS PERSPECTIVE. *8th European Conference on Space Debris,* Darmstadt, Germany, 20-23 April, 2021.

Laugen, B. T., Acur, N., Boer, H., & Frick, J. (2005). Best manufacturing practices: What do the best-performing companies do? *International Journal of Operations & Production Management*, 25(2), 131–150. DOI: 10.1108/01443570510577001

Lavuri, R., Roubaud, D., & Grebinevych, O. (2023). Sustainable consumption behaviour: Mediating role of pro-environment self-identity, attitude, and moderation role of environmental protection emotion. *Journal of Environmental Management*, 347, 119106. DOI: 10.1016/j.jenvman.2023.119106 PMID: 37806273

Lee, G.-C., Lee, S.-H., & Yoo, J. (2018). 3-D people counting with a stereo camera on gpu embedded board. *Applied Sciences (Basel, Switzerland)*, 8(11), 2017. DOI: 10.3390/app8112017

Lin Chen, Y. H.-S. (2024). Green building practices to integrate renewable energy in the construction sector: A review. *Environmental Chemistry Letters*, 22(2), 751–784. DOI: 10.1007/s10311-023-01675-2

Linneberg, M. S., Madsen, M. T., & Nielsen, J. A. (2019). Micro-level translation of corporate sustainability: When strategy meets practice in the Danish hospitality sector. *Journal of Cleaner Production*, 240, 118159. DOI: 10.1016/j.jclepro.2019.118159

Liou, J.-C. (2013). Engineering and technology challenges for active debris removal. *Progress in Propulsion Physics*, 735–748. DOI: 10.1051/eucass/201304735

Little, C., Bec, A., Moyle, B. D., & Patterson, D. (2020). Innovative methods for heritage tourism experiences: Creating windows into the past. *Journal of Heritage Tourism*, 15(1), 1–13. DOI: 10.1080/1743873X.2018.1536709

Liu, H., Azam, M., Bin Naeem, S., & Faiola, A. (2023). An overview of the capabilities of ChatGPT for medical writing and its implications for academic integrity. *Health Information and Libraries Journal*, 40(4), 440–446. DOI: 10.1111/hir.12509 PMID: 37806782

Liu, J., Mooney, H., Hull, V., Davis, S. J., Gaskell, J., Hertel, T., Lubchenco, J., Seto, K. C., Gleick, P., Kremen, C., & Li, S. (2015). Systems integration for global sustainability. *Science*, 347(6225), 1258832. DOI: 10.1126/science.1258832 PMID: 25722418

Liu, S., Kwok, Y. T., Lau, K. K. L., Ouyang, W., & Ng, E. (2020). Effectiveness of passive design strategies in responding to future climate change for residential buildings in hot and humid Hong Kong. *Energy and Building*, 228, 110469.

Liu, X., Chen, Z., Yu, Q., Fang, L., Gong, J., & Wen, C. (2023). Carbon trading systems and Credit Spreads of Highly Polluting Corporate Bonds: An Empirical Study Based on a Triple Difference Model. DOI: 10.21203/rs.3.rs-2563261/v1

Lobosco, G., Tinti, L., Magagnoli, B., Mencarini, V., Mannucci, S., & Ferrero, M. (2023). Landscape as a Palimpsest for Energy Transition: Correlations between the Spatial Development of Energy-Production Infrastructure and Climate-Mitigation Goals. *Atmosphere 2023, Vol. 14, Page 921, 14*(6), 921. DOI: 10.3390/atmos14060921

Lopes, J. M. M., Gomes, S., & Trancoso, T. (2024). Navigating the green maze: Insights for businesses on consumer decision-making and the mediating role of their environmental concerns. *Sustainability Accounting. Management and Policy Journal*, 15(4), 861–883. DOI: 10.1108/SAMPJ-07-2023-0492

Losch, A., Galli, A., & Schildknecht, T. (2024). History, concepts, and challenges of proposing an 18th UN Sustainable Development Goal related to Space. *Space Policy*, 68, 101631. DOI: 10.1016/j.spacepol.2024.101631

Lovejoy, M. P. (2011). *"Introduction." Context Providers: Conditions of Meaning in Media Arts.*

M. A. Giusti, a. M. (1993). La Certosa di Pisa a Calci. . *Pacini Editore*.

Macauley, M. K. (2015). The economics of space debris: Estimating the costs and benefits of debris mitigation. *Acta Astronautica*, 115, 160–164. DOI: 10.1016/j.actaastro.2015.05.006

Mak, A. H., & Chang, R. C. (2019). The driving and restraining forces for environmental strategy adoption in the hotel industry: A force field analysis approach. *Tourism Management*, 73(1), 48–60. DOI: 10.1016/j.tourman.2019.01.012

Ma, L., Shahbaz, P., Haq, S., & Boz, I. (2023). Exploring the Moderating Role of Environmental Education in Promoting a Clean Environment. *Sustainability (Basel)*, 15(10), 1–19. DOI: 10.3390/su15108127

Malthus, T. R. (1959). *Population: The first essay*. University of Michigan Press. DOI: 10.3998/mpub.6638

Mamdouh El Haj Assad, M. N. (2022). *Geothermal Heat Pumps: Principles and Applications*. Advances in Science and Engineering Technology International., DOI: 10.1109/ASET53988.2022.9734907

Mane, S. (2024). Theoretical Overview on Space and Sustainability. *International Journal of All Research Education and Scientific Methods*, 12(8), 2060–2066.

Manfré, L., Hirata, E., Silva, J. B., Shinohara, E. J., Giannotti, M., Larocca, A. P. C., & Quintanilha, J. (2012). An Analysis of Geospatial Technologies for Risk and Natural Disaster Management. *ISPRS International Journal of Geo-Information*, 1(2), 166–185. DOI: 10.3390/ijgi1020166

Manning, E. W. (1990). Geographic information systems and sustainable development. *Government Information Quarterly*, 7(3), 329–342. DOI: 10.1016/0740-624X(90)90028-M

Manoharan, G., Durai, S., Ashtikar, S. P., & Kumari, N. (2024). Artificial Intelligence in Marketing Applications. In Artificial Intelligence for Business (pp. 40-70). Productivity Press.

Manoharan, G., Razak, A., Rao, B. S., Singh, R., Ashtikar, S. P., & Nivedha, M. (2024). Navigating the Crescendo of Challenges in Harnessing Artificial Intelligence for Disaster Management. In Predicting Natural Disasters With AI and Machine Learning (pp. 64-94). IGI Global.

Manoharan, G., Ashtikar, S. P., Smitha, V., Sundaramoorthi, S., & Krishna, I. M. (2023). Work-Life Balance Perceptions of Women in the IT and ITeS Sectors in Kerala: A Research Study. *Journal of Pharmaceutical Negative Results*, •••, 3363–3375.

Manoharan, G., Durai, S., Rajesh, G. A., Razak, A., Rao, C. B., & Ashtikar, S. P. (2023). An investigation into the effectiveness of smart city projects by identifying the framework for measuring performance. In *Artificial Intelligence and Machine Learning in Smart City Planning* (pp. 71–84). Elsevier. DOI: 10.1016/B978-0-323-99503-0.00004-1

Manolis, E. N., & Manoli, E. N. (2021). Raising awareness of the Sustainable Development Goals through Ecological Projects in Higher Education. *Journal of Cleaner Production*, 279, 1–31. DOI: 10.1016/j.jclepro.2020.123614

Manovich, L. (2002). *The language of new media*. MIT press.

Mansour, N. (2023). Green Technology Innovation and Financial Services System: Evidence from China. *Businesses*, 3(1), 98–113. DOI: 10.3390/businesses3010008

Manurung, H., Yudoko, G., & Okdinawati, L. (2023). A conceptual framework of supply chain resilience towards sustainability through a service-dominant logic perspective. *Heliyon*, 9(3), e13901. DOI: 10.1016/j.heliyon.2023.e13901 PMID: 36915553

Marche, S. (2022). The college essay is dead. Nobody is prepared for how AI will transform academia. *Atlantic (Boston, Mass.)*, •••, 6.

María-J, P., Ricardo, I. P., & Ángeles, B. (2022). Energy transition and community participation in Portugal, Greece and Israel: Regional differences from a multi-level perspective. *Energy Research & Social Science*, 87, 102467. DOI: 10.1016/j.erss.2021.102467

Mark, M. A., Mahesti, O., & Upalat, K. (2024). Stakeholders' awareness of urban form effects on rooftop solar photovoltaic in Ghana: Implications for integrated solar energy and urban planning. *Energy for Sustainable Development*, 78, 101377. DOI: 10.1016/j.esd.2024.101377

Marra, A., Gerbino, S., Greco, A., & Fabbrocino, G. (2021). Combining integrated informative system and historical digital twin to maintain and preservation of artistic assets. *Sensors (Basel)*, 21(17), 5956. Advance online publication. DOI: 10.3390/s21175956 PMID: 34502848

Martins, C. L., & Pato, M. V. (2019). Supply chain sustainability: *A tertiary literature review.Journal of Cleaner Production*, 225, 995–1016. DOI: 10.1016/j.jclepro.2019.03.250

Masdar. (n.d.). *About Masdar City*. Retrieved from https://masdarcity.ae/en/about-masdar-city

Mashari, D. P. S., Zagloel, T. Y. M., Soesilo, T. E. B., & Maftuchah, I. (2023). A bibliometric and literature review: Alignment of green finance and carbon trading. *Sustainability (Basel)*, 15(10), 7877. DOI: 10.3390/su15107877

Masser, I. (2019). Geographic Information Systems to Spatial Data Infrastructure. In *Geographic Information Systems to Spatial Data Infrastructure*. DOI: 10.1201/9780429505904

Masser, I. (2005). *GIS worlds : creating spatial data infrastructures* (Vol. 338). ESRI Press.

Masser, I. (2019). *From Geographic information systems to Spatial data infrastructures : a global perspective* (1st ed.).

Matsumoto, T., & Yamada, Y. (2023). Multivariate Weather Derivatives for Wind Power Risk Management: Standardization Scheme and Trading Strategy. In: Caetano, N.S., Felgueiras, M.C. (eds) The 9th International Conference on Energy and Environment Research. ICEER 2022. Environmental Science and Engineering. Springer, Cham. DOI: 10.1007/978-3-031-43559-1_26

Matton, E. (2023). Avoiding the Kessler Effect: An Evaluation of the International Regulation of Space Debris. *Master's Thesis in International Law, Uppsala Universitet, Sweden.*

McDermott, D., Binan, D., Bumbaru, D., Jungeblodt, G., Odiaua, I., Soule, J., Kuban, N., Murray, C., Avrami, E., & Wilson, H. Jackson for Australia NSCES, R., Turner, M., Motta, P., Burke, S., & Colletta, T. (2017). *ICOMOS Action Plan: Cultural Heritage and Localizing the UN Sustainable Development Goals (SDGs) Prepared by Ege Yıldırım, ICOMOS Focal Point for the SDGs with contributions from Sofia Avgerinou-Kolonias, ICOMOS Sustainable Development Task Force Chair.* https://www.icomos.org/en/what-we-do/involvement-in-international-conventions/un-sustainable-development-goals-2

McFarlane, D. A., & Ogazon, A. G. (2011). The challenges of sustainability education. *Journal of Multidisciplinary Research (1947-2900), 3*(3)

McKinsey & Company. (2022, April 26). The path to net zero: Investing in carbon markets. McKinsey & Company. https://www.mckinsey.com/featured-insights/future-of-asia/the-path-to-net-zero-investing-in-carbon-markets

McLaughlin, J., & Nichols, S. (1994). Developing a national spatial data infrastructure. *Journal of Surveying Engineering*, 120(2), 62–76. DOI: 10.1061/(ASCE)0733-9453(1994)120:2(62)

Mebane, M. E., Benedetti, M., Barni, D., & Francescato, D. (2023). Promoting Climate Change Awareness with High School Students for a Sustainable Community. *Sustainability (Basel)*, 15(14), 1–15. DOI: 10.3390/su151411260

Medrado, L., & Jackson, L. (2015). Corporate non-financial disclosures: An illuminating look at the corporate social responsibility and sustainability reporting practices of hospitality and tourism firms. *Tourism and Hospitality Research*, 16(2), 116–132. DOI: 10.1177/1467358415600210

Medvedeva, A. (2015). *Space Debris Remediation: An International Relations Approach.* TU Wien Continuiting Education Center, Vienna School of International Studies.

Mei, Z., Danting, Z., & Tingfeng, X. (2024). Balancing urban energy considering economic growth and environmental sustainability through integration of renewable energy. *Sustainable Cities and Society*, 101, 105178. DOI: 10.1016/j.scs.2024.105178

Mele, P. R., Gomez, J. M., & Garay, L. (2019). To green or not to green: The influence of green marketing on consumer behaviour in the hotel industry. *Sustainability (Basel)*, 11(17), 4623. DOI: 10.3390/su11174623

Memon, A. A. Q., Osama, M., Wei, C. R., Rasool, G., Bhurgri, R. S., Siyal, D. R., & Siyal, F. J. (2024). Common Health Challenges For Foreigners In Pakistan. *Migration Letters : An International Journal of Migration Studies*, 21(S8), 131–142.

Mendoza, M. A. D., De La Hoz Franco, E., & Gómez, J. E. G. (2023). Technologies for the Preservation of Cultural Heritage—A Systematic Review of the Literature. In *Sustainability (Switzerland)* (Vol. 15, Issue 2). MDPI. DOI: 10.3390/su15021059

Menon, D. G., Sreelakshmi, S. G., & Shivdas, A. (2017). Green banking initiatives: A review of Indian banking sector. *2017 International Conference on Technological Advancements in Power and Energy (TAP Energy)*. https://doi.org/DOI: 10.1109/TAPENERGY.2017.8397303

Meshram, P., Mishra, A., & Sahu, R. (2020). Environmental impact of spent lithium ion batteries and green recycling perspectives by organic acids–A review. *Chemosphere*, 242, 125291. DOI: 10.1016/j.chemosphere.2019.125291 PMID: 31896181

Metwaly, A., Queralta, J. P., Sarker, V. K., Gia, T. N., Nasir, O., & Westerlund, T. (2019, October). Edge computing with embedded ai: Thermal image analysis for occupancy estimation in intelligent buildings. In *Proceedings of the INTelligent Embedded Systems Architectures and Applications Workshop 2019* (pp. 1-6).

Micalay-Hurtado, M. A., & Poole, R. (2022). Eco-critical language awareness for English language teaching (ELT): Promoting justice, wellbeing, and sustainability in the classroom. *Journal of World Languages*, 8(2), 371–390. DOI: 10.1515/jwl-2022-0023

Michailidis, I. (2015). Improving Energy Savings and Thermal Comfort in Large-scale Buildings via Adaptive Optimization. In Miranda, F. (Ed.), *I. &. Michailidis, Control Theory: Perspectives, Applications, and Developments* (pp. 14-1–14-22). Nova Science., DOI: 10.1159/9781634827300.ch14

Migaud, M. R. (2020). Protecting Earth's Orbital Environment: Policy Tools for Combating Space Debris. *Space Policy*, 52, 101361. DOI: 10.1016/j.spacepol.2020.101361

Migdadi, Y. (2022). Identifying the best practices in hotel green supply chain management strategy: A global study. *Journal of Quality Assurance in Hospitality & Tourism*. Advance online publication. DOI: 10.1080/1528008X.2022.2065657

Mir, A. A., & Bhat, A. A. (2022). Green banking and sustainability – a review. *Arab Gulf Journal of Scientific Research*, 40(3), 247–263. DOI: 10.1108/AGJSR-04-2022-0017

Miranda, D. D. P., Ortiz, J. A., & Cardona, J. R. (2019). Article determinants of CSR application in the hotel industry of the Colombian Caribbean. *Sustainability (Basel)*, 11(18), 5045. DOI: 10.3390/su11185045

Míriam Andrea, H. B., Jesús, S. M., Isaac, C. C., & Florentina, C. C. (2022). Is sustainability an exhausted concept? Bridging the gap from environmental awareness to emotional proficiency in science education through integral sustainability. *Heliyon*, 8(12), e12546. Advance online publication. DOI: 10.1016/j.heliyon.2022.e12546 PMID: 36593825

Mishina, N. V., Shikula, I. R., & Afanasyeva, S. A. (2020). Features of legal regulation of artificial intelligence as a guarantee of sustainable development of society. Paper presented at the *E3S Web of Conferences, 208* 06010. DOI: 10.1051/e3sconf/202020806010

Mishra, S. P., & Mohapatra, S. (2023). Anthropocene Impacts on Marine Fishers: South Odisha Coast, India. *Current Journal of Applied Science and Technology, 42*(48), 176-196. https://en.wikipedia.org/wiki/Water_management_in_Chennai#:~:text=Chennai%20receives%20about%20985%20million,required%20amount%20of%201%2C200%20mld https://en.wikipedia.org/wiki/Water_management_in_Chennai https://www.thehindu.com/news/national/tamil-nadu/cyclone-michaung-over-1600-acres-of-crops-damaged-in-ranipet/article67607625.ece

Misseyanni, A., Marouli, C., & Papadopoulou, P. (2020). How teaching affects student attitudes towards the environment and sustainability in higher education: An instructors' perspective. *European Journal of Sustainable Development*, 9(2), 172–182. DOI: 10.14207/ejsd.2020.v9n2p172

Mitoula, R., & Papavasileiou, A. (2023). Mega infrastructure projects and their contribution to sustainable development: The case of the Athens Metro. *Economic Change and Restructuring*, 56(3), 1943–1969. DOI: 10.1007/s10644-023-09493-w

Mohajan, H. (2018). Qualitative research methodology in social sciences and related subjects. *Journal of Economic Development. Environment and People*, 7(1), 23–48. DOI: 10.26458/jedep.v7i1.571

Mohamed Azeroual, T. L. (2019). Simulation tools for a smart grid and energy management for a microgrid with wind power using a multi-agent system. *Wind Engineering*, 44(6).

Mohammed, S. D. (2020). Clean development mechanism and carbon emissions in Nigeria. Sustainability Accounting. *Management and Policy Journal*, 11(3), 523–551. DOI: 10.1108/SAMPJ-05-2017-0041

Moise, M., Gil-Saura, I., & Molina, M. (2021). The importance of green practices for hotel guests: Does gender matter? *Ekonomska Istrazivanja*, 34(1), 3508–3529. DOI: 10.1080/1331677X.2021.1875863

Mo, J. L., Agnolucci, P., Jiang, M. R., & Fan, Y. (2016). The impact of Chinese carbon emission trading scheme (ETS) on low carbon energy (LCE) investment. *Energy Policy*, 89, 271–283. DOI: 10.1016/j.enpol.2015.12.002

Mokhtar, A. R. M., Genovese, A., Brint, A., & Kumar, N. (2019). Supply chain leadership: A systematic literature review and a research agenda. *International Journal of Production Economics*, 216, 255–273. DOI: 10.1016/j.ijpe.2019.04.001

Mondal, S., & Samaddar, K. (2023). Reinforcing the significance of human factor in achieving quality performance in data-driven supply chain management. *The TQM Journal*, 35(1), 183–209. DOI: 10.1108/TQM-12-2020-0303

Moons, B., Bankman, D., & Verhelst, M. (2019). *Embedded deep learning. Circuits for Always-on Neural Network Processing* (1st ed.). Springer.

Moreo, A., Cobanoglu, C., & DeMicco, F. (2007). A comparative analysis of restaurant websites and hospitality school restaurant websites. *Journal of Hospitality & Tourism Education*, 19(3), 40–47. DOI: 10.1080/10963758.2007.10696896

Morgan, F., McLeod, G., Nixon, M., Lynch, C., & Hura, M. (2018). *Gaming Space: A Game-Theoretic Methodology for Assessing the Deterrent Value of Space Control Options*. RAND Corporation., DOI: 10.7249/RR694

Morin, J., & Richard, B. (2021). Astro-Environmentalism: Towards a Polycentric Governance of Space Debris. *Global Policy*, 12(4), 568–573. DOI: 10.1111/1758-5899.12950

Mousa, S. K., & Othman, M. (2020). The impact of green human resource management practices on sustainable performance in healthcare organisations: *A conceptual framework.Journal of Cleaner Production*, 243, 118595. DOI: 10.1016/j.jclepro.2019.118595

Mowforth, M., & Munt, I. (2015). *Tourism and sustainability: Development, globalisation and new tourism in the third world.* routledge.

Mullick, S., Srinivasa, Y., Sahu, A. K., & Sata, J. T. (2019). A Comprehensive Study on Space Debris, Threats Posed by Space Debris, and Removal Techniques. SSRN *Electronic Journal.* DOI: 10.2139/ssrn.3511445

Muralidharan, E., & Pathak, S. (2018). Sustainability, transformational leadership, and social entrepreneurship. *Sustainability (Basel)*, 10(2), 567. DOI: 10.3390/su10020567

Nagariya, R., Kumar, D., & Kumar, I. (2022). Sustainable service supply chain management: From a systematic literature review to a conceptual framework for performance evaluation of service only supply chain. *Benchmarking*, 29(4), 1332–1361. DOI: 10.1108/BIJ-01-2021-0040

Naik, S., Sony, M., Antony, J., McDermott, O., Tortorella, G. L., & Jayaraman, R. (2024). Operational excellence framework for sustainability in the organisation: A design science approach. *Production Planning and Control*, 35(11), 1215–1231. DOI: 10.1080/09537287.2023.2165188

Nan, S., & Chaiprasit, K. (2023). Resilient leadership, innovation, executive incentives, and sustainable business performance: An empirical study. *Journal of Logistics. Informatics and Service Science*, 10(4), 1–19.

Naqvi, B., Rizvi, S. K. A., Hasnaoui, A., & Shao, X. (2022). Going beyond sustainability: The diversification benefits of green energy financial products. *Energy Economics*, 111, 106111. DOI: 10.1016/j.eneco.2022.106111

National Environment Agency Singapore. (n.d.). Benefits of carbon trading. Carbon Markets & Cooperation. https://www.carbonmarkets-cooperation.gov.sg/our-article-6-cooperation/benefits-of-carbon-trading/

Nedović-Budić, Z., Crompvoets, J., & Georgiadou, Y. (2011). Spatial data infrastructures in context: North and South. In *Spatial Data Infrastructures in Context*. North and South. DOI: 10.1201/b10857

Newman, C., & Cheney, T. (2023). Barriers and Gateways to Cleaning Up Earth Orbit: The Legal, Economic, and Political Dimensions of Debris Remediation. *Air and Space Law*, 48(Special Issue), 113–136. DOI: 10.54648/AILA2023034

Ng, A. W., & Tavitiyaman, P. (2020). Corporate social responsibility and sustainability initiatives of multinational hotel corporations. In: Leal Filho, W., Borges de Brito, P. & Frankenberger, F. (Eds.) International Business, Trade and Institutional Sustainability. World Sustainability Series. Springer. Doi: DOI: 10.1007/978-3-030-26759-9_1

Nobles, S., & Paganucci, L. (2015). Do digital writing tools deliver? Student perceptions of writing quality using digital tools and online writing environments. *Computers and Composition*, 38, 16–31. DOI: 10.1016/j.compcom.2015.09.001

Nomura, K., Rella, S., Merritt, H., Baltussen, M., Bird, D., Tjuka, A., & Falk, D. (2024). Tipping Points of Space Debris in Low Earth Orbit. *The International Journal of the Commons*, 18(1). Advance online publication. DOI: 10.5334/ijc.1275

Norozpour, S., & Safaei, M. (2020). An Overview on Game Theory and Its Application. *IOP Conference Series. Materials Science and Engineering*, 993(1), 012114. DOI: 10.1088/1757-899X/993/1/012114

Nozawa, W., Kurita, K., Tamaki, T., & Managi, S. (2023). To What Extent Will Space Debris Impact the Economy? *Space Policy*, 66, 101580. DOI: 10.1016/j.spacepol.2023.101580

OECD. (2020). *Space sustainability: The economics of space debris in perspective* (OECD Science, Technology, and Industry Policy Papers 87; OECD Science, Technology and Industry Policy Papers, Vol. 87). DOI: 10.1787/23074957

Omole, F. O., Olajiga, O. K., & Olatunde, T. M. (2024). Sustainable urban design: A review of eco-friendly building practices and community impact. *Engineering Science & Technology Journal*, 5(3), 1020–1030. DOI: 10.51594/estj.v5i3.955

Organización de las Naciones Unidas para la Educación, l. C. (1972). "Convención sobre la protección del patrimonio mundial, cultural y natural.". *Actas de 17a reunión.*

Oriade, A., Osinaike, A., Aduhene, K., & Wang, Y. (2021). Sustainability awareness, management practices and organisational culture in hotels: Evidence from developing countries. *International Journal of Hospitality Management*, 92, 102699. DOI: 10.1016/j.ijhm.2020.102699

Ornelas, C. (2017). Sustentabilidade, razões, práticas e impactos na indústria hoteleira. Master Dissertation, Universidade da Madeira, Portugal.

Osei, M. B., Papadopoulos, T., Acquaye, A., & Stamati, T. (2023). Improving sustainable supply chain performance through organisational culture: A competing values framework approach. *Journal of Purchasing and Supply Management*, 29(2), 100821.

Ozturkoglu, Y., Sari, F. O., & Saygili, E. (2021). A new holistic conceptual framework for sustainability-oriented hospitality innovation with Triple Bottom Line perspective. *Journal of Hospitality and Tourism Technology*, 12(1), 39–57. DOI: 10.1108/JHTT-02-2019-0022

Paidi, V., Fleyeh, H., & Nyberg, R. G. (2020). Deep learning-based vehicle occupancy detection in an open parking lot using thermal camera. *IET Intelligent Transport Systems*, 14(10), 1295–1302. DOI: 10.1049/iet-its.2019.0468

Palanca, J., Jordan, J., Bajo, J., & Botti, V. (2020). An energy-aware algorithm for electric vehicle infrastructures in smart cities. *Future Generation Computer Systems*, 108, 454–466. DOI: 10.1016/j.future.2020.03.001

Palmquist, M. (2003). A brief history of computer support for writing centers and writing-across-the-curriculum programs. *Computers and Composition*, 20(4), 395–413. DOI: 10.1016/j.compcom.2003.08.013

Palmroth, M., Tapio, J., Soucek, A., Perrels, A., Jah, M., Lönnqvist, M., Nikulainen, M., Piaulokaite, V., Seppälä, T., & Virtanen, J. (2021). Toward Sustainable Use of Space: Economic, Technological, and Legal Perspectives. *Space Policy*, 57, 101428. DOI: 10.1016/j.spacepol.2021.101428

Panse, P., & Sharma, D. K. (2023). Green Banking in India: A Sustainability Perspective. *MANTHAN Journal of Commerce and Management*, 10(1), 141–161. DOI: 10.17492/jpi.manthan.v10i1.1012308

Pantouvakis, A., & Vlachos, I. (2020). Talent and leadership effects on sustainable performance in the maritime industry. *Transportation Research Part D, Transport and Environment*, 86, 102440. DOI: 10.1016/j.trd.2020.102440

Parhamfar, M., Sadeghkhani, I., & Adeli, A. M. (2024). Towards the net zero carbon future: A review of blockchain-enabled peer-to-peer carbon trading. *Energy Science & Engineering*, 12(3), 1242–1264. DOI: 10.1002/ese3.1697

Parry, S., & Metzger, E. (2023). Barriers to learning for sustainability: A teacher perspective. *Sustainable Earth Reviews*, 6(2), 1–11. DOI: 10.1186/s42055-022-00050-3

Paulraj, A., Chen, I. J., & Blome, C. (2017). Motives and performance outcomes of sustainable supply chain management practices: A multi-theoretical perspective. *Journal of Business Ethics*, 145(2), 239–258. DOI: 10.1007/s10551-015-2857-0

Pelling, M., O'Brien, K., & Matyas, D. (2015). Adaptation and transformation. *Climatic Change*, 133(1), 113–127. DOI: 10.1007/s10584-014-1303-0

Peña, R. B. (2020). *Living Proof THE BULLITT CENTER: High Performance Building Case Study.* University of Washington Center for Integrated Design.

Perdana, S., & Vielle, M. (2022). Making the EU Carbon Border Adjustment Mechanism acceptable and climate friendly for least developed countries. *Energy Policy*, 170, 113245. DOI: 10.1016/j.enpol.2022.113245

Pereira, V., Silva, G., & Dias, A. (2021). Sustainability practices in hospitality: Case study of a luxury hotel in Arrábida Natural Park. *Sustainability (Basel)*, 13(6), 3164. DOI: 10.3390/su13063164

Perkins, M. (2023). Academic Integrity considerations of AI Large Language Models in the post-pandemic era: ChatGPT and beyond. *Journal of university teaching & learning practice,* 20, 07.

Perkins, M., Gezgin, U. B., & Roe, J. (2018). Understanding the relationship between language ability and plagiarism in non-native English speaking business students. *Journal of Academic Ethics*, 16(4), 317–328. DOI: 10.1007/s10805-018-9311-8

Petersen, S., Pedersen, T. H., Nielsen, K. U., & Knudsen, M. D. (2016). Establishing an image-based ground truth for validation of sensor data-based room occupancy detection. *Energy and Building*, 130, 787–793. DOI: 10.1016/j.enbuild.2016.09.009

Pezzey, J. (2017). *Sustainability: an interdisciplinary guide. The economics of sustainability*. Routledge.

Pham, H., & Kim, S. Y. (2019). The effects of sustainable practices and managers' leadership competences on sustainability performance of construction firms. *Sustainable Production and Consumption*, 20, 1–14. DOI: 10.1016/j.spc.2019.05.003

Pinar Mert Cuce, S. R. (2014). A comprehensive review of heat recovery systems for building applications. *Renewable & Sustainable Energy Reviews*, •••, 47.

Poirier, L., Antonio, D., Dettmann, M., Eng, T., Ganata, J., Ghosh, S., Lopez, M., Karma, R., Natekal, A., & Brinkley, C. (2024). Making plans findable, accessible, interoperable, and reusable with data infrastructure: A search engine for constructing, analyzing, and visualizing planning documents. *Environment and Planning. B, Urban Analytics and City Science*, 51(8), 1913–1930. Advance online publication. DOI: 10.1177/23998083241227471

Pommier, B., & Engel, A. (2021). Sustainability reporting in the hospitality industry. *Research in Hospitality Management*, 11(3), 173–175. DOI: 10.1080/22243534.2021.2006937

Poonuganti, S. (2023). It's Raining Rockets: Heightening State Liability for Space Pollution. *Chicago Journal of International Law*, 23(2).

Popova, R., & Schaus, V. (2018). The Legal Framework for Space Debris Remediation as a Tool for Sustainability in Outer Space. *Aerospace (Basel, Switzerland)*, 5(2), 55. DOI: 10.3390/aerospace5020055

Portugal, U. N. E. S. C. O. (2020). Objetivos do desenvolvimento sustentável. https://unescoportugal.mne.gov.pt/pt/temas/objetivos-de-desenvolvimento-sustentavel/os-17-ods

Potienko, N. (2017). Low-grade energy of the ground for civil engineering. MATEC Web of Conferences. DOI: 10.1051/matecconf/201710606026

Poyser, A. (2024). Carbon Offsets and Green Finance: Understanding Corporate Demand for Carbon Offsets and Mitigation–A Global Literature Review. Sustainable Investments in Green Finance, 139-162. DOI: DOI: 10.4018/979-8-3693-1388-6.ch009

Prabhu, G. N. (2021). GREEN BANKING PRACTICES – A Case Study on Indian Green Banking System. *International Journal of Research Publication and Reviews*, 2(3), 487–493. DOI: 10.5281/zenodo.4680016

Pradhan, N. A., Najmi, R., & Fatmi, Z. (2022). District health systems capacity to maintain healthcare service delivery in Pakistan during floods: A qualitative study. *International Journal of Disaster Risk Reduction*, 78, 103092. Advance online publication. DOI: 10.1016/j.ijdrr.2022.103092

Preziosi, M., Tourais, P., Acampora, A., Videira, N., & Merli, R. (2019). The role of environmental practices and communication on guest loyalty: Examining EU-Ecolabel in Portuguese hotels. *Journal of Cleaner Production*, 237, 117659. Advance online publication. DOI: 10.1016/j.jclepro.2019.117659

Publituris.hotelaria (2020). A prioridade do grupo Accor é lutar contra o desperdício alimentar. https://www.publiturishotelaria.pt/2020/11/25/opiniao-prioridade-do-grupo-accor-lutar-desperdicio-alimentar/

Purushothaman, S. (2019). Sustainability: Challenges in Teaching. *Ecology. Economy and Society*, 2(2), 33–36. DOI: 10.37773/ees.v2i2.73

Purwanti, A. (2023). Green Investment Appraisal: A Comprehensive Framework for Evaluating Environmental and Financial Returns. Atestasi. *Jurnal Ilmiah Akuntansi*, 6(2), 797–810.

PwC. (n.d.). Green bank deposits: A gateway to sustainable financing. In *PwC* (pp. 2–5). https://www.pwc.in/assets/pdfs/consulting/green-bank-deposits-a-gateway-to-sustainable-financing.pdf

Qadir, J. Engineering education in the era of ChatGPT: Promise and pitfalls of generative AI for education. *2023 IEEE Global Engineering Education Conference (EDUCON)*, 2023. IEEE, 1-9. DOI: 10.1109/EDUCON54358.2023.10125121

Rahaman, M. M., Islam, M. M., Polbitsyn, S. N., & Abbas, S. (2024). The electrification-social development nexus in Bangladesh: Do governmental policies matter? *Utilities Policy*, 86, 101690. DOI: 10.1016/j.jup.2023.101690

Rahman, M. (2017). The advantages and disadvantages of using qualitative and quantitative approaches and methods in language "Testing and Assessment" research: A literature review. *Journal of Education and Learning*, 6(1), 102–112. DOI: 10.5539/jel.v6n1p102

Rajabifard, A., & Williamson, I. P. (2002). Spatial Data Infrastructures: an initiative to facilitate spatial data sharing. Global environmental databases-present situation; future directions, 2.

Rajabifard, A., Binns, A., Masser, I., & Williamson, I. (2006). The role of sub-national government and the private sector in future spatial data infrastructures. *International Journal of Geographical Information Science*, 20(7), 727–741. Advance online publication. DOI: 10.1080/13658810500432224

Rajabifard, A., Feeney, M. E. F., & Williamson, I. (2003). *Spatial data infrastructures: concept, nature and SDI hierarchy*. Taylor & Francis London.

Rajabifard, A., Feeney, M. E. F., & Williamson, I. P. (2002). Future directions for SDI development. *International Journal of Applied Earth Observation and Geoinformation*, 4(1), 11–22. DOI: 10.1016/S0303-2434(02)00002-8

Rana, A. &. (2018). Thermal performance of double and triple-glazed windows Experimental results from the lab and in-situ measurements. *1st International Conference on New Horizons in Green Civil Engineering*. Victoria, BC, Canada.

Raoof, O., & Al-Raweshidy, H. (2010). Theory of Games: An Introduction. In Huang, Q. (Ed.), *Game Theory*. Sciyo., DOI: 10.5772/46930

Rastogi, M. M. (2022). A Review of Passive Design Strategies for Improving Building Energy Performance. International Journal of Multidisciplinary Innovative Research.

Rastogi, K., & Lohani, D. (2020). IoT-based indoor occupancy estimation using Edge Computing. *Procedia Computer Science*, 171, 1943–1952. DOI: 10.1016/j.procs.2020.04.208

Rattani, S. A., Dahlke, S., & Cameron, B. (2022). Cancer Care in Pakistan: A Descriptive Case Study. *Global Qualitative Nursing Research*, 9, 23333936221080988. Advance online publication. DOI: 10.1177/23333936221080988 PMID: 35237708

Raut, H., Ganesh, V. A., Nair, A. S., & Ramakrishna, S. (2011). Anti-Reflective Coatings: A Critical, In-Depth Review. *Energy & Environmental Science*, 4(10), 3779–3804. DOI: 10.1039/c1ee01297e

Rawashdeh, A., & Al-Ababneh, M. (2021). Hotel guests' perceptions of environmentally friendly practices in Jordan. *Journal of Environmental Management and Tourism*, 12(1), 107–120. DOI: 10.14505/jemt.v12.1(49).09

Recommendations, P. C. (2011). "Directive of the Prime Minister, 09/11/2011. Guidelines for the assessment and the mitigation of seismic risk of cultural heritage with reference to Italian NTC 2008.".

Rehman, A., Farooq, M., Lee, D. J., & Siddique, K. H. M. (2022). Sustainable agricultural practices for food security and ecosystem services. *Environmental Science and Pollution Research International*, 29(56), 84076–84095. DOI: 10.1007/s11356-022-23635-z PMID: 36258111

Report, B. S. (2021). Looking back at the progress we made against our sustainability strategy in 2021. https://www.sustainability.booking.com/industryinsights

Riaz, A., Riaz, N., Raza, H., & Mirza, F. (2023). Green Banking Practices: A Bibliometric Analysis and Systematic Literature Review. In *Emerald Publishing Limited eBooks* (pp. 299–317). https://doi.org/ DOI: 10.1108/978-1-80455-678-820231016

Rizvi Jafree, S., & Burhan, S. K. (2020). Health challenges of mothers with special needs children in Pakistan and the importance of integrating health social workers. *Social Work in Health Care*, 59(6), 408–429. Advance online publication. DOI: 10.1080/00981389.2020.1781738 PMID: 32614737

Rizzello, A. (2022). The Green Financing Framework Combining Innovation and Resilience: A Growing Toolbox of Green Finance Instruments. In: Green Investing. Palgrave Studies in Impact Finance, Palgrave Macmillan, Cham. DOI: 10.1007/978-3-031-08031-9_3

Roberto Stasi, F. R. (2024). Influence of cross-ventilation cooling potential on thermal comfort in high-rise buildings in a hot and humid climate. *Building and Environment*, •••, 248.

Roberts, L. D. (1992). Addressing the Problem of Orbital Space Debris: Combining International Regulatory and Liability Regimes. *Boston College International and Comparative Law Review*, XV(1), 51–73.

Robinson, J. A., & Acemoglu, D. (2012). *Why nations fail: The origins of power, prosperity and poverty*. Profile London.

Rodriguez-Gonzalez, P. T., Rico-Martinez, R., & Rico-Ramirez, V. (2020). Effect of feedback loops on the sustainability and resilience of human-ecosystems. *Ecological Modelling*, 426, 109018. DOI: 10.1016/j.ecolmodel.2020.109018

Rogalska-Marasińska, A. (2022). Reimagining Education Towards Sustainability. "Project X" – an Innovative Learning Strategy from Poland. *Studia Edukacyjne*, 66(66), 17–34. Advance online publication. DOI: 10.14746/se.2022.66.2

Rosati, F., & Faria, L. G. (2019). Addressing the SDGs in sustainability reports: The relationship with institutional factors. *Journal of Cleaner Production*, 215, 1312–1326. DOI: 10.1016/j.jclepro.2018.12.107

Roy, J. K. (2015). Green stocks: Green investment, green return. *The Financial Express*, 22, 255.

Rueda, L., Agbossou, K., Cardenas, A., Henao, N., & Kelouwani, S. (2020). A comprehensive review of approaches to building occupancy detection. *Building and Environment*, 180, 106966. DOI: 10.1016/j.buildenv.2020.106966

Ruggerio, C. A. (2021). Sustainability and sustainable development: A review of principles and definitions. *The Science of the Total Environment*, 786, 147481. DOI: 10.1016/j.scitotenv.2021.147481 PMID: 33965820

Rui, G., Shan, X., Mirzat, U., & Peter, M. (2024). Assessing the impact of urban planning policies on renewable energy: A case a China using the DID estimation model. *Heliyon*. PMID: 38463897

Ryu, S. H., & Moon, H. J. (2016). Development of an occupancy prediction model using indoor environmental data based on machine learning techniques. *Building and Environment*, 107, 1–9. DOI: 10.1016/j.buildenv.2016.06.039

Sahoo, B. P., & Singh, A. (2016). Adoption of Green Banking in India: Challenges and Prospects. In SGTB Khalsa College, *International Journal of Science and Research (IJSR)* (Vol. 5, Issue 8) [Journal-article]. https://www.ijsr.net/archive/v5i8/ART2016856.pdf

Salameh, M., & Touqan, B. (2022). Traditional Passive Design Solutions as a Key Factor for Sustainable Modern Urban Designs in the Hot, Arid Climate of the United Arab Emirates. *Buildings*, 12(11), 1811. Advance online publication. DOI: 10.3390/buildings12111811

Saleem, Z., Godman, B., Azhar, F., Kalungia, A. C., Fadare, J., Opanga, S., Markovic-Pekovic, V., Hoxha, I., Saeed, A., Al-Gethamy, M., Haseeb, A., Salman, M., Khan, A. A., Nadeem, M. U., Rehman, I. U., Qamar, M. U., Amir, A., Ikram, A., & Hassali, M. A. (2022). Progress on the national action plan of Pakistan on antimicrobial resistance (AMR): a narrative review and the implications. In *Expert Review of Anti-Infective Therapy* (Vol. 20, Issue 1). DOI: 10.1080/14787210.2021.1935238

Salem, I., Elbaz, A., Al-Alawi, A., Alkathiri, N., & Rashwan, K. (2022). Investigating the role of green hotel sustainable strategies to improve customer cognitive and affective image. *Sustainability (Basel)*, 14(6), 3545. DOI: 10.3390/su14063545

SALGADO., M. a. (2009). Re-thinking an Annotation Tool for the Museum Community Generated Content. *Nordic Design Research Conference (Nordes). Oslo: Nordic Design Research.*

Salguero, G. C., Gámez, M. A. F., Fernández, I. A., & Palomo, D. R. (2019). Competitive Intelligence and Sustainable Competitive Advantage in the Hotel Industry. *Sustainability (Basel)*, 11(6), 1597. DOI: 10.3390/su11061597

Salwa, M., & Al-Masrani, K. M.-O. (2019). Dynamic shading systems: A review of design parameters, platforms, and evaluation strategies. *Automation in Construction*, 102, 195–216. DOI: 10.1016/j.autcon.2019.01.014

Sangiorgi, I., & Schopohl, L. (2021). Why do institutional investors buy green bonds: Evidence from a survey of European asset managers. *International Review of Financial Analysis*, 75, 101738. DOI: 10.1016/j.irfa.2021.101738

Sannigrahi, A. K. (2017). Removal of Active Man-made Orbital Debris – A Great Challenge to Space Scientists. *International Journal of Scientific Research*, 6(2).

Santos, R. A., Mexas, M. P., Meirino, M. J., Sampaio, M. C., & Costa, H. G. (2020). Criteria for assessing a sustainable hotel business. *Journal of Cleaner Production*, 262, 121347. DOI: 10.1016/j.jclepro.2020.121347

Satpathy, A., Samal, A., Gupta, S., Kumar, S., Sharma, S., Manoharan, G., & Sharma, S. (2024). To Study the Sustainable Development Practices in Business and Food Industry. *Migration Letters : An International Journal of Migration Studies*, 21(S1), 743–747. DOI: 10.59670/ml.v21iS1.6400

Sauvé, S., Bernard, S., & Sloan, P. (2016). Environmental sciences, sustainable development and circular economy: Alternative concepts for trans-disciplinary research. *Environmental Development*, 17, 48–56. DOI: 10.1016/j.envdev.2015.09.002

Sayed, A. N., Bensaali, F., Himeur, Y., & Houchati, M. (2023). Edge-based real-time occupancy detection system through a non-intrusive sensing system. *Energies*, 16(5), 2388.

Schilling, T., Wyss, R., & Binder, C. R. (2018). The resilience of sustainability transitions. *Sustainability (Basel)*, 10(12), 4593. DOI: 10.3390/su10124593

Schoormann, T., Strobel, G., Möller, F., Petrik, D., & Zschech, P. (2023). Artificial intelligence for sustainability—A systematic review of information systems literature. *Communications of the Association for Information Systems*, 52(1), 8. DOI: 10.17705/1CAIS.05209

Schulze, T., & Eicker, U. (2013). Controlled natural ventilation for energy-efficient buildings. *Energy and Building*, 56, 221–232. DOI: 10.1016/j.enbuild.2012.07.044

Scott, G., & Rajabifard, A. (2017). Sustainable development and geospatial information: A strategic framework for integrating a global policy agenda into national geospatial capabilities. *Geo-Spatial Information Science*, 20(2), 59–76. DOI: 10.1080/10095020.2017.1325594

Sdg, U. (2019). Sustainable development goals. *The energy progress report.Tracking SDG*, 7, 805–814.

Sen, N., Ahuja, S., Dutta, S., & Chaudhary, K. (2024). The Application of an Emission Trade Scheme in the Indian Aviation Sector. *Air and Space Law*, 49(2), 165–190. Advance online publication. DOI: 10.54648/AILA2024020

Shabbir, H., Saeed, S., Farhan, M., Abbas, K., Rehman, M. E., Gul, F., & Basit, J. (2022). Poliomyelitis in Pakistan: Challenges to polio eradication and future prospects. *Annals of Medicine and Surgery (London)*, 80. Advance online publication. DOI: 10.1016/j.amsu.2022.104274 PMID: 36045811

Shaik, M. A., Manoharan, G., Prashanth, B., Akhil, N., Akash, A., & Reddy, T. R. S. (2022, May). Prediction of crop yield using machine learning. *In AIP Conference Proceedings (Vol. 2418, No. 1). AIP Publishing.* Sethuram, B. " Strengthening Local Governance for Cyclone Disaster Management: A Case Study of Puducherry District.

Shaikh, B. T., & Ali, N. (2023). Universal health coverage in Pakistan: Is the health system geared up to take on the challenge? *Globalization and Health*, 19(1), 4. Advance online publication. DOI: 10.1186/s12992-023-00904-1 PMID: 36635698

Shaikh, I. A. K., Kumar, C. N. S., Rohini, P., Jafersadhiq, A., Manoharan, G., & Suryanarayana, V. (2023, August). AST-Graph Convolution Network and LSTM Based Employees Behavioral and Emotional Reactions to Corporate Social Irresponsibility. In *2023 Second International Conference on Augmented Intelligence and Sustainable Systems (ICAISS)* (pp. 966-971). IEEE.

Shameem, A., Ramachandran, K. K., Sharma, A., Singh, R., Selvaraj, F. J., & Manoharan, G. (2023, May). The rising importance of AI in boosting the efficiency of online advertising in developing countries. *In 2023 3rd International Conference on Advance Computing and Innovative Technologies in Engineering (ICACITE) (pp. 1762-1766). IEEE.*

Sharar, T., Qutoshi, S. B., & Sharar, T. (2022). The level of school leadership awareness about education for sustainable development. *Journal for Educators. Teachers and Trainers*, 13(5), 358–370. DOI: 10.47750/jett.2022.13.05.032

Sharma, G. D., Verma, M., Shahbaz, M., Gupta, M., & Chopra, R. (2022). Transitioning green finance from theory to practice for renewable energy development. *Renewable Energy*, 195, 554–565. DOI: 10.1016/j.renene.2022.06.041

Sharma, M., & Choubey, A. (2021). Green banking initiatives: A qualitative study on Indian banking sector. *Environment, Development and Sustainability*, 24(1), 293–319. DOI: 10.1007/s10668-021-01426-9 PMID: 33967597

Sheer, A., Fatima, S., Owais Farooqui, M., & Ramzan Kasuri, M. (2023). A panacea to address the legal, administrative and economic aspects of space debris. *Advances in Space Research*, 72(7), 2616–2626. DOI: 10.1016/j.asr.2023.02.023

Shen, H., & Huang, N. (2022). Accounting for carbon emissions trading schemes. In *Handbook of Accounting and Sustainability* (pp. 365–380). Edward Elgar Publishing., DOI: 10.4337/9781800373518.00030

Shen, J., Li, L., Zhang, D., Dong, S., Xiang, J., & Xu, N. (2022). A Multi-Analytical Approach to Investigate the Polychrome Clay Sculpture in Qinglian Temple of Jincheng, China. *Materials (Basel)*, 15(16), 5470. Advance online publication. DOI: 10.3390/ma15165470 PMID: 36013607

Shen, N., Zhao, Y., & Deng, R. (2020). A review of carbon trading based on an evolutionary perspective. *International Journal of Climate Change Strategies and Management*, 12(5), 739–756. DOI: 10.1108/IJCCSM-11-2019-0066

Shi, L., Han, L., Yang, F., & Gao, L. (2019). The Evolution of Sustainable Development Theory: Types, Goals, and Research Prospects. *Sustainability 2019, Vol. 11, Page 7158, 11*(24), 7158. DOI: 10.3390/su11247158

Shi, H., Li, Z., Ye, H., Luo, H., Xiong, W., & Wang, X. (2021). First level 1 product results of the greenhouse gas monitoring instrument on the GaoFen-5 satellite. *IEEE Transactions on Geoscience and Remote Sensing*, 59(2), 899–914. DOI: 10.1109/TGRS.2020.2998729

Shweta Singh, S. V. (2019). *Home Automation Using Node MCU*. Firebase & IOT.

Siamak, H., Benedetto, N., Daniele, G., & Davide, A. G. (2022). Exploring the penetration of renewable energy at increasing the boundaries of the urban energy system – The PRISMI plus toolkit application to Monachil, Spain. *Sustainable Energy Technologies and Assessments*, 54, 102908. DOI: 10.1016/j.seta.2022.102908

Silva, F. (2016). Práticas ambientalmente sustentáveis nos hotéis: Perceções dos consumidores e gestores. Master Dissertation, Instituto Universitário de Lisboa (ISCTE). Lisbon, Portugal.

Singh, R. K. (2024). Impact of leadership, TQM and supply chain capabilities on sustainable supply chain performance: Moderating role of institutional pressure. *The TQM Journal*. Advance online publication. DOI: 10.1108/TQM-09-2023-0275

Sirieys, E., Gentgen, C., Milton, J., & De Weck, O. (2022). Space sustainability isn't just about space debris:On the atmospheric impact of space launches. *MIT Science Policy Review*, 3, 143–151. DOI: 10.38105/spr.whfig18hta

Sithole, M. P., & Gumede, P. R. (2022). Sustaining a tutorship programme at a university of technology: A systems approach. *Perspectives in Education*, 40(3), 224–240. DOI: 10.18820/2519593X/pie.v40.i3.15

Sjoukema, J.-W., Bregt, A. K., & Crompvoets, J. (2019). Understanding Governance Dynamics: The Governing System of Spatial Data Infrastructures. *International Journal of Spatial Data Infrastructures Research*, 14(0). Advance online publication. DOI: 10.2902/IJSDIR.V14I0.495

Skelton, C. (2014). *The implementation and education of geographic information systems in a local government for municipal planning: A case study of Dangriga, Belize.*

Slann, P. A. (2014). Space debris and the need for space traffic control. *Space Policy*, 30(1), 40–42. DOI: 10.1016/j.spacepol.2013.10.007

Snow, M. E. (2022). Patient engagement in healthcare planning and evaluation: A call for social justice. *The International Journal of Health Planning and Management*, 37(S1), 20–31. Advance online publication. DOI: 10.1002/hpm.3509 PMID: 35643909

Sorapure, M. (2003). Five principles of new media: Or, playing Lev Manovich. *Kairos*, 8(2), 26–48.

Sriyakul, T., Singsa, A., Sutduean, J., & Jermsittiparsert, K. (2019). Effect of cultural traits, leadership styles and commitment to change on supply chain operational excellence. *Journal of Computational and Theoretical Nanoscience*, 16(7), 2967–2974. DOI: 10.1166/jctn.2019.8203

Stavins, R. N. (2020). The future of U.S. carbon-pricing policy. *Environmental and Energy Policy and the Economy*, 1(1), 7–32. DOI: 10.1086/706792

Steponenaite, A., & Barakat, B. Plagiarism in AI empowered world. International Conference on Human-Computer Interaction, 2023. Springer, 434-442.

Stokel-Walker, C. (2022). AI bot ChatGPT writes smart essays-should academics worry? *Nature*. Advance online publication. DOI: 10.1038/d41586-022-04397-7

Šujanová, P. (2019). A Healthy, Energy-Efficient and Comfortable Indoor Environment, a Review. *Energies*.

Sun, C., Tie, Y., & Yu, L. (2024). How to achieve both environmental protection and firm performance improvement: Based on China's carbon emissions trading (CET) policy. *Energy Economics*, 130, 107282. DOI: 10.1016/j.eneco.2023.107282

Sun, G. H., & Hoelscher, S. H. (2023). The ChatGPT storm and what faculty can do. *Nurse Educator*, 48, 119–124. DOI: 10.1097/NNE.0000000000001390 PMID: 37043716

Sun, K., Zhao, Q., Zhang, Z., & Hu, X. (2022). Indoor occupancy measurement by the fusion of motion detection and static estimation. *Energy and Building*, 254, 111593. DOI: 10.1016/j.enbuild.2021.111593

Su, Q., & Chen, L. (2024). Carbon financial trading risk based on multidimensional analysis of data flow from the perspective of low-carbon economy. *Environment, Development and Sustainability*, •••, 1–19. DOI: 10.1007/s10668-024-05078-3

Surahman, E., & Wang, T. H. (2022). Academic dishonesty and trustworthy assessment in online learning: A systematic literature review. *Journal of Computer Assisted Learning*, 38(6), 1535–1553. DOI: 10.1111/jcal.12708

Suriyankietkaew, S. (2023). Effects of key leadership determinants on business sustainability in entrepreneurial enterprises. *Journal of Entrepreneurship in Emerging Economies*, 15(5), 885–909. DOI: 10.1108/JEEE-05-2021-0187

Svensson, G., & Wood, G. (2006). Sustainable components of leadership effectiveness in organizational performance. *Journal of Management Development*, 25(6), 522–534. DOI: 10.1108/02621710610670100

Świercz, E. Z. (2021). A Review of Heat Recovery in Ventilation. *Energies*, 14(6), 1759. DOI: 10.3390/en14061759

Sydney, O., Pankaj, L., Andres, S., & Neeraj, V. (2020). Assessment of public awareness, acceptance and attitudes towards renewable energy in Kenya. *Scientific African*, 9, e00512. DOI: 10.1016/j.sciaf.2020.e00512

Tabbi W, A.G. Olabi, Enas T.S, Montaser M, Abdul H. A, Mohammad A.A. (2024). The state of renewable energy source envelopes in urban areas. *The state of renewable energy source envelopes in urban areas*,. 21, 100581.

Tallis, J. (2015). Remediating Space Debris: Legal and Technical Barriers. *ASPJ Africa & Francophonie, First Quarter*, 80–90.

Tan, S. Y. (2012). The Practice of Integrated Design: The Case Study of Khoo Teck Puat Hospital, Singapore. Nottingham: Masters of Science in Sustainable Building Design in BCA Academy.

Tara, K., Singh, S., & Kumar, R. (2015). Green Banking for Environmental Management: A Paradigm Shift. *Current World Environment*, 10(3), 1029–1038. DOI: 10.12944/CWE.10.3.36

Tasci, A. D. A., Wei, W., & Milman, A. (2020). Uses and misuses of the case study method. *Annals of Tourism Research*, 82, 102815. Advance online publication. DOI: 10.1016/j.annals.2019.102815

Tavares, B. (2016). Práticas de sustentabilidade no setor hoteleiro: Estudo de caso do Neya Lisboa Hotel. Master Dissertation, Lisbon School of Economics and Management (ISEG). Lisbon, Portugal.

Taylor, M. W. (2007). *Orbital Debris: Technical and Legal Issues and Solutions*. Library and Archives Canada = Bibliothèque et Archives Canada.

Teoman, S., & Ulengin, F. (2018). The impact of management leadership on quality performance throughout a supply chain: An empirical study. *Total Quality Management & Business Excellence*, 29(11-12), 1427–1451. DOI: 10.1080/14783363.2016.1266244

Termeer, C. J. A. M., Dewulf, A., & Biesbroek, R. (2019). A critical assessment of the wicked problem concept: Relevance and usefulness for policy science and practice. *Policy and Society*, 38(2), 167–179. DOI: 10.1080/14494035.2019.1617971

Tien, P. W., Wei, S., Calautit, J. K., Darkwa, J., & Wood, C. (2020). A vision-based deep learning approach for the detection and prediction of occupancy heat emissions for demand-driven control solutions. *Energy and Building*, 226, 110386. DOI: 10.1016/j.enbuild.2020.110386

Tiwari, D. S. (2022). Carbon Trading: A Tool to Control Global Warming. *International Journal of Health Sciences*, (I), 5713–5722. DOI: 10.53730/ijhs.v6nS1.6246

Topinka, R. March, 2024. The software says my student cheated using AI. They say they're innocent. Who do I believe? *The Guardian*.

Toroxel, J. L., & Silva, S. M. (2024). A Review of Passive Solar Heating and Cooling Technologies Based on Bioclimatic and Vernacular Architecture. *Energies*, 17(5), 1006. DOI: 10.3390/en17051006

Tourism of Portugal. (2020). Os objetivos do desenvolvimento sustentável. https://business.turismodeportugal.pt/pt/crescer/sustentabilidade/Paginas/objetivos-desenvolvimento-sustentavel.aspx

Tourism of Portugal. (2021). Visão geral. https://www.turismodeportugal.pt/pt/Turismo_Portugal/visao_geral/Paginas/default.aspx

Tripathi, A., & Singh, M. P. (2016). Determinants of sustainable/green consumption: A review. *International Journal of Environmental Technology and Management*, 19(3-4), 316–358. DOI: 10.1504/IJETM.2016.082258

Tripathi, M. A., Tripathi, R., Effendy, F., Manoharan, G., Paul, M. J., & Aarif, M. (2023, January). An In-Depth Analysis of the Role That ML and Big Data Play in Driving Digital Marketing's Paradigm Shift. *In2023 International Conference on Computer Communication and Informatics (ICCCI)(pp.* 1-6*)*. IEEE. DOI: 10.1109/ICCCI56745.2023.10128357

Tu, C., Zang, C., Tan, Y., Zhou, Y., & Yu, C. (2022). Can information infrastructure development improve the health care environment? Evidence from China. *Frontiers in Public Health*, 10, 987391. Advance online publication. DOI: 10.3389/fpubh.2022.987391 PMID: 36091535

UAE Ministry of Climate Change and Environment. (2018). *UAE Green Growth Strategy*. Retrieved from https://www.moccae.gov.ae/assets/c11f96c6/UAE-Green-Growth-Strategy.pdf

UAE Ministry of Climate Change and Environment. (2020). *UAE National Climate Change Adaptation Program*. Retrieved from https://www.moccae.gov.ae/en/our-services/climate-change/national-climate-change-adaptation-program.aspx

UAE Ministry of Infrastructure Development. (2019). *Guidelines for climate-resilient infrastructure development*. Retrieved from https://www.infrastructure.ae/en/middleeast/infrastructure/news/2019-04-11-climate-resilient-infrastructure-guidelines-published-by-the-ministry-of-infrastructure-development-in-uae-5151725

UAE National Climate Change Adaptation Program. (n.d.). Retrieved from https://www.moccae.gov.ae/en/our-services/climate-change/national-climate-change-adaptation-program.aspx

UAE State of Environment Report. (2017). *Climate change*. Retrieved from https://www.ead.gov.ae/en/our-services/environmental-resources/climate-change.aspx

Ulle, R. S., & Kumar, A. S. (2014). A review on total quality leadership in TQM practices-industrial management and organizations. *International Journal of Emerging Research in Management &Technology*, 3(5), 152–155.

UN. (2015). *Transforming Our World: The 2030 Agenda for Sustainable Development* (p. 41). DOI: 10.1891/9780826190123.ap02

UN. (2018). National geospatial data and information systems. In *United Nations Committee of Experts on Global Geospatial Information Management,United Nations Economic and Social Council,E/C.20/2018/16/Add.1*. DOI: 10.1093/oxfordhb/9780199560103.003.0007

UNESCO. (1972). *Convention Concerning The Protection Of The World Cultural And Natural Heritage*. 1–58.

Unesco. (2023). *Operational Guidelines for the Implementation of the World Heritage Convention UNITED NATIONS EDUCATIONAL, SCIENTIFIC AND CULTURAL ORGANIZATION INTERGOVERNMENTAL COMMITTEE FOR THE PROTECTION OF THE WORLD CULTURAL AND NATURAL HERITAGE WORLD HERITAGE CENTRE*. https://whc.unesco.org/fr/orientations

UN-GGIM. (2020). *Virtual Regional Seminar on Operationalizing the Integrated Geospatial Information Framework in Asia and the Pacific*.

United Nations Conference on Trade and Development. (2023). Capital markets and sustainable finance. In World investment report (pp. 205–237). DOI: 10.18356/9789210027847c009

United Nations Environment Programme (UNEP). (2020). *Making peace with nature: A scientific blueprint to tackle the climate, biodiversity, and pollution emergencies*.

United Nations. (1995). *Human Development Report 1995*. Human Development Reports.

Urrutia-Azcona, K., Usobiaga-Ferrer, E., De Agustín-Camacho, P., Molina-Costa, P., Benedito-Bordonau, M., & Flores-Abascal, I. (2021). ENER-BI: Integrating Energy and Spatial Data for Cities' Decarbonisation Planning. *Sustainability 2021, Vol. 13, Page 383, 13*(1), 383. DOI: 10.3390/su13010383

Van Wynsberghe, A. (2021). Sustainable AI: AI for sustainability and the sustainability of AI. *AI and Ethics*, 1(3), 213–218. DOI: 10.1007/s43681-021-00043-6

Vaswani, A., Shazeer, N., Parmar, N., Uszkoreit, J., Jones, L., Gomez, A. N., Kaiser, Ł., & Polosukhin, I. (2017). Attention is all you need. *Advances in Neural Information Processing Systems*, 30.

Veckalne, R., & Tambovceva, T. (2022). The Role of Digital Transformation in Education in Promoting Sustainable Development. *Virtual Economics*, 5(4), 65–86. DOI: 10.34021/ve.2022.05.04(4)

Vera, M. L., Torres, W. R., Galli, C. I., Chagnes, A., & Flexer, V. (2023). Environmental impact of direct lithium extraction from brines. *Nature Reviews. Earth & Environment*, 4(3), 149–165. DOI: 10.1038/s43017-022-00387-5

Victor, O. E., Aziz, N. A., & Jaffar, A. R. (2015). Privatization of Electricity Service Delivery in Developing Nations: Issues and Challenges. *International Journal of Built Environment and Sustainability*, 2(3). Advance online publication. DOI: 10.11113/ijbes.v2.n3.82

Wahab, I. A., Ismail, L. H., & Kadir, A. (2019). Opening Design and Position Effect on Building Natural Stack Effect and Cross Ventilation. *International Journal of Engine Research*, 5(1), 13–22.

Wahl, F., Milenkovic, M., & Amft, O. A distributed PIR-based approach for estimating people count in office environments. 2012 IEEE 15th International Conference on Computational Science and Engineering, 2012. IEEE, 640-647.

Wali Shah, M., Ali, I., Mahmood, A., Ashar Awan, M., Ahmad, N., Syed Hamid Arshad, H., Pakhtunkhwa, K., & Scholar, P. (2024). *Safeguarding Endangered Built Heritage In Khyber Pakhtunkhwa-Pakistan: A Digital Restoration Approach. 21*(S7), 437–449. www.migrationletters.com

Wang, F. (2020). Why public health needs GIS: a methodological overview. In *Annals of GIS* (Vol. 26, Issue 1). DOI: 10.1080/19475683.2019.1702099

Wang, C., Zhang, Y., Zhou, Y., Sun, S., Zhang, H., & Wang, Y. (2023). Automatic detection of indoor occupancy based on improved YOLOv5 model. *Neural Computing & Applications*, 35(3), 2575–2599. DOI: 10.1007/s00521-022-07730-3 PMID: 36068815

Wang, S., Burnett, J., & Chong, H. (1999). Experimental validation of CO2-based occupancy detection for demand-controlled ventilation. *Indoor and Built Environment*, 8(6), 377–391. DOI: 10.1177/1420326X9900800605

Wang, S., Huang, X., Liu, P., Zhang, M., Biljecki, F., Hu, T., Fu, X., Liu, L., Liu, X., Wang, R., Huang, Y., Yan, J., Jiang, J., Chukwu, M., Reza Naghedi, S., Hemmati, M., Shao, Y., Jia, N., Xiao, Z., & Bao, S. (2024). Mapping the landscape and roadmap of geospatial artificial intelligence (GeoAI) in quantitative human geography: An extensive systematic review. *International Journal of Applied Earth Observation and Geoinformation*, 128, 103734. DOI: 10.1016/j.jag.2024.103734

Wang, T. (2016). A Liability and Insurance Regime for Space Debris Mitigation. *Science & Global Security*, 24(1), 22–36. DOI: 10.1080/08929882.2016.1127036

Wawrzyniak, D. (2023). Animal husbandry and sustainable agriculture: Is animal welfare (only) an issue of sustainability of agricultural production or a separate issue on its own? *Animal*, 17, 100880. DOI: 10.1016/j.animal.2023.100880 PMID: 37793705

WCED - World Commission on Environmental and Development. (1987). *Our Common Future*. Oxford University Press.

WCED. S. W. S. (1987). *World commission on environment and development. Our common future. 17*(1), 1–91. https://idl-bnc-idrc.dspacedirect.org/bitstream/10625/152/1/WCED_v17_doc149.pdf

Weinzierl, M. (2018). Space, the Final Economic Frontier. *The Journal of Economic Perspectives*, 32(2), 173–192. DOI: 10.1257/jep.32.2.173

Wei, P., Zhou, J., Ren, X., & Huynh, L. D. T. (2024). Financialisation of the European Union Emissions Trading System European Union Emissions Trading System and its influencing factors in quantiles. *International Journal of Finance & Economics*. Advance online publication. DOI: 10.1002/ijfe.2950

Weizenbaum, J. (1966). ELIZA—A computer program for the study of natural language communication between man and machine. *Communications of the ACM*, 9(1), 36–45. DOI: 10.1145/365153.365168

Wen, F., Zhao, L., He, S., & Yang, G. (2020). Asymmetric relationship between carbon emission trading market and stock market: Evidences from China. *Energy Economics*, 91, 104850. DOI: 10.1016/j.eneco.2020.104850

Wilcox, H. (2023). 2023. Cheating Aussie student fails uni exam after being caught using artificial intelligence chatbot to write essay-now Australia's top universities are considering a bizarre solution to stop it happening again. DailyMail., Retrieved February 2, 2023, from.

Williamsson, J., Schaad, G., & Sandoff, A. (2022). Building legitimacy for sustainable business schools: Using the business model concept when teaching corporate sustainability. *Journal of Cleaner Production*, 367(13311), 1–8. DOI: 10.1016/j.jclepro.2022.133116

Wilson, S. (1993). *The Aesthetics and Practice of Designing Interactive Computer Events Introduction : Design Challenges of Interactive Multimedia.*

Wilson, A. R., & Vasile, M. (2023). The space sustainability paradox. *Journal of Cleaner Production*, 423, 138869. DOI: 10.1016/j.jclepro.2023.138869

Winter, J., Zhai, J., & Cotton, D. R. E. (2022). Teaching environmental sustainability in China: Opportunities and challenges for business and economics faculty in higher education. *Environmental Education Research*, 28(2), 318–332. DOI: 10.1080/13504622.2021.2012560

Wollny, S., Schneider, J., Di Mitri, D., Weidlich, J., Rittberger, M., & Drachsler, H. (2021). Are we there yet?-a systematic literature review on chatbots in education. *Frontiers in Artificial Intelligence*, 4, 654924. DOI: 10.3389/frai.2021.654924 PMID: 34337392

Wong, I. (2017). A review of daylighting design and implementation in buildings. *Renewable & Sustainable Energy Reviews*, 74, 959–968. DOI: 10.1016/j.rser.2017.03.061

Wood, D., Rathnasabapathy, M., Stober, K. J., & Menon, P. (2024). Challenges and progress in applying space technology in support of the sustainable development goals. *Acta Astronautica*, 219, 678–692. DOI: 10.1016/j.actaastro.2024.03.064

World Bank. (2014). *Social development: putting people first.* https://documents.worldbank.org/curated/en/703901468147574810/Social-development-putting-people-first World Bank. (2015). *Bonds for Sustainable Development.* www.worldbank.org/debtsecurities

Wosk, J. (1991). Postmodern Currents: Art and Artists in the Age of Electronic Media by Margot Lovejoy. *Technology and Culture*, 32(1), 138–140.

Wronski, P., Koetsenruijter, J., Ose, D., Paulus, J., Szecsenyi, J., & Wensing, M. (2021). Healthcare planning across healthcare sectors in Baden-Wuerttemberg, Germany: A stakeholder online survey to identify indicators. *BMC Health Services Research*, 21(1), 510. Advance online publication. DOI: 10.1186/s12913-021-06514-0 PMID: 34039315

Wu, C., Raghavendra, R., Gupta, U., Acun, B., Ardalani, N., Maeng, K., Chang, G., Aga, F., Huang, J., & Bai, C. (2022). Sustainable ai: Environmental implications, challenges and opportunities. *Proceedings of Machine Learning and Systems*, 4, 795–813.

Wu, M., Zhao, K., & Fils-Aime, F. (2022). Response rates of online surveys in published research: A meta-analysis. *Computers in Human Behavior Reports*, 7, 100206. Advance online publication. DOI: 10.1016/j.chbr.2022.100206

Xianguo, Z., Yanan, Z., & Zhiyong, C. (2021). Development and research of rural renewable energy management and ecological management information system under the background of beautiful rural revitalization strategy. *Sustainable Computing : Informatics and Systems*, 30, 100553. DOI: 10.1016/j.suscom.2021.100553

Xiao, J., Li, G., Zhu, B., Xie, L., Hu, Y., & Huang, J. (2021). Evaluating the impact of carbon emissions trading scheme on Chinese firms' total factor productivity. *Journal of Cleaner Production*, 306, 127104. DOI: 10.1016/j.jclepro.2021.127104

Xue, W. (2024). The influence of environmental and social criteria in green finance decision-making: Insights and trends. *Economic Change and Restructuring*, 57(3), 110. DOI: 10.1007/s10644-024-09686-x

Yang, W. B., Yen, Y. N., & Cheng, H. M. (2015). Digitalized preservation and presentation of historical building-taking traditional temples and dougong as examples. *ISPRS Annals of the Photogrammetry, Remote Sensing and Spatial Information Sciences, 2*(5W3), 371–376. DOI: 10.5194/isprsannals-II-5-W3-371-2015

Yan, S., Seng, B. J. J., Kwan, Y. H., Tan, C. S., Quah, J. H. M., Thumboo, J., & Low, L. L. (2019). Identifying heterogeneous health profiles of primary care utilizers and their differential healthcare utilization and mortality - A retrospective cohort study. *BMC Family Practice*, 20(1), 54. Advance online publication. DOI: 10.1186/s12875-019-0939-2 PMID: 31014231

Yan, Y. (2023). Application of the principle of common but differentiated responsibility and respective capabilities to the passive mitigation and active removal of space debris. *Acta Astronautica*, 209, 117–131. DOI: 10.1016/j.actaastro.2023.04.033

Yan, Z., & Yunqi, L. (2024). Statistical Evaluation of Sustainable Urban Planning: Integrating Renewable Energy Sources, Energy-Efficient Buildings, and Climate Resilience Measures. *Sustainable Cities and Society*, 101, 105160. DOI: 10.1016/j.scs.2023.105160

Yin, R. K. (2014). *Case study research: Design and methods*. Sage Publications.

Young, J. (2015). Towards zero energy buildings: lessons learned from the BedZED development.

Yueyan, C., Maoli, T., Jianxin, X., Linda, Z., & Jinli, Y. (2024). Data-driven approaches to achieve energy transition: Statistical analysis of demand-side management and smart decision making in urban energy hub networks. *Sustainable Cities and Society*, 101, 105150. DOI: 10.1016/j.scs.2023.105150

Yu, F., Zhang, G., Zhao, F., Wang, X., Liu, H., Lin, P., & Chen, Y. (2023). Improved YOLO-v5 model for boosting face mask recognition accuracy on heterogeneous IoT computing platforms. *Internet of Things : Engineering Cyber Physical Human Systems*, 23, 100881. DOI: 10.1016/j.iot.2023.100881

Yun, J., & Lee, S.-S. (2014). Human movement detection and identification using pyroelectric infrared sensors. *Sensors (Basel)*, 14(5), 8057–8081. DOI: 10.3390/s140508057 PMID: 24803195

Yu, W., Jacobs, M. A., Chavez, R., & Yang, J. (2019). Dynamism, disruption orientation, and resilience in the supply chain and the impacts on financial performance: A dynamic capabilities perspective. *International Journal of Production Economics*, 218, 352–362. DOI: 10.1016/j.ijpe.2019.07.013

Zabolotnyy, S., & Wasilewski, M. (2019). The concept of financial sustainability measurement: A case of food companies from Northern Europe. *Sustainability (Basel)*, 11(18), 5139. DOI: 10.3390/su11185139

Zaidi, S. A. H., Shahbaz, M., Hou, F., & Abbas, Q. (2021). Sustainability challenges in public health sector procurement: An application of interpretative structural modelling. *Socio-Economic Planning Sciences*, 77, 101028. Advance online publication. DOI: 10.1016/j.seps.2021.101028

Zemouri, S., Magoni, D., Zemouri, A., Gkoufas, Y., Katrinis, K., & Murphy, J. (2018, September). An edge computing approach to explore indoor environmental sensor data for occupancy measurement in office spaces. In 2018 IEEE International Smart Cities Conference (ISC2) (pp. 1-8). IEEE.

Zemouri, S., Gkoufas, Y., & Murphy, J. (2019, August). A machine learning approach to indoor occupancy detection using non-intrusive environmental sensor data. In *Proceedings of the 3rd international conference on big data and internet of things* (pp. 70-74).

Zhaiting, Y., & Youwei, J. (2024). Quantifying resilient urban energy systems: Statistical analysis of climate adaptation, renewable integration, and socioeconomic dynamics. *Sustainable Cities and Society*, 101, 105153. DOI: 10.1016/j.scs.2023.105153

Zhang, X., Zhi, Y., Xu, J., & Han, L. (2022). Digital Protection and Utilization of Architectural Heritage Using Knowledge Visualization. *Buildings*, 12(10), 1604. Advance online publication. DOI: 10.3390/buildings12101604

Zhang, Y. (2022). An Analysis of the Regulations on Environmental Damage to Outer Space Caused by Space Debris. *Science of Law Journal*, 1(1), 19–26.

Zhao, J., & Gómez Fariñas, B. (2023). Artificial intelligence and sustainable decisions. *European Business Organization Law Review*, 24(1), 1–39. DOI: 10.1007/s40804-022-00262-2

Zhong, J., & Pei, J. (2024). Carbon border adjustment mechanism: A systematic literature review of the latest developments. *Climate Policy*, 24(2), 228–242. DOI: 10.1080/14693062.2023.2190074

Zhou, H., & Li, L. (2020). The impact of supply chain practices and quality management on firm performance: Evidence from China's small and medium manufacturing enterprises. *International Journal of Production Economics*, 230, 107816. DOI: 10.1016/j.ijpe.2020.107816

About the Contributors

Farhan Ujager, combines expertise in Artificial Intelligence (AI) and computing with a commitment to promoting sustainability, recognizing its essential connection to all aspects of society. With a certification in Carbon Literacy, Dr. Farhan leverages modern technologies like AI and the Internet of Things (IoT) to address sustainability challenges, particularly in smart healthcare. Through practical projects and scholarly contributions, Dr. Farhan demonstrates how technology can advance the UN Sustainable Development Goals (SDGs) while fostering awareness of sustainability's importance. Actively participating in panel discussions and advocacy efforts, Dr. Farhan inspires others in academia and beyond to join the collective journey toward a sustainable and equitable future

***

Munir Ahmad is a seasoned professional in the realm of Spatial Data Infrastructure (SDI), Geo-Information Productions, Information Systems, and Information Governance, boasting over 25 years of dedicated experience in the field. With a PhD in Computer Science, Dr. Ahmad's expertise spans Spatial Data Production, Management, Processing, Analysis, Visualization, and Quality Control. Throughout his career, Dr. Ahmad has been deeply involved in the development and deployment of SDI systems specially in the context of Pakistan, leveraging his proficiency in Spatial Database Design, Web, Mobile & Desktop GIS, and Geo Web Services Architecture. His contributions to Volunteered Geographic Information (VGI) and Open Source Geoportal & Metadata Portal have significantly enriched the geospatial community. As a trainer and researcher, Dr. Ahmad has authored over 50 publications, advancing the industry's knowledge base and fostering innovation in Geo-Tech, Data Governance, and Information Infrastructure, and Emerging Technologies. His commitment to Research and Development (R&D) is evident in his role as a dedicated educator and mentor in the field.

Tariq Akhtar is a specialist in Mechatronics with a strong focus on AI and Data Science, holding an MSc in Data Science and Artificial Intelligence from the University of East London. With a diverse background in mechatronics, Tariq has extensive experience in integrating mechanical, electronic, and computational systems. His research centers on AI-driven solutions for energy efficiency, particularly in utilizing deep learning and IoT technologies for smart building management and occupancy detection. Tariq has contributed to several publications, including co-authoring the book *"Exploring Pillars of Sustainability for Modern Age Improvements"* and publishing in the IEEE with *"Occupancy Detection for HVAC Systems Using IoT Edge Computing and Vision-Based Image Processing."* He is currently engaged in a governmental project focused on enhancing building energy efficiency through advanced AI solutions.

Asmat Ali is PhD in Remote Sensing & GIS. He is Director of Cartography & GIS at Survey of Pakistan. He is also Project Director Feasibility Study for Establishment of National Spatial Data Infrastructure (NSDI) for Pakistan. In, 1998 he earned Professional Master Degree in Geoinformatics from ITC, University of Twente, Enschede, Netherlands. He got MSc Degree in Geo-information Management from the same university in 2008. Title of his MSc thesis was, POTENTIAL OF PUBIC PRIVATE PARTNERSHIP FOR NSDI IMPLEMENTATION IN PAKISTAN. In 2022, he got PhD degree in Remote Sensing & GIS from PMAS-Arid Agriculture University Rawalpindi, Pakistan. Title of his dissertation was SPATIAL DATA INFRASTRUCTURE AS THE MEANS TO ASSEMBLE GEOGRAPHIC INFORMATION NECESSARY FOR EFFECTIVE AGRICULTURAL POLICIES IN PAKISTAN. He has several research publications on SDI, GIS, Remote Sensing, & Land Administration. He is on the Reviewer Board of MDPI ISPRS International Journal of Geo-Information (IF 3.4) and Information Development (IF 1.9). In 2008, SDI Asia-Pacific identified him as FOCAL POINT FOR SDI DEVELOPMENT IN PAKISTAN. GSDI awarded him as SDI IMPLEMENTER FROM PAKISTAN in 2016.

Charu Banga is an associate professor in finance at Rochester Institute of Technology Dubai. With over 19 years of experience in teaching, research, industry, and consultancy, her areas of interest are corporate finance, business valuations, sustainable finance, and market microstructure. She is a Ph.D. (Finance) from the University of Delhi, India; a Chartered Financial Analyst (CFA), a Certified Financial Risk Manager from GARP, USA, and a senior fellow of Higher Education Academy (UK).

Balaji Gopalan is an Assistant Professor in the Department of Decision Sciences at the CMS B School, Bangalore (India). His area of specialization is Production and Operations Management, Logistics and Supply Chain Management and Value Co-creation.

Diliza Madikiza based in the Research, Innovation, Postgraduate Studies and Commercialisation section of the University of South Africa. He has research interests in Legislative Affairs, Sustainability, Communication and Media Studies. He has been involved in the training of parliamentarians in the SADC region and has undertaken consulting assignments for the South African government. Diliza Madikiza holds postgraduate degrees in International Business, and Telecommunications Policy from the University of Liverpool and the University of South Africa respectively.

Geetha Manoharan is currently working in Telangana as an assistant professor at SR University. She is the university-level PhD program coordinator and has also been given the additional responsibility of In Charge Director of Publications and Patents under the Research Division at SR University.

CS. Priyanka Mathur, Assistant Professor at Prestige Institute of Management and Research, Indore holds an Academic experience of 8 years with an Industry experience of 2 years. She is a Company Secretary and MBA in Finance.

Harsh Purohit is a Professor of Personal Finance at the largest residential university for women's education in India Banasthali - Vidyapith and the Dean-FMS. Prof. Purohit has authored over 65 articles in reputed journals and over 70 students have been awarded PhD in his supervision. He is the Co-Investigator of the NITI Aayog's prestigious Atal Incubation Centre at the University and has recently conceptualized the framework of 'Bharatiya Model of Financial and Digital Literacy' and 'Work-Life Natural Fusion' which have potential to bring another revolution.

Yoganànthan S is an Assistant Professor in the Department of Decision Sciences at the CMS Business School, Bangalore (India). His areas of specialization are Production and Operations Management, Logistics and Supply Chain Management, Sustainability and Value Co-creation.

Ravishankar S Ulle is an Assistant Professor in the Department of Decision Sciences at the CMS Business School, Bangalore (India). His areas of specialization are Production and Operations Management, Logistics and Supply Chain Management, Sustainability and Value Co-creation.

Mohd Firrdhaus Bin Mohd Sahabuddin is a Professional Architect registered with the Board of Architects Malaysia (LAM), a member of the education committee for the Malaysian Institute of Architects (PAM) and a graduate member of the Royal Incorporation of Architects Scotland (RIAS). He is also a graduate member of the Chartered Institute of Building Services Engineers (CIBSE) and the American Society of Heating, Refrigerating and Air-Conditioning Engineers (ASHRAE). He gained his Master's degree in Advanced Sustainable Design from the University of Edinburgh and was later awarded the 'Andrew Grant Award' for the best dissertation in the University for the year. He then pursued his study at a PhD level at the University of Strathclyde, Glasgow with the British Government Sponsorship under the Commonwealth Scholarship Tenable in the United Kingdom. He worked with the Public Works Department (PWD) of Malaysia from 2005 to 2021. His last position was as the head of the Architectural Forensic Unit in the Architectural Branch, Public Works Department of Malaysia. Currently, he is holding a position as a Senior Lecturer at the Department of Architecture, Faculty of Built Environment, University of Malaya. His research work focuses on the areas of sustainable architecture and indoor environmental quality (IEQ) such as thermal comfort, visual comfort, indoor air quality, and building ventilation.

Faina Saleem is a qualified Architectural Engineer and has spent more than 3 years in the building and construction sector. She is currently studying for Ph.D. at the Faculty of Built Environment of the University of Malaya, Malaysia. In her Ph.D. research, she is exploring the entanglement of technology and building efficiency in the built environment, considering both sustainable energy and building performance of the Inner city of Multan (ancient cite), Pakistan. She holds MSc in Architectural Engineering from Vrije University Brussels (VUB), Belgium, and BSc in Building and Architectural Engineering from Bahaudin Zakariya University (BZU). Ms. Faina is working as Project Manager at Vision Developers and Builders (VDB) She is also working with Dr. Engr. Kamran Shabbir as a part time Research Assistant where they are trying to explore different efficient, sustainable construction materials. Previously, she worked with Defense Housing Authority (DHA), Multan as an internee and as a Visiting Lecture at Bahauddin Zakariya University (BZU), Multan.

Maryam Saleem is a lecturer (BZU, Pakistan) specializing in animation and digital arts, focusing on integrating new media technologies and AI into art and design education. She is also a PhD researcher (USM, Malaysia) exploring the digital preservation of architectural heritage in Pakistan using tools like 3D mapping, VR, and AR. Her teaching and research emphasize a multidisciplinary and sustainable approach to education and cultural heritage conservation.

Nidhi Solanki is an assistant professor with 18 years of experience with banks, financial institution and Academics. She has experience if Financial industry for 11 years and 7 years of experience in academics. She has specialization in Behavioral finance as her research work.

Farhan Ujager combines expertise in Artificial Intelligence (AI) and computing with a commitment to promoting sustainability, recognizing its essential connection to all aspects of society. With a certification in Carbon Literacy, Dr. Farhan leverages modern technologies like AI and the Internet of Things (IoT) to address sustainability challenges, particularly in smart healthcare. Through practical projects and scholarly contributions, Dr. Farhan demonstrates how technology can advance the UN Sustainable Development Goals (SDGs) while fostering awareness of sustainability's importance. Actively participating in panel discussions and advocacy efforts, Dr. Farhan inspires others in academia and beyond to join the collective journey toward a sustainable and equitable future

Zubia Waqar holds an MSc in Public Health from the UK and an MSc in Health Professions Education. Currently a PhD Scholar in Dental Health Sciences at Jinnah Sindh Medical University (JSMU), she has over 11 years of experience in academia. Her expertise spans general & dental public health and Health Professions Education, with a strong focus on research and curriculum development. Dr. Waqar is currently working as a medical educationist at Indus Hospital & Health Network that is aiming to become Indus University of Health Sciences. She has been actively involved in faculty development and mentorship, contributing to the advancement of health education. She is dedicated to enhancing healthcare practices through innovative teaching strategies and research initiatives.

Ihab Wattar brings over 30 years of extensive National and International experience in both industry and academia, with a rich background in senior management and engineering across diverse sectors. He holds a Doctor of Engineering degree from Cleveland State University, complemented by a Bachelor's and Master of Science in Electrical Engineering from Southern Illinois University. Throughout his dynamic career, Dr. Wattar has held positions such as assistant professor at Cleveland State University since 2021 and assistant professor at Al Ghurair University in Dubai from 2020 to 2021. He also served as a Proposal/Bidding Manager at ABB from 2012 to 2018, Deputy General Manager at ITAS Group in 2012, and Chief Operating Officer at Syrian Qatari Holding. Notably, he played a pivotal role in United Nations Development Programs (UNDP) from 2003 to 2008, overseeing multimillion-dollar projects. Before these roles, Dr. Wattar excelled as a Senior Application Engineer at Intel in the Chicago area. In addition to his professional endeavors, he has shared his expertise as an Adjunct Faculty Member at Cleveland State University, Kalamoon University in Damascus, Robert Morris University, and Southern Illinois University at Carbondale. Dr. Wattar's research interests span various fields, including Voice Recognition, Chaotic Vibration Signal Analysis, fault diagnosis of rotating machinery, and Machine Learning. His contributions to academic research are evident in his authored articles and the numerous accolades he has received from companies he's collaborated with.

Seema Yadav has academic qualifications of M. Sc. (Chemistry), M Ed (RIE), UGC-NET, and PhD in Education. Presently she is working as an Assistant Professor, in the Department of Education at The Bhopal School of Education, Bhopal. She has almost 18 years of teaching experience. Her area of specialization is educational psychology, environmental education, and curriculum development.

Index

Symbols

A

B

C

D

E

T

U

V

www.ingramcontent.com/pod-product-compliance
Lightning Source LLC
LaVergne TN
LVHW062109300926
849161LV00010B/2411